# X Marks the Spot

# X Marks the Spot

## The Lost Inheritance of Mathematics

Richard Garfinkle
David Garfinkle

CRC Press is an imprint of the
Taylor & Francis Group, an **informa** business
AN A K PETERS BOOK

First edition published 2021
by CRC Press
6000 Broken Sound Parkway NW, Suite 300, Boca Raton, FL 33487-2742

and by CRC Press
2 Park Square, Milton Park, Abingdon, Oxon, OX14 4RN



CRC Press is an imprint of Taylor & Francis Group, LLC



**Library of Congress Cataloging-in-Publication Data**

Names: Garfinkle, David, 1958- author. | Garfinkle, Richard, author.
Title: X marks the spot : the lost inheritance of mathematics / David Garfinkle, Richard Garfinkle.
Description: First edition. | Boca Raton : AK Peters/CRC Press, 2021. |
Includes bibliographical references and index.
Identifiers: LCCN 2020037191 (print) | LCCN 2020037192 (ebook) | ISBN 9780367187064 (hardback) |
ISBN 9780367187040 (paperback) | ISBN 9780429197758 (ebook)
Subjects: LCSH: Mathematics--History. | Mathematics--Study and teaching.
Classification: LCC QA21 .G26 2021 (print) | LCC QA21 (ebook) | DDC 510.9--dc23
LC record available at https://lccn.loc.gov/2020037191
LC ebook record available at https://lccn.loc.gov/2020037192

ISBN: 9780367187040 (pbk)
ISBN: 9780367187064 (hbk)
ISBN: 9780429197758 (ebk)

Typeset in Palatino
by Deanta Global Publishing Services Chennai India

# Contents

## Part II Theory in Practice

## Part III Toolkit of the Theoretical Universe

## *List of Figures*

# *Preface*

Most books on math are written from the vantage point of mathematicians. This book aims at something different: to show math both from that point of view and that of the users of mathematics. Seen from the overlap of these perspectives, math is revealed as a multipurpose toolkit with its roots in the description of nature.

The tools in this toolkit are the rightful inheritance of all human beings: not just the mathematicians who made the tools or the scientists, engineers, accountants, etc. who use them.

In our first book, *Three Steps to the Universe*, we talked about how scientists think and work. Our vehicle was an examination of gravity, astrophysics, black holes, dark matter, and dark energy.

In that book, we followed a precept common to science popularizations: minimize the math. There is an assumption that mathematics drives people away from the cool interesting parts of science.

After mulling this over, we came to the conclusion that there's something wrong here. Mathematics is part of the shared inheritance of human thinking. It should not be excluded or shunned or set aside as the province of some select group of either gifted or nerdy people, depending on how one wishes to look at it.

In explaining science, we had found that a useful question to ask, and answer, is "how do they know that?" that is, what combination of experimental and theoretical techniques allows scientists to make confident pronouncements about nature.

Similarly, we have found that the understanding of mathematics can be promoted by asking, and answering, the question "where does this come from?" Most mathematics was abstracted and generalized from previous mathematics. And the previous mathematics was designed with a specific purpose in mind: usually as a tool to describe nature.

But the abstraction gives the mathematics an alien feel, which makes it harder to understand and harder to learn. By unraveling the abstraction process through the "where does this come from?" inquiry, that understanding can be retrieved.

Furthermore, mathematics is stated in a very precise language that is specifically designed to leave no room at all for ambiguity or error. But this precise language also gives mathematics its alien feel, and can be an impediment to learning. By asking, and answering questions of the form "what were they trying to say here?" and "why did they feel the need to say it in this way?" more mathematical understanding can be retrieved.

We set out to write a book that would help people reclaim more of their share of the human heritage of mathematical thinking, and thereby to see the world around them in a more vibrant, alive, and measured way.

This book is the result of that effort.

# *Authors*

### David Garfinkle

David Garfinkle was born in 1958 and wanted to be a physicist ever since his first year of high school. He got a bachelor's degree from Princeton University in 1980 and a PhD from The University of Chicago in 1985. Since 1991 he has been a physics professor at Oakland University in Michigan.

David is the author of over 100 articles in physics journals. His main areas of research are black holes, spacetime singularities, and gravitational radiation. He performs computer simulations of gravitational collapse to resolve questions about black holes and singularities.

David was named a Fellow of the American Physical Society (APS) with the citation reading "for his numerous contributions to a wide variety of topics in relativity and semi-classical gravity."

### Richard Garfinkle

For the better part of his early life, Richard Garfinkle thought he wanted to be a mathematician. He went so far as to spend four years studying math at the University of Chicago before discovering that he really wanted to be a writer. He has had several science fiction and fantasy novels published. His first *Celestial Matters* (Tor 1996) won the Compton Crook award for best first novel. Richard lives in Chicago with his wife and children.

David and Richard have written *Three Steps to the Universe* (U. of Chicago Press, 2008) a book on black holes and dark matter.

# 1

# *Why This Book?*

If we said that this was a math book, how many people would run screaming into the night rather than read it?

If we said that this was a book about how mathematicians think, how many would toss it aside because who cares how they think?

But if we said that this is a book about something mysterious, invisible, and barely understood that is used to control and manipulate your life, what would you do?

And if we said that this is a book about a rightful inheritance, a wealth of tools for remaking the world as you see fit, that for many of you was lost through no fault of your own, wouldn't you want it back?

Even if that inheritance was stored in a box with a complicated lock and the name "Mathematics" on it?

Math is the common heritage of all people, a power over the world that early in life we have a sense of and a desire for.

Consider:

> The first complaint a child makes is "I don't want to!"
> The second is "It's not fair!"

The first comes from human nature, the second from mathematics. Without a sense of equality, how can we have the wounded fury that rises from seeing someone else have more than we do?

Once we get past this primal arithmetic judgment, many of us lose this basic grasp of mathematical principles. We are soon taught that math is a mysterious other way of thinking, that the people who practice it are strange, aliens dwelling among us. But in that second cry of a resentful child can be seen how deeply math is ingrained in our everyday thinking. The aliens are not just among us; they're in our heads! (cue fleeing crowds).

But what is this lost inheritance, and why is it so tightly bound up with basic human thinking?

Consider the following process of thought that everyone follows:

*Step 1: Observation.* A person looks out at the world and sees something.

*Step 2: Internalization.* A thought of that thing is created in the person's mind.

*Step 3: Imagination.* The person manipulates that thought, changing it or placing it into a story so that a different thought related to the initial thought is generated.

*Step 4: Externalization.* The person takes the latter thought and tries to treat the real-world object we started with as if it had to act according to the thought we imagined.

In most cases, this process does not work. We cannot, just because we imagine a witty dialogue with another person, make that person play the right part, saying their lines at our behest and accepting our conclusions when we have made them. We cannot, just because we envy them, fly like birds. We cannot make clouds edible just because they look like marshmallows. We cannot force the world to conform to our imaginations.

This process of thought, of attempted control of reality by act of conception, is sometimes called magical thinking and it does not, in general, do what we want.

But there is a region of thought and action where we can make this process work. It is a region that must be entered with care and navigated with precision. In this region, we:

1. Observe with great care, measuring accurately and properly the characteristics of the real-world object.
2. Internalize those characteristics into a mental object that works as a model of the physical object.
3. Transform this internalized mental object using processes of thought which preserve the truth and accuracy of our modeling to produce other thoughts that we:
4. Apply out in the real world using new knowledge of the object.

Done correctly, this more controlled and delicate use of magical thinking actually works. It allows for a cycle of interaction between world and mind in which mind takes from the world, alters what it takes, and then puts it back in order to change the world.

In this way, we can find the weights of buildings, the heights of mountains, the distances of stars, the odds of winning and losing games, and the reason written words appear on a computer screen as we tap keys.

This region, this strange region of thought where magic works, is mathematics.

It may seem that a region of magic where you have to work at it and take care isn't magical enough. After all, what's the point of wishing for something if it takes ages to prove whether or not you can have your wish? But it seems to us worth the effort. It's mind-boggling that there can be such a correspondence between the real world and our minds, and that yes, one may have to work at it, but at least some of our wishes can be granted if we learn the proper ways and are willing to take care with how we formulate our wishes.

This is the lost legacy of mathematics.

To be fair, there are two perspectives on this process of removing from the world, working in mind, and putting back: the mathematical, which largely focuses on the work of imagination, and the scientific, which is equally concerned with the back-and-forth between the worlds.

Thus, understanding math as an inheritance to be recovered requires not just looking at mathematics, but at science as well. We wrote a book, *Three Steps to the Universe* (University of Chicago Press, 2008), in which we tried to explicate the way scientists think in the course of discussing some of the stranger aspects of modern astrophysics.

Unlike that book, the book in your hands does not have the flare and flash of the sun, the destructive images of black holes, nor the hidden picturelessness of dark matter and dark energy. This book is almost completely concerned with things that go on in the human mind. But those things are closer to hand than the material objects we discussed in our first book.

Look below and count the dots:

.  .  .  .  .  .  .  .  .  .
1  2  3  4  5  6  7  8  9  10

The dots are spots of ink on a page, or spots of light on a screen. The numbers printed below the dots are also patterns of ink or light. But the counting, that's a thing in your head, a purely theoretical thing.

If you read *Three Steps to the Universe* (and, of course, we selfishly think you should), the word "theoretical" will set off bells in your minds. We talked about scientists seeing the world as made of three universes:

> The perceived universe, containing what is actually observed.
> The detected universe, containing what can be discerned by experimentation.

The theoretical universe, containing the explanations for what is perceived and detected as well as the means by which predictions of later perceptions and detections are crafted.

Mathematics is one of the vital elements that make up the theoretical universe of the sciences, particularly in physics. Math makes that universe rigorous. This sounds kind of dull, but it is the rigorousness, the exactitude, and the care with which mathematical ideas are transformed from one to another that preserves the truth that enters the theoretical universe from the detected, so that what emerges at the end of the process is as true as what came in.

That's actually pretty amazing. A path of thinking that preserves truth is unusual, to say the least. Truth is easy to lose in the forest of associations, emotions, memories, desires, fears, images, and catchphrases that fill up people's minds.

How can this preservation of truth within the bewildering forests of mind work?

We'll be looking into that a lot in this book, but let's start by looking at the theoretical universe.

This is the universe where our understanding of things lies. In its shallowest regions, there are names of things. We look at something and think "chair." That's a theory of what the object is. Looking closer, we might find out that it's a papier-mâché model of a chair or a picture of a chair or that it actually is a chair, but it's too small for us to sit in.

We examine and we test and we refine our theories based on reality. But we can also use our theories to discern what the nature of the objects that are real are. It's a chair; I can sit in it. You sit down and something snaps; low comedy ensues. We look more closely and see that the wood of one of the legs is rotten. We refine our theory further. I might be able to sit in the chair, depending on how sturdy it is.

And now we hit science and math. "How sturdy is that?" is a scientific question with a mathematical answer. In order to answer it, we need to define sturdiness, measure whatever contributes to sturdiness, and calculate an answer to the question. That answer has to be a helpful one.

We need a measure of sturdiness that will allow us to know what we can safely pile on the chair (starting with us). The uses to which we will put the answer help us form the question properly and guide us through making the theoretical and detecting processes we can use to find that answer.

This backward creation from needed answer to means of calculation and needed experiments is one of the ways that math and science are created and one of the reasons they actually fit the world of mind and the world of fact together.

The point is that the deeper we want to look into the real universe, the deeper we need to look into the theoretical universe. Inherited estates in the latter give us estates in the former. Gain ground in theory, gain it in reality.

Mathematicians reading the above may object to the subordination of math to science that is implied. That's because mathematicians and physicists are like siblings sharing the same work. They know a lot about each other, and while they work together well, they are also strongly aware of how different they are each from the other, and they are very insistent on those differences, sometimes to the point of loud and occasionally silly arguments (much to the amusement of other family members).

Seriously, there is a fundamental difference between scientific and mathematical thinking, and while it will show up throughout this book, we won't get to it until quite late (Chapter 12: The Smith and the Knight). Before we can explore those differences, we need to survey the world of mathematics in its assumptions and applications.

Why read this book? We've admitted that the flare and flash of new discovery isn't here. The cool stuff from science books isn't going to jump off the pages. So what about the subject of mathematics itself?

It is the exploration of a world that is simultaneously the shadow of our world and that casts our world as a shadow. It is a reflection that reveals the thing reflected. It is intimately bound up with reality in our minds, and to a great extent we cannot ever leave it. We can learn to know this vast estate or hide in one small room wondering what the noise is around us.

Unfortunately, to most people, mathematics is something that was forced upon them in school without any of this connection to reality. It was an arbitrary set of symbols and manipulations without meaning or purpose.

It's as if one were taught grammar without ever knowing that nouns, verbs, adjectives, adverbs, articles, and conjunctions were words. Think what it would be like to be taught all about "sentences" "phrases," and "clauses" as abstract, empty concepts, and that only if you learned all of grammar would someone come along and tell you that this was about language.

It's no wonder that so many people eventually give up on mathematics, either because it seems to have even less to do with their lives than science, or because they slammed up against a wall in learning math and decided they could not do it.

Both of these reasons for abandonment exist because of aspects of how math is taught, and both are unfortunate because they alienate those who do not learn math and make aliens of those who do.

Why read this book? Because math is so ingrained in human thought that it yells from the lips of children, and yet this part of the human mental heritage, this wealth of mental tools and understanding, is given up without a fight to a small group of people who, because they possess this heritage, are seen as paradoxically inhuman.

The people who succeed in learning the math are those who can most easily understand it as it is currently being taught and who have a talent for solving the kinds of problems that are currently used to test mathematical ability. But the teaching of math has changed over time and the problems given are not fixed.

As recently as 300 years ago, it was almost universal for visual artists in Europe to be given rigorous education in geometry so that they could map and deform the three-dimensional world they saw around them to the two-dimensional canvas in ways that fooled the eye. In consequence, they coupled an artistic depth of perception with a mathematical depth of detection, allowing them to walk every day through a world of beauty and symmetry that most people cannot even dream of.

Now those courses are taught rarely and the specialized mathematics of artists (such as curvilinear perspective) have fallen away from both artistic and mathematical education. The artists still see beauty, but they do not see how that beauty holds together. They have a hard time taking mental hold of the large-scale arrangements of the world, and a hard time transforming it in their minds in ways that preserve and reveal the unseen parts of reality. Art is weaker for this loss.

Mathematics is a common human heritage, a wealth of ways of thinking which are deeply dug into our minds and our lives in ways unnoticed by most. They hide in plain sight in the matter of fairness; in the questions of money and cost; in the odds of winning games and surviving; in the shapes of our homes; the traffic patterns on our roads; the growth of flowers; the spread of fires; the distribution of galaxies; the processes of weather; and the weird, brightly lit boxes on which we work, listen to music, watch videos, play games, and write books.

Math is inescapable and is the common property of humanity. Why give it up just because of problems in learning it? And why let ignorance of mathematics lead to bad gambles, bad purchases, bad game play, bad choices of where to build houses, bad scientific understanding, and endless calls to tech support? The distribution of math need not be unfair. The child need not cry out.

To reach this common heritage, we have to dig pretty deep into the ways people think. As we do so, a lot of common ideas that are used carelessly in everyday life will be transformed into their careful, rigorous counterparts, and then from those beginnings we will construct more ideas, seemingly farther removed from reality, that have made it possible for modern physicists to discover even deeper realities.

We'll start with looking for the source of that primal cry, the mathematical meaning of fairness.

# Part I

# The Roots of Mathematics

Most people when they think of mathematics at all see formulas, equations, geometric shapes, and nowadays, perhaps, computers. These seem like alien writings or esoterica of some mysterious ancient order. (To be fair, the Pythagoreans were a mysterious ancient order, but never mind that now.)

Most people don't see any connection between these things and their lives until math somehow drops in on them in the form of statistics or science reporting or the actuarial tables predicting their deaths.

It is the processes of mathematical thinking, the ways of mind by which math is created and used, which bridges the gap between things that impact people and the mathematics underlying them.

Ordinary mathematical education is a process of selection, of trying to find those who learn these ways of thought most easily. Please note this: the process selects for those who learn them most easily. Not those who can learn them, just those who have the easiest time.

The process is a combination of subtlety and brute force. The brute force is giving problems that some will learn to solve and others will not without help. The help is given in techniques of problem-solving. The subtlety is that those who learn mathematical thinking most easily begin to see deeper into what they are presented. They ask questions and find answers from their teachers, books, or other sources. Those who do not see the interest lose the interest and are taught to do the problems mindlessly. These people, regrettably, cannot wait to stop doing math.

In order to unlock the treasures of mathematical inheritance, it is necessary to understand the processes of thought that are normally selected for rather than taught. Once one understands and can use these ways of thought, all of math opens up.

Each of these processes is a refinement of an everyday way of thinking, a stretching out and a sharpening and sometimes a twisting round of commonly held faculties, but they are no more alien to humanity than a spear is alien to a stick.

# 2

## *Sticks and Stones*

Mathematics is not real. It is an illusion we substitute for reality, a theoretical mirroring.

Accepting this can be difficult because the most basic mathematical ways of thought are so deeply ingrained in the way people think that they often seem realer than reality. Two questions, two basic mathematical queries, are dug in so deep into modern human thinking that it is hard to see them as anything more than fundamentals of human curiosity. They also seem so rooted in the factual world around us that it seems alien to consider them unreal:

> "How many...?"
>
> "How much ...?"

We are so used to these questions that we rarely notice that we are asking them. When we go to a store and look at the price of something, the words "How much?" do not even form in our minds. We ask and answer the question when we look at the labels. We look at clocks to find how much time is left before we are done working. We mark calendars, wanting to know how many days are left until we can go on vacation. We count our money to see if we have enough, and if not we measure our credit limits.

Odometers on cars count the miles or kilometers we travel so that we can figure out when to get the cars serviced. Some people count their possessions to see if they have more than their neighbors. Children with candy count obsessively to see how much longer they have until they have subtracted it all. Dieters count calories in that selfsame candy, fearing what it will add.

> How many? What number of things?
>
> How much? What amount of stuff?

In order to see the basic mathematical thought processes, we need to shine a light upon and examine these basic mathematical thoughts.

These earliest roots of mathematics, these first treasures of our inheritance are bequeathed from prehistory. We cannot discern how and when they entered our ways of thought. Nor can we glean much of help from a historical context that led to their arising.

But that's all right. We don't really need to know all of that because this is not a history of mathematics. Our goal is to learn mathematical thinking, so it does not really matter to us where, when, how, or even ironically how many times these first thoughts arose. We will begin our discussion of them, therefore, not with confidence, documentation and proof, but with speculation and story. Complain about the lack of scholarship if you will, but you know what they say, "Sticks and stones ..."

## The Unnumbered World

Before we can understand the unreality of these most basic of mathematical questions, we need to look at the real world. Any student of any mental discipline will tell you that this is a task that can take a lifetime before a single grain of the world is exhausted. We will not take all that time, but then we do not plan to look upon pure reality. Instead, we need to examine those aspects of reality that "How much?" and "How many?" mirror. To do this, we need to remove for a moment one single illusion. We need to imagine the world without numbers, and reflecting it upon our imaginations see the reality.

Seeing this numberless world is hard on our minds because to do so we have to abandon a number of things that we are taught to be concerned with: how much, how many, how big, how long, and so on. These are habits of thought taught in childhood and absorbed into thought until they are nearly instinctual.

Try this. Look around you at each object you see in turn. As you shift from one object to another, see if you can ignore the questions in your mind, the comparisons about which one is bigger than the other, which is closer, which brighter. Keep doing that until you get to the point where you wonder how long you've been doing this or how long you should keep doing this or how long it is until lunch. If you can go without even thinking about amounts of time, then you've got a good handle on the unnumbered world. Even if you can't do this at all, in the process of trying you should get a sense of just how much you think about numbers without even thinking about numbers. Even if you cannot see no numbers, you will see how much you see numbers.

The unnumbered world is a way of looking at the observable universe that involves taking away part of the detected universe and seeing what really lies beneath.

**In looking at the unnumbered world, we can get hints of the earliest experiments that created the detected universe.**

*Experiments. How can you call these experiments? They're just people looking at things.*

**And thinking about what they see in relation to a theoretical structure.**

*Where's the apparatus? Where are the tools?*

[*Note*: For the most part, we two authors speak with one unified voice, but from time to time (like now) we find it useful to engage in dialogue. When we do so, **Richard will speak in boldface** (as he does in real life) and *David will speak in italics.*]

Back to these earliest experiments in comparison. They were probably done only with eyes, hands, and mind. Take two objects. Put them next to each other. Comparisons will arise in mind just by the apposition, the nearness of each to the other. This one's bigger. This one's prettier. This one's older. This one's harder. This one's redder. This one's brighter.

The objects themselves are in the observable universe, but the comparisons are experiments. The results of those experiments (the green rock is bigger than the red stick, but the red stick is redder) belong to the detected universe.

The ideas we are using to compare with – bigger, prettier, older, harder, and so on – belong in the theoretical universe. They are mathematical in nature, but they are not numbers. They are probably precursors to our concepts of number, and they fit neatly into our senses. Brighter, louder, hotter, rougher, more chocolaty, stinkier. These concepts have inherent in them the attributes of "more" and "less," but they do not yet comprise a theoretical object by which we can say *how much* more or less. There is a big mental leap from "bigger" to "ten meters longer and three meters wider." The commonplace change from "bigger" to "how much bigger" was likely once a leap of genius, a brilliant change in thinking made once or many times which created our modern awareness of the numbered world.

**Here we find the first great treasure of our inheritance, a treasure as common and vital to our existences as clothing.**

*Uh, how do you know the discoverer wasn't naked at the time? Or to put it another way, maybe the conception of number is a genetic mutation from pre-human times, and all Homo sapiens are hardwired with a rudimentary concept of number.*

**Well that could be. Good thing that we are doing storytelling rather than anthropology. However, for our purposes, the main thing is the difference between using the concept of number and not using that concept. Look at the gap between the numbered world and the unnumbered world, how vastly, startlingly unlike they are. Consider that we are all born into the unnumbered world. Most of us move into the numbered world fairly easily because we are taught to see that world, taught so well that the unnumbered world becomes the alien place.**

There is no leap in the learning of mathematics that is so radical as that first jump. If that can be internalized, nearly anything can.

But how come this leap is so easily taken?

As with so many acts of genius, one can make an analysis after the fact that removes the brilliance of the initial leap. In this case, the jump can be reduced to two not-so-dramatic steps, thanks to things lying around all over the place: stones and sticks.

## Stones: Counting

Take a pile of small stones (if you don't actually have any, imagine them). Now take a group of things you want to count (preferably small things like jelly beans or pencils). Pick up each one of those things in turn and put it down elsewhere. Each time you do that, pick up one of the stones and put it in a separate pile from the first lot of stones. When you run out of objects you wanted to count, the new pile of stones will have the same number of stones in it as you had objects.

Wait, it gets better. Suppose you do all of that without yet having any concept of number. Even without the abstract notion of numbers, you can use the pile of stones as a practical counting tool. Suppose you need to have the same number of some other kind of object as you had of the first kind. For example, suppose the first group of objects was all left shoes and a new group all right shoes. Pick up one of the objects in the second group and put it aside. Each time you do this, take one stone away from the pile of stones you made using the first group of objects until you run out of stones. When you do so, you will have the same number of second objects as you did first objects.

*All right, so what? You needed a pile of rocks to do that trick? Who's going to have a pile of rocks with them when they go to the shoe store?*

**As you can see, David is currently playing the straight man or stooge. Don't worry, we'll swap these roles around as we go along.**

**You can do that trick with a pile of rocks in your mind. Heaps of stones without any stones, that's what counting numbers are.**

There's the leap and there's the legacy. Whoever came up with this took the numberness out of a pile of rocks and made a mental object from it. That object is number, or rather counting number. No other characteristic of the pile of stones goes with number. There is no rocky texture, no weight, no color, no little clumps of moss attached to the sides. The only thing kept in mind for imagination to work on is the number. Thanks to this, we can carry around the ability to pile up and pull down piles and so make things equal without having to lug a single pebble.

There are two useful ways to carry these piles in mind: words and symbols. We don't propose to get into a philosophical debate about the distinction between them. We'd be on this topic for ages and never get back to the pile of rocks. Let's just stick with: a word is something you can say, a symbol is something that needs to be depicted.

Words for numbers are many:

> One, two, three, uno, dos, tres, un, deux, trois, eins, zwei, drei, etc.

As are symbols shown in Figure 2.1.

By using the words one can speak the news. "Three hundred and twelve sheep went out to graze this morning and three hundred and six came back. Let's go look for the coyotes."

Or with symbols one can write the message:

**Sheep Out: 312**

**Sheep In: 306**

**Hunting party. Now!**

The power and efficiency of this process, of creating rock piles in the mind, is found in this example. It takes a lot of work to tell if one pile of rocks is bigger than another. But just look at 312 and 306 with a mind taught Arabic numbers and the ideas of less than and greater than and you know the sheep are gone.

But why is this so? Why does a change in numbers mean a change in reality? Isn't that magical thinking?

Yes, but it's the kind we were talking about before, the magic that works. It works because rocks don't just appear and vanish in reality, and neither do sheep. The number of things we count remains the number of things we had unless something happens in reality.

*Coyotes.*

**Exactly.** If this were not true, if in the real world the number of rocks in the pile changed haphazardly, or if sheep were spontaneously created and destroyed in the air, then counting would not be a magical process that works. It would not fit reality.

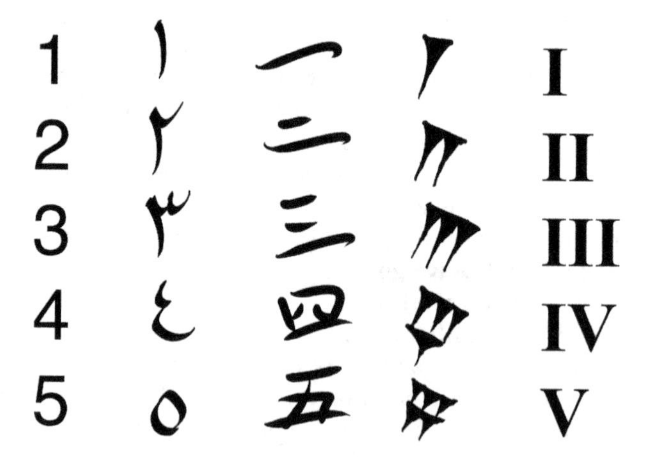


**FIGURE 2.1**
Symbols for Numbers

Because the world leaves number of things unchanged unless there's a physical cause, we can use counting as the kind of imagination that works as a stand in for reality. Furthermore, we can extend beyond counting. We can by careful use of imagination learn more than we did before. We can count my sheep, your sheep, and the sheep of that strange guy who lives over the hill. We can add the numbers up and get the same number of sheep we would have if we put them all together and counted them together. That's serious magic. We counted once, then added, and don't need to count again after the addition. And if we do count again and find the numbers off, we're back to the coyotes or to the wandering sheep, or to the strange guy over the hill.

Every mathematical process ever developed falls into one of two categories. Either it's like counting in that it directly touches the real world, turning reality into imagination and back again, or it's like addition in that it takes objects of imagination (like numbers) and creates other objects of imagination (more numbers) in such a way that the mental objects produced are as true to reality as the ones one started with.

But the mental objects created are not all created equal. Some are easier to work with than others. Both number words and number symbols have to deal with a complication that is unique in language, having an infinite number of things to talk about. Long before the concept of infinity became part of mathematics, the mental reality of it became a practical problem. Early on in number use, one might need only a small set of numbers and markers – Yan, Tan, Tethera – and tally markers for counting animals. Eventually, if one starts counting large numbers of things (stars, grains of sand, excuses), one needs more and more number words and symbols.

In order to solve this, all numbering systems have ways of adding on more numbers, be it the simple process of putting one stroke next to another (such as is done in tallying) or the addition of a new number place representing a higher order of magnitude (as is done in the Arabic numerals we use today, so that we can have 1 10 100 1000 10,000 and so forth).

A history of the various different representations can be interesting, and others have covered it. But – as we will delve into later – mathematics is primarily a tool kit. When dealing with tools, the question is not "what different forms does this tool take?" but "how well do those forms work?" Is a long handled knife better for chopping? Is a slotted spoon good for soup eating? We quickly find that some ways of representing or speaking numbers work better than others. The system we use today is extraordinarily efficient and pretty simple.

Consider trying to add two numbers using a tally system. Suppose one has counted three hundred and twelve sheep, and then put them together with another herd of two hundred twenty-nine sheep. Adding them together with tallies is eye-blurring (Figure 2.2).

But in Arabic column-style addition, one simply does

$$\begin{array}{r} 312 \\ \underline{229} \\ 541 \end{array}$$

Furthermore, in carrying out this process of addition, one need never do more than add two single-digit numbers at one time. Pardon the belaboring to follow, but it will show the efficiency of the system.

To add the two numbers above, we start at the right side in the ones place. We add 2 and 9 to get 11. Put the 1 down in the ones place below the line and carry the 1 to the tens place. In the tens place, add 1 and 1 to get 2, then add 2 to get 4; put that down in the tens place. Then in the hundreds place, add 3 and 2 and get 5. This gives us the result we were after.

**FIGURE 2.2**
Tallies

*Boring, but quick.*

If you compare that to the piles of stones or to the tallies or to Roman numerals or to any other system we have ever had, it is astonishingly easy to do.

It gets even better with multiplication, an immensely difficult task in many number systems. In the system we use, one never needs to do anything more complicated than multiply any two single-digit numbers.

It is for this reason that children have been taught addition and multiplication tables. After all, if you know the product of any single-digit number times any other single-digit number, you can do calculations at speeds that would have astonished people using earlier number systems. This form of arithmetic is a brilliantly effective mental tool.

*Although learning it bores kids to tears.*

Unfortunately, addition and multiplication tables are usually not taught in a way that makes it clear what they are being taught for. There has been a tendency to treat them as unmotivated but demanded exercises. Generations of children slogged through times tables not knowing how easy they had it. After all, remembering $5 \times 7 = 35$ is easier than carrying around a bag of rocks.

As with so much of the inheritance of mathematics, both the giver and the receiver treated the legacy as lead, not gold.

Herein lies one of the clearest examples of how one can be given and yet driven away from the lost inheritance. Counting is easy to take in, addition is a little more work, multiplication more work still. We have now the most efficient system for doing these things ever created, yet it is taught in a way that is simultaneously the best and the worst, the best because it focuses on exactly what work the student needs to do, the worst because the rocks are gone. Because the number system is nothing but marks on paper to the children learning. There is too little of the world, too little of the stone in the multiplication table.

Strangely enough, the stone counting that underlies the counting numbers became in the 19th century the basis for a deeper understanding of mathematics called **set theory**, and in the 1960s this in turn became the basis for a bewildering method of mathematical teaching called "the New Math." Unfortunately, this attempt to teach the gold made the learning even more leaden.

Set theory was invented by 19th century German mathematician Georg Cantor who had a delightful sideways way of looking at things. Cantor had an impressive ability to see things simply and then create ideas from that simplicity that burrow down into the depths of things. His simplest idea was to make a mathematical object from having a bunch of things together and thinking about them as one grouping. He called this idea the **set**. Sets recover the piles of stones so long abandoned and then go beyond them to a deeper mathematics.

Cantor conceived of the concept that any grouping of objects would be a set. The objects are called **elements** or **members** of the set. So, for example, one could have a set of all the shoes one had ever owned, or a set containing one lemon, one potato, and the Empire State Building. The first set mentioned would be written {A | A is a shoe I have owned}, which would be read "The set of all A's where A is a shoe I have owned." The second would be {lemon, potato, Empire State Building} and would be read, "The set containing a lemon, a potato, and the Empire State Building."

Cantor then took the stone counting process above and formalized it into what is called a one-to-one mapping, declaring that two sets had the same number of elements if one could match them element to element, using up all the elements of each. The set

> {lemon, potato, Empire State Building}
> and the set
> {Fred, Adele, Ginger} have a one-to-one mapping if you pair lemon with Fred, potato with Adele, and Empire State Building with Ginger. This is not the only such one-to-one mapping, since one could match potato to Fred, Empire State Building to Adele, and lemon to Ginger.

This really is the pile of rocks method of counting. A one-to-one mapping picks up one thing from each set and puts it aside. If you run out of members of each set at the same time, they have the same number of elements.

When you have two sets, there are things you can do with them to make new sets. These processes are mathematically exact forms of basic things one might do with piles of stuff and they relate directly to the basic operations of arithmetic. First, you can take the **union** of the sets, which is the set containing all the elements of either set. So the union of

> {lemon, potato, Empire State Building}
> and
> {Fred, Adele, Ginger}
> is
> {lemon, potato, Empire State Building, Fred, Adele, Ginger}

Union therefore is a matter of taking one pile of things and sticking it in with another pile.

You can take the **intersection** of two sets, which is the set of all elements that belong to both sets. So the intersection of

> {lemon, potato, Empire State Building}
> and
> {Fred, Adele, Ginger, lemon}
> is
> {lemon}

Intersection is more like the answer to a question than it is a physical operation. Intersection asks which things are in effect being counted twice when you look at both sets.

Notice that if there are no common elements of two sets, then the union of the sets has as many elements as the number of elements in the first set plus the number of elements in the second set.

> {lemon, potato, Empire State Building} has three elements.
> {Fred, Adele, Ginger} has three elements.
> {lemon, potato, Empire State Building, Fred, Adele, Ginger} has six elements.

This points to how one can develop arithmetic from set theory, in effect by formally recreating the process of counting this pile then counting that pile. This was very important when set theory was invented, since it helped establish it as a branch of math that provided an understanding of numbers deeper than the study of numbers.

This is one of the paradoxes of mathematical inheritance. One can develop a concept, use and explore it for years, centuries, or millennia, and then go back and look at it again with the benefit of that experience and see the simplicity in a whole new light. Only after all those tallies and rock piles and accountant balance sheets and censuses could someone steeped in the advances of millennia of mathematics look at the roots of counting to see a new math inside it.

Unfortunately, in the 1960s, the explanation I just gave for how to get addition from set theory was taught to elementary school children, many of whom floundered in these same depths. The creators of the New Math had assumed that giving deeper understanding was always the best way to teach.

*Remember, always teach kids to swim in the deepest parts of the ocean, during a storm, while sharks and orcas are having a dispute about who's on top of the food chain.*

So have we contradicted ourselves? Isn't this the approach we were advocating?

No, because despite being back in the stone age of counting, set theory is a very formal and sophisticated piece of mathematics. It is one thing to say to a child that these numbers and this arithmetic are all about counting piles of stuff, it is another entirely to first give the child all the mathematical apparatus of set theory and then develop arithmetic from that.

New Math also sought to teach students not to be fooled by the symbols, to know the numbers as numbers instead of as their representations. Again, an important deep idea, but hard to grasp when one still has difficulty dealing with the simple aspects of numbers.

Here's the concept that was taught: the symbol we use for a given number is an ordered list of digits, so one hundred fifty six can be seen as the three digits 156. To figure out what number this is, we take 1 times 100 plus 5 times 10 plus 6 times 1, or $1 \times 10^2 + 5 \times 10^1 + 6 \times 10^0$. The symbol 156 is called the **base 10 representation** of the number.

Any other whole number can also be used as a base to represent the same number. For example, in base 2 (also called binary), this same number is

10011100

which means

$1 \times 2^7 + 0 \times 2^6 + 0 \times 2^5 + 1 \times 2^4 + 1 \times 2^3 + 1 \times 2^2 + 0 \times 2^1 + 0 \times 2^0$ or

$1 \times 128 + 0 \times 64 + 0 \times 32 + 1 \times 16 + 1 \times 8 + 1 \times 4 + 0 \times 2 + 0 \times 1$

So two different symbols, 156 and 10011100, can mean the same number. A person who learns this is freed from the notion that the symbol, which is an idea in concrete form (since you can draw the symbol), and the number (which is an idea purely of mind) are the same. The person is also free to use different representations for the same number, using whichever is most convenient. Binary may not look convenient, but as we'll see later, computers can't get along without it.

New Math was invented with the best of intentions. A student who learns set theory and multi-base mathematics has a deeper and fuller understanding of numbers than one who doesn't. For that matter, elementary math before New Math was severely lacking in any explanation, relying instead on repetitive drills. Unfortunately, the deeper understanding came to very few, and the repetitive drills are actually necessary in order for the student to be able to be comfortable with arithmetic as a set of tools.

A student comfortable with the tools will have an easier time later learning the deeper meaning of them. On the other hand, a student bored with the drills will not practice and will not seek any understanding when it is offered. This paradox between practice and understanding lies at the heart of the difficulties in mathematical education, and we will return to it in the final chapter.

This difficulty is unfortunately one of the causes of the loss of the inheritance. What do you teach and how do you teach it? And how, for those who need the drill, do you get them to understand the value of what they are learning?

The paradox of New Math is that it would have made a good later step in learning about numbers. Going from numbers to set theory can make the numbers more real and bring their connection to reality into sharper focus.

For now, back to prehistory.

## Sticks: Measuring

Stones tell you "how many." What about "how much?"

There is another mental leap involved in going from how many to how much. How many is definite, clear, sharply marked out by counting numbers used to represent individual objects.

1. apple
2. apples
3. apples

etc.

The one-to-one correspondence between apples and stones in a pile and thence to whole numbers is easy to follow once you have the idea of numbers at all. But "how much" is a fuzzier idea, drawing on our ideas of big and small, of relative size.

If our idea of number comes from stones, then it is not obvious that bigness is susceptible to number. Numbers, as abstracted from stones, are solid, fixed things, clear, and set out. You can hold them in your hand, pile them up, and carry them away in your mind. They don't seem at first blush to have anything to do with the mushy idea of muchness. It is hard to conceive a way to transform the theoretical universe process of counting into a process that allows one to work with size.

But, as we will see later, the adaptation of a mathematical tool to a new area of thought and application by taking the new concept and transforming it in some way so that it fits the old concept is a vital mathematical act. To substitute what you already know how to do for what you do not is the essential ladder on which math has been built up.

Exactly how this particular change was effected is unknown, and indeed the process of changing from much to many is so ingrained in our thinking that we rarely notice it. But there is a remnant of it, a process used consciously in modern scientific thought that may point all the way back to the first crossing over. This process is the creation of units.

What is involved here is breaking the bigness of something up into standard-sized pieces and then counting the pieces. This is the critical step. Rather than finding a natural "one thing" to count (such as an apple), an artificial "one thing" (such as a meter stick) is created and then the artificial things, the sticks, are counted as if they were stones.

There is a phrase in use in mathematics, *thus reducing this to the previous problem*. It means find a way to turn what you can't do into what you already can do and then do that. Making a thing that represents one of something, like the meter stick, gives us a thing to count. We have reduced the problem of the stick to the problem of the stone.

Or have we? We have not completely done so, as we will see later. It's more that we've used the mental object of the stone and the processes it uses (counting, addition, and so on) to give us a place to begin when we work with sticks. "Many" shows us the way to "much."

But before we can accept this stony stick, we need to know that we've actually caught the idea of muchness in it. After all, it's not enough to imagine something and say it means something else. We need to make sure that stick counting really fits our real-world experience of big and small.

Our sense of bigness and smallness is blatant in our sense of sight. We see things as large or small, as taking up a lot or a little space. If we try to create an artificial unit to represent this idea of extent in space, we are after something that in our eyes shows an extent between one thing and another. Lying around in the world are just such objects, real-world things that seem to embody extent: sticks.

A stick has two ends; the "length" of the stick means the bigness between one end and the other. Each stick is a natural artificial unit. Pick up any stick and you can use it to find the length of something (more or less, as we will see below) by putting the stick against the object, with one end of the stick at the beginning of the length you are measuring. If the stick is shorter than the object, move the stick so that its beginning is where its end was before. Repeat until you run out of object, counting as you go (Figure 2.3).

1 stick long

2 sticks long

3 sticks long

4 sticks long

not a whole stick.

So the length of the thing is more than 4 but less than 5 sticks. Call it about 4 sticks and a bit more.

The critical difference between counting and measuring can be found in the phrases "between 4 and 5 sticks" and "4 and a bit." Counting is always definite, done as it is with whole numbers. Measuring involves a lot of more-or-lesses.

This stick-based measuring process is the place where the detected universe of physics started, the Big Bang of detection in a piece of wood. This radical change in thinking comes about because where before we were carrying around as numbers things we had created, piles of rocks, flocks of sheep, and so on, here we are taking out of things in the

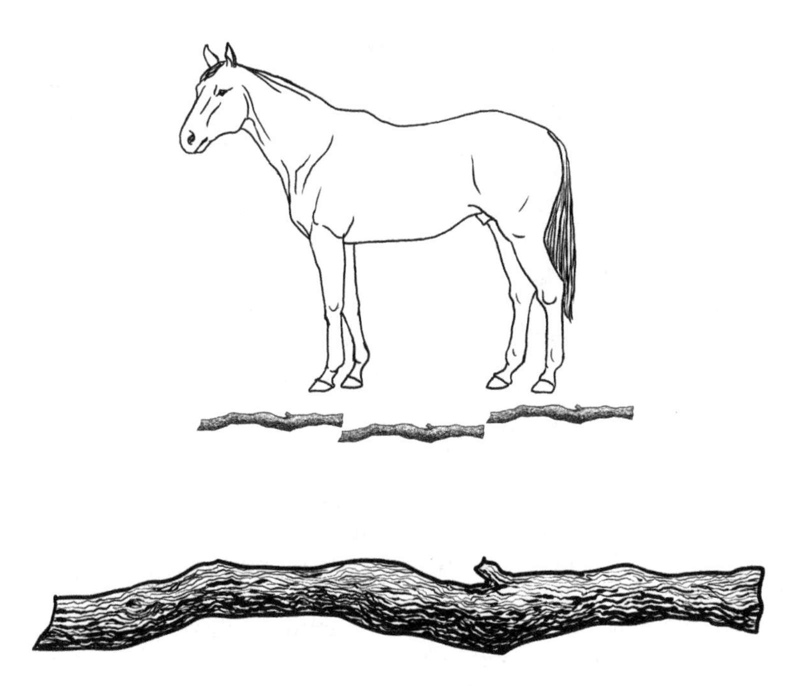

**FIGURE 2.3**
Measuring with Sticks

world something inherent in them, how big they are. We are delving into the world, prodding it as it were.

All that is needed to make a stick is a thing with two ends where the distance between the ends does not change. (Well, does not change much. Some of the sticks are a little sloppy.) There are two kinds of sticks one might measure with, personal and universal.

Personal sticks take their ends from some characteristic of the measurer. For example, the *foot* originally meant the stick made by using the back of the heel as one end and the tip of the most sticking-out toe as the other. The *foot* was useful because by walking carefully, heel to toe, one could pace out a distance. Related to the foot is the *pace,* which is the distance a particular person covers in a normal step. The *pace* is a lot more variable than a *foot* since people do not cover the same distance in each step. The *inch* comes from thumb widths, the *cubit* from fingertip to elbow, the *span* from pinky tip to thumb tip in a stretched-out hand, the *hand* from one edge of the palm to the other, including the closed thumb, and so on.

Universal sticks take their ends from a standard stick chosen as a reference purpose and resorted to by all who wish to use this stick. Whenever a new copy of the universal stick needs to be made, it is measured against the standard stick, and if it is too long, it is cut down to size. An extreme version of this can be found in the Greek myth of Procrustes, a dubious host who whenever he had a guest stay overnight at his inn measured him on his bed. If the guest was too short for the bed, the guest would be stretched on a rack until he fit; if too long, his extremities would be cut down. Needless to say, this exercise in data fudging did not fit the standards of the Greek scientific community of the time and Procrustes was killed by one of the few ancient heroes known for intelligence, Theseus.

Personal sticks are useful if one is making measurements for oneself and never needs to share them. Universal sticks are much more valuable, since one person can do the measurement and another person do the work that comes from the measurement. (This is particularly important in pursuits like architecture, where if the architect's foot is size 6 and the builder's foot is size 12, a house created by personal measurements will be considerably larger than intended.) This is why some personal sticks (like the foot) have been replaced by universal sticks of the same name (like the foot).

The difficulty with universal sticks comes from the fact that they need by definition to refer back to a single source stick that must be unvarying. This is harder than it sounds because errors tend to crop up and accumulate.

Suppose we have a meter stick and someone comes and makes a stick from it, but they are off by 1 centimeter. They really have a 99-centimeter stick. Suppose someone else makes a meter stick from that one that is also 1 centimeter short of the stick it was measured from. It will be 98 centimeters. After only five copies, one could have a "meter stick" that is only 95 centimeters, which can be the difference between comfortable ceilings and bashing your head on the lintel every time you come into the kitchen.

Solutions to this problem became more and more sophisticated. It used to be that the official meter was defined as a fraction of the earth's circumference (a problem to measure, to say the least). Later an official meter stick was created and stored in a temperature-controlled environment (the controls were necessary because objects can grow or shrink if the environment is too hot or too cold).

The current solution makes use of the fact that nature really does have standard sticks, albeit very small ones: all atoms of a given type have certain identical properties and can be used to define our sticks, though in an indirect manner. This indirect manner makes use of the fact that nature also has universal constants, such as the speed of light. The unit of time we call a second is defined as a certain number of vibrations of a particular type of atom, as measured in an atomic clock. The unit of distance called a meter is defined as the distance light travels in a particular fraction of a second. (Thus, we have both a universal stick and a universal clock.) Another constant of nature, Planck's constant, which is connected with quantum mechanics, is used to define a measure of mass.

Regardless of what kind of stick it is, mere possession of a stick lets us measure length, but only up to the accuracy of the stick. Here is another of those crucial concepts that is explained casually, but if taught with a little more care would make some of math and much of science far more accessible. The accuracy of a measurement is only as good as the accuracy of the means of measurement.

But accuracy is related to truth. Remember that mathematical imagination preserves truth. It doesn't create it. It makes sure that what you produce from mathematical operations is as true as what you put in. Consider the following:

$$2+3=5.$$

That's true when counting. But when measuring you need to be more careful. Something 2 meters long measured with a meter stick and something else 3 meters long measured with a meter stick might, if you stuck one on top of the other be more than 6 meters long if your 2 meters was really 2 and ¾ meters and your 3 meters was really 3 and 1/3 meters. If the accuracy of your measurements can be off by as much as whole stick then it's not really 2 + 3 = 5. It's (2 + something between 0 and 1) + (3 + something between 0 and 1) = 5 + something between 0 and 2. In other words, your answer could be anything between 5 and 7.

In this case, we had to be careful with the truth we started with to make sure we ended up with the real truth. This is why all real scientific experiments have estimates of error, to make clear where the answers can actually lie. More about this is given in later chapters.

Let's see if we can do better with our sticks. With a stick we can do 4 and a bit, but can't be more accurate than that. What can we do if we need to measure something that is smaller than the stick by which we measure?

There is a simple solution that leads to the first expansion of the concept of number. (A lot more of these expansions will crop up as we go along.) Up until this point, we have only used whole numbers: this many stones, this many sticks. But to find out how big something is, we sometimes have to use more than a single stick. We need a big stick and a little stick, such as a foot and an inch, for example. A common practice is to measure something with as many big sticks as we can and then measure with little sticks the bit which is more than the last whole big stick. So we might say that someone is 5 feet 4 inches tall. We get this by measuring their height in feet to get 5 and a bit more. Instead of sticking the foot measure at the 5 foot mark, we put down an inch measure and use that instead, going up inch by inch to get 4 of those.

This looks like we have a measurement consisting of two numbers – 5 feet, 4 inches – each of which has its own unit. That doesn't help much if we cannot in some way connect the foot unit and the inch unit. Since the foot unit is longer than the inch unit, we can measure the foot in inches and get 12 inches = 1 foot. This means that we can use our foot measure, then our inch measure, and convert as follows: (5 feet × 12 inches per foot) + 4 inches = 64 inches. This ability to convert between two units that measure the same thing (length, in this case) is vital, but simple. One uses one stick against the other, or sometimes a third smaller stick against both to create the factors needed to substitute one unit for another. Conversion between units that measure the same thing is one of those mathematical processes that once learned is automatic in one's thinking.

But now we need to stop and look more carefully. We've actually gone right past several vital concepts. The first is that of equality. You remember, the thing kids yell about, fairness. Consider this:

"It's not fair. He got twelve inches, and I only got one foot. Twelve is more than one, so he got more than me."

And, of course, one has to explain that 12 inches and 1 foot are the same things, that one can be substituted for the other whenever one wants. This is a critical concept. Equality means interchangeability. So much of math consists of swapping in one thing for another thing that is for the practical purposes we are concerned with the same thing. Grasping the idea of substitution is like holding the key to the theoretical universe because by substituting one thing for another thing we can progress throughout theory.

One must be careful with substitution because while one can mathematically substitute equal measures, in reality we cannot swap things out if they are made of different materials. One foot of wood and a foot-long sandwich cannot be substituted either in building or in lunch.

Substitution can also confuse people if they focus on the wrong aspect of measurement; if, for example, they think that the number ("5" in "5 feet," for example) is the important thing, not the number connected to the unit. 5 feet = 60 inches does not mean that 5 = 60. Number and unit together have meaning, not number alone. We will return to this topic in depth later.

12 inches = 1 foot

Means that twelve one-inch things stuck end to end will be a foot long. If we wanted to, we could keep our measurements in whole numbers by only using the smaller stick. This can get very inconvenient – if, for example, you want to measure the distance to another star and you insist on measuring in inches. The numbers get large and unwieldy.

Besides, there is another piece of mathematics that comes from looking again at the simple statement

$$12 \text{ inches} = 1 \text{ foot}$$

We have seen this as a process of assembly: put 12 small sticks together to make a big stick. But we can also look at it in the opposite direction. A 1-foot stick can be broken up into 12 little pieces, each an inch long. So an inch is a piece of a foot. How big a piece? One-twelfth (1/12).

This is where fractions come from. One thing is a piece of another thing. We want to know how big a piece, and for this we create ratios: 1 foot is 12 inches; 1/12 foot is an inch.

We have used the example of feet and inches because they make for clearer teachings on fractions. But they're bad for measurements and a nuisance for calculations. Let's go metric.

In the metric system every unit, every stick has bigger and smaller variants of itself, other units derived from it that measure larger or smaller scales of the small quantity. Each variant is named by putting a prefix like deci-, centi-, mega-, or giga- in front of the name of the base unit. The ratio between versions is always a power of ten.

In particular, let us consider the meter.

A decimeter is 1/10 of a meter.

A centimeter is 1/10 of a decimeter, hence 1/100 of a meter.

A millimeter is 1/10 of a centimeter, hence 1/1000 of a meter.

Let us suppose we want to measure a metal bar down to millimeter accuracy so that if we stick another bar on it, the total length will be off by at most 2 millimeters.

We would first use a meter stick, ending up with, say, 3 meters and a bit. Then we would use a decimeter stick on the bit, getting 4 decimeters and a smaller bit. Centimeters follow, producing 8 centimeters and an even smaller bit, and finally we carefully tweeze out 7 millimeters with perhaps a really small bit remaining. How long is the bar?

$$3\,\text{meters} + 4\,\text{decimeters} + 8\,\text{centimeters} + 7\,\text{millimeters}$$

$$= 3 + 4/10 + 8/100 + 7/1000\,\text{meters}$$

The funny thing about tenths, hundredths, and thousandths is that we have a convenient shorthand for them: decimal place value.

> The length is 3.487 meters
> = 34.87 decimeters = 348.7 centimeters = 3487 millimeters, depending on how you want to express it.

This is, in part, where fractions and decimals come from, from a scientific desire for accuracy in measurement and a human desire for ease of notation and calculation. This is also why the metric system is easy to use. Its measurements can be turned directly into one of our simplest-to-use forms of numeric notation.

One other thing. It is easy to make a single real-world measuring instrument that combines multiple sticks, multiple scales of measurement in one device. Simply use smaller sticks to make marks on a bigger stick. Decimeter, centimeter, and millimeter marks can be placed on a meter stick so that one instrument can do all the measurements above.

## Multiplying the Stick

Looked at edge-on, a stick seems only good for one thing, measuring length. But do not discount so quickly the powers of a simple length of wood. There are a number of things that we think of as qualitatively different that the stick can measure on its own. The length, width, and height of an object are three different measurements that can be made with that stick. We can also measure the distance between things, again with that single stick.

We got on to measuring with sticks by starting from an intuitive idea of bigger and smaller and creating something that embodied the idea of "this big," but we've only approached one form of bigness and our eyes see a lot of different kinds of sizes. We know that the stick shown in Figure 2.4 is bigger than that in Figure 2.5 and we can show this with experiments using stick measurement.

But which of these is bigger? (Figure 2.6)

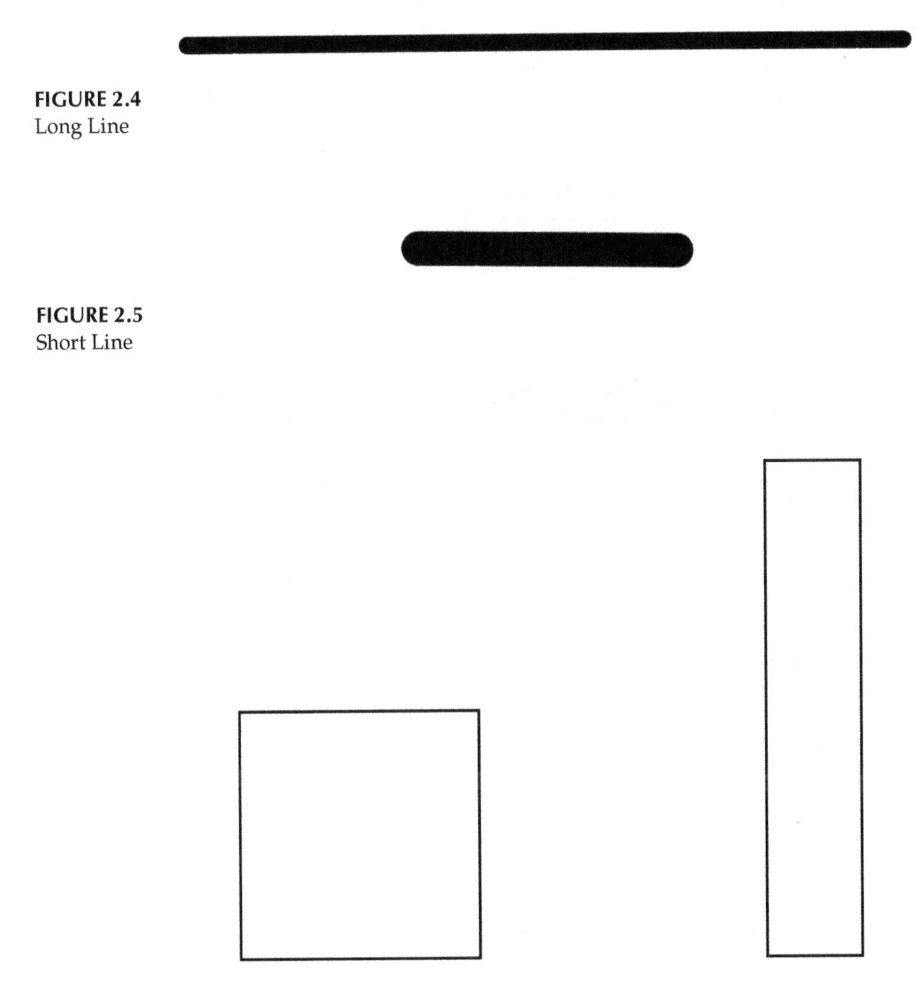

**FIGURE 2.4**
Long Line

**FIGURE 2.5**
Short Line

**FIGURE 2.6**
Square and Rectangle

Even though we have an intuitive understanding of the meaning of this question, we don't know yet how to take our idea of this kind of bigness and make it something we can really work with. Often when confronted with a new problem, people try old solutions first, hoping somehow to make the known work upon the unknown. We can use the sticks we have to measure the sides of these shapes. We quickly discover that the square is shorter but wider than the rectangle. This means that by one measurement (length) the square is smaller, and by another (width) it's bigger. This does not help us find the answer.

*OK. Now we have to tell them why anyone would care to measure those shapes.*

**Why? How about people crying "not fair"?** Let's consider land. Suppose two people own plots of land of these sizes and shapes (All right, not these exact sizes. You couldn't plant more than a few beans on these pieces.). Is it a fair exchange to trade one piece of land for the other? Are they equal?

To answer that we need a means of measuring the pieces of land that reflects the relevant bigness, which brings us back to the question before us. If we can't find a way of measuring this, then, as is traditional when humans can't answer questions, weapons will be drawn. To keep the knives in their sheaths, we need to know which is bigger.

We have come to the next jump in thinking. Up until this point our measurements have been done with sticks. Here we still need sticks, but instead of using them directly, we're going to assemble them into something that seems to have the same kind of bigness as the shapes do (Figure 2.7).

We will measure with a square made of one unit on a side. How can we use this? The same way we did with the stick, but a little trickier. We break up the areas we want to measure into little squares and count the little squares (Figure 2.8).

The square is made of 25 little squares and the rectangle of 20 little squares. Pretty, but does it prove that the rectangle is smaller than the square?

**FIGURE 2.7**
Square Made of Sticks

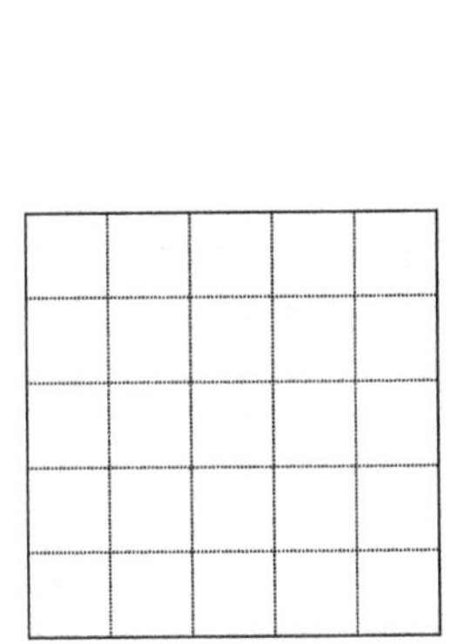

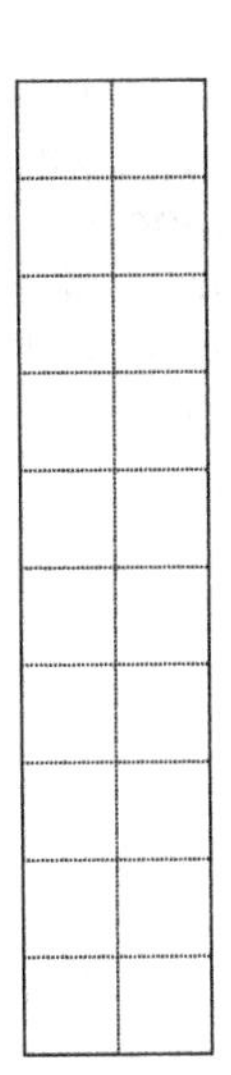

**FIGURE 2.8**
Gridded Square and Rectangle

It does, because we can cut up the rectangle into those 20 little squares and stick them on top of corresponding pieces in the square, with 5 squares left over. That's one-to-one mapping. We just made little square stones to measure area with.

Let's pause again because we just went past a piece of magic.

We created out of sticks a new way of measuring, we measured using it, and then we found a way to test to see if our measurements were accurate. We did this by breaking up what we were measuring into objects that were the same size. We then said that since they are the same size, they are interchangeable, as far as this measurement is concerned. We then counted the squares and saw that one pile had more than the other, therefore it was the larger pile, therefore the corresponding initial shape we started with was the larger one. We went out of the world of fact into the detected universe using the square measurements, did calculations in the theoretical universe, and then returned with the results we needed to know, and hopefully prevented stabbings.

These little squares we measured with are called square centimeters or centimeters$^2$ or cm$^2$ for reasons that will become clear soon.

We've found something more than a way to measure area, we've discovered a new mathematical operation: multiplication. Before, when counting stones and sticks, we discovered in passing addition and subtraction (more things, more rocks, fewer things, fewer rocks). Here we've found something else. Look again at the two figures in the illustration.

$$5 \text{ cm} \times 5 \text{ cm} = 25 \text{ cm}^2$$

$$2 \text{ cm} \times 10 \text{ cm} = 20 \text{ cm}^2$$

That looks like multiplication, and indeed it would not be surprising to learn (although impossible to know) that this kind of stick-by-stick leap in thinking, this need to create area from length, is where the idea and the process of multiplication came from.

Actually there's more here than just the multiplication of numbers, there's multiplication of units. When using addition and subtraction, the units of what you add or subtract have to be the same. You can't add apples to inches, you have to add inches to inches. After all, what's the meaning of 5 apples + 3 inches? You can't just say it's 8. 8 whats?

With multiplication, on the other hand, new units are created. Centimeters times centimeters gives square centimeters. We've made one kind of counting thing (a square) out of another kind (a stick) and from it extended the concept of measurement. But we've done something else, something subtler, which is much more important than the fairness of whose farm is bigger. We've developed the idea of taking units and from them making new units.

We'll expand this soon, but the thing is that the leap in the mental creation of tools (units give rise to new units) is helped along by the visual assist of making a square from the sides of the square.

The mental process of making the square as a unit of measurement opens the mind to the possibility of creating units, and shows that multiplication is a way to actually do this creation. Once a new way of doing this is created, it will be used over and over in a variety of different fashions. The leap from unit as an object found in nature (stick, foot, etc.) to an object created by manipulating the units found in nature (multiplying stick by stick) opens the mind to the concept that units need not be fixed things derived directly from experience.

Rather, units are created mental objects that can be crafted out of other units to fit a need. From this point onward, physics created units as it needed them and created them based on arguments analogous to the one we used before to figure out the square as the means of measuring. This one step started a very long journey in physics which is still going on.

Let's take another of those steps. We can repeat the process that made area to deal with another concept of bigness, space taken up. No longer are we walking a line or drawing figures on the ground. Now we're dealing with things like us, with length and width and height. The same process that gave us square centimeters can give us little cubes to measure volume (Figure 2.9).

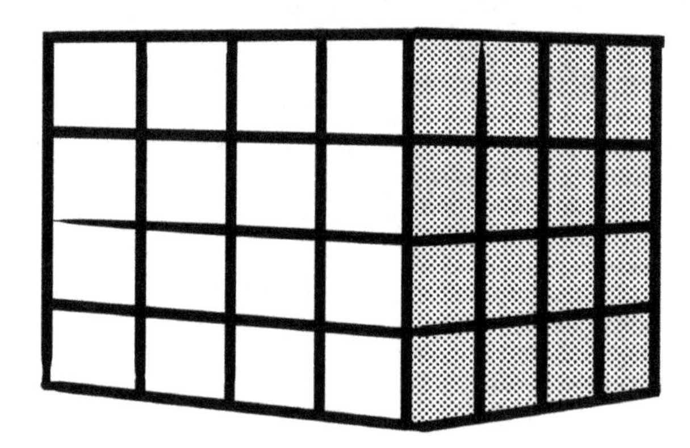

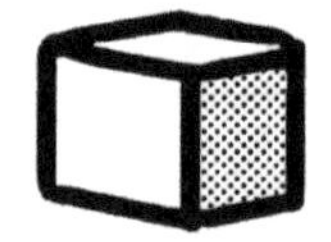

**FIGURE 2.9**
Centimeter Cube and Gridded Box

In this case we multiply centimeters by centimeters by centimeters to get cubic centimeters.

*Now we have to tell them what this helps us to be fair about. We don't measure land in cubes.*

**How about drinks?** How much has he had to drink? Three mugs. And I've had four glasses. So obviously I have to buy him a drink to even things out. But wait a minute. How much is in one of those mugs? These little cubes give us an answer. 250 cubic centimeters for each of his three mugs and my four glasses only held 100 each, so he's drunk 750 cubic centimeters and I've drunk only 400. He owes me three and a half drinks.

We have, of course, another means of measuring drinks, a more direct means. We make a standard cup and fill things up using that cup. How many cups does it take to fill something up? Two cups to a pint, two pints to a quart, four quarts to a gallon, et cetera. These units of volume are created not from measuring size but from pouring water.

*Wait, those units are done by experiment, not math.*

**Not quite.** We're still using the idea of measuring things by taking a standard unit and seeing how much of whatever we're measuring it takes up. In this case, instead of a stick we're using a cup. But the basic stick to measuring idea is there in the chug-a-lug mentality.

We now have two different ways of measuring volume: one created by physical tools that hold things (cups), and another by measurement of sides (cubic centimeters). But we have the idea that they represent the same kind of measurement, space taken up. We have two means for the same end. If they are the same kinds of units, we need a way to translate from one of these units to another the same way we went from inches to feet.

Again let's pause. We have entered the theoretical universe from two different places in the detected universe, measurement by sticks and measurement by cups. But we can see that we have reached the same theoretical object, a measurement of volume. If we can create the means of translation between these methods, we can use these tools interchangeably, measuring by stick or cup as we need.

There's an easy way to do the translation. We take a little $cm^3$ cube and fill it with water. However, much fills up that cup is the cup volume of a $cm^3$. It's a little tiny cup, too small to be a convenient measure on its own. We can make a bigger unit. By experiment, we can see that 1000 of these little cups is around a quart, and a quart is about the size of a wine bottle, so let's use that as a practical measure. Call it a liter (the above is completely nonhistorical, but we've got our bottle and our relationship). This gives us a simple way to translate.

$$1\,\text{cm}^3 = 1/1000 \text{ of a liter} = 1 \text{ milliliter.}$$

Again these are straightforward ways of going from one way of thinking about a concept (volume as length times width times height) to another (volume as amount a container will hold). But once one is used to this substitution of concepts and implementations, most of mathematics opens up. This is another basic idea that points much deeper toward a mental versatility in considering the theoretical and the practical.

## Tricks with Sticks

Area and volume are natural extensions of length measurement, but they answer only some kind of how much. Objects in the real world have many more muches that are observable. To start with, when we pick something up, there is more than size in our hands; there's heft, weight, mass.

"How much?" thus might need to answer "How heavy?"

In order to do this measurement, we need a stick and a stone. The stick is not used in its simply sticklike way, but needs to be made part of a special piece of instrumentation (Figure 2.10).

Weights on both sides pull down. The heavier side pulls down harder, and therefore falls farther. When the stick lies flat and even the weights in both sides are the same. Here is our symbol for fairness, here is the idea of justice, here in the weighing in the balance can be found images of salvation, damnation, rightness, wealth, and care (Figure 2.11).

Using carved stones or lumps of baked clay or pieces of metal cut so that they balance equally, we can make multiple sizes of weights and measure the "how much heaviness" of things, using the different weights in a manner analogous to our use of sticks of many sizes.

With an almost casual ease, now we transfer an idea (using bigger and smaller things added together to measure things with greater precision) from one use (sticks and length) to another (rocks and mass). We took our numbers from sticks of different sizes with known ratios of length. We now do the same thing with objects of different weights that have known ratios of weight. We can add up pounds and ounces just the way we did feet and inches, and of course we can flee swiftly to the arithmetically easier pastures of the metric system using grams, milligrams, and kilograms.

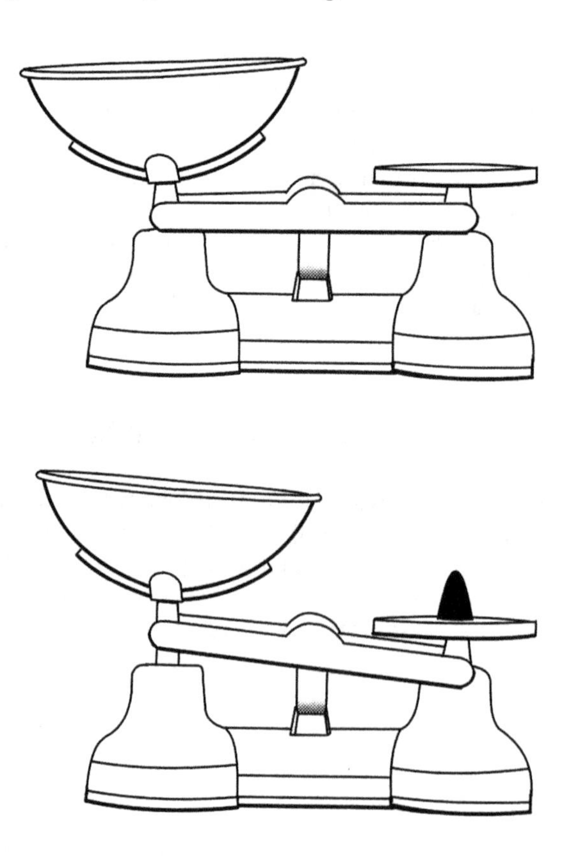

**FIGURE 2.10**
Balance Scale

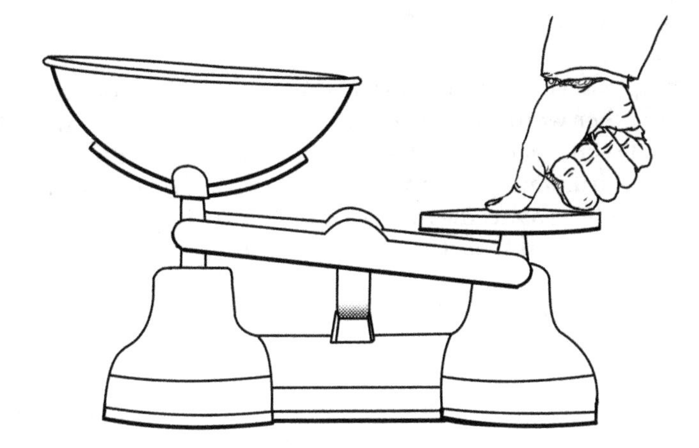

**FIGURE 2.11**
Scales as Symbols

Two kilogram rocks and four hectogram rocks and six decagram rocks and nine gram rocks means a mass of 2469 grams or 2.469 kilograms.

Look how fast that went compared to the time it took us to get to the various different size-measures. Is it because weight is a simpler concept to measure? No, it's because we've done this trick before. We've already solved most of the basic problems in creating and manipulating a muchness. All the concerns of accuracy, error, multiple units, comparison, and so on are the same. Therefore, once we've solved the question of how to do the basic act of measurement here, we've reduced the problem of turning those measurements into usable numbers into problems we've already solved.

So from now on all we need to do to be able to bring the entire power of numbers and arithmetic to bear on a physical concept is to properly create a unit for it that accurately measures it.

The balance scale gives us two different kinds of units which for a long time were indistinguishable because we were Earth-bound: weight and mass. Weight might be described as how heavy something feels and mass as how much stuff there is in it. When dealing with solids and liquids at the Earth's surface, the two seem to be the same thing, but move away from the Earth, and one discovers that weight can vary while mass stays the same. (Even on Earth weight can vary a little, because the force of gravity is not precisely the same at two different places on the Earth's surface.) This is because weight is really a measure of force (which we will discuss below), which depends on the interaction between two objects, where mass is a more basic thing, a "how much" that is intrinsic to the object. But this does cause a complication in the two prevalent measuring systems. A pound is a measure of weight and a gram is a measure of mass, so the standard conversion:

$$1 \text{ kilogram} = 2.2 \text{ pounds}$$

is not a true equation. The units on the left are not the same as the units on the right (apples and oranges, as they say). A true statement would be that on the Earth's surface, an object with a mass of 1 kilogram has a weight of approximately 2.2 pounds.

The scales, however, are a simple stick trick compared with the next one. The scales measure something you can hold in your hand. What about something that seems to pass by you, but you can never catch hold of? How do you measure mortality and the passing of the seasons?

Time is something that we perceive to exist because things do not seem to stay the same as "time" passes. In other words, we believe in time because of perceived change. On the other hand, the whole concept of change is rooted in our idea of time. From a philosophical perspective, this is frustratingly circular. But the first people who wanted to measure time didn't care about philosophy; they were too busy growing food. And as for circles, the universe was shoving them in their faces. The day turned in a circle as the Sun wheeled overhead. In the night sky, the stars turned in circles. The year turned with the seasons. Circles, circles, circles. We'll come back to them later in the discussion of geometry and use them to map the courses of the skies.

For now, notice that each of these circles gives us a way of measuring time, since each circle represents a series of changes through a cycle of positions (the Sun rises in the east and sets in the west) and characteristics (snow falls in the winter, flowers bloom in the spring).

One of these circles, the year, provides its own rough measurement of time in the turning of the seasons, but it also provides a stick-worthy method of measuring the positions of other natural timekeeping devices, the Sun and the planets. There is a tool made of two sticks which we can use to track position in the sky. That tool is the angle (Figure 2.12).

The separation of the sticks in the tool is the measurement of the angle. Closer together, smaller angle; farther apart, wider angle. The angle from the horizon upward and the angle from east to west can be used to measure the changing positions in the sky of the timekeepers above and can be used to turn the solar system into what we would now think of as a clock to measure the passage of the years as the angles change and cycle (Figure 2.13).

For a smaller circle, the day, there is a simpler use of a stick that can measure time. Place a stick upright in the ground on a clear day and watch the shadow of the stick move and change as the sun rises, passes through the sky, and then sets. By this, can the hours of the day be measured?

What about the night? Or overcast days? For that, another kind of stick can be used: a candlestick. Thick candles marked with even lines like a ruler can be used to measure the passage of time by simply setting them alight and burning down through the passage of time.

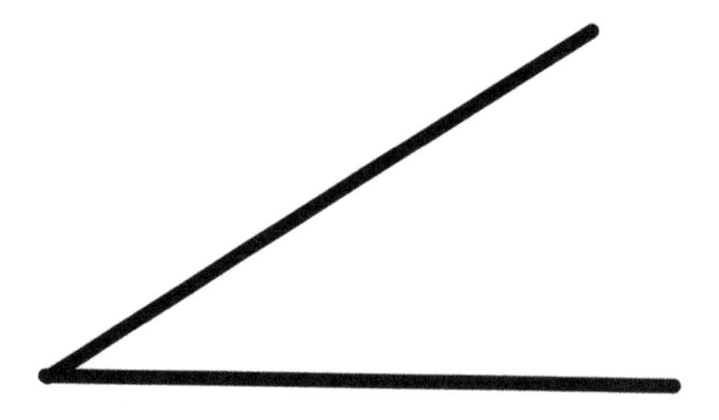

**FIGURE 2.12**
Angle

**FIGURE 2.13**
Astrolabe

From these and other things, we have evolved distressingly irregular and decidedly Earth-centric measurements of time. These are personal units for our planet, the feet, cubits, and spans of time:

*Year* – One passage of the Earth around the Sun, approximately 365 days. Each planet has its own year.

*Month* – One passage of the moon around the Earth. Well, kind of. We adjust the months to fit the year so the months and the moon don't harmonize anymore. Each world with moons can have a different kind of month for each moon. (Don't ask what month it is if you're on Jupiter.)

*Day* – One full turning of the Earth or, more precisely, one cycle of the Sun's position in the sky. Each world that turns has its own day.

*Hour* – Arbitrary division of the day, 1/24th of a day.

*Minute* – Arbitrary division of the hour, 1/60th of an hour.

*Second* – Arbitrary division of the minute, 1/60th of a minute.

Once we're down to seconds, we can start throwing on metric prefixes to get nice, well-behaved units that are divisible by tens: millisecond (1/1000 of a second), microsecond (1/1,000,000 of a second), nanosecond (1/1,000,000,000 of a second).

We have much better ways to measure time these days than sticks and clocks, but they all turn on the same basic idea. Take something that happens regularly and make a tick every time that happening happens. This means that to measure time, we really have to have a sense of what we mean by happens regularly. We have a need, a need for speed.

## Rate of Change

We know time through change and change through time; so along with a sense of time passing, we have a sense of how fast things change. The seasons turn slowly, ice melts quickly in hot weather, horses run fast, turtles are slow. In all of these, there is an intuitive concept: how much change happens in how much time.

This is easiest to see in speed of movement. We perceive a person who runs 10 meters in 5 seconds as moving at the same speed as a person who runs 20 meters in 10 seconds. Indeed, if these two people start at the same time and run side by side, they will maintain the same relative distance for as long as they are running. Of course, once the person who covers the 10 meters in 5 seconds stops, the other will pull away, but never mind that. We don't care where they end up at the moment, we only care how fast they're going.

This brings us to the last of our four basic mathematical operations, division. A person covering 10 meters in 5 seconds we say is moving at a rate of 10 meters/5 seconds ("ten meters divided by five seconds") = 2 meters/second ("two meters per second"). The person covering 20 meters in 10 seconds is moving at a rate of 20 meters/10 seconds = 2 meters/second. Our detection that the two rates are equal (since running side by side, the two runners keep pace with each other) is mirrored in the operation we just invented.

But why is this operation justified? We made two independent measurements, one of distance (10 meters) and one of time (5 seconds). If we line these two measurements up one above the other as shown in Figure 2.14 and then break each of them up into a number of pieces equal to the number we are dividing by (in this case five for the 5 seconds), (Figure 2.15) we end up with two sets: one of distance sticks (2 meters long each), and one of time sticks (1 second each). These two sets have a one-to-one correspondence. The first two meters correspond to the first second, the second two meters to the second second, and so on. There are therefore 2 meters for every second or 2 meters per second.

Notice that just as multiplication creates new units, so does division. Speed we measure in distance/time ("distance divided by time" or "distance over time"). In general,

**FIGURE 2.14**
Running in Time

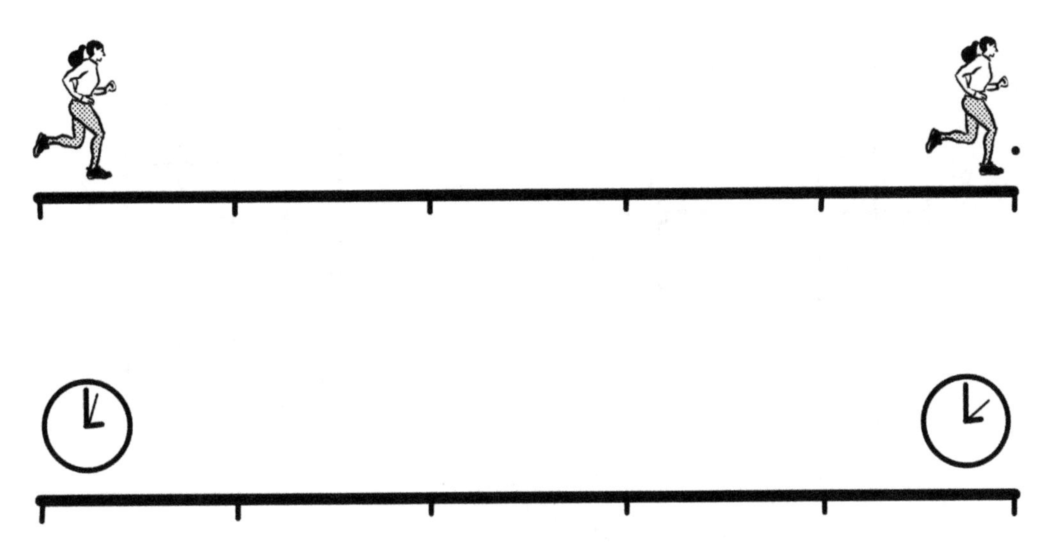

**FIGURE 2.15**
Divisions of Running in Time

we measure the rate of change of something as the thing being changed/the time it takes to change. If a growing child puts on 1 kilogram of mass each month, we would say it was gaining mass at the rate of 1 kilogram/month, or approximately 0.000385 grams/second.

Now we've reached something a little uncomfortable to our thinking. We have the idea that division and multiplication work in opposite directions, that if you multiply by something and then divide by that same thing, you should get back what you started. How does that work here?

Since both processes create new units, we should be able to multiply and then divide or vice versa and somehow get back to our original units.

Let's try it. A person is covering 20 meters in 10 seconds, giving a speed of 2 meters/second. In the course of 10 seconds, that person will cover a distance of 2 meters/second × 10 seconds = 20 meters second/second. If we say that second/second is no unit at all, then the distance covered is 20 meters. In other words, if we treat division of a unit by the same unit as making that unit disappear, then all our calculations will work out.

While this sounds like a cheap conjuring trick, it isn't. If we think of division of units as representing a rate per unit of something happening, then multiplying that by how many of the happenings there are should remove the entire consideration of happenings and restore the things we began with.

Consider the following: Somebody receives a shipment of ostrich eggs in crates. There are 25 eggs/crate and 16 crates. The number of eggs is 25 eggs/crate × 16 crates = 25 × 16 eggs crate/crate = 400 eggs. The crates, as they say in fractions, cancel.

This makes sense when you divide and then multiply. What if you multiply and then divide? We created multiplication from a need for area. Let's take our old friend, the 2 cm × 10 cm rectangle, which gives us 20 cm$^2$. If we then divide this by 4 cm, we would end up with 20 cm$^2$/4 cm = 5 cm.

What does this mean? It means if we took the 20 little squares and made from them a rectangle, one of whose sides was 4 cm, then the other side would be 5 cm, which is correct (Figure 2.16).

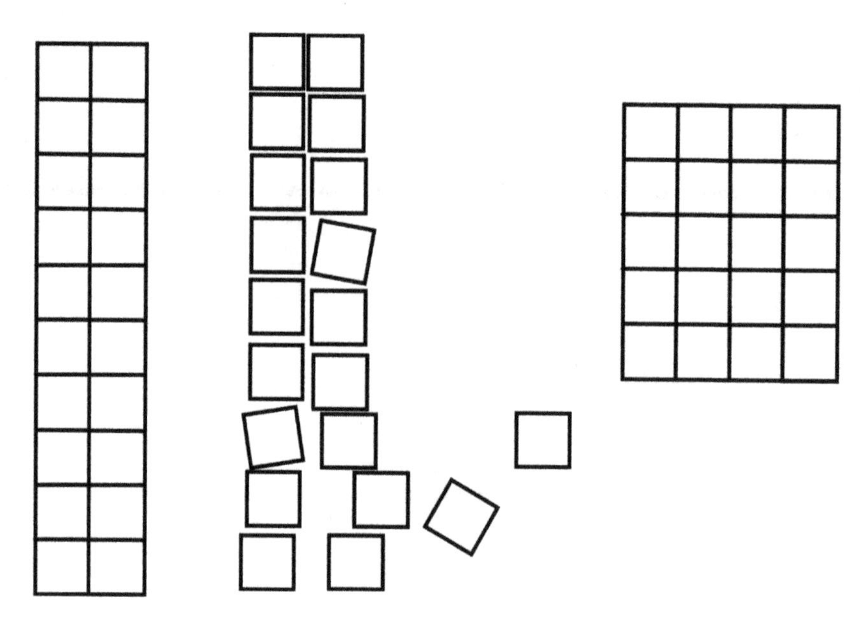

**FIGURE 2.16**
Multiplication with Boxes

## Units Multiply Like Rabbits

Using multiplication and division, we can create arbitrary sets of units. We could have something that was $mass^{15}$ × $distance^{5}/time^{3}$. But just because we can do so doesn't mean that the units make sense for any real-world quantity.

Physicists have created new units when they needed them. Most can be made from mass, distance, and time. Indeed, we're about to jump through millennia of physics. We can do so because the mental tools we created from sticks and the four operations of addition, subtraction, multiplication, and division, the elementary school math, are all that is needed to understand a great deal about the way physicists measure the universe.

Here are a few of the vital units concerned with motion. Later on we'll talk about those dealing with heat and with electricity and a few other odds and ends of the universe.

Speed, as we noted, is distance/time. But that only gives us a real sense of how the object moves if the speed is constant. What about when speed changes?

For this we need the rate of change of a rate of change. This is called **acceleration** and is measured in distance/time/time again, or $distance/time^{2}$. Acceleration is a quantity we can feel, usually in our guts on a roller coaster.

Force, we also have a sense of, is the push or punch of something. Newton's second law of motion says that force is equal to mass times acceleration, so the units of force are mass × $distance/time^{2}$. If done metrically as kilograms × $meters/seconds^{2}$, the unit is called a **Newton**.

**Momentum** is something we have a partial sense of. If we are pushing on something big but slow-moving, we can sort of feel that the effort to stop it is the same as that needed to stop something small but fast-moving. Momentum is mass times speed, or mass × distance/time.

If we push something across a distance, like pushing a box across a floor, doing a lot of work, we end up feeling tired. We have a sense of having used something up in exerting force. This quantity we're using is called **work**, also called **energy**, and is measured in force times distance or mass × distance$^2$ / time$^2$. The standard metric unit is a Joule, which is 1 kg meter$^2$/second$^2$.

The rate at which energy is being used to perform work is called **power** and is measured in mass × distance$^2$/time$^3$. If done metrically as Joule/second – that is, kilograms × meters$^2$/seconds$^3$ – the unit is called a **Watt**. Light bulbs are measured in Watts. The higher the wattage, the more the power consumption (because the bulb is either brighter or less efficient).

Underlying all of the above discussion is a subtle point that is easy to miss. In physics, numbers are meaningless without units. The value 5 has no meaning except in counting, but even then you need to know what you are counting. 5.2678 has no meaning at all. It could stand for any amount of anything if the right stick was provided. As we will see later, the meaninglessness of numbers is vital to the versatility of mathematics. Right now let's focus on how to give numbers meaning.

When units are provided, numbers have meaning. Five apples is something, something you can eat, where 5 is not – not only not edible, not anything. Likewise, 5 meters/second means something, a slow-moving car perhaps. Similarly, 10 kilograms might mean something if we knew what it was 10 kilograms of. It might be a large turkey or a small bomb.

This meaningfulness when units are provided implies that two things with the same units are at least comparable, even if they are not the same thing. We can ask which is bigger when we have the same units.

But suppose we use two different processes of measurement and calculation and end up with the same units for the quantities arrived at. We are saying something real when we discover that two seemingly different quantities are measured in the same units. We are saying that somehow they are the same, which leads us to ask in what way are they the same. We saw this a bit in the discussion of volume. Now let's crank it up.

### The volume?

**The energy.** We defined energy in terms of work done over a distance. But many other things come out to units of energy. For example, an object moving at a given speed has **kinetic energy** equal to ½ of its mass times the square of its velocity. This has the units of energy, but why should we think of it as energy?

An elevator carrying an object up against the force of gravity does work equal to the force of gravity times the distance traveled (that is, the height reached). The object carried up is said to have **potential energy** equal to the force of gravity times that height. Drop the object out of the building and it will start moving downward faster and faster because of the force of gravity. Its potential energy will decrease as the object falls (since its height is getting lower), but its kinetic energy will increase (since its speed is increasing). By the time it hits the ground, its kinetic energy will be equal to the potential energy it had at the top (less some that has turned into heat, which we won't get into here). The energy of work (which was the energy we first defined) became potential energy which then became kinetic energy. A real transformation occurred, suggested by this equivalence of units.

This idea of transformation of one kind of thing measured according to units into another kind of thing measured in the same units is called a **conservation law**. The principle illustrated above is called the **conservation of energy**. Energy is neither created nor destroyed; it is only transformed from one kind of energy to another.

A conservation law translates directly into a mathematical structure, an equation. If a quantity such as energy is conserved, then the measurement of the quantity before an action and the measurement afterward have to be equal. If they aren't, then something happened to that quantity.

Going back to the kinetic and potential energy, if you do the experiment above in a place without atmosphere, kinetic and potential energy will balance. But if you do it where there is an atmosphere, you find some energy missing at the end. Where did it go? It went into heat created as the object fell through the resistant air. The object and the air heated up. Therefore, heat is a kind of energy.

This equivalence of real-world phenomena in the equivalence of units is one of the demonstrations that math works as a tool for physics. The apparatus of units is called **dimensional analysis**, and it is a deeply rooted theoretical tool that proves its worth time and again in showing when things mean the same thing.

Sometimes, however, the value in sticks lies in using two different ones to approach the same problem. Sometimes you need to look at different aspects of things that are connected.

Let us consider mass and volume. They are not based on the same sticks, but each in their own way represents how much of something there is. When dealing with solids and liquids, we notice that there is a correlation between them. There's an old saying, "A pint's a pound the world around." What this means is that a pint of water weighs one pound (at sea level on Earth).

Mass and volume do not correlate outside of a particular material (a pint of gold weighs many pounds), and they do not correlate perfectly, since heat and cold can change volume. But materials do have a value called **density**, which is mass per volume, that is, mass/distance$^3$, and it expresses the relation between these two different ideas of "how much stuff."

Density figures in one of the classic science/math stories that have been told over and over again. It concerns Archimedes, who was famous in ancient Syracuse as a mathematician and weapons designer (A Roman soldier killed him, possibly for being too good at the latter). The story goes that the king of Syracuse had had a new crown made, but he worried that the maker may have substituted an equal weight of silver for some of the gold the king had given him to make the crown. The mass of the crown was right, but how to tell if it was not fully gold?

Supposedly, while sloshing around in a bath, Archimedes realized two things. First, that silver, being of lower density than gold, would take up more volume; and second, that an easy way to test this would be to put the crown in a container of water and see how high the water level rose, since this would show the change in volume. If the water level rose higher than it would have for that volume of gold (volume of water + volume of gold), it would mean the crown had been made of some inferior material (volume of water + volume silver).

We'll leave out the image of Archimedes running through the streets naked, shouting "Eureka," which means "I've found it."

*I wonder what the people of Syracuse thought he'd found.*

We've delved pretty far into the detected universe with sticks and stones and in the process built up a decent foundation in the theoretical universe. We have counting, measuring, whole numbers, fractions, decimals, addition, subtraction, multiplication, division, units, and dimensional analysis. Not a bad legacy from a rock and a branch. And not a bad legacy from the math taught in elementary school.

But mathematicians do much more than work at the border between the detected and the theoretical universe. They build and explore, often taking ideas that start out bound to the detected universe and breaking them out to create purely theoretical objects.

This may sound odd if you think of math as the toolmaker to the sciences. But it is in these explorations that mathematicians find things that end up being of use in surprising ways. Mathematicians in their exploration/building of the theoretical universe need to go beyond the real world in order to create what might reveal even more about it.

Sticks and stones point out beyond these boundaries to –

## Infinity

Numbers begin in sticks and stones. But what do you do when you need more stones than you can have? Or if the stick needs to be longer than it possibly can? In short, what happens when the numbers themselves push farther out than the world can handle, out to –

**Infinity.**

Infinity implies a word of drama that stirs the imagination because people have a hard time imagining it. It is a concept that arises naturally in mathematics. We saw it before in the simple question of how to keep counting.

> 1 2 3 4 5 …
> Keep counting.
> 4,000,001 4,000,002, 4,000,003, …

No, you can't just skip to the end. There is no end. There is no highest number. You can always add one and get to a next number. The process of counting never needs to stop. In the act of counting stones, humans created the process of counting. It's a simple process, easy to state. If you have a number, add one to it and you get the next number. But since this process takes a number and produces a number, it can then be applied to the number you just got to make yet another number, and so on, *ad infinitum.*

It's certainly possible to make the argument that only in theory does this process never end. In reality, no matter how many people and machines count for however long you set them to do the task, they'll eventually all give up, break down, or die of boredom, and then the counting will stop. It could go on, but it doesn't, so what does it matter? All it means is that we need never run out of numbers. It does not, in this view, mean that we have to worry about the unending character of the process of counting.

Except that counting is itself a thing of theory, and if we want to understand our theories, we sometimes have to make theoretical objects that are about other theories. It's not enough to populate the theoretical universe with the objects we create directly from the other two universes. We have to connect the ideas within it. After all, we created number by doing just that, by realizing that counting one kind of thing is like counting another kind, and that measuring one kind of thing is like measuring another. We need to make the theoretical universe robust and alive. Otherwise, we'll never be able to go beyond the theories we started with, and that defeats the entire purpose of the theoretical universe, to grow and change and give us new ideas to test in the observable and detected universes.

Do we have a way to get hold of this elusive concept, the infinite, using ideas already developed?

We actually have several such ideas, and they lead to different infinities. We'll start with the apparently simplest: set theory. If we follow Cantor, we can make the set {all counting numbers}, which we can write as {0, 1, 2, 3, …}. The three dots at the end mean keep going.

Then we can ask the naive question, how many elements are there in this set? After all, one of the reasons we have sets is so we can count their elements. Here's a set, count it.

Two sets, as said before, have the same number of elements if we can create a one-to-one correspondence between the elements of the first set and the elements of the second.

Just looking at this set above, it would seem that there's no way it can have a meaningful answer to "how many are in it," since if there were some definite number of elements (let's say 8, for example), we would be able to find a one-to-one correspondence between a set with eight elements and the set of counting numbers. **Here's a set with eight elements:**

{! @ # $ % ^ & *}

*Are you spelling out your cuss words so that the children can't hear you?*

**They do enough cussing without my help.**

Let's try to make a one-to-one correspondence with the set of counting numbers.

{! @ # $ % ^ & *}

{1 2 3 4 5 6 7 8 9 10 …}

Nope 9 10 and everything after it have no correspondents in the set of dingbats above. And it won't work with any other set we can number. So can we create a meaningful answer to how many? Let's push on a bit.

We need a few more set theory concepts before we can answer this. To start with, we need the concept of **subset**. One set is said to be a subset of another if every element of the first set belongs to the second set. So

{fred, wilma} is a subset of {wilma, barney, betty, dino, fred}

Every set is a subset of itself, and the empty set {} (the set with no elements in it) is a subset of every set. A **proper subset** is a subset of a set that is not the whole set, so

{fred, wilma} is a proper subset of {wilma, barney, betty, dino, fred}, but
{wilma, barney, betty, dino, fred} is a subset that is not proper because it contains all the elements of the set.

It looks obvious that the number of elements in a proper subset is less than the number of elements in the whole set. Let's have a bit of mathematical notation so that we can check to make sure of this. One way of writing the number of elements in a set A is $|A|$. So

$|\{fred, wilma\}| = 2$ (read this as "the number of elements in the set fred, wilma is two")
and
$|\{wilma, barney, betty, dino, fred\}| = 5$.

$2 < 5$, which fits our idea that proper subsets always have fewer elements than the set they are proper subsets of. *Note*: This is an example, not a proof. We'll talk about proof in the next chapter.

But let's take that set {all counting numbers} and take a proper subset of it: {all even numbers greater than or equal to 0}. This is a proper subset, since each nonnegative even number is a counting number. Let's write these sets out a little:

{0 1 2 3 4 5 6 7 8 9 10 11...}
{0 2 4 6 8 10 12 14 16 18 20 22...}

Line them up and you find a one-to-one correspondence. Each number in the set of counting numbers corresponds to two times that number in the set of nonnegative even numbers. We said that two sets have the same number of elements if there was a one-to-one correspondence between them, but we also said that proper subsets should have fewer elements than the sets they are proper subsets of. Something's clearly gone wrong here.

*Maybe it's the mathematicians! Why are they messing up such obvious ideas as counting?*

**It is precisely because these ideas are obvious that they need to be examined.** Obvious things are often wrong or naive. They need to be tried and tested and refined. That's what math and science are all about; and sometimes in the trial surprising ideas and truths emerge like the shining metal that can be extracted from the black rock of silver ore. Sometimes the legacy is like inheriting a mine. There's a lot of wealth in there, but you've got to do work to get it out.

The processes mathematics uses to transform elements of the detected universe require their own study and exploration. It's not enough to do what works. You have to understand the workings of it. Sometimes to do that, you push the workings beyond their normal bounds and see what happens. Infinite sets, while not real, are easy to create, and infinite counting is a natural consequence of that. While at first sight it may seem like a step away from real-world useful mathematics (more about this later), infinity is still only a single step away.

Cantor created the refinement of counting for infinite sets in a characteristically simple way. He flipped the idea around. Rather than saying that there was something wrong with there being as many even numbers as there were counting numbers, he said the characteristic definition of an infinite set was one that had proper subsets with the same number of elements as the whole set. In effect, he said that the idea we were comfortable with – smaller-set-fewer-elements – was the nature of the finite, and that the infinite is revealed in those sets where this was not so.

By the way, while because of infinite sets we have lost the idea that every subset has fewer elements than its "superset," it's still true that every subset has at most as many elements as its superset.

Written out in symbols this is: If A and B are sets, then

if $A \subseteq B$ (if "A is a subset of B"),
then
$|A| \leq |B|$ ("the number of elements in A is less than or equal to the number of elements in B").
And if A is finite,
$|A| < |B|$("the number of elements in A is less than the number of elements in B").

What number is this number of counting numbers? It can't be any of the counting numbers since they all stand for finite sets. It can't be a fraction or decimal since those are for measuring, not counting. We need to create a whole new number with its own symbol.

The symbol that was picked for this was $\aleph_0$ (read as "aleph null" or "aleph naught"). The first letter of the Hebrew alphabet was used, probably because Roman and Greek letters had been flogged to within an inch of their font sizes for the last 2000 years and it was time to start whacking some other typography.

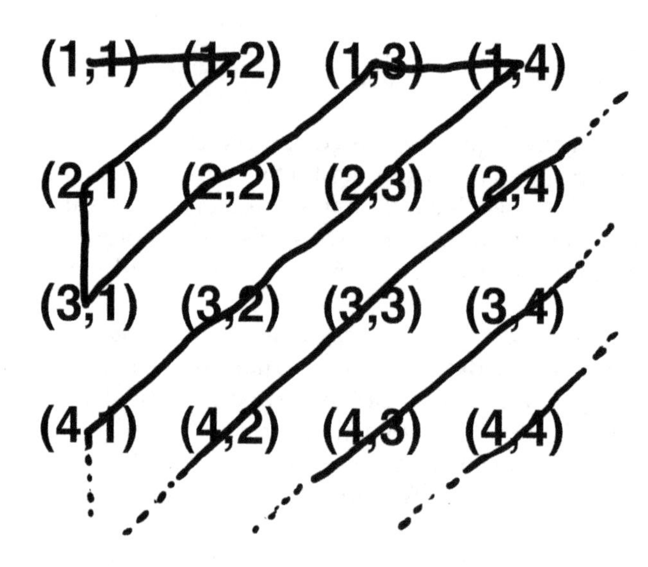


**FIGURE 2.17**
Cantor's Snake

Rather than stick a lot of $\aleph_0$'s into the text, we'd like to introduce another term for this: **countably infinite**. A set is countably infinite if it has the same number of elements as the counting numbers – in other words, if it can be put into one-to-one correspondence with the counting numbers. (Just to confuse the issue, a set is called **countable** if it is either finite or countably infinite.)

One useful thing about countably infinite sets is that you can list their elements on infinitely long pieces of paper. And you can make one-to-one correspondences by putting two columns on that infinite piece of paper. Here's a listing that shows there are as many positive and negative numbers as there are counting numbers:

| Counting number | Set Element |
|---|---|
| 0 | 0 |
| 1 | 1 |
| 2 | -1 |
| 3 | 2 |
| 4 | -2 |
| 5 | 3 |
| 6 | -3 |
| etc. | |

Cantor proved several surprising results about infinite sets using a few sneaky tricks. He proved that there are as many rational numbers as positive integers with a drawing. The rational numbers are those numbers that can be expressed as fractions of whole numbers. Cantor created a picture that showed the process of counting all the fractions. He laid them out on a square spreadsheet with numerators for columns and denominators for rows.

The bent line drawn in Figure 2.17 is actually the one-to-one correspondence. Here it is written out, at least to start with:

| Counting number | Set Element |
|---|---|
| 0 | 1/1 |
| 1 | 2/1 |
| 2 | 1/2 |
| 3 | 1/3 |
| 4 | 2/2 |
| 5 | 3/1 |
| 6 | 4/1 |

This actually overcounts the rational numbers, since each fraction is counted many times (1/1 = 2/2, etc.). This is not a problem, since this shows that the set we are really looking at is not the rational numbers but the set of pairs of numbers, and the rational numbers are a subset of that set. So the set of pairs of numbers is countably infinite, which means the rational numbers are at most countably infinite. But the counting numbers are a subset of the rational numbers, so the rational numbers are at least countably infinite. A number that is at most a given number and at least that same number must be that same number.

*That came off a little glib.*

**Sorry, that was math proof speak.** It tends to be fast at exactly those places where explanations are needed. All the above really says is that

Number of rational numbers ≤ Number of pairs of numbers

Number of counting numbers ≤ Number of rational numbers

And Cantor's snake-like proof shows that

Number of counting numbers ≤ Number of pairs of numbers

So the number of rational numbers is no more than one number and no less than another. But the number it is no more than is the same as the number it is no less than, so the number of rational numbers is equal to both those numbers.

The symbolic forms for what we've just been saying use letters for each set. N is the counting numbers, Z is the integers, Q is the rational numbers.

$$|N| = |Z| = |Q| = \aleph_0$$

Cantor's next trick is far more impressive and became the basis of a kind of mathematical argument called **diagonalization,** which, as we will see later on in the discussion of logic, was eventually used to blow people's minds. Here's a small foretaste.

So far it looks like all the infinite sets we know of are countably infinite. Maybe $\aleph_0$ is the only infinite number?

Nope, says Cantor. Observe. Cantor pulls the real numbers out of his hat. Later on we'll talk about where the real numbers came from. For now, let's just say that they are all numbers representable in decimal form. This includes all the rational numbers and a lot more besides. The set of real numbers is called R. We'll take a subset of that set, all the numbers between 0 and 1 (usually written [0, 1], sometimes called the **unit interval**).

Well, says the magician, we have here an infinite set of numbers. Suppose it is countably infinite. Then we can list its elements. Come up on the stage and examine the box. Notice the solid lines in the spread sheet. Kick the rows and columns to ensure there are no holes or false bottoms. Now pick all those numbers and put them down in one of those infinite lists. Any order will do, just lay them out one at a time and fill up the list.

| | |
|---|---|
| 1 | 0.6111222233333344444 |
| 2 | 0.39312333888891 |
| 3 | 0.141592654… |

etc.

Any order will do. Shuffle the cards and put them down as you want.

Now, you say you've used up every card in the deck, every real number between 0 and 1 is down on that list, not one is missing. Examine the interval, check to make sure. You're certain you've got them all.

Watch my hands. I'm going to make a number in the interval that's not on your list.

When I say make a number, I mean it. We're going to construct a number from your list by going down the diagonal (Figure 2.18).

Here's how we do it. We're going to make the number digit by digit. Let's call this number we're making "a." For the first digit to the right of the decimal point, we look at the first decimal place of the first number in the list. In this case, that number is 6. We add 1 to that number to get 7 and that becomes the first digit of a.

> An innocent sounding letter, isn't it?
> So far we have a = 0.7.

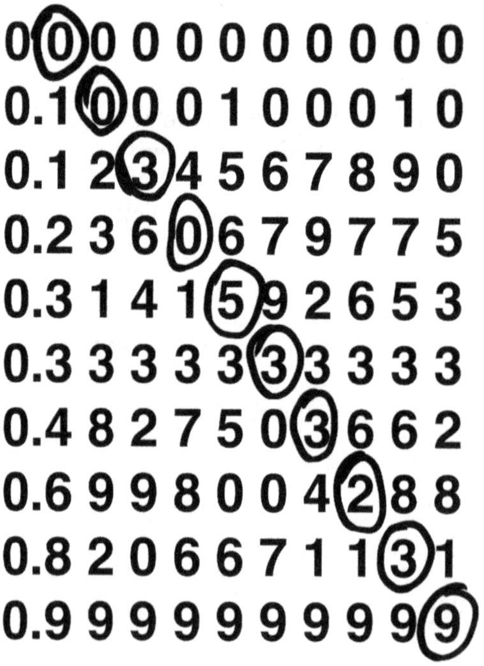

0 0 0 0 0 0 0 0 0 0 0
0.1 0 0 0 1 0 0 0 1 0
0.1 2 3 4 5 6 7 8 9 0
0.2 3 6 0 6 7 9 7 7 5
0.3 1 4 1 5 9 2 6 5 3
0.3 3 3 3 3 3 3 3 3 3
0.4 8 2 7 5 0 3 6 6 2
0.6 9 9 8 0 0 4 2 8 8
0.8 2 0 6 6 7 1 1 3 1
0.9 9 9 9 9 9 9 9 9 9

**FIGURE 2.18**
Cantor's Diagonalization

Now we go to the second digit of the second number, which is 9. We add 1 to that which gives us 10. We only care about the ones digit of that, so we get 0.

> a = 0.70
> The third digit of the third number gives us 1, to which we add 1, giving us 2.
> Now a = 0.702.

We keep going down the list, finding the nth digit of the nth entry and make 1 plus that digit the nth digit of a.

> Once we're done –
> *Done? How can you be done?*

**Okay, an infinite number of steps are necessary.** But we got the counting numbers by doing an infinite number of steps. There will be more about this doing an infinite number of things in a little while.

Once we're done, we have the number a, which is clearly a number between 0 and 1. Since we put all of those numbers on the list, a must be somewhere on the list. That means we can find the row a is on. Call this row n.

But in constructing a, we made sure that the nth digit of a was different from the nth digit of whatever number is on the nth row. Therefore, a cannot be the nth number for any number n.

We've pulled an uncounted number out of the interval. You didn't use them all up. (Smug magician's smile.)

Well, you might say, just add it to the list, stick it in somewhere. But we can do the same process again to the new list and get another number you haven't counted. Stick the rabbits in the hutch, we can always pull more out of the hat. You can't count the interval, therefore you can't count the real numbers.

> All right. So how many real numbers are there?
> If $|N| = \aleph_0$, is $|R| = \aleph_1$?
> *$\aleph_1$? What does that mean?*

We named $\aleph_0$ to be the first infinite number. $\aleph_1$ is meant to be the second infinite number.

So is the number of real numbers the second infinite number?

Disturbingly, the answer is maybe. In fact, it can't be determined if it is or isn't. We'll come back to this later when we talk about logic and undecidables; we'll also reuse Cantor's magic trick.

Since we can't say which infinite number it is, a term for the number of real numbers was coined. It is (cue thunder and lightning): ***the Power of the Continuum!*** (We're not kidding. It really is called that.) If you need a shorthand symbol for this number, it's written as a C (c in a gothic font).

So what's the point of all this? We're talking about counting infinities beyond infinities. How does that help us?

It stretches our idea of number, our concept of counting, our notion of aggregation (that is, sets), and our ideas of correspondence beyond the confines of our everyday use. It opens our minds and leads us to see that not only are there "more things in heaven and earth," there are more things even than that.

But that's an argument from aesthetics, a sense of the mathematical beauty of this. How does it help us? Is the worth of this legacy of infinity only to be found in the prettiness of its results?

No, there are two human-practical purposes in this understanding. First while the universe might not have an infinite number of things in it, we can construct infinite sets in our imaginations by considering variations and possibilities, by wondering if there are an infinite number of possible colors or an infinite number of variations in the history of the universe, or an infinite number of possible positions for objects and so on. There are potentially infinite sets in our conception of our world.

The second use is found in the breaking of concepts, in the realization that perhaps even number is not as simple as we think, that our reliance on stones might be naive and we might need to cast a mistrustful jaundiced eye upon them.

Enough of infinite stones, what about infinite sticks? Even if we count to infinity, surely we don't measure out to infinity.

Actually, artists have been doing that for centuries. They use infinite distance to create perspective lines (Figure 2.19).

And where did they get this idea? From the geometers who had been casually throwing around infinite sticks since before Euclid wrote the *Elements*. Consider this simple statement: parallel lines never meet.

This "never" means that no matter how far you go along these parallel lines in either direction, they will not touch. Geometers had **rays** which start off at a finite place and go off to infinity, **lines** which go off to infinity going forward or back along the line, **planes** which go off to infinity along a flat, um, plane, and **space** which went off to infinity in all directions (all directions that they knew about, at least).

They did not really mean that there was an infinity these sticks went to, but rather that they could go out arbitrarily far in the specified directions.

But those artists' lines seem to meet. Surely they aren't parallel.

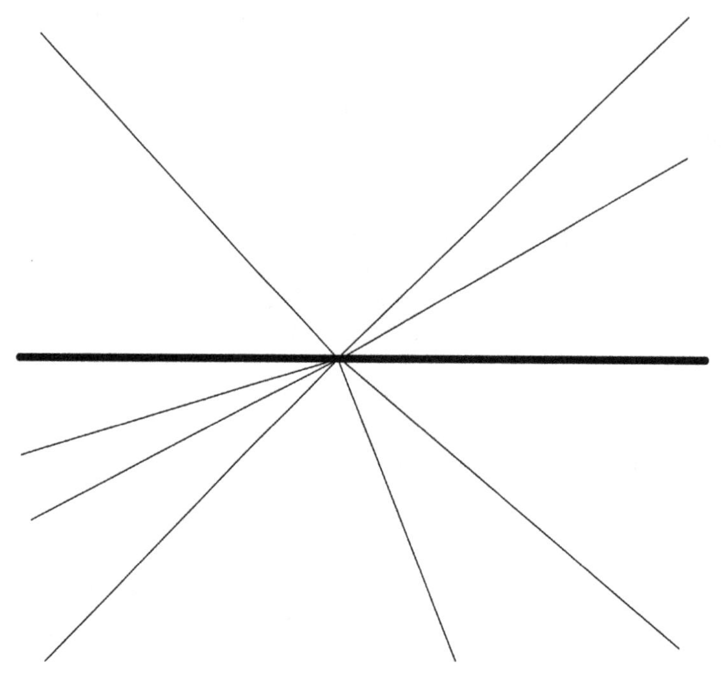

**FIGURE 2.19**
Perspective Lines and Vanishing Point

No, the ones drawn are not really parallel, but they represent parallel lines.

In Figure 2.20, there are two parallel lines seen from above.

What those lines look like if you are standing between them is shown in Figure 2.21.

The appearance of infinity in this case looks finite and creates the concept of a point at infinity, the **vanishing point**, where parallel lines seem to cross.

Infinite sticks are so deeply built into geometry that talking about them is part of the geometric mindset, which we will cover later. It is a curious thing that while infinite counting came very late and is covered in controversy, infinite measuring was simply subsumed in the underlying assumptions of the visually minded.

This part of the legacy can be found in any picture done with this kind of perspective. It was created by mathematically minded artists and is used to create beauty and to map the heavens. It is an infinity that is so commonly used that we walk right by it, perhaps admiring its contribution to the prettiness of a landscape.

There is one more kind of infinity which we have touched on before: an infinite process, doing the same thing over and over and over an infinite number of times and seeing what the result is. On the face of it, this seems nonsensical. Process and change are things we connect with time, and, as has been pointed out by far wiser heads, none of us has forever.

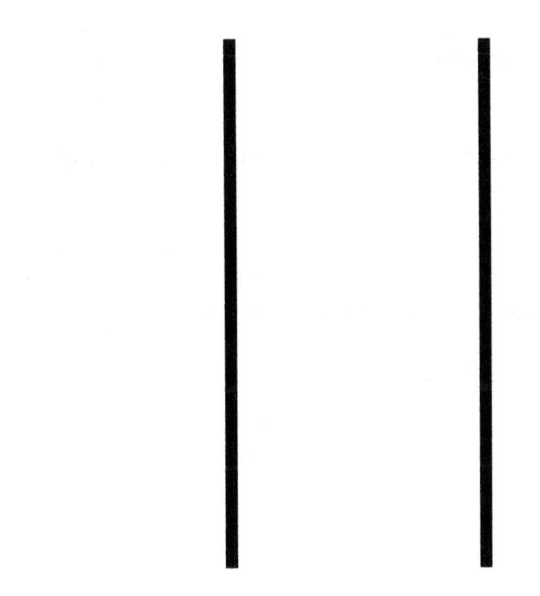

**FIGURE 2.20**
Parallel Lines

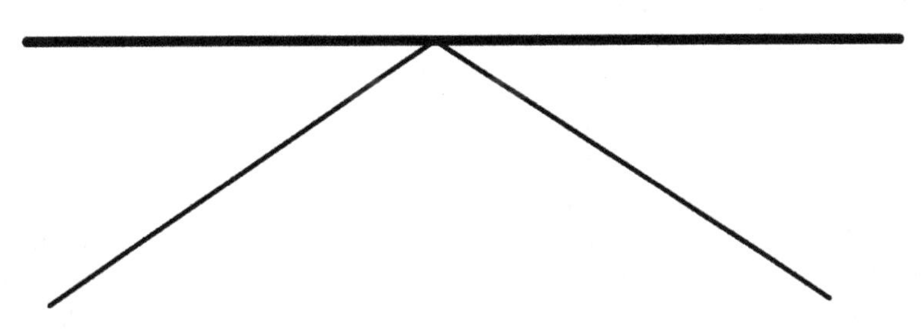

**FIGURE 2.21**
Vanishing Point Perspective

But we've already sort of seen one infinite process. In theory, our measurement of length can be an infinite process wherein we start with sticks of a certain length, then use sticks 1/10 that length to get more accuracy, then we can go to 1/100, 1/1000, etc. In theory, we could keep this going forever to get a perfect measurement. In practice, there's a lower limit to the exactness of our sticks, but we don't have to worry about that in the theoretical universe.

Besides, sometimes we want to look at a process not in terms of its practicality but in terms of its long-term effect. Suppose we look at a process and, by seeing the way it is going, figure out what would happen if one kept doing it over and over. Then we can, as it were, skip to the end without doing the middle parts. We can summarize the infinite process into a single process.

> Here's a simple example.
> Start with 0. Put it into a set.
> Add 1. Put that into the same set.
> Add 1 to that. Put that into the same set.
> Keep going. What do you end up with?

{0, 1, 2, 3, ...} We end up with N, the counting numbers. We can see where we're going, so we just go there. This is part of mathematical thinking that skips all intermediate steps if the results of each step can be completely determined and if the result of accumulating the steps can be determined.

Let us consider the process of dividing a stick up into smaller sticks. How long is each smaller stick, and how many sticks do we end up with?

> If we divide a stick in 2, we have two sticks, each 1/2 the length of original.
> If we divide a stick in 3, we have three sticks, each 1/3 the length of original.
> If we divide a stick in 4, we have four sticks, each 1/4 the length of original.

We can look at this as two sequences of numbers:

> 2, 3, 4, ...
> and
> 1/2, 1/3, 1/4, ...

In the long run, the first sequence grows to infinity and the second shrinks to 0. We would have an infinite number of sticks, each of 0 length. In geometry, that would be said as "a line segment (a stick) is made up of an infinite number of points (geometric objects with no extent in any direction)."

In analysis, this would be rendered as "the limit as n goes to infinity of 1/n = 0." We'll come back to limits in the chapter on calculus.

Look again at this process. We can not only tell where it is going to be at the end, we can tell where we will be at any step of the way. We can drop in on the 1,000,067th step and say 1,000,067 sticks, each 1/1,000,067 the length of the original.

We have taken an overview of the process, and can drop in and out anywhere we wish from beginning to infinite end without having to do the step-by-step work. This way of looking down on ongoing processes, of swooping in casually and taking what we need, is also part of the legacy, an aristocratic aerie from which we oversee the labor of ongoing work and take what we need or want from it.

*That sounds really lazy and obnoxious.*

**It would only be obnoxious if there were actually people doing the work of the process and we were really stealing their labor.** As for lazy, we'll come back to that in a later chapter.

> Indeed, we will return to all these infinities many times.
> *How many times?*
> **Countably many.**
> *How many?*
> **Okay, we'll return to the infinite a finite number of times.**

## The Numbered World

What have we made from sticks and stones?

A world where numbers are given to anything we can think of, where births carry height and weight, where houses are advertised by their areas and hotels list cost and travel times to desirable locations, where price is always asked, where timepieces are carried on wrists and belts and synchronized with networks, houses are numbered and people are counted, and where a measure of not-yet-grown coffee can be bought and sold for prices in a dozen different currencies.

It is a world where not only are things counted and measured, but where we are tempted to count and measure, where a certain kind of reality attaches to numerical valuation. How much? How many? What percentage? What are the poll results? What are the statistics? Give me the odds. Tell me the costs. Predict the profits. Make it real by putting it into numbers. Cast the first stone. Take a stick to it.

The numbered world is an image of the real world in the theoretical universe. We walk in that world as much as we do in the observable universe, but we don't pay proper attention to how the numbers come about and what they mean. We often walk carelessly, crossing traffic without looking both ways, not paying attention to the warnings and signposts, not knowing the rules of the road, not knowing how to number safely and to watch out for reckless statisticians and street theater people juggling numbers and vamping us with their figures.

Seeing the numbered world is not enough. We need to know the how and why of it, the ways used to make this world of sticks and stones before we can live safely in it.

*So that's why we're doing this? Fear of the numbered world?*

**No. That's just the warning.** The numbered world is like a power tool: safety is the first consideration. The benefits of it come from careful use. The numbered world helps people every day when they use it right. If you know and understand the numbers attached to things, you can better judge. You can decide on prices, on area, on travel times. You can know if you are driving too slow or your heart is beating too fast.

You can have a house made to your specifications, a meal cooked to the desired doneness. You can have tools that work within acceptable tolerances, computers that take care of a number of annoying tasks. You can inherit reality from the abstraction of the numbered world and know if the reality is real by knowing if the numbers are right.

The numbered world is the central estate of mathematical inheritance, as old as sticks and stones but as up-to-date as the software on computers and the prices in a grocery store.

The rules for living in it are simple ones, but regrettably not taught. Something is given a measurement. How? In *Three Steps to the Universe*, we said that the most vital question for people hearing about science was "How do they know that?" In the numbered world, the question is: "Where did that number come from?" Was it measured directly? Was it calculated from direct measurement? Were the calculations used those that preserve truth? Was care taken with accuracy? And if the number has no apparent source, did somebody just make it up? Lots of numbers and measurements are not connected to reality, they're just announced with authoritative voices.

The numbered world is a tool we live in that lets us reach out with our eyes and our hands and grasp and judge without having to grasp and measure ourselves, provided those that have placed the numbers have done so in a reliable fashion.

But what is reliable? How can one know which numbers are created rightly, which sloppily, and which are made up? For that we need to look at how mathematicians think about and help make this image of the world. It's time to get inside their heads.

# 3

# *Abstraction, Mistrust, and Laziness*

Mathematical thinking is built upon three processes: Abstraction, Mistrust, and Laziness. Abstraction is the process of taking something real and forming an idea from it (abstraction also means stealing). Mathematicians steal number, shape, and process from the real world and build their part of the theoretical universe from it.

Mistrust is the process of not taking someone's word for something, demanding proof instead. Mathematicians have the strictest standard of proof consequently the greatest mistrust.

And Laziness means avoiding unnecessary work. Mathematicians try to answer a single big question so that *other* people can do the work of answering a million small questions using that single big answer. Mathematicians love Laziness so much that they've crafted an entire aesthetic of it. Elegance, it's called.

*Other people, like scientists and engineers?*

**Ahem, yes, well. Perhaps. Oh, look over here at the elaborated discussion.**

## Abstraction

Abstraction is one of humanity's basic mental processes. We use it moment by moment whenever we perceive things and get an idea about what they might be (that looks like Charlie, that smells like chicken soup, that sounds like the worst music of all time). We use it in deeper form when we take a characteristic out of something and connect it in our minds with a description (red hair, rough surface, brilliant idea, bad example).

The mathematical process of abstraction is a more precise and elaborated version of this everyday theft of worldly perception into thought. When abstracting from the world, we do two things. First, we ignore nearly everything about the object we are abstracting; we focus our attention on a small set of its characteristics. This ignoring is vital to the whole process.

Look at this stick (Figure 3.1).

We can find the length of the stick by looking only at the endpoints and using another stick to measure it by. We can then say this stick is 1.26 (or whatever) meters long. Now look again, and see what we are leaving out: color, shape (except for endpoint-to-endpoint length), material, age, smell, taste, how easily it would burn, etc. We're ignoring all the cellular characteristics of the stick's makeup, the life processes of the tree that caused the stick to form as it did, the atomic and subatomic composition of the stick, any bugs that happen to be living in it, any diseases it's harboring.

And that's just ignoring the stick as a particular object.

We're also ignoring everything around the stick, the stick's place in the ecosystem of the woods it came from, the place of those woods in the country, the country in the world. We're also paying no attention to the stick's artistic qualities: whether the stick makes a

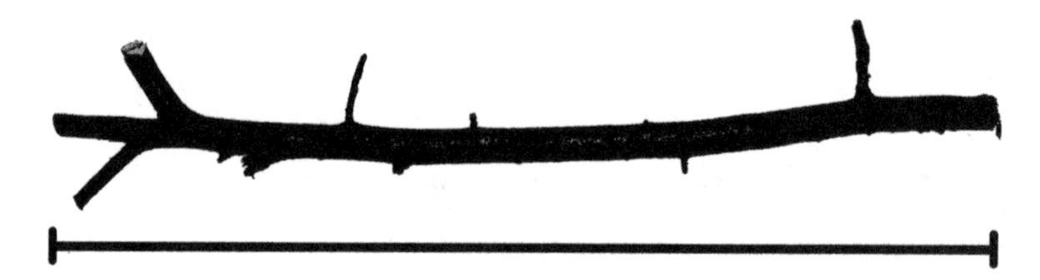

**FIGURE 3.1**
Stick and Line

**FIGURE 3.2**
Drawn Circle

good metaphor for the act of measurement, whether a beautiful haiku could be written about it, what might be made from it, what colors of paint would work if one were painting it, and so on.

This ignoring is vital because there's just too much stuff to work with for anything in the universe. From the smallest subatomic particle to the entirety of the cosmos, we are immersed in too much information. In order to work on it in our minds, we need to be indifferent to most aspects of it. The crucial thing in mathematical – and therefore scientific – abstraction is to discern what characteristics of the object matter for the mental work you need to do upon the object.

At this point, everybody in every profession and walk of life can start yelling at us that they too also have to abstract properly, and that's correct. Visual artists need to abstract appearances; lawyers need to abstract legal categories and characteristics; cooks need to abstract taste and texture. So what's different in mathematical and scientific abstraction?

What's different is precision in the definition of what concept we are taking the object into. If we are precise and clear in what we are taking away, then we can know how right we are in asserting that we are abstracting a certain quality. Consider the following shape (Figure 3.2).

We abstract this into the idea "circle," but in fact it isn't a circle. If we look closely, we see much non-circularity about it (Figure 3.3).

**FIGURE 3.3**
Arc from Drawn Circle

We abstract the drawing into a circle, knowing how much it succeeds and fails in fitting our abstraction. We can do that only if we have precise definitions of the mental objects we are abstracting into.

*Definition: A circle is the set of all points in a plane at a given distance (the radius) from a given point (the center).*

Math requires definition before abstraction. Using this definition, we can look at the drawn circle and say that it is, on the scale we are looking at, approximately a circle. If we go to a smaller scale, as we saw, then it fails to be a circle. As long as we know under what circumstances our mathematical abstraction works, we can use it. This is vital for all math and science. We need to be aware of the limits of the utility of our mathematical ideas. The real world has no circles, but it has a lot of things that, in a given range of scales, can be approximated as circles. As long as we are dealing with those ranges and as long as our concern with the object involves its shape, we can abstract these objects into circles.

The critical element here is how much you can rely on the abstraction. Consider the following poetic rendition from one of the authors:

> My lady's eyes are storm-tossed seas.
> Her hair the sunset's rays.

If I proceeded to clarify that the poet is speaking about coloration only – and even in that, there are many sea colors not present in the poet's wife's eyes and her hair does not give off ultraviolet radiation – going into all that would damage the quality of the poetry (such as it is). But if the poet actually needed to abstract the colors in order, for example, to use them on a computer (where colors are numerically specified), then one would have to forego the poetry and be exact in definition.

The reliability of the abstraction depends on how clear and precise the idea is and how well you can match the object to the idea. Clarity and precision of ideas are embodied in mathematics in the concept of definition.

Mathematical definition is not the same thing as dictionary definition. A word in a language is a fuzzy thing, covering a variety of ideas and flowing and changing over time, so that words that at one time meant the same thing can become polar opposites. For example, "awful" and "awesome" both used to mean the same effect: the engendering of awe in the viewer. But over time the scary aspects of awe accumulated into awful and the majestic and grand aspects into awesome. The words split apart.

A mathematical definition must be fixed and exact. Any changes in the definition over time must be done deliberately in order to make the definition cover exactly the ideas it is meant to cover. In order to do that, definitions must be built up from other definitions in a clear manner.

Let's consider the definition of a circle given above: *A circle is the set of all points in a plane at a given distance (the radius) from a given point (the center).*

For this definition to make sense, we have to know the meanings of set, point, plane, and distance. Lacking clear meanings for these, a circle could be anything. You can play the old parlor game where you substitute words and make whatever appeals to you out of it.

*A circle is the set of all in a given (the radius) from a given (the center).*

Hilarity can ensue, if desired.

Rather than go through the entire process of filling in the correct definitions, let's look at what happens if we change one word in the definition: plane.

Suppose we substitute the word "line" for "plane."

We end up with *the set of all points in a line a given distance (the radius) from a given point (the center).* This ends up being just two points (Figure 3.4).

Suppose instead we substitute "space" for "plane." We have *the set of all points in space at a given distance (the radius) from a given point (the center).* This is a sphere (Figure 3.5).

Suppose we go back to the circle and change "at a given distance" to "at most a given distance": *the set of all points in a plane at most a given distance (the radius) from a given point (the center).*

This gives us a solid disk (Figure 3.6).

If in everyday life you showed someone the circle above and the disk above and asked, "Are these circles?" he or she would say yes. A mathematician, if being careful, would say the first is; the second isn't.

Similarly (although harder to illustrate), we can do the "at most" change to the definition of a sphere and get a ball. In everyday life, sphere and ball are synonyms, but in math they aren't.

It gets even messier. If the definition is *the set of all points in space whose distance is less than a given distance (the radius) from a given point (the center),* we have what's called an "open" ball. If it's "less than or equal to," we have a "closed" ball. The difference here is that a closed ball includes the sphere of the given radius, but the open ball does not.

What difference does that make? you might sensibly ask. The best we can say at the moment is that an open ball can contain every point leading up to a point on the sphere, but not the point itself. So an open ball does not contain the obvious endpoints of a number of processes (such as moving out from the center toward the surface). That doesn't help much now, but it will later when we talk about the mathematics of limits.

**FIGURE 3.4**
Equidistant Points on a Line

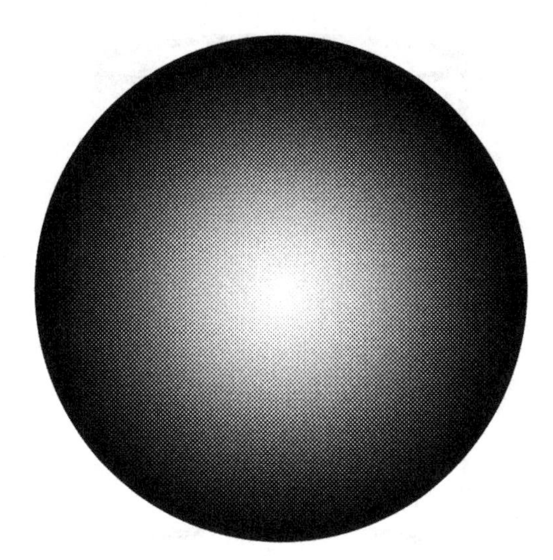

**FIGURE 3.5**
Sphere

**FIGURE 3.6**
Disk

From this it should be clear how precision of definition helps us in seeing whether or not something fits a definition well enough for us to abstract the object into the definition.

Is this a circle? (Figure 3.7).

No. Its points are not all in a plane, and they are not all at a given distance from the center. But if you didn't need to worry about the up and down direction, you could treat it as a circle.

Is this a circle? (Figure 3.8)

No, its points lie in a range of distances. But if you didn't need to be precise, you could treat it as a circle.

Is this a circle? (Figure 3.9)

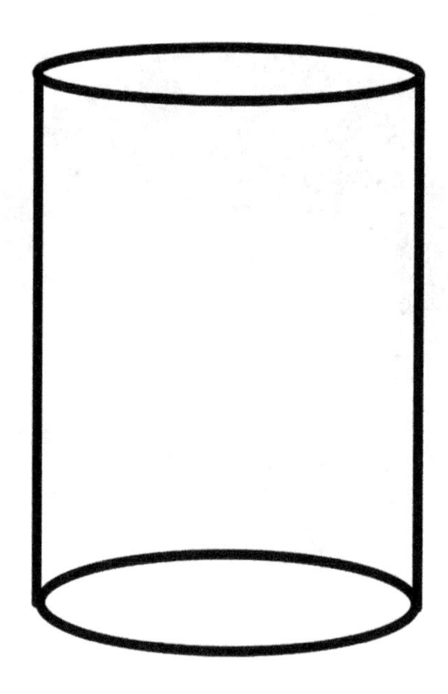

**FIGURE 3.7**
Cylinder

**FIGURE 3.8**
Annulus

No, because it does not have *all* points in the plane at the given distance, etc. But if you only need to check the behavior of this part of the circle or the behavior of individual points, you can treat it as a circle.

All right, you say, enough already. We'll treat it as a circle. What good does that do us? What's the point of this process of abstraction?

The point is that mathematical abstractions are not isolated ideas. They exist in a world of interconnected concepts and processes that allow us to work upon them and then gain an understanding of the abstracted object that was not there before the abstracting.

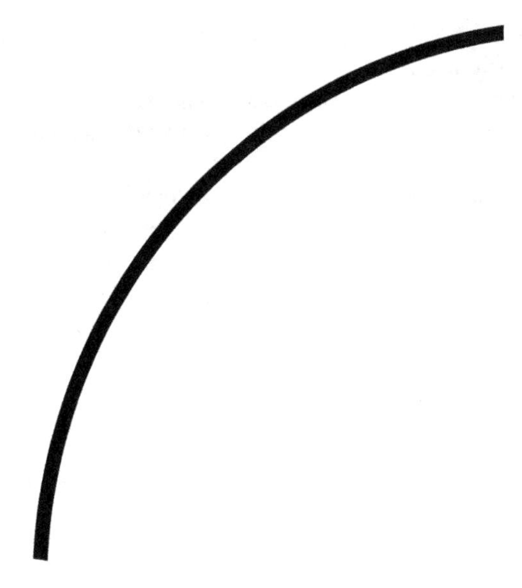

**FIGURE 3.9**
Arc

Let us suppose one can abstract an object in the real world as a ball. Suppose further that by measurement one can determine that the object has a radius of 0.5 meters. Suppose that chemical tests determine that it is made of gold. Ignoring the question of taxes owed once one gets the thing home, there is the matter of how much mass it has (and therefore how strong a crane one needs to lift it). It's too heavy to easily put it directly on a balance scale. So how can we determine its mass?

The abstraction of the object is a ball 0.5 meters in radius made of gold. Gold has an approximate density of 19,320 kg/m$^3$. This abstraction does not float alone in the theoretical universe; it abuts with two useful equations:

$$m = \rho V \left(\text{mass} = \text{density} \times \text{volume}\right)$$

and

the volume of a ball is

$$V = 4\pi r^3/3 \left(\text{Volume} = 4 \times \text{pi} \times \text{the cube of the radius divided by } 3\right)$$

Putting in these values gives us $V = 0.5236\ m^3$.

Plugging this value for V into the first equation gives us

$$m = 10{,}115.92\ \text{kg}$$

At the time of this writing, gold was selling for about \$50,000/kg. So the current value of this regrettably hypothetical ball of gold is about \$500 million. So you can probably afford to rent the means of moving it.

It is the interconnection of theoretical objects that makes the theoretical universe work. This is most clear in the case of mathematical formulas. If two formulas refer to the same abstraction, then that abstraction can be used in both formulas in a way that allows one to discern new understandings in the theoretical universe and then come back to the real world with them and apply the results.

*This is the basic use of math in science. We abstract from reality to math by measurement, employ mathematical tools to the measurements, and produce results that should be as accurate as the measurements we started with.*

But how can we know that these theoretical structures and connections, these formulas, shapes, functions, spaces, and all the other mathematical apparatus are reliable? After all, people can just make things up. We could announce that the average size of women's feet times the number of dishes of saganaki (a Greek dish of flaming cheese) sold in Chicago in 1998 determines the mean density of the now-downgraded planetary body Pluto. I could even write it in a formula:

$$\rho_{\text{pluto}} = A\left(F_w\right)S_{\text{Chicago},1998}$$

But you have no reason to accept the truth of this. This brings us to the second fundamental property of mathematical thinking.

## Mistrust

Just take my word for it. I'm your friend, aren't I?

You know it's true. We're of the same ethnic background/religion/nation/social club.

It's true. I'm yelling it, aren't I?!!!!?

Believe me or else!

Believe me because I'm pretty/cool/an actor/a singer/a politician.

Surely you know I'm honest. Just look at my face. A smart person like you couldn't possibly be fooled. Just sign here – no, don't bother reading it.

Hey, just look at my cover. That's all you need to judge me.

Look at the pretty pictures in my presentation. Everything's right with pretty pictures.

How dare you challenge my integrity! You can tell I'm an old school, reliable, just look at the size of my mustache.

Trust from person to person must be earned by acting in a trustworthy fashion. But careful mathematicians never trust each other. They never trust themselves. They don't trust anything mathematical they can't prove.

No problem. I've got five witnesses and the best forensic evidence money can buy.

I got proof right here – a sworn statement from a person of unimpeachable integrity.

You want proof. Drink this. 100 proof.

That's not good enough. Mathematical mistrust is so deep and so harsh that it has a standard of proof unmatched anywhere else. Unfortunately, it can't be used anywhere else because it is proof that never touches the real world.

To get a clear sense of where mathematical mistrust lies on the scale of mistrust, we will examine different standards of proof wherein each higher standard looks down with condescension or, at best, sorrowful compassion on those lower on the scale.

The everyday standard of trust is that if someone the listener has no reason to mistrust says something that sounds believable, the listener will tend to believe it. Listeners will also tend to believe what they have apparent direct experience of. Advertisers have learned to take advantage of both of these, as have stage magicians.

Moving up a notch from everyday standards, we come to law.

The legal standard of mistrust varies, depending on the court case. Two common standards are: "the preponderance of the evidence" and "beyond a reasonable doubt." In the former, the question is whether the evidence presented makes it more likely that one side is right than the other. In the latter, the question is not whether there exists room for *any* doubt as to what is true, but whether the circumstances would cause a *reasonable person* (whatever that is supposed to mean) to doubt. (For example, "My evil twin did it" is usually not cause for reasonable doubt unless the person demonstrably has an evil twin.) Because they are schooled in these standards, lawyers tend to feel themselves better judges of the truth than most people.

Climbing a step up the ladder of mistrust, we reach science.

The scientific standard of mistrust is a combination of the mathematical, which we will get to soon, and the experimental. If a result can be tested in multiple ways by different experimenters using different equipment, and if the results each time conform within acceptable error ranges to the predictions, then the theory behind those predictions is considered more likely to be true. We won't go too far into this subject because we covered it in our first book. But in summary, scientists test their ideas repeatedly from multiple directions and are always on the lookout for anything that could prove them wrong.

The everyday and legal standards of mistrust are founded in human reliability. The scientific standard is founded on the reliability of repeated actions in the real world.

The mathematical standard of mistrust is not connected to the real world. It has no interest in evidence or testing. It uses a process that boils down to this: if one starts with a set of premises and uses those premises through processes that preserve truth (that is, processes of reasoning that never go from true premises to false conclusions) and one comes to a set of conclusions, one has proven not that the conclusions are true but that given that the premises are true, the conclusions will be true.

## Premises

Consider the matter of the ball of gold discussed above. The mathematical proof of the process would be: given a ball of gold 0.5 meters in radius and given that the density of gold is 19,320 kg/m$^3$ and given that the price of gold is \$50,000.00/kg, we can establish using the following formulas:

$$m = \rho V \left(\text{mass} = \text{density} \times \text{volume}\right)$$

$$V = 4\pi r^3/3$$

$$\text{price} = \text{number of unit masses} \times \text{price/unit mass}$$

that

$$m = 10{,}115.92 \text{ kg}$$

and

$$\text{price} = \$505{,}796{,}000.00$$

This does not mean we have proven that we now have more than $500 million worth gold because the premise that we actually have such a ball of gold is false.

In this definition of mathematical proof, there are areas of concern: the premises themselves and the idea of processes of reasoning that preserve truth.

Premises fall into three groups (except they're really two groups and the distinctions between those two are artificial, but we'll keep the words for these groups and try to explain them): axioms, postulates, and hypotheses.

In ancient Greek an axiom was something that was blindingly obvious, so obvious that no one could possibly deny it. For example, one of Euclid's axioms for geometry said that given any two points, there is a unique line containing those two points. This says both that such a line exists and that there is no more than one such line.

The problem is that several of these blindingly obvious truths turned out to be not as true as they thought, particularly those dealing with the shape of space, as we will see later. As a result, the word "axiom" has now come to mean the fundamental premises of what you are doing, the underlying assumptions on which all else is built. Nowadays, every mathematical object needs a set of axioms so that you know what you are talking about.

After axioms are postulates. These are also underlying ideas, but they do not have the cachet of being blindingly obvious. The distinction between axioms and postulates is now nonexistent. Indeed, before the change of meaning in the axiom, one of Euclid's axioms was downgraded to a postulate. It's called the parallel postulate, and as we will see later, it became the root of new kinds of geometry. The parallel postulate says that given a line and a point not on that line (Figure 3.10) there is one and only one line passing through

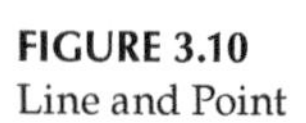

**FIGURE 3.10**
Line and Point

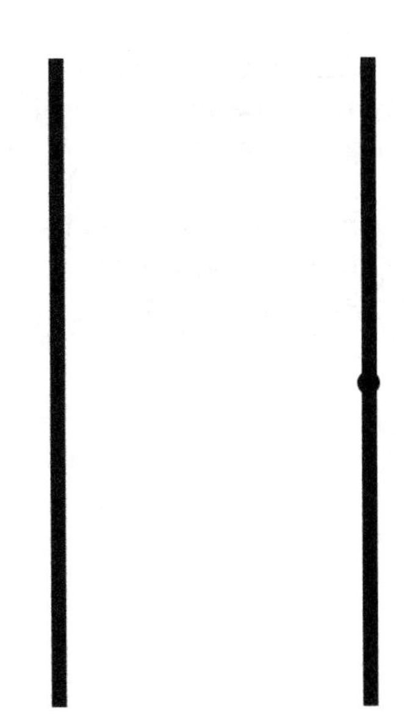

**FIGURE 3.11**
Parallel Lines

that point that is parallel to the first line. (Parallel lines are lines that lie in the same plane but never intersect.) (Figure 3.11).

This seemed obvious to Euclid because if you look at a plane, it's true. If you tilt that second line even a little, it will eventually intersect with the first line. Refuting this idea involves bending space. In the real world, gravity takes care of that for us.

So axioms and postulates are foundational premises that apply whenever one is working in a particular area of mathematics.

Hypotheses, in contrast, are premises that are brought up for particular problems. They are created and discarded in each circumstance. Say we have one problem with a premise that a variable $y = 5x + 3$, and in another problem $y = 3x - 2$.

While these two problems both rely on the axioms of algebra to solve them, each problem has its own premises for the relation between y and x. Unless it is specifically said that these two problems are about the same y and x, the hypothesis of the first problem has nothing to do with that of the second and the x of the first need not be the x of the second.

Keeping track of axioms/postulates and hypotheses is vital in any use of mathematics because they tell you what you are talking about. You have to understand what you're talking about before you start talking, which is a big difference between math and most other aspects of human thought and conversation.

## Methods of Reasoning in Proofs

Care in choice of premises is, however, not only present in math. Law and science both (it is to be hoped) use such care. It is the use of truth-preserving reasoning that distinguishes math. The means of truth-preserving reasoning are fairly easy to understand; indeed, most will seem obvious. The main thing is not what is allowed in this kind of reasoning; it's what is *not* allowed. Analogy is not permitted; stories are not allowed. Metaphor is forbidden. Poetry, imagery, all the things that make language persuasive and beautiful are kept out.

Although math, as we shall see, has its own ideas of beauty, and mathematicians are as susceptible as anyone else to the grace and flow of art and image, they keep it away from their work, for the same reason that a biochemist keeps pastrami on rye out of her clean room. No matter how much she loves the sandwich, she doesn't want the mustard messing up her work.

So what is allowed in mathematical reasoning? What are these obviously truth-preserving methods?

First, a couple of classics that have fancy old names: *modus ponens* and *modus tollens*.

*Modus ponens* says that if it is true that the truth of A implies the truth of B, and A is true, then B is true. If it is true that if someone hits Fred then Fred will become angry, and it is true that Margie hit Fred, then Fred will be angry.

*Modus tollens* says that if the truth of A implies the truth of B, and B is false, then A is also false. If it is true that if someone hits Fred then Fred will become angry, and Fred is not angry, then no one hit Fred.

Related to *modus ponens* is the specification. If something is true for everything in a set, then it is true for each individual thing in that set. The classic example is: all men are mortal; Socrates is a man; therefore, Socrates is mortal.

There is also generalization. If it can be proven that something holds for each individual member of a set, then it holds for all members of that set. So if Fred is not smart, Barney is not smart, and Dino is not smart, then every member of {Fred, Barney, Dino} is not smart.

There are some more such methods of reasoning that are even more obvious than the basics:

If A is true and B is true, then statements A and B are true. Fred is dumb. Barney is dumb. Therefore, Fred is dumb and Barney is dumb.

The converse also holds. If A and B are true, then A is true and B is true.

And one more blindingly obvious one: you can repeat something true later in the argument as many times as you need to.

Let's move on to things that have to do with numbers.

There's a symbol we've been using casually without actually saying what it means. That symbol is =. What does equality mean in math? In arithmetic, it means that two things represent the same number. When we say $3 + 2 = 5$, we mean 3 rocks with 2 more rocks makes 5 rocks. When we say $3 + x = 5$, we mean that x can only be whatever number added to 3 gives 5; hence $x = 2$.

Most of the math people are taught in elementary and high school and the beginnings of college revolve around the = sign. And indeed only four characteristics of the = sign come into play in most of the problems people are given. Three of them have names; the fourth is a deduced principle.

The three named properties are:

*Reflexive*: For any number x, $x = x$. This sounds like a brainlessly obvious characteristic, but if we replace = with <, then we find a statement that is absolutely false.

*Symmetric*: For any numbers x and y, if $x = y$, then $y = x$. Again, obvious for equality, but false if = is replaced with <.

*Transitive*: For any numbers x, y, and z, if $x = y$ and $y = z$, then $x = z$. In this case, it's still true if we replace = with <, but not true if we replace = with ≠ (which means "does not equal").

Any relation between things that is reflexive, symmetric, and transitive is called an equivalence relation. Equivalence relations satisfy our sense of what it means for things to be like each other. Same size, same color, same shape, same taste in music, and so on. These all seem = like to our minds. What separates the numerical = from these is the fourth property, the one that lets us do all the calculations.

If $x = y$, then if you do the same numerical operations to x and to y, the results will still be equal. Do the same thing to the left and right sides of an equation, and you will preserve truth.

So if $x = y$, then $x + 5 = y + 5$, $36x = 36y$, $x + x = y + x$, $2^x = 2^y$, etc.

Choosing what to do to both sides of an equation determines how easily one can solve an equation. The following may cause flashbacks for people who had trouble in algebra. Don't panic.

Suppose $5x = 3x + 6$, and we are asked to solve for x, which means to have x alone on one side of the equation and on the other side have no x's whatsoever.

We could do anything we want on both sides of this equation. We could take the square root of both sides, multiply them by 1,000,000.8637, divide by $\sin(y + 42)$. But none of these would do any good. We need x alone on one side. So we'll subtract 3x from both sides, which gives

$$5x - 3x = 3x + 6 - 3x$$

Now we're going to do the next two steps quickly and then slowly because the first time what we're doing will look reasonable but have no justification behind it. The second time we'll justify our actions.

$2x = 6$

$x = 3$

Now for the slow.

$5x - 3x = 3x + 6 - 3x$

There is an axiom of addition called the commutative law, which says that

$a + b = b + a$

So we will replace $3x + 6 - 3x$ with $3x - 3x + 6$. What justifies the replacement? The transitive property of =. We know

$5x - 3x = 3x + 6 - 3x$

and

$3x + 6 - 3x = 3x - 3x + 6$

So

$5x - 3x = 3x - 3x + 6$

But if we consider what addition and subtraction mean (piling up rocks) whatever value 3x is, if we take away that same value, we end up with 0. We can therefore replace $3x - 3x$ with 0.

$5x - 3x = 0 + 6$

An axiom of addition says that 0 plus any number = that number. So

$5x - 3x = 6$

Now, there is an axiom that connects addition and multiplication called the distributive law, which says:

$a(b + c) = ab + ac$

$(b + c)a = ba + ca$

So

$5x - 3x = (5 - 3)x$

$(5 - 3)x = 6$

$2x = 6$

So

$x = 3$

The scary thing is that we still slurred over a few of those steps and would have had to digress and prove even more results before results, but we didn't want to create too much madness in one place. We prefer to spread them out, to take an egalitarian approach to the propagation of madness.

Mathematicians almost never solve problems like this. They spend their time trying to prove much more general results for reasons of laziness (see below). The proofs in the previous chapter about the number of elements in various sets are the kinds of things they do.

The lazy results proven by mathematicians are usually called theorems. Once a theorem has been proven, it can be used in any other proof so long as the premises of the theorem are also premises of the new proof. For example, the fact that there are more real numbers than integers can be used in any proof involving the number of real numbers and integers.

Theorems tend to be big, muscular results that stand on their own. Accompanying them are two smaller kinds of proofs, the smaller siblings of theorems: lemmas and corollaries.

Lemmas are results that are useful in proving other theorems but don't have any other use. If math were cooking, Lemmas would be spice mixtures or marinades or sauces. Sometimes they are difficult to make; and they are critical to dishes, but you wouldn't want to eat them straight.

Corollaries are small results that follow easily and swiftly from theorems. Their proofs are too fast and easy to be accounted for the heavy work of theorem proof. Corollaries are like sandwiches made from leftovers (they said, pushing the cooking metaphor to its limits), made of the same tasty ingredients, but less work to concoct since someone else already did the work of cooking the filling.

## Kinds of Proof

The direct generation of conclusions from premises is not the only kind of proof, nor even the most commonly used. There are a number of different ways mathematicians go about their business, and these methods are at least as revelatory as the method of direct reasoning:

*Proof by Contradiction.* A staggeringly popular form of proof, this is based on *modus tollens.* Suppose you want to prove that a given statement is false, for example, "There exists a highest integer." First, you say, let's suppose this is true. What follows from it? If you can, by direct reasoning, find that something false follows from this premise, then you know the premise is false. Because we are using truth-preserving methods in which truth leads only to truth, if something leads to falsehood, it must be false. So suppose there is the highest integer. Call it a.

It is an axiom of arithmetic that if a is an integer, a + 1 is also an integer. It is also axiomatic that a + 1 > a. Hence, a cannot be the highest integer, which means that our initial premise, which required that a be the highest, is false, and there is no highest integer.

Cantor's diagonalization proof (see Chapter 2) is another example of this kind of proof.

*Proof by Counterexample.* People have a tendency to think that something is proven true if you can find a lot of examples of it. But that's just evidence, and math doesn't care about evidence. Math does say that something is proven false if you can find a single example that is supposed to fit it and doesn't. As a simple example: "All integers are even." Hey, what about 1? Nuts, We're wrong. Proof by counterexample sometimes involves careful construction of the offending object, and sometimes they're just staring you in the face.

*Proof by Exhaustion.* If a hypothesis can be broken down into a finite set of possibilities and each possibility proven in turn, then the entire hypothesis has been proven. Until the advent of computers, Proof by Exhaustion usually involved a small set of possibilities. With computers, that limitation has exploded. Some people are nervous about this. (We'd come up with an example of this proof method, but we're too tired.)

There are other correct methods that will show up as we go along. There are also false proof methods which do not work in math but which can be persuasive. Part of the legacy of math involves learning to mistrust these methods:

*Proof by Intimidation*: Big strong person armed with big fists or gratuitous weapons "suggests" that you agree with them.

*Proof by Seduction*: Highly attractive person with gratuitous speech and gestures "suggests" that you agree with them.

*Proof by Technobabble*: Person speaks rapidly and confidently, using words and concepts you are not familiar with, and then ends with "that's clear, of course."

*Proof by Examples*: Page after page of examples that fit the premises with no mention of any that don't fit.

*Proof by Beauty*: The idea is too beautiful not to be true. This one does catch mathematicians sometimes because of the elegance of the ideas.

## Elegance

The mathematical idea of elegance is a little hard to put into words. In an elegant proof, the ideas fit together smoothly and cleverly; they flow and sparkle. The ideas themselves reveal and illuminate all around them and all within them. The trouble with trying to bring this concept across is that beauty is very much in the eye of the beholder. People who see this as beautiful already see it. Those who don't … well, it's like arguing over any aesthetic. It would get loud and be pointless.

If you looked at Cantor's proofs presented in the last chapter and felt a warm glow, or were amused, or you objected to the picture of the snake and the diagonal line, then you might have an appreciation for mathematical elegance. If not, don't worry about it. The important thing to note about elegance isn't whether you get it or not but to understand that it matters to mathematicians, that it is one of the things that can guide what they look at and try to prove and how they go about trying to prove things. Nobody is completely immune to the lure of beauty, and the desire for clean, bright, clever proofs is one of the lures of mathematicians.

It can also be a trap. Mathematicians enamored of certain aesthetics have been known to disregard other possibilities. Concepts such as fractals were once called pathological (i.e., diseased) because they did not conform to the aesthetics of the time. Only by opening up to a broader sense of beauty and elegance were they brought within mathematical aesthetics.

One lure however has been vital to the evolution of math, the desire to do all the work at once.

## Laziness

There are two kinds of laziness: not wanting to do any work, and wanting to work as efficiently as possible. All mathematicians fall into the second group. Some – you know who you are – also belong to the first. We're not going to talk about them.

The laziness involved here is the kind that wants to solve a problem once and for all. It's the laziness of a toolmaker seeing the need for a new tool so that one needs to no longer break one's back and heart in wasted effort. It is creative laziness.

Here are a couple of examples from kinds of equations that show up all the time.

A common kind of equation is the "linear" equation. (We'll see why it's called that in chapter 5.) They usually have the form $y = mx + b$. If one has values for m, x, and b, finding y is a quick calculation. But suppose one has values for y, m, and b, as in this case:

$$11 = 3x + 2$$

In order to solve this, one does the following:

Subtract 2 from both sides.

$$9 = 3x$$

Divide both sides by 3.

$$3 = x$$

It is possible to come up with a single formula which solves all such equations by dealing with the basic equation $y = mx + b$ so that one can solve it without knowing any of the values.

$$y = mx + b$$

Subtract b from both sides.

$$y - b = mx$$

If m = 0, then there is no solution unless y = b, in which case it doesn't matter what x is.

If m does not equal 0, the solution is

$$(y - b)/m = x$$

Given any equation in the form y = mx + b, one can put in the y, b, and m values to the above equation and get the answer. This is not that impressive; after all, linear equations need only a few steps to solve. How about something a little harder?

$$0 = x^2 + 4x + 2$$

Solving this involves a process called completing the square, which is based on a little bit of algebra:

$$(x + a)^2 = x^2 + 2ax + a^2$$

from which we can deduce that $(x + a)^2 - a^2 = x^2 + 2ax$.

We can use this to transform the right side of the equation above by looking at $x^2 + 4x$ and turning it into $x^2 + 2(2x)$, which we can replace with $(x + 2)^2 - 2^2$.

So

$$0 = x^2 + 4x + 2 \text{ turns into}$$

$$0 = (x + 2)^2 - 2^2 + 2$$

which since 2 squared is 4 turns into

$$0 = (x + 2)^2 - 4 + 2$$

$$0 = (x + 2)^2 - 2$$

or

$$2 = (x + 2)^2$$

If we take the square root of both sides, we get

$$\pm\sqrt{2} = x + 2$$

or

$$x = -2 \pm \sqrt{2}$$

The process of completing the square is already a general method, a decent tool for solving these problems, but it can be made even lazier. Let's take the general equation $0 = ax^2 + bx + c$. If a is 0, we are dealing with a linear equation, which, as the mathematical jargon says, reduces to the previous problem.

We do not know the source of the following joke, but it has made the rounds in various forms.

*Question*: How does a mathematician put out a fire, when given an empty bucket and a sink?

*Answer*: He uses the sink to fill the bucket with water and then pours the water from the bucket on the fire to put it out.

*Question*: How does a mathematician put out a fire, given a bucket filled with water and a sink?

*Answer*: He pours the water from the bucket into the sink, thus reducing it to the previous problem.

If a is not 0, we transform the equation by dividing both sides by a, producing:

$$0 = x^2 + (b/a)x + c/a$$

We can now complete the square, replacing $x^2 + (b/a)x$ with its equivalent in completing the square. In order to complete the square, we must take ½ of whatever is multiplied by x (in this case, we take ½ of b/a, which is b/2a), then squaring x + b/2a and subtracting the square of b/2a. This gives $(x + b/2a)^2 - b^2/4a^2$.

So the equation becomes

$$0 = (x + b/2a)^2 - b^2/4a^2 + c/a$$

Subtracting $-b^2/4a^2 + c/a$ from both sides gives

$$b^2/4a^2 - c/a = (x + b/2a)^2$$

A little finagling with fractions makes the left side $(b^2 - 4ac)/4a^2$. So the equation becomes

$$(b^2 - 4ac)/4a^2 = (x + b/2a)^2$$

We take the square root of both sides and end up with

$$\pm\sqrt{(b^2 - 4ac)}/2a = x + b/2a$$

Subtract b/2a from both sides, and we end up with the famous quadratic formula:

$$x = \left(-b \pm \sqrt{\left(b^2 - 4ac\right)}\right) / 2a$$

A little messy to calculate for a human, but a computer can easily be programmed to do this over and over as long as it is supplied with values for a, b, and c. And if you think that doesn't matter, we'll run back into the quadratic formula in the discussion of motion, where it is used to determine how far an object shot from a cannon will fly and how long it will take before it hits. In times of war, this formula matters a lot, and fast calculation has been the difference between life and death, certainly between hitting and missing.

It is this desire to solve a problem once and for all and then go have lunch that is one of the motivating factors in the expansion of mathematics. It's not just trying to solve an equation finally; it is trying to find theorems that are so universal as to be applicable in arbitrary numbers of situations, of devising methods of problem-solving that can be used broadly and deeply, of defining things in terms so general but so focused on the particular needs of a situation that the resulting results can be applied anywhere the definitions apply.

The principle here is that every individual problem is a special case of a general problem. In order to solve the individual, we first solve the general and then stick the particulars of the individual problem into the general solution. It may be more work to solve the general problem this time. But the next time you come up against another individual form of the general problem, you can solve it immediately by using the general solution you came up with. And the time after that you can do it again. And if you teach it to your students, they can do the same, and they can pass it on, and it can spread out into human knowledge, making less work for generations to come.

Furthermore, if you come up with a general solution, you or somebody else might later come up with a better general solution based on that general solution. Completing the square came before the quadratic formula. The former was a good general solution; the latter is a great one, unlikely to be improved upon, since it reduces an entire class of equations to a single process of calculation that can be done easily by a computer.

Once you have a sufficiently general solution, you often discover that it can be used in situations you did not anticipate. Probability, when it was created, existed to help gamblers beat the house. No one at the time anticipated it could be used to answer fundamental questions about the workings of the universe. But that is happening right now in quantum mechanics.

Entire fields of math have been created by finding a way to solve a small group of problems (calculus, as we will see, arose from two kinds of problems) and then expanding outward from those solutions.

Mathematical laziness allows one person to devote a lifetime to make work easier for everyone who comes after them.

## Putting the Three Together

So what goes on in a mind that practices abstraction, mistrust, and laziness?

Most of the time, just regular everyday stuff.

But when such a mind looks at a kind of problem that seems to be an instance of something more general, it takes up a different approach. It turns the problem over, looking for the aspects that are relevant to the problem, seeking what can be abstracted. It especially seeks to find relevant aspects for which there are already tools so that it does not need to do unnecessary work.

The mind then turns over the relevancies, seeking how to connect them to the needed solution using means that preserve truth and are hopefully not too complicated to implement. It works to pare away the unnecessary in order to make as general a solution as it can.

The goal is to make a tool that can be used again and again, not just by the creator of the tool, but by anyone willing to learn to use it. In this way mathematicians' work is meant to spread to minds beyond theirs and last longer than their lives. They seek to add to the tool kits humanity uses when dealing with the real world by creating tools that are not themselves of the real world, but can be applied back to the real world.

The crucial process shows the interplay between those dealing with factual problems (such as scientists) and the mathematicians making tools for them.

In later chapters, we will see how this back and forth between reality and mathematics has evolved both math and science.

# 4

# *Algebra, Geometry, Analysis: The Mathematical Mindsets*

All mathematicians employ abstraction, mistrust, and laziness, but they do not all think alike. Some people abstract and manipulate qualities of things, some people abstract and manipulate the appearances of things, and some people abstract and manipulate the ways things work.

In the realm of imagination these are largely aesthetic differences, but in the region of mathematics they represent three different approaches to abstracting and transforming the world through theory. These approaches see the world differently and therefore they abstract differently. Each way of thinking is most comfortable with creating and manipulating different kinds of mental objects.

These three different ways of thinking and working have given birth to the three fundamental branches of mathematics, each of which tends to attract the people who are most comfortable with its thinking.

The three are algebra, which comes from considerations of number and operations, and which abstracts qualities; geometry, which arises from consideration of shape and space, and which abstracts appearance; and analysis, which is sparked by concepts of process and behavior, and which abstracts how things happen. At various times in the history of math, each has had an ascendancy over the others. In such times one branch has been deemed the fundamental mathematical way of thinking with the other two as adjuncts, helpers to the main.

Right now we live in a highly algebraic time, with the assistance rendered from analysis. Geometry is left far behind in consideration, though its child branch of mathematics, topology, has taken over much of its previous place. The dominance has changed before and probably will again. The rise of fractal geometry and the use of computer graphics have to a certain extent resulted in a revival of geometry. The rise of computer-based analysis as a tool in the sciences has pushed analysis up as well. But for now, for reasons dealt with in a later chapter, algebra rules the abstracted roost.

A feeling for each kind of thinking can be found in examining a little more closely what kinds of things each tends to abstract and what kinds of theorems each tries to prove.

*Algebra* abstracts objects and operations on those objects. Numbers are abstracted objects (from sticks and stones), and addition and multiplication are abstracted operations (from heaping up rocks and making new sticks). Algebraic theorems try to establish universal characteristics of those operations on those objects. The commutative law of addition, $a + b = b + a$, is such an algebraic statement, as is the quadratic formula.

*Geometry* abstracts shapes and properties of those shapes. Circles are abstracted shapes (from round things). That the area of a circle is $\pi r^2$ where r is the radius of the circle is a property of that shape (how much extent is inside the round thing). Geometric theorems tend to establish what the properties of shapes are.

*Analysis* abstracts processes into mathematical functions and the behavior of processes into the study of those functions. Falling is a process. Abstracting one aspect of the process into a function produces Newton's universal law of gravitation: $F = GMm/r^2$ where F is the force of gravitation, G is a constant, M and m represent the masses of the objects attracted to each other, and r is the distance between them. Theorems in analysis tend to establish universal principles about kinds of functions.

Any mathematical object can be looked at from each of these perspectives. A circle can be an object to an algebraist (as a kind of object defined by a batch of numbers representing center and radius), a shape to a geometer (because it is a shape), or the process of drawing a circle to an analyst (who sees it as a function by which to generate each point of the circle).

It is easy to fall into the trap of thinking of these views as disparate and disconnected, but they are not. They are only ways of thinking. Some people have an easier time with one than another, and some people can move freely between them and combine them casually. As we will see later, some of the most important and useful math has been created by people who abstracted the same objects in more than one of these ways and then connected the abstractions into single unified wholes that had implications and uses in multiple branches.

These ways of thinking are not confined to mathematicians. Everyone uses all of them to a greater or lesser degree and with greater or lesser facility.

We each possess an everyday algebra of abstracted objects and operations (fork + food = forkful of food which can be eaten).

We each walk through a geometry built from our sense impressions, the shape of the world that comes from sight and sound, smell, and touch.

We each take note and care (hopefully) of the processes of our life, the beating of our hearts, the pace of our breaths, our growth and aging, the turning of time measured in moons, seasons, and bill payments.

And each of us is more or less comfortable with each of these ways of thinking. We tend to rely on the ways that we have the easiest time with and shy away from those that are too much work for us personally.

The mathematical versions of these processes are created from these everyday ways of thought by applying a refiner's fire made of abstraction, mistrust, and laziness to produce algebra, geometry, and analysis.

The refinement itself is done by learning to be careful of meaning and definition in the process of abstraction (What is a circle?); learning to mistrust both other people's words and one's own opinions (It's got to be a circle. It's round, isn't it?); learning to seek and generalize the most universal characteristics of the object (What can we carry over from circles when we stretch them out into ellipses?); and finally, practice.

Mathematical thinking is developed by applying math to real-world-derived problems and concerns. As with anything one learns, the more one practices, the more comfortable one becomes with it and the more capable one is of using it. This is where people's talents and proclivities come into play. It's easiest to understand and to practice the ways of thought one is most inclined toward. Each of us inclines more or less to one of the underlying mental processes that can be refined by thought and practice.

Many people were turned away from learning math because they were confronted with processes they were not at ease with and so could not refine. They needed teaching methods that emphasized and empathized with their own ways of thinking. We'll talk more about this in the final chapter.

But it's a mistake to focus only on what one is comfortable with and ignore the other ways of thinking. There is a lot to be gained in studying something one lacks the talent for (for example, Richard finally learned to sing at the age of 46). Difficulty of learning leads to having to take more care, and the more care one puts in to learning something, the more one will understand its limitations.

The ways one learns easily can be used to help one learn the other ways. If one can see the shape of the circle and understand it that way, then one can use the shape to help learn the process and the object. If one can understand the process of functions, one can come to understand shapes by seeing how the graphs of the functions become those shapes.

Using this complementarity of the processes, one can use each to help the other and eventually learn whatever one wants.

*Wait a minute. That sounds like we're saying that the math that was created was the math that mathematicians were comfortable with. But physicists tend to think that much of math was first developed because it was needed for a correct description of Nature. (At any rate, that's the part of math that we physicists are most interested in). More generally, why would anyone who is not a mathematician care to learn the math that mathematicians felt comfortable developing?*

**It's not that they developed what they were comfortable with. They developed what was needed in ways they were comfortable developing them. They transformed normal thinking into mathematical thinking, and they created the mathematical tools to change the world using processes that are just refinements of normal ways of thinking.**

*So to learn math you need to give up the way you think about things and instead think the way that mathematicians do?*

**Not give up. Refine with care. If you look at abstraction, mistrust, and laziness, they are all aspects of taking care with one's thoughts. "Be careful with that." "Watch where you're going." "Put everything away where it belongs." These are all parental admonitions, things we teach children so they won't run around heedless of how things need to be done.**

And what happens when they do this? They start seeing the world more clearly, discovering that there are more things around that can be interesting and that there's more that can be done with the world and that the world is full of more toys than they could find in the biggest shopping mall.

And just as a child who learns the rules to a game discovers that the game is more interesting than the little pieces used to play it, so a person who comes to learn mathematical thinking sees the objects, shapes, and processes of the world as more interesting than someone who simply lives through what's happening and has no idea how or why. One of the hidden secrets of mathematics is that while it is done for serious purposes, what's going on in the minds of mathematicians is play.

If you look at each of the aspects of the three realms of mathematical thinking and watch how one thought leads to another, you'll see how the care of abstraction, mistrust, and laziness are the means by which mathematics plays with the ideas of that realm.

We began math by piling up stones. Now let's see what happens to those who abstract the piles and the piling up and using them place abstracted stone upon abstracted stone to build algebra's mansions and playgrounds.

## Algebra

2 + 1 = 3

A stone cold, simple number statement. Take 2, add 1, the result is 3. From this statement, it's easy to make a simple question:

2 + [ ] = 3. What goes in the box?

Questions like this show up on kid's activity menus in family restaurants, along with "find all the fish in this picture" and "how many words can you make out of COLESLAW?" These problems are so nonthreatening that restaurants are willing to dump them in the laps of 5-year-olds (along with more sugar than should be consumed by a small city).

And yet, the number of people who have acquired math phobia from the following question is startling:

Solve for x: 2 + x = 3

All that's happened in transitioning from placemat/menu to Algebra I class is the letter "x" used in the place of [ ] and "solve for x" instead of "what goes in the box?" How is it that this simple transformation has become the most serious stumbling block in learning math? Why is it that the words "algebra" and "variable" have sent millions of children screaming into the night?

**I have a difficult time answering this, since I confess to having an algebraic mindset.**

*Not me. I'm just a simple-minded physicist. We're either not supposed to have any mathematical mindset, or to owe our allegiance to geometry or to calculus, which in the classification of this chapter is part of the analysis mindset.*

**Don't worry. Later we'll get to complain about the algebraic mindset.**

There seems to be a strong emotional difference between the box and the x among those who do not have the algebraic mindset. One possibility is that the box feels like a space for a missing number where the x seems to be a whole new thing, a jump into a new level of abstraction. Another possibility is that it isn't the x that's troubling but the process of solving that causes the uneasiness. After all, a box in a puzzle can be addressed by trial and error: one can just try some number in the box and see if it works. But the x comes with some formal rules for maneuvers that one is supposed to make to get the x by itself on one side of the equals sign and a number on the other side.

The strange thing is that this is a much less extreme abstraction than the first one that went from stones to numbers. Number is a radical concept; variable is just a box with a label on it. However, number has things it can directly connect to in the real world, the counting and measuring of stones and sticks. Variables are abstractions that go from one thing in mind to another thing in mind without touching the world.

Therein lies the leap. Number is always touching the world – how much of this, how many of that – but variable is not. It is a second level of thought removed from the world, because in a real sense the variable is not so much the answer to the question but the question itself.

Let's look again at "Solve for x: 2 + x = 3"

Another way to look at this is as the question, "What number added to 2 gives 3?" From this perspective, x itself is the question. "x = 1" then becomes a way of saying that 1 is the answer to the question "What number added to 2 gives 3?"

From this point of view, x is just an incredibly compact way of phrasing the question. That may sound trivial, but it's actually important. The algebraic mindset is capable of shrinking big ideas into small, compact forms and then manipulating them. The fact that the question shrinks down neatly makes it easier for the mind to work with.

The algebraic mindset works by abstracting things into generic objects that represent things like the abstracted thing. The generic objects have no symbolic or representational character. There's nothing about letters used as symbols that carry with it a sense of number. An algebraist could say "let x be a number" or "let x be a potato" with equal willingness.

This kind of mental jump seems difficult. In most cases when we symbolize something, we like the symbols to have some connection to the thing it represents. The simplest form of this is drawing pictures. The drawing is a symbol for the thing drawn.

But in reality the human mind has a phenomenal ability to create and connect symbols with objects without any real connection. We call this ability language. The word "fire" has nothing in common with the reality of fire (It's not red, doesn't give off heat, doesn't sparkle, doesn't incorporate oxygen into fuel in an exothermic reaction, doesn't actually cause words next to it to burn, etc.). It's just a placeholder in our minds that evokes whatever our idea of fire is.

This process of unconnected symbolism can be used to contain extremely complex concepts that if unfolded would take books and books to describe. Just consider the density of words like love, government, art, anger, happiness, and so on.

In many respects, algebra is just drawing on our fundamental ability to craft and learn language. So where does the problem lie?

It lies in the fact that when we make a word, we tend to keep its meaning. Once we have the word fire, we connect it to a bundle of associations, all the fires we've ever seen or heard about, the feeling of being near fire, fear of burning, and so on.

We may change what fire the word fire refers to. In one sentence we might be talking about a fire in a fireplace, the next we might mean a forest fire, the third a metaphoric fire of passion. But in all cases these associate in our minds with the word fire.

In algebra the variables we use do not carry over association from one usage to the next. x may mean one thing in one problem (2 + x = 3) and something else entirely in another (x is the potato in the sack with the largest mass). There's no connection. They aren't even the same kind of object.

In computer programming, this is called localization, which means that the same word can have completely different, unconnected kinds and meanings.

The strange thing is that there are linguistic objects that satisfy this kind of memoryless, typeless questioning, and if algebra were taught using these linguistic objects, it would probably be easier for most people. These words are pronouns.

Consider the word "it." It can refer to anything of any kind, from the most abstract (it can even be the idea of abstraction) to the most concrete (like the broken off hunk of concrete on a street corner). Writers know that using the word it can be tricky because people can lose track of its meaning from sentence to sentence, especially if one lets it change what it refers to.

It's raining on my car.

It got wet.
It got in through the car window.
It should have been shut.

If each "it" here (rain, car, and window) was given a pronoun all its own (x, y, and z), then it might be easier to follow this:

x's raining on my car [y].
y got wet.
x got in through the car window [z].
z should have been shut.

But this does not sound like the questions we were talking about before.

There's another kind of pronoun, an interrogative pronoun, "What?"

What got wet?
My car.
Let y be what got wet.
What is y?

Later on, in a different statement, the word "what?" or its stand-ins x, y, or z can be used again, just as one use of the word "what?" or the word "it" does not commit us to keeping the same object associated with the pronoun. We can change the meaning as we need.

The principle of use of variables in algebra is that one should not change variable meanings inside a single calculation or proof. Once you assign the meaning of x, keep it that same meaning until you are done with this particular bit of work.

Let's go over this again, from a slightly different perspective. We've just seen that one can use variables as pronouns. But why would one want to? What's the advantage in this kind of empty symbolism?

The leap from box to variable, from explicit long question to implicit short one with its own pronouns, is the heart of the algebraic mindset. Indeed, what distinguishes the algebraic mindset from the others is that it makes such leaps easily. It can take an object and, without any cues from the object itself (like shape or color or smell), generalize its characteristics to create a generic substitute for that object, the variable. From that representation, that implicit question, it can seek answers not just about a particular instance of the object (the answer to a single question) but all instances of objects like it (answers to all questions of a certain type).

Let's take a look at that again a little more slowly. As was noted in the discussion on abstraction, the human mind is always making substitutions of thought for real object. Why are variables different? Because they lack memorial value. In most cases, when the mind substitutes abstraction for reality, it uses something that connects in the memory of the person to the thing substituted for. We use symbols and appearances that remind us of the thing to be recalled. This is the entire idea of a memorial for something, a means of recalling.

But algebra rejects this, seeking instead a bland memoryless substitution. It does this for the benefit of laziness.

Consider the following equation:

$$\text{apples} = \text{granny smiths} + \text{jonathans}$$

This equation contains obvious memoria, but it reminds us, unsurprisingly, of apples. It would be hard in our minds to have apples mean oranges, granny smiths mean navels, and jonathans mean valencias.

Whereas the equation

$$a = b + c$$

has no memorial value whatsoever and therefore works just as well for apples as oranges.

Note that there is in the sciences a compromise between these two views, where a small amount of memorial value is permitted in the choice of letters. So Newton's second law of motion is usually written

$$F = ma$$

F(orce) = m(ass) times a(cceleration).

There will be more about this later.

Generic algebraic symbolism is a powerful mental action, a strong act of generalization. It allows one to discern truths of great generality because it does not care what the objects are, it only wants to know and show how they work together.

But as with all abstraction, it loses things along the way. The poetry of 2 (love) and of 3 (parents and child) disappears into "x." (Of course, the fact that there is a poetry of individual numbers says a lot about what poets can work with – and some have done things with "x" as well. ) This caution against loss is true in all areas of thought but especially important in algebra, where anything can be abstracted, as we shall see.

Let's return to our stone-based starting point.

$$2 + 1 = 3$$

How many things are there to abstract? Let's count them, shall we?

2 (1 abstraction)
\+ (2 abstractions)
1 (3 abstractions)
= (4 abstractions)
3 (5 abstractions)

Five abstractions (*cue lightning, thunder, and the Count from Sesame Street*).

The abstractions of + and = we will look at later. For now, let's look at abstracting numbers, pulling up those stone and stick values into things as yet undiscerned.

### Here it is: Value Abstraction. Let's bring him out again, x the unknown, Variable Extraordinaire!

Our little equation 2 + 1 = 3 has three values. Does this mean we have to abstract the 2, 1, and 3 into x? No, not if you think about x as a question. x + 1 = 3 is not the same problem as 2 + x = 3, nor the same as 2 + 1 = x.

On the other hand, if one looks at them from the algebraic mindset, these questions are very similar. Each of these equations has one blank space and that space has one number filling it (2, 1, and 3, respectively). These correspond to questions of the form: "What number added to (given number) gives (other given number)?" or "What number do you get when you add (given number) to (other given number)?"

The basic task of most math problems people are confronted with is this principle of having a blank space that asks a single question wherein work needs to be done to provide an answer. How much money will I have in my bank account next month? How many eggs do I need to buy if I'm making pancakes three times next week and scrambled eggs twice? How long should I expect the drive between Chicago and Detroit to take, assuming there's no construction (*there's always construction*) and I leave at nine in the morning on a Sunday?

These are all problems of one variable, problems in which a single hole needs to be filled with a single number. Solving problems like this is the bread and butter of everyday math. The problems may end up being immensely complicated and incredibly difficult to solve. They may even be eye-blurring messes like

$$3\sin\left(e^{\cos(x)}\right) = 52.6666667842 - x + x^4$$

but ultimately they all boil down to seeking a single number that satisfies an equation, a number that if you substitute it for the variable produces a true statement about numbers.

$$1 + x = 3$$

This is a question.

Substitute 2 for x.

$$1 + 2 = 3$$

This is a statement.

This is a little subtle, but it has a grammatical equivalent.

You went where?

The "where" in the above is a variable, so the above is a question.

I went to New York.

New York is a value substituted for where to produce a statement.

Mathematical equations lack question marks, so it's not easy to tell the questions from the statements.

We talked about the process of solving these kinds of equations in the last chapter. The principle is that you transform the question into a statement by doing the same thing to both sides of the equation until you end up with something of the form:

$$x = (\text{number})$$

Then to check that you got it right, you substitute this value for x into your original question to produce the statement that is true.

Most facility with this kind of algebra consists of figuring out what operations to perform on both sides of the equation to help the process along. But there is another element, a hidden algebra even in this single variable world where sometimes one introduces other variables just to make things easier. Consider the following equation.

$$x^4 - 2x^2 + 1 = 0$$

At the moment we don't have a way to solve this, since all we've dealt with so far are linear and quadratic equations. But we can change the above equation by making a further substitution if we create a new variable out of whole cloth defined by $y = x^2$.

Why would we do this? Because sometimes in order to answer a difficult question, you transform the question into two questions that you answer in succession, using the answer of one to help you answer the other.

How do I bake a cake?

First you make a cake batter.

Then how do I bake the batter?

The making of the batter is a problem embedded in the question of baking the cake. Implicit in it is the question, how do I make the batter? In effect, the above $y = x^2$ is a way of removing the question of making the batter and concentrating on the baking. If you don't see that yet, watch.

We can substitute y for each occurrence of $x^2$ in our earlier equation. We end up with

$$y^2 - 2y + 1 = 0$$

We can do this, since $y^2 = yy = x^2x^2 = x^4$.

In effect, we've come up with an intermediate question. Instead of asking what x is, we ask what y is.

But this equation in y is a quadratic and it has only one solution, $y = 1$. Now that we know the answer to the middle question, "what's y?" we can go back and ask what x is, using the equation $y = x^2$.

$$1 = x^2$$

So x is 1 or –1.

Notice that sometimes the answer to a question is not a single answer but a list of possibilities.

This process of making further substitutions until the problem is simplified enough to begin the process of solution is a hallmark of laziness, and when done with variables it is a strength of the algebraic mindset.

Generally, the people who are best at solving equations are those who are sneakiest at their substitutions. For them, coming up with a clever substitution is a form of mental play.

It's possible to create or find an arbitrary number of single variable problems of many kinds, some easy to solve, some hard. They form good practice, but they do get boring once you know how a problem is to be done and can do so with facility.

Let's go back to our original equation (we'll do this a lot in this section). Earlier, we abstracted only one variable out of our $2 + 1 = 3$. Something very different happens if we abstract both the 2 and the 1 into separate variables to get $x + y = 3$. This is an entirely different sort of question, requiring a closer look at the meaning of "variable."

If we go back to $2 + x = 3$ and translate it as "What number added to 2 gives 3?" we see again that a variable is a question. "What something or somethings of a particular kind will make the statement true?" – something or somethings because sometimes, as we saw before, even a one-variable question can have more than one answer.

$x^2 = 1$ has two answers. x can equal 1 or x can equal –1.

$0x = 0$ has an infinite number of answers. Since any number times zero is 0, any number makes this equation true.

$3 < x \leq 7$ also has an infinite number of answers, if we're talking about real numbers (all the numbers greater than 3 and less than or equal to 7), but only a finite number if we are talking about integers {4, 5, 6, 7}.

This last points out in mathematical language that you have to know what kind of answer you're talking about when you ask a question. Unfortunately, math texts tend to describe this using phrases like "the set the variable ranges over," which means what's the list of values a variable can take.

This is one of the aspects of variables that often confuses people, not because the idea is difficult but because the explanation, while exact, is not edifying. (There will be more about this difficulty in math texts in the final chapter.)

If we take a different perspective, the idea is not hard to understand. Once again let's play (trademarked term for fill-in-the-blank game).

My *(noun)* has been *(verb in past tense)* by your *(adjective) (adjective) (gerund) (noun).*

To fill in these blanks you must supply words of the particular kind specified for each space. We'll leave this to you readers.

Just as noun is a category of answers to fill in the blank and verb is a different category, so "real number," "integer," "integer between 3 and 7," "even number," and "multiple of 17" are categories of answer. And just as the above blanked sentence only makes sense if the answers come from the appropriate sets of possible words (such as gerunds), so solutions to math problems only make sense when the answers come from appropriate sets of possible answers (such as the real numbers).

All right, but how does that help us with $x + y = 3$?

Let's suppose that x and y are questions about nonnegative integers. We can then rephrase the question as "What pairs of whole numbers greater than or equal to 0 add up to 3?" This question we can answer with a table.

| x | y |
|---|---|
| 0 | 3 |
| 1 | 2 |
| 2 | 1 |
| 3 | 0 |

This shows us that what we have answered is not "What are x and y?" but "What is the relationship between x and y?" And the answer to that is "They add up to 3, so tell me a value for one of them and I can tell you the value for another."

This might sound like we're giving only partial answers, but in reality these dependence relationships are vital to the use of math as a tool in science. Newton's Second Law of Motion states that $F = ma$. The force on an object is equal to the mass of the object times the acceleration the object undergoes. Give me any two of these and I'll give you the third. Physics ardently seeks relations like this, laws that can be turned around in any direction to produce needed information from given information whatever its form.

Establishing these kinds of dependence relationships between physical quantities explicitly tells us a great deal about the real world and allows us to make direct predictions for the consequences of various actions. It is in these apparently partially solved problems that discovery and prediction are made.

All well and good, but while $F = ma$ is a statement about the nature of all matter in the universe and a predictor of the paths of pool balls and planets, $x + y = 3$ is just some silly little thing we pulled out of our false-bottomed top hat. It doesn't reveal much of anything at all.

That's true, but suppose we sweeten the deal and give you two equations for the price of one. Yes, good people, call in now and we'll send you not just

$x + y = 3$, I'll also throw in

$$3x - y = 1$$

Remember when we said that two equations don't have to have anything to do with each other, that the x in one doesn't have to mean the same as the x in the other? That's true, but it's equally true that they can mean the same if you want them to.

**It's as if I said, "What's black and white and red all over?"**

*A newspaper. A nu–*

**Never mind. The point is that the above has a lot of possible answers. If I then said "What is a four-legged mammal?" you'd say:**

*I'd throw a field guide to quadrupeds at your head.*

**But if I said that both of those questions refer to the same creature ...**

*Aha. A blushing zebra!*

**Yes.**

The point is that we can connect our questions together by saying that the answers need to be the same. We can declare the whats to be the same it.

Going back to our two equations, this means we are expressing two relationships between the same two values: (x and y add up to 3) and (3 times x then take away y giving 1). If x and y range over real numbers, then each of these equations gives us an infinite number of pair of values. Is there some way to combine these two relationships and find out if any pair fits both equations?

If we couldn't, algebra would never have made it into high school textbooks.

$$x + y = 3$$

We go back to the principle that if you do the same thing to both sides of an equation, you preserve truth. We can subtract x from both sides of this equation and get

$$y = 3 - x$$

Still, you might object: Earlier we only did that with numbers. How do we know that we can do that with variables?

Here's how. Whatever number x ends up being, we would be able to subtract it from both sides. As a result, we can subtract x from both sides. It's like reconstructing a crime in a mystery story. We don't know who the killer is, but whoever he or she was had to be standing in this place in order to fire the shot.

Back to the equations: since y in the second equation is the same as y in the first equation, we can substitute 3 – x for y in the second equation, since y is always equal to 3 – x regardless of what x is.

*Hang on. We'd better give a decent explanation of what we mean by substitute. Otherwise, our readers might think it's like when the regular teacher is off for the day and we try to trick this substitute into letting us play baccarat in the classroom while downloading dubious videos into the school's network.*

**OK. This substitution is a simple application of equality. If two things are equal, you can always swap one for the other in an equation.**

It's like making change. You can always trade two 5's for a 10. In this case we can swap 3 – x for y since they are, as it were, the same value. This is not the conventional meaning of the word substitute which implies something being used in place of something else perhaps not doing as good a job (like sugar substitutes). In mathematics, substitute has a different meaning, that of swapping in something that is actually the same even though it may not look the same.

This transforms

$$3x - y = 1$$

into

$$3x - (3 - x) = 1$$

which becomes

$$3x - 3 + x = 1$$

Then we add 3x and x and end up with

$$4x - 3 = 1$$

We can then add 3 to both sides and get

$$4x = 4$$

Dividing both sides by 4 gives

$$x = 1$$

x now has a single contrived value. And we can find y by going back to our earlier equation, y = 3 – x

which yields y = 3 – 1 = 2

from which we can get

$$3 = 2 + 1$$

At this point, you may discern a pattern: we could solve one equation, with one variable, two equations with two variables. Does the pattern continue? The following statement needs to have a lot of disclaimers and warning statements in front of and in the middle of it to be mathematically accurate, but we'll weasel on that because the idea comes across more clearly without them.

In general, if you have the same number of equations and variables, solving those equations together will give explicit values for all the variables. So five equations in five variables will in general give individual values to all five variables. Or in other words, if you don't know 5 things, answering 5 questions about the interrelationships between those 5 things will give you the answers to all of them.

The basic process by which one does this is to take one of the equations and partially solve it so that one expresses one of the variables in terms of the others. For example, if one had $12x + 4y - 2z + 10 = 0$, we could do the following:

$$12x + 4y - 2z + 10 = 0$$

$$12x + 4y + 10 = 2z$$

$$6x + 2y + 5 = z$$

We now have z expressed in terms of x and y. If we substitute the expression on the left for each occurrence of z in all the other equations, we will effectively have removed z as a concern and will have one fewer unknown to solve for. In effect, we sacrificed one equation to remove one unknown. If we continue this process, we will end up with one equation in one unknown which we can then solve normally.

*We'd better also tell them how to get the values for all the other unknowns.*

**OK**. We just substitute back through the equations we used to eliminate the unknowns. Suppose we originally had three equations in three unknowns (x, y, and z). And in eliminating unknowns, we ended up with

$$z = 6x + 2y + 5$$

$$y = -5x + 10$$

and finally

$$x = 1$$

In order to find y, we substitute 1 for x in $y = -5x + 10$, getting $y = -5 + 10 = 5$.

Then we substitute 1 for x and 5 for y in the top equation, getting $z = 6 + 10 + 5 = 21$.

Now, let's stick on a disclaimer.

Assuming the equations specify different relationships between the variables, then the solutions will be specific values. This disclaimer addresses the difficulty that two different-looking equations can have the same meaning:

$$x + y = 3$$

and

$$1000x + 1000y = 3000$$

are the same relationship,
as is

$$x + \sin(x) - \sin(x) + y = 3$$

So if the equations do not all mean different relationships, then we really don't have as many equations as we thought we did.

*It's a case of redundancy, of saying the same thing again and again, of unnecessary reiteration, of trying to get away with a really old joke, a running gag, a …*

**OK, that's enough of that.**

*(for now)*

All of the other disclaimers are similar to the above. They clarify exactly under which circumstances the basic principle above is true. From this point on, we could fill a series of textbooks on methods for solving multiple equations in multiple variables. A lot of people have put in a lot of work creating the means to do so. It seems a shame to ignore them. But given the scope of this book, we can't get into every mathematical process, particularly given how many books there are on those subjects. From this perspective, the small set of problems above tells us most of what we need to know about variables in equations.

If all algebra was the ability to create variables in equations as questions and combine them to create answers to multiple questions based on the interrelations between known and unknown quantities, it would still be one of the most powerful mental tools ever created. By determining relations between quantities and placing those relations together in the form of simultaneous equations, we can pull out numbers no one has any means of measuring. Algebra lets us keep the number of sticks we use small while simultaneously increasing without bound the number of things we can effectively measure by calculation.

If that was all, algebra could take its bows and retire to the Middle East where it was born. But algebra is more ambitious than that. It started looking beyond variables a while ago.

The truth of the matter is that x isn't all that impressive any more. His act has gotten old hat. We take him for granted. We'll just give him jobs as an extra. Let's try to bring on something more exciting, more *au courant*.

*Hey, I didn't say that.*

**Sorry, it's a manual of style thing: foreign phrases in italics.**

*Great. I'm being ordered around by a book I'm not even writing.*

**It's OK. We'll use both our voices (bold italics) to announce the next topic.**

***Introducing on our algebraic stage: +, the binary operand that walks like addition***

Going back to 2 + 1 = 3, let's have a look at +, our old friend addition, rock star since the stone age. What can we abstract from him?

At first glance, this question looks really strange. Addition is putting one pile of stones with another pile of stones and seeing how many stones you have. What is there to abstract?

From the rock perspective, very little. But if we examine the behavior of addition, we discover many properties that are useful and abstractable. To do this, we use what addition is and from it discern how it works, and then we abstract from how it works.

*We'd better say that again, only a little less abstractly.*

**OK, let's try it this way:** we have a tendency to look at objects in terms of what they are, but we also have to pay attention to what they do and how they do what they do. A windmill may be a picturesque thing, but it's also a device that takes the movement of the wind and uses sails or propellers to create a turning motion that can be used to turn millstones to grind grain into flour, or to power pumps to keep water levels low, or to simply charge up batteries with electricity. What makes a windmill a windmill isn't what it is, but what it does and how it does it. This makes it possible to create windmills of different forms and purposes and to refine the concept beyond wind power to other kinds of power (water wheels and windmills work on very similar principles).

Let's look at addition in a similar light. Addition is a process that takes two things and from them makes a third thing. A thing that does this, that takes two objects and makes a new one from them, is called an **operation**.

Whenever we abstract something, we do so in the hopes of isolating and exploiting certain properties of that thing in order to employ them elsewhere than where they started. We hope to find that the single thing we started with is really an example from a large class of things. Windmills are a kind of motor. Hearthfire is a kind of fire. Addition is a kind of operation.

So what properties does the operation of addition have?

Before we dive into that, we're going to use variables in a slightly different way than we did before. We are going to use them as placeholders that allow us to say an infinite number of things in one statement.

Let's look at the theorem, for any a, $a + 1 > a$. This is a shorthand for the infinite number of statements:

$0 + 1 > 0$

$1 + 1 > 1$

$2 + 1 > 2$

and so on.

The use of the placeholder variable a allows us to write this infinite number of statements lazily in one line. But remember we're dealing with mathematical laziness in which shorthand reveals new truths. The general statement not only allows us to create an arbitrary number of explicit statements (as seen above), but also lets us generalize again.

We can say $a + 2 > a + 1$ and $a + 1 > a$; therefore, $a + 2 > a$. We can go even further and create another general rule that if $b > 0$, $a + b > a$. These lazy statements become theorems we can draw upon in later proofs and problems.

If you want to think in terms of variables as questions, these are like rhetorical questions (sometimes called **dummy variables**). They are phrased in the form of a question, but they really make a statement. So as we'll see in a bit, instead of asking which values of a and b satisfy $a + b = b + a$, the equation is really saying that for any values of a and b, $a + b = b + a$. The equation exists not to be solved, but to reveal the characteristics of the operation.

Let's get back to addition and its properties, which we will state in this kind of shorthand.

First and foremost, if we have two numbers – call them a and b – then $a + b$ is also a number. This property is called **closure**. This may sound obvious, but there are ways to combine things that produce things unlike what you started with, such as sand combined with heat creates computer chips.

This property, that if and b are numbers, then $a + b$ is also a number, means that you can keep doing addition over and over again to the results of an earlier addition. You can add some other number c to $a + b$ and get $(a + b) + c$, which will again be a number.

This leads us to the second property of addition, which is called the **associative** property. It says that $(a + b) + c = a + (b + c)$. That is, if we first add a and b and then add c, we get the same result as if we first add b and c and then add a to that. Although this sounds obvious, not all operations are associative. Subtraction isn't:

$(a - b) - c$ is not the same as $a - (b - c)$
$(6 - 4) - 2 = 2 - 2 = 0$
but $6 - (4 - 2) = 6 - 2 = 4$

The third property is called the **commutative** property. This says that $a + b = b + a$. Again, this is obvious for addition, but false for subtraction.

The next property is called **existence of identity**. This says that there is a unique number – call it 0 – for which $a + 0 = 0 + a = a$ for any number a. Again, it's obvious for addition but false for subtraction, since there is no number b such that $a - b = b - a = a$ for every a.

The last property we'll look at is called **existence of inverses**. This says that for every number a, there is a special number $-a$ such that $a + -a = -a + a = 0$.

*So we've got a bunch of properties of addition. So what?*

First, notice the kinds of properties we abstracted. We did not abstract properties of individual effects of the operations, such as $2 + 1 = 3$. We abstracted properties that apply to all applications of the operation (such as the associative property), or that specify the existence of individual objects that have universal properties (such as the identity), or that specify universal existences in relation to the property (such as the existence of inverses). In short, we took out a listing of properties that characterize the overarching behavior of the operation, not its individual results. We took the operation out of the operation.

And that's where the algebraic mind comes in. In a transcendent act of laziness, algebraic mind thinks as follows:

> *Whatever I can prove about addition that uses only these properties will be true not just of addition but of every other operation that is enough like addition to share these properties. What's more, these operations don't even have to be operations on numbers; they can be operations on the members of any set.*

Here's a couple of proofs doing this.

We will prove that an operation with the above properties cannot have more than one identity. There cannot be two distinct numbers – call them 0 and O – such that

$$a + 0 = 0 + a = a$$

and

$$a + O = O + a = a$$

The proof is simple. If both 0 and O are identities, then

$$0 = 0 + O$$

and

$$0 + O = O$$

Therefore, 0 = O.

Notice that this proof requires only the existence of identities.

We've shown in these few steps that any operation on any set that is enough like addition to have an identity has a unique identity.

We will next prove that for an operation with inverses, every number has one and only one inverse. This time assume that there is a number a with two inverses, −a and ~a.

So a + −a = −a + a = 0 and

$$a + \sim a = \sim a + a = 0$$

We will add as follows:

$$(-a + a) + \sim a = 0 + \sim a = \sim a$$

but (−a + a) + ~a = −a + (a + ~a) because of the associative property and

−a + (a + ~a) = −a + 0 = −a.

So

chaining all of these equations together gives

$$\sim a = 0 + \sim a = (-a + a) + \sim a = -a + (a + \sim a) = -a + 0 = -a$$

So any operation that is associative, has identity, and has inverses has unique inverses.

We'll go more fully into this in a later chapter. The point here is not so much the proofs as the abstraction of the attributes of addition from the nature of addition itself. Algebraists call a set with an operation that satisfies the rules laid down above an **abelian group**. Whatever they can prove about abelian groups holds not just for addition but for anything enough like addition.

*Well, that's a definite distinction between any of the mathematical mindsets and the physicist's mindset: mathematicians are perfectly happy to deduce general results from abstract principles. Physicists like to see examples. Let's give our readers an example of something that is "enough like addition."*

**Okay, here's one. If we are dealing with positive real numbers, then multiplication is like addition.**

*OK, now that's just going to confuse our readers. Especially after we told them that multiplication comes from sticks by sticks, whereas addition comes from stone counting. And we told them that multiplication makes new units, whereas addition keeps the old units.*

**All of that is true from the perspective of analysis, but algebra doesn't care about origins or units. It only looks at the properties of the operations.**

So let's take a look at multiplication.

Closure: if a and b are numbers, ab is a number.

Associative: (ab)c = a(bc).

Commutative: ab = ba.

Possession of identity: There is a special number 1 such that a1 = 1a = a.

*Right. So 1 is the new 0.*

Possession of inverses: For every positive real number a, there is another positive real number 1/a such that a(1/a) = (1/a)a = 1.

Note that if we did multiplication in the integers, this would not work; ½ is not an integer, therefore 2 does not have an inverse in the integers, nor does any other integer except 1. Furthermore, if we included all real numbers, it wouldn't work either, since we can't do 1/0.

So the positive real numbers under multiplication form an abelian group. And it's actually even more extreme than that. From the perspective of algebra, the real numbers under addition and the positive real numbers under multiplication are in many ways the same abelian group.

*Right, that's the craziest thing about the mathematical mindset! And especially the algebraic mindset: we can take two things that are completely different, and as long as they have the same structure, the mathematicians will consider them the same.*

**Right. We're not talking about the real numbers and the positive real numbers with all their properties and characteristics.**

We don't at the moment care about sticks and stones, about the meaning of 1 and 1/2. We don't want to have meanings for these numbers or most of their interrelationships. We've left all that behind in the abstraction of the operation. It might seem ridiculous, but we will discern certain deep properties of an object by abandoning most of the characteristics of that object. This is the normal process of abstraction, where we only take what we need for what we're working on. We're carrying it out on entire number systems and looking only at certain properties of the operations in those systems.

We're talking only about the way the real numbers behave under addition and the way the positive real numbers behave under multiplication. We don't care about the fact that one of them has the shape of a line and the other the shape of a ray. Nor do we care about the way they are ordered, nor their meanings for measurement, nor any other characteristic. We're only looking at one thing and seeing it as the same as another thing.

It's like two different pieces of music that have the same rhythm. Everything else about the pieces can be completely different, but the beats are the same and therefore you can impose the lyrics of one upon the other.

In the same way, we can map the actions of addition in the real numbers into multiplication in the positive real numbers because the underlying beat (the way the operations work) are the same.

This map is a special one-to-one correspondence called an **isomorphism**. We'll talk about them in more detail in later chapters. The central idea is that an isomorphism is a way of associating one element of one set with another element of another set in such a way that if you operate on elements of the first set to get a new element, that new element is the same one you would get if you first mapped the elements and then operated on the results.

Rather than take this apart too much, let's look at the particular case we're dealing with. We're going to make a map from (R, +) (real numbers with addition as their operation) to ($R_+$, .) (positive real numbers with multiplication as their operation).

The map takes a real number a and associates it with the positive real number $10^a$. So, for example, the real number 2 goes to the positive real number 100 (which is 10 to the second power) and the real number −3 goes to 0.001 (which is 10 to the −3 power). This is a one-to-one correspondence, and on its own it proves that there as many positive real numbers as there are real numbers (since we've made a one-to-one correspondence between the sets).

What makes this an isomorphism is that it sends the operation of addition into the operation of multiplication.

Let's consider a + b in the real numbers. The mapping sends this to $10^{a+b}$. But $10^{a+b} = 10^a 10^b$. This is the same as multiplying what a maps to (that is, $10^a$) and what b maps to ($10^b$).

For example, $10^3$ (which is 1000) is equal to $10^2 10^1$ (100 × 10).

So this map preserves the structure of the operations, sending addition into multiplication. It preserves the structure of every one of the characteristics we described before for addition and multiplication. Closure is obvious, but look:

$$(a+b)+c = a+(b+c)$$

maps to

$10^{(a+b)+c} = 10^{(a+b)}10^c = (10^a 10^b)\,10^c = 10^a(10^b 10^c) = 10^a 10^{b+c} = 10^{a+(b+c)}$. So the associative property maps to the associative property.

And

$$a+b = b+a$$

maps to

$10^{a+b} = 10^a 10^b = 10^b 10^a = 10^{b+a}$. The commutative maps to the commutative.

$10^0 = 1$, so the identity maps to the identity.

And $a + -a = 0$ maps to

$10^{a+-a} = 10^a 10^{-a} = 10^a\,(1/10^a) = 1$. So inverses map to inverses.

Another way to look at this is to see that from the algebraic perspective, operations are like rules to a game that let you take two things and from them make a third. It is possible to make game pieces that look completely different and game rules that sound different, but which end up being the same game because the moves in one game mimic precisely the moves in another, even if the way they are mimicked is a little brain twisting. This happens a lot in computer games that use the same underlying software (called the game engine) but have different appearances for objects on the screen.

An algebraist can look at addition and multiplication, which when you perform them do not work like each other at all, and see the ways they are the same.

This act of abstraction of operation may look like it has no way back to the real world, that it's nothing but a mathematician's game, and hardly the inheritance of anyone else. But just as one example, consider the role of symmetry in particle physics. Physical measurements come out the same no matter where they are done, what time they are done, and what the spatial orientation of the apparatus. As shown by Einstein, even a uniform motion of the observer leaves things unchanged. Now consider all the possible changes in position, time, orientation, or velocity of the observer as the elements in our group, and consider composition (doing one change after the other) as the "multiplication" operation. This group is called the Poincare group, and the fact that all of our fundamental theories have to be invariant under the Poincare group forms the basis of particle physics. Furthermore, having gotten used to the Poincare group, physicists devised other symmetry groups that form the basis of the electromagnetic, weak and strong forces. So all of our deepest theories of the universe are tied up in this abstract group theory.

This is something that could not have been realized had mathematicians not explored the theoretical groundwork and left this field behind them. This is one of the most important truths of math, that sometimes it is done in explicit response to a real-world problem and sometimes it is done by mathematicians thinking mathematically and expanding the understanding of math because they are interested in the expansion. And while the first usually provides needed answers, the second often surprises by supplying them as well.

Let's consider what's happened in the abstractions we've done so far. We've taken

$2 + 1 = 3$ and turned it into

$a + b = c$, where we no longer need to know what the values of a, b, and c are as long as we know that they work right with whatever + we've got. And we no longer need to know what + is as long as we know that it acts correctly.

The process of looking at operations and seeing their properties, of taking the action out of its context and seeing when things are similar, is a vital one in everyday conversation. People are often fooled by the words in arguments without paying attention to the action underlying the argument.

For example, few people (except the pickle obsessed) will be fooled by

> "If you put all of your pickles into my magic pickle barrel, you'll be able to pull out twice as many pickles as you started with without anyone having to create new pickles."

But a lot will be tricked by.

"If you put your money in our fund, it'll double in no time, guaranteed."

That's the same operation, and if you abstract it from the first, you might ask, "Where do the pickles come from?"

If you do it from the second, you might ask, "Where does the money come from?"

Same operation, same question, but without the distraction of the wrong green objects.

We're not done with our original equation. One more abstraction can be performed.

***And now our final performer. We give you the champion of justice and fair play, the mighty = !***

We know what we mean by =. The same number of stones, the same length of sticks. We understand equals. But we knew what addition was also, and the algebraists wouldn't leave that alone. Just as they want to know what we can prove about addition that applies beyond addition, they want to know what can be proven about equality when it's a different kind of equality. Just as they abstracted the properties of addition, so they do with equality.

We've seen the three vital properties before:

Reflexive: $a = a$.

Symmetric: If $a = b$, then $b = a$.

Transitive: If $a = b$ and $b = c$, then $a = c$.

A relation that satisfies these properties is called an **equivalence relation**. The first thing we need to examine is –

*OK, hold up a minute. They're not going to know what we mean by "relation" (i.e., not the same as you are my "relation" because we are brothers).*

**Right, not that kind of relation.**

A relation is something that is true or not true about two things. For example, "is the brother of" is a relation, as in David *is the brother of* Richard.

*OK, again it sounds like we're talking about us. We have to define our terms in order to be clear what we're talking about.*

**You really want a definition rather than an explanation?**

*Yes.*

**You'll be sorry. Here we go.**

A relation R is a set of ordered pairs. If an ordered pair (a,b) belongs to the set R, then a is said to be in the relation R with b, often written aRb.

*Well that's about as clear as mud.*

**I warned you.**

We're dealing with an area where the definitions are exact, but abstract and uninformative. We'll come back to this definition in a later chapter when we've got better groundwork.

**Will you let me get back to the intuitive explanation and save the exactness for later?**

*All right, just as long as you clean up your intuitive mess. We don't want too much of that stuff on the pure math.*

**Fine. Yeesh, like there's no intuition in math.**

A relation is a characteristic that two things have relative to each other. There are lots of these in everyday life. "Bigger than" is a relation. In truth "big" is not a concept that makes sense on its own; only "bigger" does. We always need to know what we are saying something is bigger than rather than just saying it's big. A human is huge compared to a cell, but a planet is much bigger than a human and a galaxy bigger than a planet.

Here are just a few relations that we deal with commonly: bigger, brighter, happier, richer, more expensive. These are a kind of relation called **orderings**. They compare one thing to another and place them in an order. One object is higher up in the order than the other, which is lower down. Orderings are abstractions of "more" and "less" that we use every day. We have a sense that we can apply these ideas to anything that we can somehow see the more-or-lessness of. This is one of those places where the algebraic mindset is commonplace, just as it is in counting and measuring. And it is common for the same reason. 'More' and 'less' we can hold in our hands the way we can hold sticks and stones. We abstract into them, but we do not lose our grips on them.

Here are some other relations: likes, loves, hates, knows about. These are everyday relations that do not have to have any mathematical properties we are interested in. Nevertheless, they are relations.

And here's a few of what we're talking about when we abstract =: is the same shape as, is the same size as, is the same color as. These are **equivalence relations**. They share all three of the properties we laid down for =. They are equals-like, and anything we can prove about = using only the reflexive, symmetric, and transitive properties hold for them as well.

There are also relations that seem like they might be equivalence relations but that fail in one or more properties.

Let's take "is a brother of." This is not an equivalence relation, because if a is the brother of b and b is female, then b is the sister of a, not the brother of a. So "is a brother of" is not symmetric.

What if we substitute "is a sibling of"? Then we might have an equivalence relation if we mean "has both the same parents as." But if we allow sibling in this case to mean "shares a parent with," then the relation need not be transitive. If you have two people who have children together but each of whom has had children with other people, then a (child of parent 1) can be the sibling of b (child of both parents) and b can be the sibling of c (child of parent 2), but a and c are not siblings. Furthermore, even if we restrict ourselves to "has both the same parents as," we need to be talking about biological parents, since adopted children can have siblings from birth family and adoptive family without those siblings being each other's siblings. We are left with "has both the same biological parents as" if we want to make an equivalence relation.

That may have sounded overcomplicated, but it illustrates the necessity for exact definition if one wants to be able to discern things with math. The same everyday word, "sibling," either is or is not an equivalence relation depending on what we are talking about. Most people would casually slip between meanings, calling their biological and step- and adopted siblings (assuming they have them) all siblings. And most people would not say they are their own siblings. For everyday use they would be correct, but for mathematical use more care would be needed.

Equivalence relations are not so abstract as operations, since there are a lot of them in everyday life: same height, same weight, same mass, same taste in fusion cuisine.

Notice a common thread. The mathematics of equivalence relations is the mathematics of different kinds of sameness. What then is the purpose of abstracting an equivalence relation $\equiv$? It is to treat things that are the same as far as $\equiv$ is concerned as if they were the same. As far as equality of mass is concerned, all objects with the same mass are the same. We don't need to know anything else about them. We can talk about a 1 kilogram object without knowing its shape, color, taste in fusion cuisine, taste in fusion jazz, tendency to undergo nuclear fusion, or anything else about it as long as the only thing we are concerned with is its mass.

Another equivalence relation of relevance to earlier discussions in this section is **is isomorphic to.** One algebraic structure is isomorphic to another if there is an isomorphism from the first to the second. That this is an equivalence relation is a little difficult to prove at the moment, so we'll defer it until a later chapter. *Darn right we will. We haven't even explained what an algebraic structure is.* We're mentioning it here because this relation formalizes the idea that two isomorphic algebraic structures are effectively the same thing, but only as far as their mentioned algebraic characteristics are concerned. In other words, in so far as $(R, +)$ and $(R_+, .)$ are the same, their isomorphic structures are the same. They are two sets with the same number of elements that algebraically work the same. As far as the algebra is concerned, they are effectively interchangeable, but only as far as that.

We've been talking about relations as if they existed in isolation, but to be more careful we have to talk about the set of objects that can have this relation to each other. We tend to use = for numbers. "is a sibling of" we use for human beings. But we can define them on different sets. We can restrict = to the integers, and we can expand "is a sibling of" to any animal.

So to be careful, we talk about an equivalence relation on a set, meaning that at the moment we are only looking at the elements of that set when we check for equivalence.

We can use an equivalence relation on a set to divide up the elements of the set we are talking about and sort them into **equivalence classes**. These are subsets of the original set, each of which contains all and only all those items that are equivalent to each other.

In essence, what we are doing is looking at each object in turn and putting it into a box with every other object that is equivalent to it according to the particular equivalence relation we are working with. If we were doing equivalence in mass, we would have a box containing everything that has a mass of 1 gram, another box for everything that has a mass of 5 grams, a box for 68.327 kilograms, and so on. If we were doing our "has both the same biological parents as" relation, we would have boxes for each batch of biological siblings.

So far there's no math here. But just as we could prove properties of all operations that were enough like addition, we can prove properties of all equivalence relations. For a start, we can prove that every object in the set we are dividing up will belong to one and only one equivalence class.

Here are two definitions to formalize what we're talking about. First, if we have an equivalence relation ≡, then an **equivalence class** of a set A under this relation is a subset B of A such that if a and b are members of B, then a ≡ b, and if a is a member of B and a ≡ b, then b is a member of B. In other words, every element in an equivalence class is equivalent to every other element in that same class and every object that is equivalent to an element of such a class is a member of that class.

Second, a **partition** of a set A into equivalence classes is a list of equivalence classes such that every element of A belongs to one of the equivalence classes in the list. In other words, a partition into equivalence classes divides up the set into subsets, each of which is an equivalence class.

We've been a bit casual. Technically, one needs to specify what relation the equivalence classes are equivalence classes under. After all, dividing the set of all people into "is a sibling of" classes produces one partition but dividing the same setup into "has the same name as" classes produces a different one. But if we are only talking about one equivalence relation at a time, we tend to be a little sloppy and not note the relation as we talk about the classes.

So let's prove that every element of a set belongs to one and only one equivalence class in the partition.

We know it belongs to one equivalence class since we defined that in the partition. To prove that it can't belong to more than one, we'll do proof by contradiction.

Suppose that an element a of the set belongs to two different equivalence classes B and C. Now consider an element b of B. Since it shares an equivalence class with a, it must be that a ≡ b. But if a ≡ b and, as we said, a is a member of C, then by the definition of an equivalence class, b must be a member of C. In other words, every member of B must be a member of C and vice versa, which means that B and C must be the same set.

*Hold it. We haven't yet proven that equivalence classes are even possible. We just defined them and used the definitions. Why should anyone believe we can do this dividing up?*

**That comes from the properties of equivalence relations.**

Let's do a little symbolism. Call [a] the equivalence class containing a. There must be such a class because a ≡ a (reflexive property). Suppose a ≡ b. Then the [a] must contain b. But since if a ≡ b, then b ≡ a (symmetric property), it follows that [b] must contain a. And if there is an element c in [a], it must also be in [b] because if c is in [a], a ≡ c, but if a ≡ c and a ≡ b, then b ≡ c (transitive property). This means that every element in [a] is also an element in [b] and vice versa. Two sets with the same elements are the same set, so if a ≡ b, [a] is the same set as [b] and if [a] is the same set as [b], a ≡ b.

What's the use of dividing into equivalence classes? It means that for purposes of the equivalence relation, we can treat all elements in the same class as the same object. This is particularly important for relations like "is the same shape as." It allows us to treat all squares of the same size as if they were the same square. We don't have to worry about where the square is or whether it's balanced on a point or lying on its back after drinking too much Flatland booze. As far as "is the same shape as" is concerned, they're the same square.

But that's where we started when we first abstracted things. We said we were only going to care about certain aspects of the objects and in effect treat all others as irrelevant. We did so by focusing on one aspect and saying that all things that have the same value for that aspect are one. In creating equivalence relations and classes, we've formalized the very process we started with and now we can do so with greater clarity and exactness. We can even say whether or not we can mathematically abstract a property based on whether or not possession of that property is an equivalence relation.

We have now abstracted object, operation, and relation, leaving 2 + 1 = 3 a confused and empty husk. Its initial jump from the real world to the theoretical universe was but a small one compared to the ways we have pulled it up and around.

The algebraic mindset is astonishingly effective once you get the knack of realizing that anything can be abstracted, that nothing in any statement of any kind need be concrete. You can always abstract again and again. This is not necessarily a good idea, as we will see later, but grasping this process, this freedom from saying that a thing must be a certain way and act a certain way, liberates the mind, while the cautions of mathematics keep one from jumping too far in that liberation.

## Geometry

*Disclaimer*: Neither of us started out with a geometric mindset, and even now we have only a fairly simplistic one. In Richard's case, it's a secondhand mindset developed by being married to a visual artist (Alessandra Kelley, who as a minor sideline to her painting did many of the illustrations for this book) and from writing computer animation software. David does research on Einstein's theory of general relativity, which is a geometric theory of space, time, and gravity. However, David concentrates on those aspects of the theory that can be addressed using differential equations and computer simulations. (Though David's graduate school mentors have brilliant geometrical insights, he is not able to emulate them.)

So what is this shape thing we were talking about in the short description of the geometric mindset?

In many respects, shape is an illusion, an artifact of the scale on which we live, and our reliance on sight as a sense. On the subatomic and atomic scale, there is no shape, only emptiness and particles. On the molecular and cellular levels, there are shapes, but they aren't what's important and there is no sight. Touch matters there, what's next to what and what can interact with what. The only thing shape does there is get in the way of or allow interaction. The bends and twists of molecules can prevent or allow two molecules to connect or fail to connect. In this respect, shapes are used as keys and locks.

On the levels between macrocellular and ecosystem scale, shape does matter, because there is a need to sense and discern things at a distance. But shape is not the only means of telling what something is. Smell, taste, and sound are at least as important as the discerning of shape. So much for scales smaller than us.

On larger scales, particularly planetary and upward, shape does not matter much, mass matters. Because most of the interaction is gravitational and gravity is a very boring sculptor, it tends to make things into balls. Spin can stretch things out and pressure creates churning prominences from stars and all that, but basically objects become spherical. On this scale, the only interesting shapes are in the arrangements of things, the curdling of galaxies and the threading chaos of planetary rings and asteroid belts.

We, on the animal scale, living in an ecosystem, evolved to depend primarily on sight. We're pretty good at this vision stuff. We're not up to the raptors for distance, nor the owls for night sight. We don't have the sonar advantages of bats and cetaceans. But we can see things and make associations based on what we see.

But we don't really see shape. We see color and shading. Our color range is limited to visible light, which is too bad. It would be nice to see the X-ray sky or the pulsations of our microwave ovens. Our color receptors work when things are sufficiently bright. Shading we can see in near darkness, and it is from shading that we discern shape. Take a look at Figures 4.1 and 4.2.

Both are the same internal object on a computer, a sphere. The first looks like a disk because it lacks shading. The second looks like a sphere because it is shaded.

We have two eyes which give us binocular vision, allowing for a sense of depth. Refer to Figures 4.3 and 4.4. Here are two pictures, one rendered as if seen in monocular vision (looks like a drawing), and one showing what binocular vision is like. Figure 4.4, composed of two images, shows apparent depth.

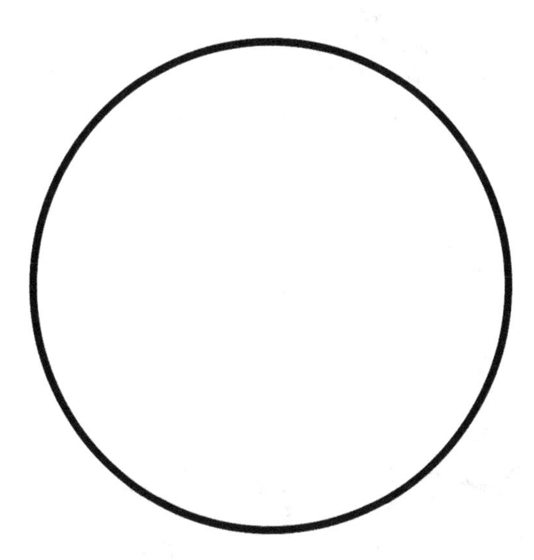

**FIGURE 4.1**
Unshaded Sphere

**FIGURE 4.2**
Shaded Sphere

**FIGURE 4.3**
Monocular View

**FIGURE 4.4**
Binocular View

We evolved to take advantage of this reliance on sight, and our minds learn to recognize things by sight. While we don't really *see* shapes, we do *recognize* shapes. Or to put it another way, the raw data from our retinas are processed by our optic nerves and brains using sophisticated algorithms to give the postprocessed signal of shape. Shapes are things of mind, the images in thought of what our sense of sight shows us. They are basic to our thinking, the memoria for the sights of the world: person, building, sun, bird. These are words connected in our minds with shapes we see. But the shapes themselves are already abstractions from the perceptions formed in our minds from the light that hits our eyes.

This is one of the most basic of abstraction process: light → idea of what you are seeing. This process we call visual recognition, and it seems to be hard-wired into our minds.

But different things can happen in different minds when they see things.

For a lot of people, recognition is all that occurs. They see a shape that seems to be a person. They recognize.

"Hey, that's Fred!"

And what stays in their minds is the thought "Fred," not the image they saw, nor the shape abstracted from that image.

Others see Fred and they keep Fred's shape as a thought in mind. This tendency of mind to hold on to shape past the point of recognition is highly useful in the visual arts and in the geometric mindset. Some people have a talent for this, automatically holding on to the shape. But it can be trained into those of us who had the habit of simply letting shapes go once we had an idea of what we were looking at.

Here are a couple of practices Alessandra suggests for learning to hold on to the shape as you look at it.

For a simple shape, create a mental connect-the-dots that marks important parts of the shape and creates a simple mental framework to conjoin them in your imagination. Practice this with things like people's faces.

For complex shapes, find a simple part of the shape that you can create a connect-the-dots in. Make this the focus for remembering the shape so that the connect-the-dots carries the full image with it. (You may need to extend the connect-the-dots beyond the focal area to catch other parts of the shape.)

It's a useful skill in many walks of life and worth the practice, as it makes one's memory of the world more vivid and accurate. But while this holding on to shape is vital to the geometric mindset, it is not itself that mindset. The geometric mindset sees the shape and wonders about it, tries to discern it, take it apart, abstract it. It spins it in space, twists it around, and does things to it.

The geometric mindset can be seen in a woodworker who puts a length of wood on a lathe, sets it spinning, and, by touching here, here, and here with a chisel while it spins, turns it into a table leg. It can be seen in the painter who looks at a scene, sees where everything is, and, back in the studio, transforms the arrangement of people and buildings and creates a whole new image from it (Figure 4.5).

In short, the geometric mindset treats the shapes of the world as a treasure trove of raw materials. It mines the world for shape and arrangement and from them builds the theoretical universe of how things can be shaped and arranged. And once it has done that, it can be used with real-world labor and tools to reshape parts of the world into statues, buildings, cities, and all the other physical creations of humanity.

In many ways, the geometric mindset is the most obviously effective of the three because it has brought us all these clearly visible objects that we live and work in.

But how does one transform basic awareness of shape into geometry?

## Sticks Are Lines

The geometric mindset begins with the stick. In Greek, geometry means "Earth Measure," and the Ancient Greeks measured the world with sticks. This method of measurement has some problems which we'll get to in Chapter 8, but for now let's stick with the sticks.

Measurement as we've seen is done endpoint to endpoint with the stick. If you keep only the number that results, you are abandoning the geometric mindset for the algebraic and/or analytic. Geometry sees not only the tool and act of measurement, but also the ideal of the measuring instrument, the perfection of linearity that comes from taking the real stick away from the stick (Figures 4.6 and 4.7).

**FIGURE 4.5**
Cityscape: Upriver by Alessandra Kelley

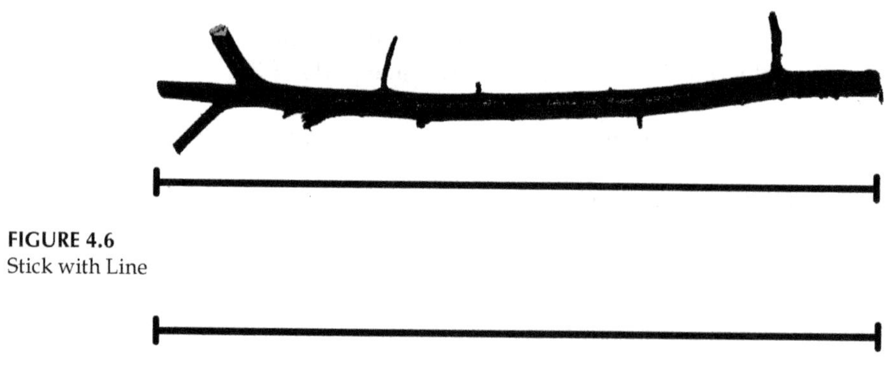

**FIGURE 4.6**
Stick with Line

**FIGURE 4.7**
Line

Geometry abstracts shapes from the real world into ideas of shapes in the theoretical universe. This process starts with sticks and lines. Look at that stick. It really does not look like a line. It's gnarled and twisted and the only thing that makes it line-like is it's sort of straight and you can lay down its endpoints on something to measure the thing.

The line as a tool of measurement is the idealized form of the stick. It is sharp and clean. It does not deviate or wobble. It goes from point to point, and the distance between those points can be traced along the line. You can mark it up as you want, making any of your measuring sticks fit on the line. It's perfect.

Of course, it doesn't exist in the real world. There are no objects with that perfection of straightness and exactness of length. But then, numbers don't exist in the real world, so why should we worry that the theoretical object is, well, theoretical? Besides, while the line does not exist, we can do a good job of faking it, provided we say how small a scale we want to measure down to. We can smooth and sand and sharpen and make things that are pretty straight and sharp and clean and marked up (Figure 4.8).

Notice how little the ruler resembles a natural stick. The ruler is made as an embodiment of the idealized stick. The perfect line is used as a guide to make real-world tools. This idealization of a shape and then use of that theoretical pure shape as a guide for real-world creation is one of the oldest uses of geometry. Think about it for a moment. In many ways, the first thought of this was as great an act of genius as the first abstraction of number from pile, from recognition to create an ideal that cannot truly be found in the world and then to make a thing in the world that more resembles the ideal than the thing recognized in the first place. It is an act of brilliance so commonplace now that we do not even notice we are using it. But it is the paradigm of the uses of mathematics. Take from the world, imagine in mind, and then reshape the world to meet the imagination.

The geometric mindset starts from here and then can visualize what the real-world form of one of its shapes is and then push that form out into reality, so that using lines one can lay out floor plans, (Figure 4.9) put in frames of wood, (Figure 4.10) and from it create a building (Figure 4.11).

**FIGURE 4.8**
A Good Ruler

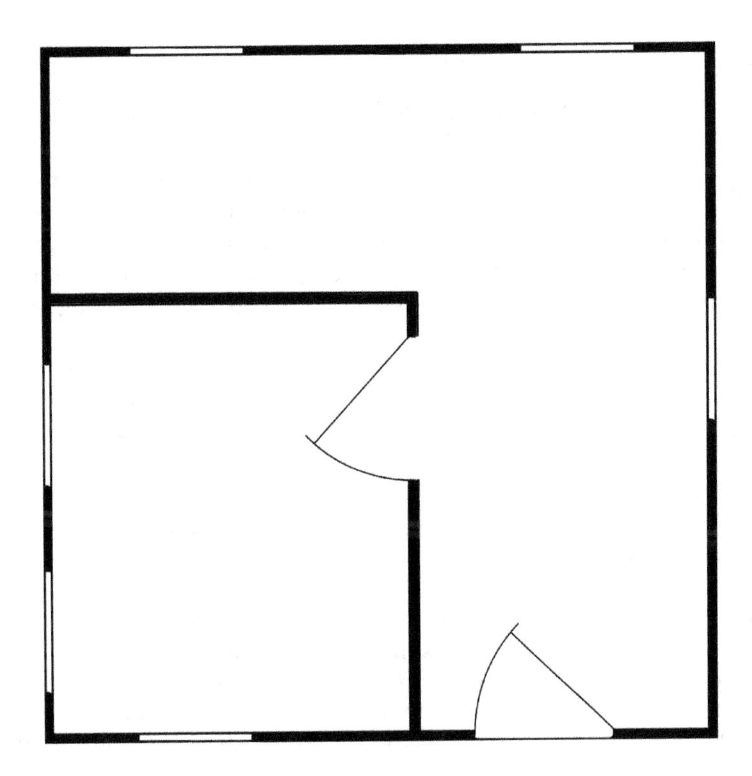

**FIGURE 4.9**
Floor Plan

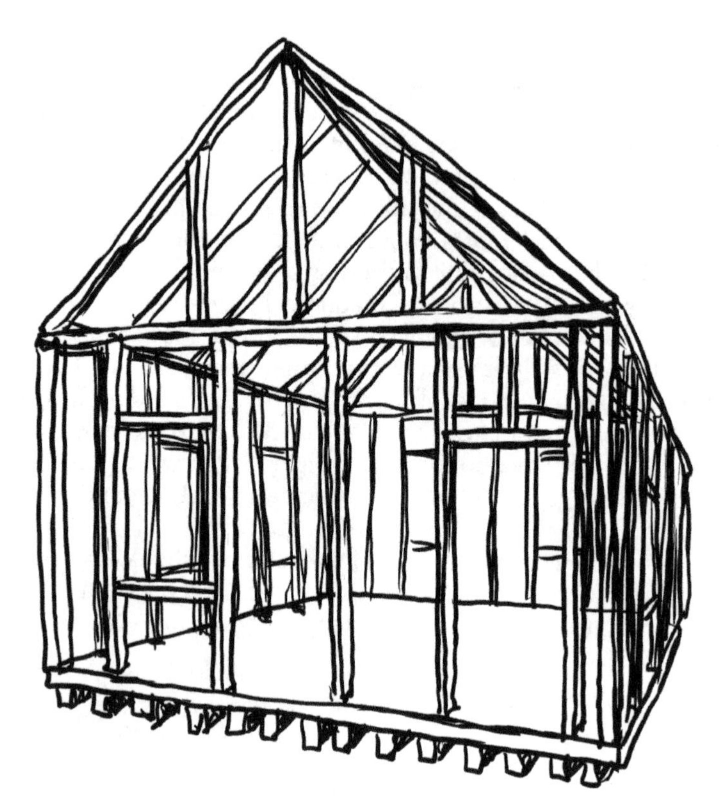

**FIGURE 4.10**
Skeleton of a House

But we're getting ahead of ourselves. We've only got lines at the moment, and haven't said how to make them.

Lines come from drawing straight between the endpoints of sticks. But we haven't said what "endpoints," "drawing," or "straight" mean. Non-mathematicians who have the geometric mindset would answer that they know what drawing and straight are and they can see what the endpoints are. This doesn't help mathematicians with the geometric mindset. They need to know these things with the same level of mistrust that algebraists need to know how numbers work.

Funnily enough, the "know it when they see it" method of describing shape used to be what geometers used as the basis of their mistrust. That's where Euclid's axioms came from. The ancient geometers looked at their ideas of lines and pulled out the characteristics they saw, making them the axioms of geometry. They turned their awareness on the lathe of their mistrust and created statements of obvious truth. (They weren't true in the real world, but they are close enough for now.) We won't give all the axioms, focusing only on the most relevant one for lines:

> **Two points determine a line.** Given two points, there is one and only one line that contains those two points. That was enough for Euclid because he knew what a line looked like and he knew what he meant by points. He knew with geometric thinking what he was talking about. Only many centuries further down the line, as we will see in a later chapter, did mathematicians wonder if Euclid knew what he was

**FIGURE 4.11**
House

> talking about. Indeed, they wondered if he knew what he was seeing, or even if he saw what he knew.

For the moment we're going to pretend he did and not make more exact definitions, for the simple reason that geometry developed for many centuries using this view and it worked as a mathematical model because it was accurate enough for the uses to which it was put. We know a **line** when we see it, a straightness extending infinitely in both directions containing all the points that lie on it.

A few other useful concepts.

> A **point** is an infinitesimal dot. It does not extend at all in any direction.
>
> A **line segment** is a piece of a line that is finite in length.
>
> A **ray** is a piece of a line that extends infinitely in one direction (Figure 4.12).

Points, lines, line segments, and rays are the building blocks of geometry.

*Hey! What about circles, ellipses, hyperbolas, parabolas, all those other shapes?*

**Euclid made them by playing games of definition and drawing with lines and points.**

Before we dive into those, we need to take a step back, as an artist would to look at her drawing, and look at drawing itself.

Drawing is an invisible act of geometric meaning that yet leaves the visible behind it. It does not have a mathematical conception, but is vital to the geometric mindset.

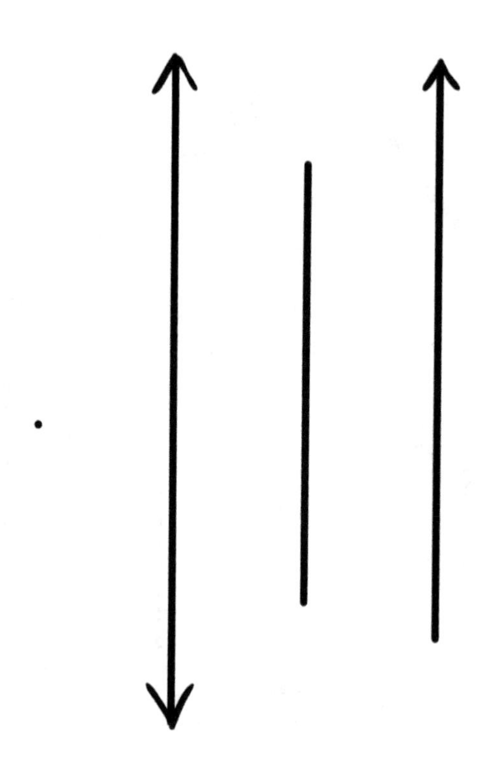

**FIGURE 4.12**
Point, Line Segment, and Ray

So draw up a chair (Figure 4.13).

Thanks. Let's have a look at drawing.

In the standard geometric conception, space consists of an infinite number of points, each at a single definite location. (We'll worry about the location of the locations in the chapter on analytic geometry.) Shifting a bit from this perspective to a more algebraic view, every shape is a subset of this set of points. But most of these shapes themselves contain an infinite number of points (lines, rays, and line segments). Listing the points in a shape is an exact but totally impractical method of knowing what subset one is dealing with.

Besides, this is geometry, not algebra. We need to know the shape of the shape. We need, because shape comes to us from sight, to know what the shapes look like.

Fortunately, there is a physical process that resembles the process of marking out a subset of the set of all locations. Rather than listing the points as the elements of a set, we'll tag the points we're looking at. We'll make them a different color from the points we don't want to consider. Physically one does this by spreading something like a mixture of dirt and oil or ground clay or graphite over it. In other words, we draw the shape by coloring it.

Various definitions of geometric shapes amount to directions for drawing the specified shape. They even imply the possibility of kinds of tools for the drawing.

Let's return to an old friend, the circle. A circle is the set of all points a given distance (called the radius) from a given point (called the center).

This definition leads almost directly to the process by which an image of the circle may be created in the real world. To draw a circle, take a stick as long as the radius, put one end of the stick at the center, and turn the stick around until the other end of the stick has touched each and every point it can. If the other end of the stick has some means of drawing, a circle will be left behind.

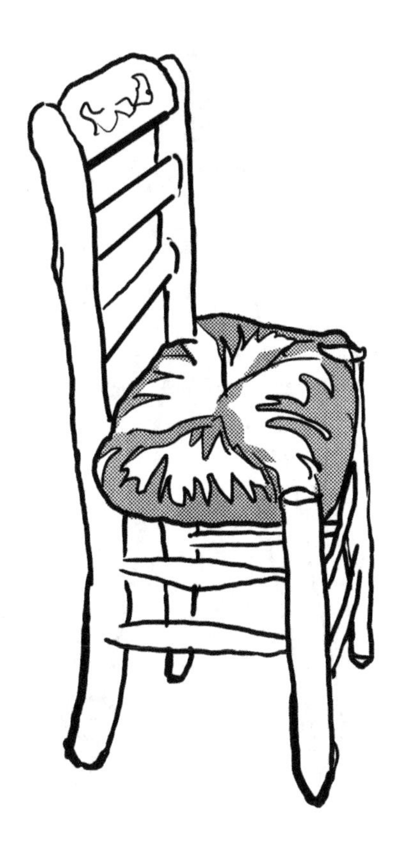

**FIGURE 4.13**
Chair

The practical problem with this instruction is that you need to make a new stick for each different radius circle you want to draw.

Fortunately, a tool can be created that amounts to a stick whose size you can control and which has a pin at one end and a pencil on the other so that you can stick one end in a center and draw with the other end. This tool is called a **compass** (Figure 4.14).

Classical geometry used only two drawing tools: the straightedge (an unmarked ruler) and the compass, which allowed the drawing of lines and circles. They did a lot using only these tools. We'll cover this later when we discuss the **constructible numbers**.

But beyond that let's look more closely at the intimate way drawing turns definition into appearance. The circle contains its own drawing instructions; in many ways nearly every geometric shape does. Geometry has sometimes been described as the process of taking inaccurate drawings and producing accurate theorems. But another way to look at that is that the geometric mindset can translate theorems which are statements into shapes for the eye to see by making those theorems into drawing instructions. That's pretty amazing. Two artists given the same everyday description of something will produce different drawings of it based on their personal styles and interests. Two geometers given the same definition of a shape will produce remarkably similar drawings, inaccurate though they are.

Think about that for a moment. Just as a number can be used to carry the conception of how much or how many from place to place and mind to mind and allow one person to recreate amounts that another person measured out perhaps hundreds of years before in

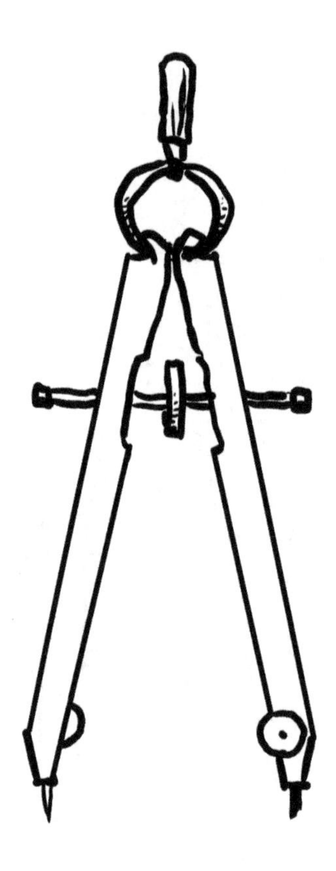

**FIGURE 4.14**
Compass

distant lands, so geometric definitions allow for the carrying of shape across space and time in human minds and also for the creation of startlingly similar pieces of reality.

This ability is what makes architecture possible, and engineering. The ability to transport shape in definition and instructions means that we can rebuild and alter the world according to definition and theorem because we can, for example, transport a circle from one place to another by specifying a center and a radius.

We will have made an identical circle.

*Stop. What do you mean identical? What equivalence relation are you talking about?*

## When Are Shapes the Same?

Drawing creates shapes. Since sets are only the same if they have the exact same elements, two shapes would only be the same if they covered the exact same sets of points. This does not fit our intuitive sense of sameness in shape and indeed does not fit the usage described above for transport of shape in mind. Obviously then, the algebraic idea of same sets is not the equivalence we want to use in geometry.

Geometers came up with two related equivalence relations, either of which qualifies in some way as "same shape." These two relations are **congruence** and **similarity**.

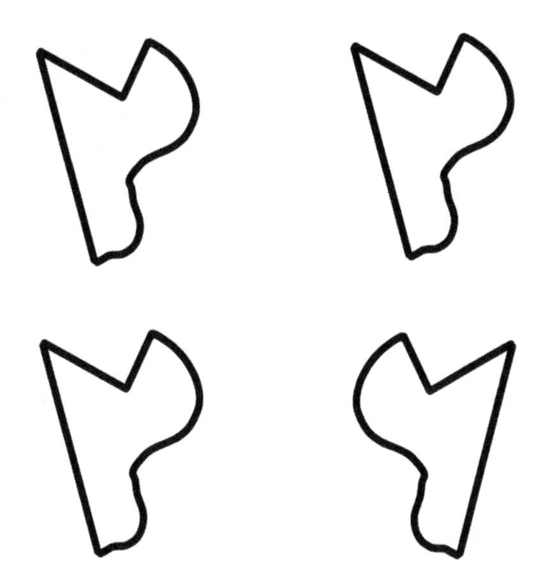

**FIGURE 4.15**
Congruent Shapes

Two shapes are **congruent** if by moving one shape around in space without distorting it, one can make it cover the exact same set of points as the other shape. In other words, congruence is an equivalence relation that does not care about positioning (Figure 4.15).

The shapes shown in Figure 4.16 are not congruent.

Notice those last shapes. They look alike to our eyes, except they aren't the same size. We would like to say that they are also in some sense the same shape. This is what similarity is for. Two shapes are **similar** if one shape can be moved around and made uniformly bigger or smaller in order to cover the same set of points as the other.

Intuitively, we can see that both congruence and similarity are equivalence relations.

For **congruence**:

*Reflexivity*: Each shape is certainly congruent to itself since we can move it a distance of 0 and lay it over itself.

*Symmetry*: If we can move shape A onto shape B, we can move shape B onto shape A.

*Transitivity*: If we can move shape A onto shape B and shape B onto shape C and then if we do the first motion followed by the second motion, we will have moved A onto C.

For **similarity**: It's the same as the above except we also might be making the shape bigger or smaller.

With some work, we could prove this to algebraic satisfaction. Geometry, however, is a bit more practical in its laziness. There are some shapes that get used over and over again in geometry and a decent amount of time was spent creating proofs for when two shapes of these kinds are congruent or similar. Two circles are congruent if they have the same radius, since all you need to do is move one so that it has the same center as the other and behold the same set of points. Any two circles are similar, since if you put one at the same center as the other and then change its radius (which amounts to shrinking or enlarging it) to be equal to the other, you end up with the other circle.

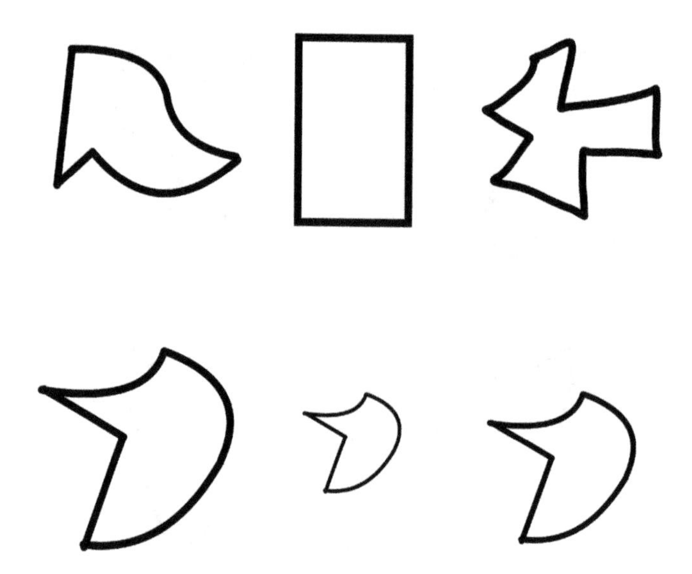

**FIGURE 4.16**
Noncongruent Shapes

A number of tests were created to determine if two triangles were congruent. These always involved comparing the lengths of sides and the sizes of angles. These tests were formulated into rules put in a shorthand of "S" and "A." For example, the SSS rule says that if two triangles have corresponding sides of equal length, they are congruent. The SAS rule says that if two triangles have two sides and the angle between them is equal, they are congruent. The ASA rule says that if they have two angles and the side between them is equal, they are congruent.

*Note*: Geometry students often try to use a rule based on one angle and the next two sides equal as a test for congruence. This test fails and the students often feel like the acronym for this.

Two triangles are similar if they satisfy the AAA rule, which is not always keep a full tank of gas, but if the angles are equal they are similar.

One more thing to notice about these tests. They are methods of determining if shapes are the same, but they are built up from tests that measure equality of length and angle using our friend the sticks. This is part of the process of building the theoretical universe and making sure it works. If you create measurement of length and angle, then you can create and prove the workings of congruence and similarity and those equivalences will be as accurate as your underlying measurements. We construct one set of mathematical means out of another, and by preserving truth in our constructions, we make sure that the result is as accurate as the initial values.

## Shapes as Tools

The geometers did not just create shapes that looked pretty to them. They made shapes that were useful. Circles were used to help mark the paths of the Sun, stars, and (using a combination of circles) planets in the sky. Lines were used to measure distances.

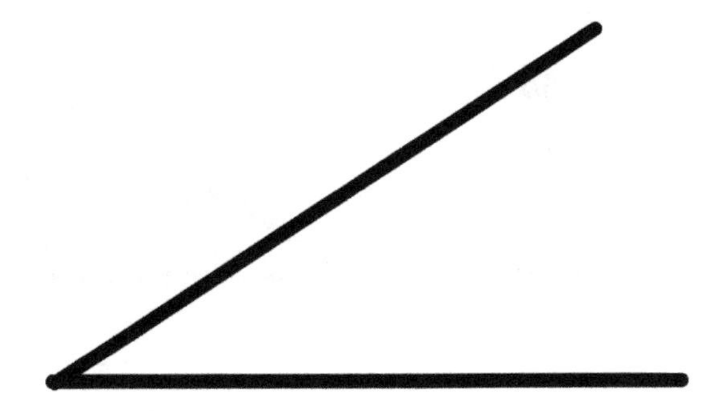

**FIGURE 4.17**
Angle

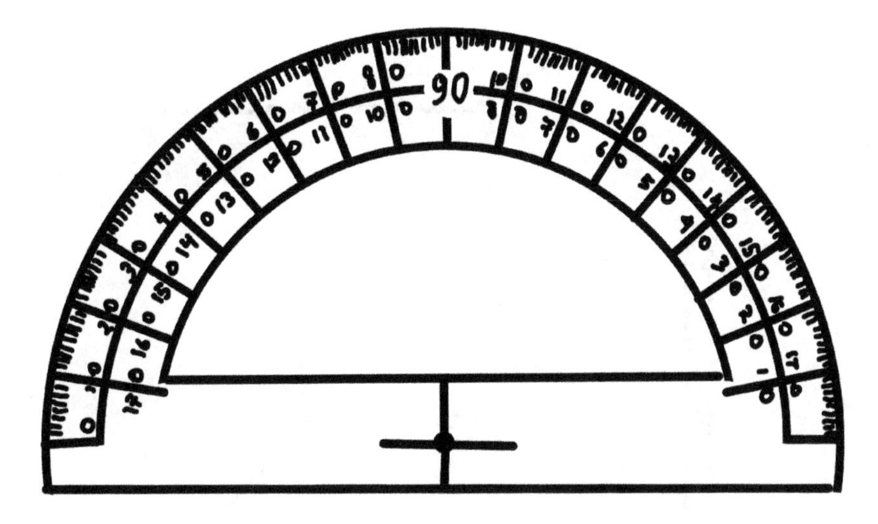


**FIGURE 4.18**
Protractor

Two shapes ended up being stunningly useful: **rectangles** and **triangles**, both of which contain the word "angle," which, as we said before, is an object made from taking two sticks and putting one end of one stick on top of one end of another and looking at the gap between the sticks (Figure 4.17).

If one holds one stick stationary and moves the other stick around, keeping the ends touching, the moving stick traces out a circle. This means that a circle can be used as a measuring device to determine the size of an angle. Half circles called **protractors** are often used for this purpose (Figure 4.18).

The everyday measurements of this method divide a circle into 360° ("360 degrees") and says that an angle is so-and-so-many degrees (Figure 4.19).

One particular angle, the one where one stick lies along the ground and the other points straight up, exercised a peculiar fascination. It is an angle measured at 90° and was given its own special name, the **right angle**. This doesn't mean that all the other angles are wrong, only misguided. They just need a little straightening out and they too can follow the right way.

Right angles have a clear physical reason for existing. Take a pole and set it upright (that is, at right angles to the ground). It is being pulled down by gravity, but it's being pulled

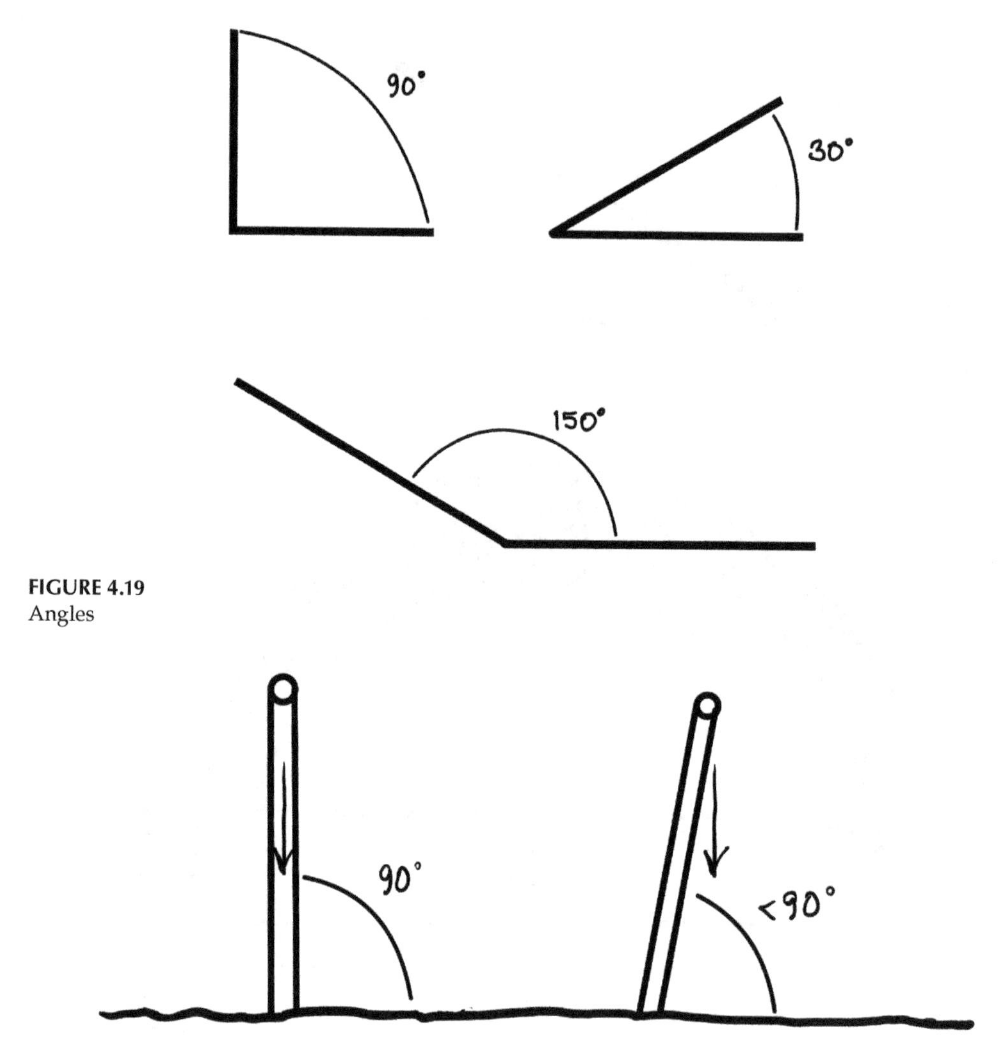


**FIGURE 4.19**
Angles

**FIGURE 4.20**
Right Angle Using Poles

straight down, which means that it's not going to fall over. If you have it at another angle, then the top is being pulled straight down from where it is, and unless propped up it's going to fall over (Figure 4.20).

Take two sticks of equal length, then two other sticks of equal length, and arrange them as shown in Figure 4.21.

This shape, the rectangle, a shape with four right angles is extremely useful in building things. We use it for floors, walls, doors, windows, and roofs. It is easy to figure out how much ground you need to lay out a rectangle, how much wood or stone to make a wall, and so on. All you need to do is multiply the length of the stick along one side with the length of the stick at right angles to it and you'll have the area.

Using angles, triangles, and rectangles, we can create elaborate amalgams of shapes that can be used as designs for buildings. But because we're using these shapes, we can calculate the area needed and the materials required, and our requirements will be as accurate as our creation.

**FIGURE 4.21**
Rectangle

*Hmm ... that depends on what we mean by "same shape." If "same shape" means "congruent," then our model of a house would have to be the same size as the house. That doesn't sound very useful.*

**Right, in this case "same shape" doesn't mean "congruent." It means "similar."**

Suppose you made a small-scale model of the house you wanted to make, or a scale drawing. A **scale model** is a three-dimensional model of something created with such care that each shape in the model is geometrically similar to the shape you really want in the house, and all the shapes are smaller by the same factor (called the **scale**). A scale drawing is a two-dimensional drawing that does the same job. For a house, you need a lot of scale drawings since you need one for each floor and wall.

If you did a scale drawing where 1 cm on the drawing represents 1 meter in the real world, then everything you are making in the real world would be 100 times bigger than the corresponding shape drawing. A scale drawing like this is a tool that allows you to discern the size of the real objects you plan to make and how much material you need for them. All you need to do is measure the drawing and use it to figure out how much material you need in real life. If one wall is 3 cm × 5 cm on the drawing, then the wall in the real world will be 3 m × 5 m. This means that the drawing can be used on its own to determine what is needed to build the actual house. This use of geometry made drawings into real and vital tools for architecture.

While rectangles are useful in architecture, right triangles (that is, triangles with one right angle) spawned a whole branch of mathematics and have allowed people to reach out and measure the stars. I refer to the one of the great stumbling blocks in math education: **Trigonometry**.

*Cower before it, ye that are not Right!*

Let's take a look at a right triangle (Figure 4.22).

A standard theorem of geometry says that if you add up the angles of a triangle, you get 180°. Since one of the angles of the triangle is 90°, the other two angles must add up to 90°.

Notice, by the way, that this right triangle is exactly half of the rectangle whose sides are a and b. So its area is ab/2 (Figure 4.23).

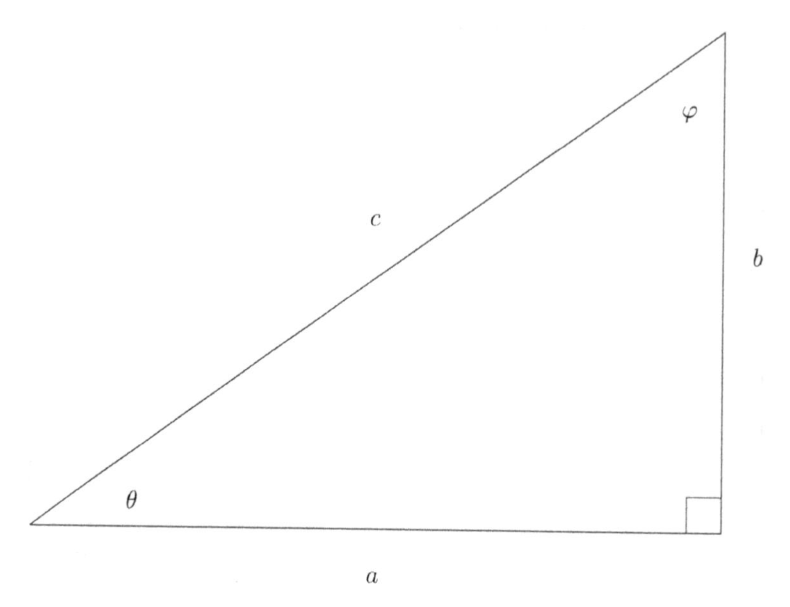


**FIGURE 4.22**
Right Triangle with Sides and Angles Labeled

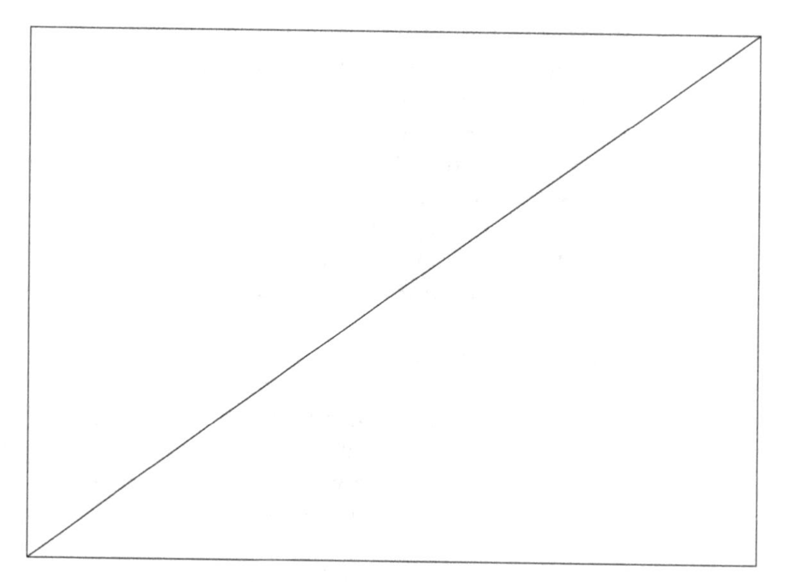

**FIGURE 4.23**
Rectangle Divided into Right Triangles

One of the most famous and useful theorems in geometry is called the **Pythagorean theorem**, named after the ancient Greek philosopher Pythagoras, who supposedly proved it. Pythagoras who lived in the sixth century BCE, while counted in the pantheon of mathematicians, was more like a prophet and a founder of religion. It's hard to determine exactly what he did because his followers tended to attribute things to him.

*A mathematical religion. I'll bet that caught on big.*

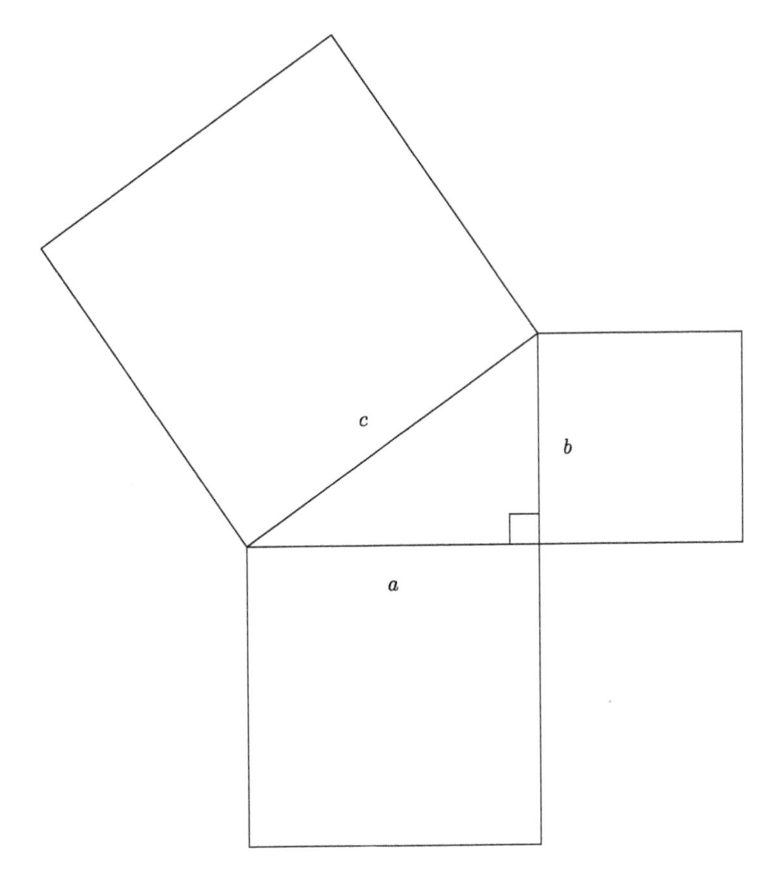


FIGURE 4.24
Squares on the Triangle

**Look, just because these days people don't found religions based on math, music, and not eating beans because they have human souls in them doesn't mean it wasn't a viable spiritual system.**

*Beans, you say.*

**Ahem.**

The Pythagorean theorem says that in a right triangle, the sum of the areas of the squares on the two sides opposite the non-right angles is equal to the square on the **hypotenuse** (the side opposite the right angle). Or in algebraic terms,

$$a^2 + b^2 = c^2$$

The phrase is "square on," not "square of," because the square Pythagoras was talking about was the actual square shape constructed on the side mentioned (Figure 4.24).

To prove this and also by way of showing what a geometric proof can look like, let us consider the square whose sides are length a + b (Figure 4.25).

If we mark the points on the square where the length is a, we can divide the square up into two squares and two rectangles (Figure 4.26).

The squares have areas $a^2$ and $b^2$, and each rectangle has an area of ab. The total area of the big square is $a^2 + b^2 + 2ab$.

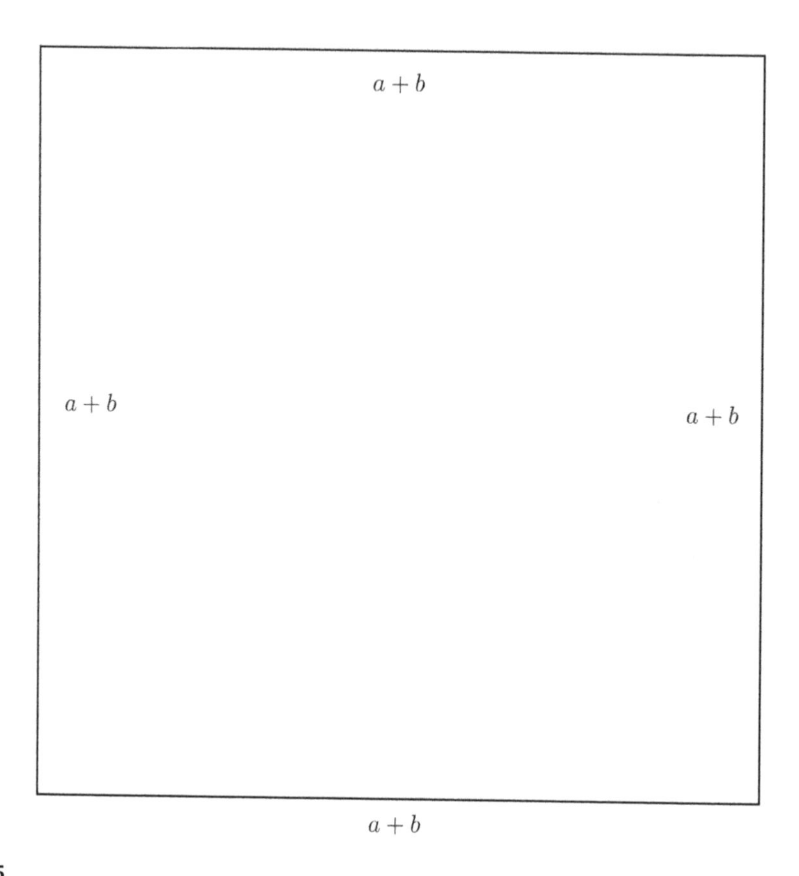


**FIGURE 4.25**
Square

We can also take this same big square and divide it up into four right triangles that are copies of the triangle we started with and whatever area is left over when you've taken these right triangles. That these triangles are all congruent to our starting triangle can be proven by applying the SAS test using the a side, the right angle, and the b side. Since all four triangles here have sides a and b and the right angle between them, they are all congruent. But if we look at the shape we've created inside our big square, we can see that the area left over is a square whose sides are length c (Figure 4.27).

The area of this square is then the area of the interior square plus the area of the four triangles. The interior square is $c^2$ and each of the four triangles has an area of ab/2, so the four together have an area of 2ab. So the entire square has an area of $c^2 + 2ab$. Since the two big squares are the same big square, just divided up differently, we can equate the areas:

$$a^2 + b^2 + 2ab = c^2 + 2ab$$

Subtracting 2ab from both sides gives

$$a^2 + b^2 = c^2$$

Notice that implicit in this is the real-world observation that two congruent objects have the same area. In other words, if you move something around, it stays the same size.

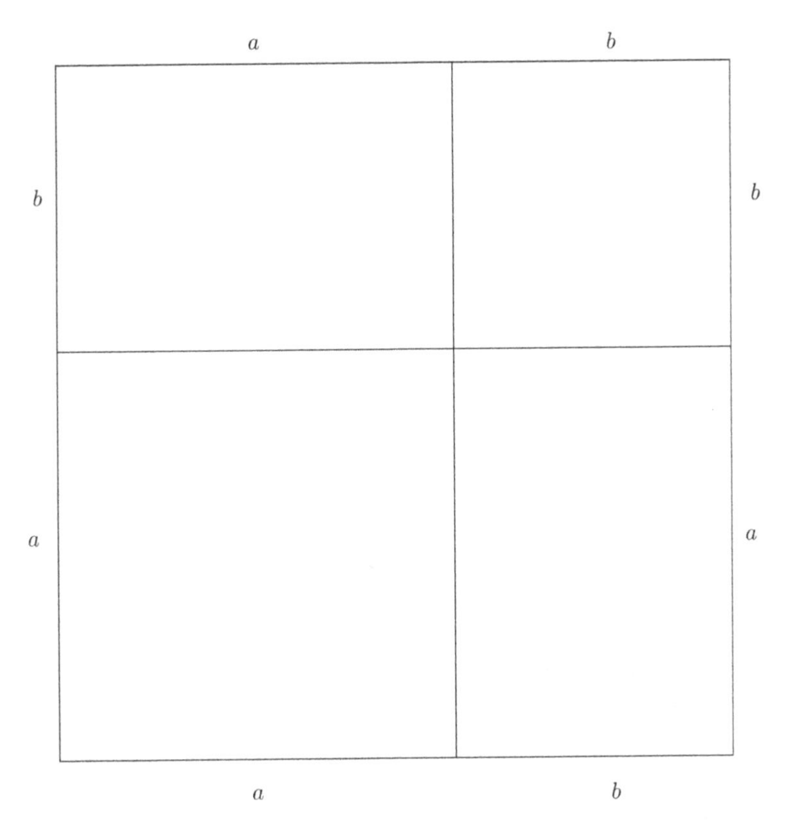


**FIGURE 4.26**
Square Divided into Rectangles and Squares

Ironically, given general relativity, this is not completely accurate. This is one of those situations where obvious observation turns out to be only approximately true.

There is something more important here than just the triangles. Notice that the side c is the line connecting the two points on the triangle that are not at the right angle. The Pythagorean theorem tells us the length of that side. But that side is the line segment connecting those points, and therefore its length is the distance between those two points.

*Back up. You've used distance and have not defined it.*

**I just did.** In classical geometry the distance between two points is the length of the unique line segment connecting those points. Therefore, the distance between these points can be determined without measuring it, assuming we know the lengths of a and b. We'll use this a great deal in the chapter on analytic geometry.

If you look again at the right triangle in Figure 4.28, you will notice five variables: a, b, c, $\theta$, and $\varphi$. These five are not independent. That is, we can't change them willy-nilly. They have interconnections, relations that restrict the values we can assign them (these are called dependence relations). We know that $\theta + \varphi = 90°$ and that $a^2 + b^2 = c^2$. But if we look closer, we can see another dependence relationship. Two triangles are congruent if they have the same sides. Since a and b determine c, two right triangles are congruent if we specify two of the sides and if we say whether either of those sides is the hypotenuse. This means that the triangle is fixed if we give only two values. It's also fixed if we give the value of a or b and either $\theta$ or $\varphi$. In Figure 4.28, notice that we've given a and $\theta$, and they together determine b.

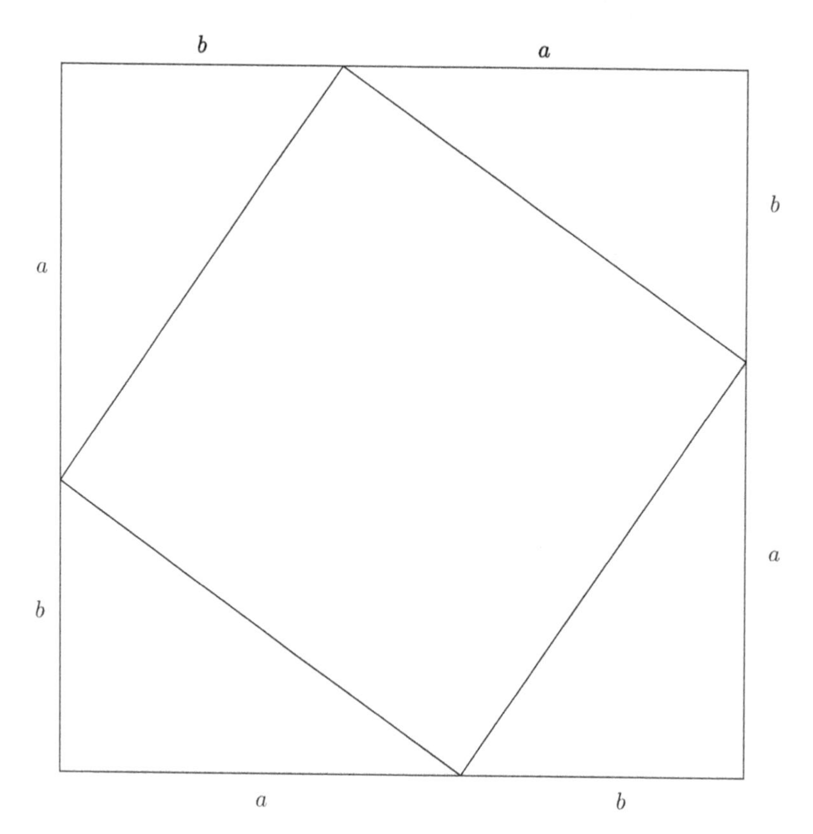


**FIGURE 4.27**
Square Divided into Square and Triangles

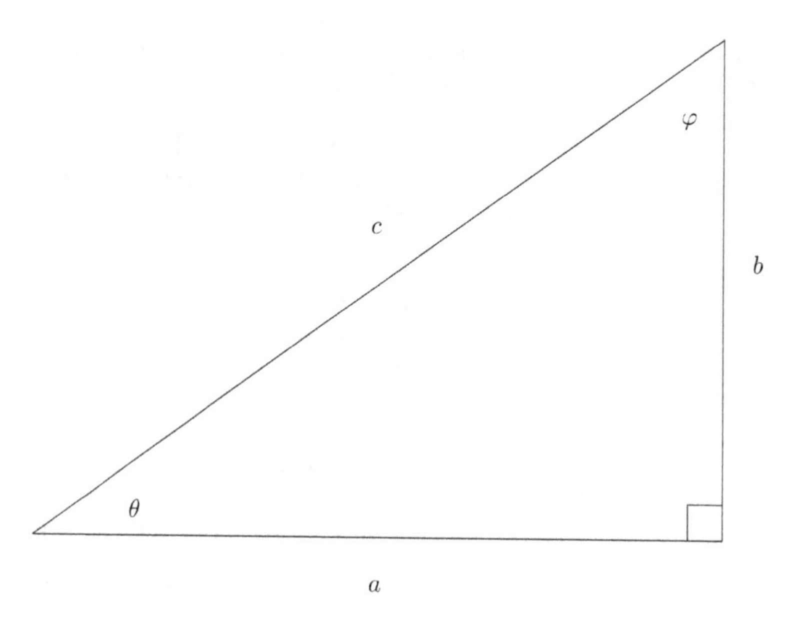


**FIGURE 4.28**
Triangle with Angles Marked

So why does this matter? Because one of the most basic uses of mathematics is to measure as little as possible in order to calculate any other needful values. The less information we need to determine a right triangle, the more information we can get out of it relative to the amount we put in. This is another example of laziness. Get as much output as you can for as little input as you need to provide.

In any case, there is clearly some interconnection between the angle θ and the side a that allows us to determine b. Let's look more closely at the triangle as drawn. Suppose we doubled the lengths of the sides. We would have a triangle similar to the initial one, so the angles would be the same. That means that whatever the interrelationship between a and θ is that determines b, the same interrelationship between 2a and θ gives 2b. This hints at some kind of ratio between the sides.

In fact, if we take a right triangle and divide one side by the length of the hypotenuse, (a/c), this value will be the same for any triangle similar to this triangle. Weird as it sounds, this means that this value (which is determined by dividing one length by another) depends only on the angle that is opposite the side a, not on the lengths that were used to calculate it. This ratio is called the **sine** of the angle, and is usually defined as follows:

sine of θ – written sin(θ)– is equal to the length of the side opposite the angle θ divided by the length of the hypotenuse of the triangle.

Sine is the first of the **trigonometric functions**. ("Trigon" means three angles, so trigonometric means measurement of three-angled things. In other words, it just means measuring triangles.)

There are two other important trigonometric functions: the **cosine** and the **tangent**. There are also three unimportant ones which we're not going to bother with.

The cosine of θ – written cos(θ) – is equal to the length of the side touching (or adjacent to) the angle θ divided by the length of the hypotenuse. So in the triangle we have above, it's b/c.

The tangent of θ – written tan(θ) – is equal to the length of the side opposite the angle θ divided by the length of the side touching (or adjacent to) the angle θ (a/b in the triangle in Figure 4.28).

We'll analyze these more in the section on analytic thinking. (Coming up in just a couple of pages. Stay tuned, analysis fans.)

Trigonometric values could be calculated by making triangles and calculating ratios, but there are better methods (now made much faster using computers). Before computers, people would make tables of the answers so that they could be looked up when needed.

*Unfortunately, many people think of trigonometry as an irrelevant subject used to annoy high school students. We'd better explain why and how people used a table of sines, cosines, and tangents for practical purposes.*

Suppose you need to know how tall some object inconvenient to measure is, like a tall tree or a mountain. 'Cause, you know, measuring high things can be a problem.

*Oh, because of that annoying falling thing people talk about, or yell about while it's happening.*

Suppose further that you know how far away the object is from where you are.

Set up a stick where you are with one end in the ground so that one can look from the ground through the top of the stick to the top of the object (Figure 4.29).

Draw an imaginary line from the top of the object to the top of the stick and then continue that line down to the ground (a piece of string can be used there). You can measure the angle made between this line and the ground. Find the tangent of that angle on a table or using a calculator. (*Not available in all time periods; check your local anachronism authorities for misuse of temporally misplaced hardware.*) The height of the object will equal that tangent times the distance to the object. Distance as a ground measurement is somewhat easier to determine than height.

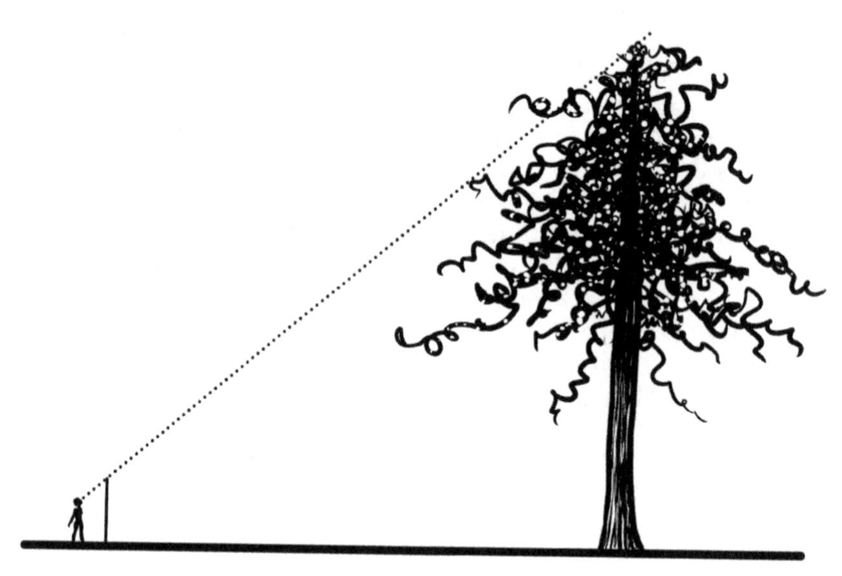

**FIGURE 4.29**
Measuring a Tree

The height is the opposite side of an imaginary triangle you are creating, and the distance to it is the adjacent side.

Since tangent = opposite/adjacent,
opposite = tangent times adjacent.

It's not just heights that are inconvenient to measure. Large distances are more difficult and time consuming to measure than small distances. Trigonometry tells us that if we have a right triangle with one small side and one large side, then measurement of the small side and the angle allows us to conclude the size of the large side, thus saving us the time and effort we would have had to expend in directly measuring the large side. For this reason, trigonometry has been an essential tool in surveying.

It is also an essential tool in astronomy. Even in ancient times, Eratosthenes used this method to measure the size of the Earth. Here, the small side of the triangle was the measured distance between two cities and the angle was found by measuring the shadow of a vertical stick. More recently, astronomers use this method to measure the distance to the stars. Here the "small" side of the triangle is given by the distance between the Earth and the Sun, and the angle is measured using the method of parallax (how far the image seems to shift during the year as measured against a background of distant stars).

Trigonometry has been a tool of surveyors and navigators since its inception, but it has also been a nuisance to learn. While its functions are determined from the sides of the triangle, they really only depend on the angles. This marks a boundary point in geometric thinking because the use of the tool of the triangle is not found in the triangle, but in analyzing the properties of its use.

To go on from here, we have to leave geometric thinking behind, a thing that has happened time and again. More than any other branch of mathematics, geometric thinking has been transformed and subsumed into the other branches. This is unfortunate, because now many people who might once have been deemed mathematicians (architects, builders, painters, surveyors) are no longer so considered, and their voices and views are silent in the discussions of what math is and what it needs to do. This is part of a general phenomenon (which we will touch on in detail later) of the difference in point of view between the mathematicians and the users of mathematics.

This is a case where mathematicians themselves, as a by-product of their emphasis on abstraction, end up casting out part of the inheritance and demeaning its value. In the long run, we think this trend will pass, but in the meantime aspects of geometric thinking are being preserved by the users of mathematics and could be brought back at a later time.

But there's no need to follow the current fashion. The ability to switch one's view in mind, to reshape things with thought, and to carefully preserve measurement and scale and see shapes as like or unlike, the basic tools of geometry, can still be practiced and still be useful in the everyday careful observation of the world and detection of its forms.

These ways of thinking are subtly present in our minds. Our sense of rightness in shape, of what looks good and beautiful, is often geometrically informed. Our eyes see more sharply when our minds know what they are looking at and can judge thereby. Our minds play more with what we see when we can hold in them the shape of what we remember and can alter it in our honed imaginations.

Besides, as we will see later, there are new shapes to see and new geometries that are waking up and blooming forth.

## Analysis

In geometric and algebraic mindsets, the mind focuses on objects, seeking to understand them. One can think of the analytic mindset as looking instead at process. It's like attending to the baking, not the cake. This can sound odd if you want to eat, but not if you want others to be able to eat.

To a certain extent, the treatment of analysis in this chapter is very preliminary. Much of analysis is concerned with abstracting and extending the tools that were used to put calculus on a properly mathematically rigorous footing. Treating that part of analysis will have to wait until we do calculus. For now, we will concentrate on the idea of a function.

Let's look again at $2 + 1 = 3$. In the algebra section, we seemed to exhaust all the possible abstractions of this equation. But there is a way of looking at it we neglected. We did not examine the process of calculation inherent in this equation. Here it is as a step-by-step procedure:

Step 1: Take the number 2.

Step 2: Perform the action of adding 1 to the number from Step 1.

Step 3: Write down the result of Step 2.

Even for something as simple as adding 1, writing this process out takes a lot of space, rather like writing down a recipe. In most cookbooks, a single recipe occupies an entire page or more. A more compact form for calculation procedures would be highly useful, particularly if one wanted to display something more complicated than the simple addition above such as:

Step 1: Take two numbers.
Step 2: Square the first number
Step 3: Cube the second number.
Step 4: Multiply the cube in Step 3 by the square in Step 2.
Step 5: Multiply the result of Step 4 by 25.
Step 6: Multiply the first number by the second number.
Step 7: Multiply the result of Step 6 by 16.
Step 8: Add the result of Step 7 to the result of Step 5.
Step 9: Add 9 to the result of Step 8.
Step 10: Write down the result of Step 9.

A very compact way of writing down processes like this has been developed. Unfortunately, it produces things that look like equations and which lose the entire sense of a process.

This form is called a "**function**." Functions are the objects that the analytical mindset has the easiest time seeing and working with.

The analytical is a mindset that looks less at "what?" and more at "how?" How do we make the cake? How do we calculate a value? "How" is one of those questions that everyone implicitly asks when they need to learn to do something.

The analytical mindset is less interested in individual hows than it is in the how of the how. It wants to abstract how into forms that can be grasped and analyzed, hence its need to shrink processes down to compact forms in order to hold them easily in mind and mess around with them.

The compact form of a function begins with a name for the function. Often the name is just a single letter, like "f" for example.

The name of the function is then followed by a list of what are called **parameters** or **arguments.**

*This is definitely a difference between the mindsets of mathematicians and physicists. Mathematicians like to use "f," whereas physicists like to use "f(x)." More generally, mathematicians like to use a completely abstract notation where any letter can stand for anything, and you specify things with an initial definition ("let f be a map from the real numbers to the real numbers"), while physicists like a notation where looking at the notation tells us what kind of mathematical object we're dealing with.*

The list consists of dummy variables representing the things that the process operates on. For example, a function that takes a single number and adds 1 to it has only a single parameter, a variable representing the number that is to be added to. Argument lists are usually enclosed in parentheses, with commas separating the parameters. After this comes an equals sign. This equals sign means that the function is *defined* to be what follows. The other side of this is usually a formula involving the parameters of the function.

For example, the function for the process "add 1" is:

$$f(x) = x + 1$$

This is read "f of x equals x plus 1," which, as I said, looks like an equation. It isn't. It's actually a definition. The function f is defined as follows:

Step 1: Take a number. This number will take the place of every instance of the variable x in the function definition.

Step 2: Replace the variable x with the number in step 1 wherever it appears in the function definition.

Step 3: Carry out the function calculation (in this case adding 1).

Step 4: Write down the result of step 3.

Suppose one wanted to apply this function to the number 8. One would first take the function definition

$$f(x) = x + 1$$

and substitute the number 8 for every occurrence of the variable x, getting

$$f(8) = 8 + 1$$

This is read "f of 8 equals 8 plus 1."

The next step is simple in concept, easy to write down, but confusing in meaning. We then add 8 and 1 and get 9. So we write down

$$f(8) = 9$$

This is an equation, not a definition. This equation tells us that the value of the function f when applied to the number 8 is equal to the number 9. This is unlike the earlier

$$f(x) = x + 1$$

which we used to define what the process of the function is.

The confusion that comes from this notation has thrown off many people, which is regrettable.

*Regrettable? People losing their mathematical heritage because the notation is confusing? Isn't that like losing your house because of a confusing mortgage contract?*

**I'm afraid so.**

Let's go over this again, more slowly.

When you write the function out with the parameters as variables, you have a definition of the function. When you specify values for the parameters you have an equation derived from the function.

*Okay. That's better.*

**Only kind of, because the definition can also be used as an equation.** Just as if you have $y = x + 3$, you can substitute $x + 3$ into another equation involving y, so you can substitute the definition of f(x) (in this case $x + 1$) for f(x) in any equation. If you had

$$y = f(x) + 7$$

you would be perfectly justified in writing

$$y = x + 1 + 7 = x + 8$$

as an intermediate step in trying to solve the equation above.

*So what you're saying is that you have to be careful about whether you are dealing with definitions or equations, but you can use the definitions as if they were equations.*

**Yes.**

*Go on.*

Let's go through the process of turning a process into a function definition using our more complex example.

Step 1: Take two numbers. (call them x and y)

Step 2: Square the first number. ($x^2$)

Step 3: Cube the second number. ($y^3$)

Step 4: Multiply the cube in step 3 by the square in step 2. ($y^3x^2$)

Step 5: Multiply the result of step 4 by 25. ($25y^3x^2$)

Step 6: Multiply the first number by the second number. (xy)

Step 7: Multiply the result of Step 6 by 16. (16xy)

Step 8: Add the result of Step 7 to the result of Step 5. ($25y^3x^2 + 16xy$)

Step 9: Add 9 to the result of Step 8. ($25y^3x^2 + 16xy + 9$)

We would write this as $f(x,y) = 25y^3x^2 + 16xy + 9$

Which would be read as "f of x and y equals 25 y-cubed x-squared plus 16 xy plus 9."

If we needed to know the value of this function when x is 2 and y is 1, we would write f(2,1) = 25 times 1 cubed times 2 squared + 16 times 2 times 1 + 9. This is 25 times 4 + 32 + 9 = 100 + 32 + 9 = 141.

It helps, we find, to think of a function like a machine, a factory that takes in the raw materials of its parameter values and produces a finished product. This will be especially useful when we get to the chapter on computers, which are real machines that do this with functions. The problem with this metaphor is that the compact mathematical notation for functions does not fit this concept.

Why use this notation that looks like equations? Ironically, it's because, as noted above, the functions can then be used in equations, substituted for their definitions.

If we go back to $f(x) = x + 1$, we can write the equation

$$f(x) = 4$$

and ask what values of x satisfy this equation. That is, what numbers can we substitute for x in the definition of f(x) to get a value of 4?

This is another annoying shift. When we defined f(x), x was a dummy variable a place holder, nothing more. But when we write

$f(x) = 4$ and ask for what values of x satisfy this equation, we've turned x into an actual variable, a question seeking an answer.

To solve this, we swap in the definition of f in the equation $f(x) = 4$ to get

$$x + 1 = 4$$

This yields $x = 3$. This tells us that the function f takes the value 4 when its parameter is 3. We'll make this process more visual in the chapter on analytic geometry.

This sounds like it's only a notational trick, but it's actually crucial to the analytical mindset. The purpose of functions looks like an artificial trick to create compact forms of calculations. But there's a missing element in all the above. The functions we've so far created are meaningless. They are nothing more than compact calculation machines. But the functions analysts look for and try to find the formulas for are meaningful. For example, the force of gravity on an object of mass m, exerted by another object of mass M if the distances between the objects is r, is

$$F(m,M,r) = GMm/r^2$$

We've seen this as an equation, but it's also a function that produces a value for the force, given the parameters specified above. This function along with being a means of calculating force is also a means of saying what is needful to calculate the force. Only those three parameters, just m, M, and r, that's enough information out of an entire universe full of abstractable objects and values to find the gravitational force one object exerts upon another. The existence and rightness of this function is a statement about the processes of the universe – and it's compact enough to fit in a single simple formula. Furthermore, it's a statement that acts as a machine to calculate an infinite number of values for the force that will help you determine how objects move. And your results will be as accurate as the initial measurements of mass and distance.

## Building Machines

If one looks at functions as machines or machine parts, one can see that one can build complex functions out of simple ones. For example, one might take three functions:

$$f(x) = x + 1$$

$$g(x) = 2x + 1$$

$$h(x) = x/6$$

and make a new function by multiplying them:

$$j(x) = f(x)g(x)h(x) = (x+1)(2x+1)x/6$$

One could make a completely different new function by adding them.

$$k(x) = f(x) + g(x) + h(x) = x + 1 + 2x + 1 + x/6$$

Notice that the process of adding these functions produces something with a messy calculation. We can simplify this process by noticing that the steps from these functions can be combined by a little reshuffling.

$$k(x) = x + 1 + 2x + 1 + x/6 = x(1 + 2 + 1/6) + 1 + 1 = x(3 + 1/6) + 2$$

Thus, instead of

Step 1: Take a number.
Step 2: Take the number in Step 1 and add 1 to it.
Step 3: Take the number in Step 1 and multiply it by 2.
Step 4: Take the number in Step 3 and add 1 to it.
Step 5: Take the number in Step 1 and divide it by 6.
Step 6: Add the numbers in Steps 2 and 4.
Step 7: Add the numbers in Steps 5 and 6.
Step 8: Display the number in Step 7.

We have the shorter process:

Step 1: Take a number.
Step 2: Take the number in Step 1 and multiply it by 3 + 1/6.
Step 3: Take the number in Step 2 and add 2 to it.
Step 4: Display the number in Step 3.

This creates an important point. We deem two functions equal if they produce the same values from the same parameter lists. Even though the procedures are different, the ultimate result is what we care about. This sounds paradoxical. If analysis is the mathematics of process, why would mathematicians deem processes the same that don't work the same?

The answer comes in another mathematical name for a function: a "map." Recall that each point on a real map corresponds to a particular place. Mathematicians think of a function as a map that associates each "point" in its "parameter space" with the resulting number that the function produces from those parameters.

In this sense, a function can be thought of as an equivalence class of procedures. All procedures that take the same parameter values to produce the same results are the same function. This is not the formal mathematical definition (which is a variant of the ugly definition of relation in the last chapter), but it is closer to the reality of mathematical work. This leads mathematicians and scientists to try to find easy to calculate forms of functions. Thus, does laziness find its way in.

Along with the obvious processes of combining functions noted above, there is also a more subtle and, for a lot of people, confusing way to combine two functions. This process is called **composition**. The composition of two functions is the function that arises from applying one function, taking the results of that, and applying the other function to those results. This amounts to taking a number, running it through the machinery of one function, and then taking the result of that and running it through the machinery of another function.

Looked at as a series of steps, there's nothing strange here. After all, in making a cake, one would first mix the ingredients into a batter (one function), then bake the batter into a cake (second function), then frost the cake (third function), then eat the cake (fourth function), then digest the cake (fifth function) .... And we'll ignore any functions that come thereafter.

Mathematically, it's easy to see this way of combining functions, as long as one is not looking at the function notation.

Consider the function "take a number and square it," and another function "take a number and add 2 to it."

If we do the first function and then the second, we've taken a number, squared it, and then added 2. If we did that to the number 3, we would have $3^2 + 2 = 9 + 2 = 11$.

If we do the second function, then the first, we've taken a number, added 2 to it, and then squared it. If we did that to the number 3, we would have $(3 + 2)^2 = 5^2 = 25$.

This shows that it matters what order you do the functions in. This is obvious with the cake process as well. After all, if you eat the cake and then you frost it, well.... Let's not think about that too much, okay?

Notationally, if we have two functions,

$$f(x) = x^2$$

and

$$g(x) = x + 1$$

then doing f first and then doing g is written as g(f(x)) (read "g of f of x"). We calculate this by taking the definition of g(x) and substituting f(x) for every occurrence of x, which gives $g(f(x)) = f(x) + 1 = x^2 + 1$.

Doing g first and then f gives f(g(x)) (read "f of g of x"), which gives

$$f(g(x)) = (g(x))^2 = (x+1)^2$$

There are two confusing aspects to this. First of all, we're really changing the meaning of x each time we do one of these compositions. We've got serious "it" problems. Notice that we did not have these problems when we added or multiplied functions. The parameters stayed the same. In composition, we are substituting one function as a parameter value for another. This is one of those things that one simply has to get used to. This can be done by taking care. If you get confused, the main thing to do is to write the steps out carefully, keeping track of each act of substitution with an explicit statement.

f(g(x)) = f(x) with g(x) put in each place where x occurs in the definition of f. Make note of each of these places, do the substitutions, then put the whole of it together.

Composition also has a visual confusion. The function we do first is the one that's inside, not the one we reach first in reading. f(g(x)) means do g first, then do f, but in reading left to right our minds see do f first, then g.

There have been notational attempts to solve this, but none of them really help much and we're not going to dig deeply into them.

The difficulty many people encounter is that the function notation is compact, useful in calculation, and easy to use in equations, but only if one is comfortable with it. If not, it can trip people up repeatedly.

This is another of those unfortunate learning difficulties. The analytical ability to roll up a complex operation into a single mental object and work with that entity in combination with others and so produce new understanding is made possible by the ability to compact the notation. It's like the difference between having a single word for something and having to spout three pages of text every time you want to talk about it. But the word only works if the person you're talking to understands its meaning, which might be those three pages of text. The analytical mindset is comfortable shifting back and forth between the compact and the long-form description of a process.

Because of this ability to shift back and forth, the analytical mind is harder to confuse on certain points. If you misuse the word, the analyst can open it up and show the misuse. It's important to be able to shift back and forth between $F = ma$ and force is equal to mass times acceleration. If you learn to slide between the formula and the meaning, the formulas will acquire meaning. This is a useful skill for anyone to learn, because of the simultaneous ability to easily use processes and to open them up and fix them if something goes wrong.

## Function Behavior and Unknown Functions

The analytic mindset does not just want to know what functions are, but also how they behave and what determines how they behave. Determination of function behavior is largely a matter of asking the question: how does this function act as the parameter values change?

Starting with a simple case, consider a function $f(x) = mx + b$, where m and b are values that have been picked already.

This kind of function, a linear function, either grows at a constant rate or shrinks at a constant rate, depending on whether or not m is positive. If m is 0, the function is $f(x) = b$, having a constant value. Constant functions couldn't care less what their parameter values are.

For another example, the function $f(x) = x^2$ is never negative and its values grow as x gets away from 0 either in the positive or negative direction.

The function $g(x) = x^3 - x$ is negative if x is between 0 and 1 and then it becomes positive at a faster rate than $f(x) = x^2$.

As a more physical example,

$$F(m,M,r) = GMm/r^2$$

acts linearly in M or m – that is, it grows at a constant rate as m or M increases – but it shrinks at much faster rate as r increases. What this means is that in terms of the force of gravity, distance matters more than mass.

There are a number of tools for exploring function behavior. Most of them are easiest to explain using analytic geometry, so because we're lazy we'll wait until that chapter to discuss them.

Similarly, much of the motivation for analysis comes from calculus, so we're going to postpone a lot of that until that chapter. But here's a quick preview: physicists look for laws that describe how things change. Suppose we want to describe the trajectory of an object by giving its position x at all times t. That is, we are looking for a function x(t) that describes a motion of the object. Oddly enough the laws of physics are not the functions x(t) but rather equations, called differential equations and written in the language of calculus, whose solutions are the appropriate functions x(t). It is the solving of differential equations (or even determining that those equations have solutions) that forms much of the work and motivation for analysis. This too will be taken up in a later chapter.

Analysis is the branch of mathematical thinking that most directly connects the scientific study of the world to the theoretical universe of mathematics. It is the behavior of the world, how it works is the goal of scientific investigation, and analysis is the mathematics of how. Cultivating the analytical mindset opens up all scientific research to the person who cultivates it. In our book *Three Steps to the Universe,* we said the most important question for a layperson reading about science was "How do they know that?" Analysis is the mathematical side of the answer to that question just as experiment is the physical side.

The people on both sides of that question, the scientists and the mathematical analysts, are like two groups of people living on opposite sides of the same river, swimming in the same waters, boating in the same boats, fishing for the same fish.

*But we're different because we live on this side of the river, not that one.*

**That's what they say, too.**

*Hey, do you mean we're arguing so much because we're close neighbors?*

There'll be a lot more of this in later chapters.

## Logic

**Well, that wraps up the branches of mathematical thinking.**

*No, it doesn't.*

**I said ...**

*And I said, no it doesn't. We haven't talked about logic.*

**What about it? Logic is only sort of a branch of mathematics. It's at least as much a part of philosophy.**

*Yes, and analysis is at least as much a part of physics, and besides, we've already pointed out that one famous mathematician was really a philosopher. Or don't you want to admit how close some aspects of mathematical thinking are to philosophy?*

**Hey. We're different. We live on this side of the river of logic. Not that one.**

*(Sarcastic glare)*

**All right. We'll talk about the way logicians think, which is not the same as logical thinking.**

Let's start with a joke. We heard this one many years (and it was probably old then). We don't know where it comes from, and apologize for the lack of attribution.

Once upon a time there was a beautiful castle called Mathematics. In the basement of this castle lived some giant spiders called Logicians who spent all their time spinning webs in the basement. One day a flood came through that destroyed all the webs. The spiders set to spinning as fast as they could, because they thought their webs were holding up the castle.

But what is logic?

Apart from a source of at least one alien culture on a certain TV show, I mean.

Logic is the abstraction of mistrust.

Logic takes the statements used in creating proofs and abstracts them to discern and analyze the processes by which truth is discerned.

Logic does this first by abstracting the concept of a statement, allowing there to be variables that range over the set of statements. We saw a little of this in discussing acceptable proof forms where we used A and B for generic statements. For example, one might say A = "Fred is a terrible driver," B = "Charlie is a good driver."

A and B are **propositional variables** (variables whose values can be propositions).

Logic also abstracts some, but by no means all, of the connections that can exist between statements. Unfortunately, logic does not have a standard set of symbols for these connections. We'll use some that are convenient to produce on a word processor:

& represents "and." So A & B represents the statement, "Fred is a terrible driver and Charlie is a good driver."

| represents "or." So A |B represents the statement, "Fred is a terrible driver or Charlie is a good driver." This is called the "**inclusive or.**" It means A is true or B is true or both A & B are true. There's also the "**exclusive or,**" which means A is true or B is true but not both. We're not going to worry about that one at the moment even though the everyday use of the word "or" is the exclusive or.

~ represents "not." It changes a single statement rather than connecting two statements. ~A represents the statement, "Fred is not a terrible driver."

→ represents "if ... then ..." So A→B represents the statement, "If Fred is a terrible driver then Charlie is a good driver." Another way to read A→B is "A implies B."

↔ represents "if and only if." So A↔B represents the statement, "Fred is a terrible driver if and only if Charlie is a good driver."

Logic that is based on this symbolic manipulation is called **symbolic logic** and was created in the mid-nineteenth century. Logic without symbols has been around a lot longer than that, as might be deduced from reading ancient Greek and Roman philosophers. The particular form of symbolic logic that uses propositional variables and these connectives is called the **propositional calculus**. (No relation to the calculus that scares people in college. We'll be dealing with that later.)

An early pioneer of symbolic logic was a 19th-century mathematician and clergyman by the name of Charles Dodgson, better known as Lewis Carroll. He's far better known these days as the author of *Alice in Wonderland* and *Through the Looking Glass*. It is an interesting philosophical question as to whether his interest in logic spurred his skill with nonsense or vice versa.

For sentimental reasons, we would like to use his symbolism, but we won't, because it's really bad. His logic books are more eye-blurring than the goofiest passages in *Through the Looking Glass*.

Using symbolic logic, it is possible to put several of the forms of proof discussed earlier into symbolic form. For example, modus ponens is (A & (A→B)) → B which can be read "A and (A implies B) implies B." Modus tollens is (~B & (A→B)) → ~A.

Learning to read these statements is a matter of decoding a kind of weird rebus. One starts out translating each symbol in turn, so as to produce a statement that sounds like English or whatever one's native language is. Eventually, one learns to simply read symbolic logic directly.

Some of the standard proof forms require more sophisticated symbols. Specification from the general cannot be done with propositional calculus, so an expanded idea of statement was created. The idea was to take statements and divide them into **objects** and **object variables**, which are the things the statements are about, and **predicates** and **predicate variables**, which are what the statement says about the objects it's about. For example, instead of having "Fred is a terrible driver," we have an object, f = Fred, and a predicate, D = "is a terrible driver." The statement "Fred is a terrible driver" is then written Df, which can be read, "the predicate 'is a terrible driver' is true for the object Fred," or more simply, "Fred is a terrible driver."

This mirrors to some extent the grammatical structure of a standard indicative sentence, which is usually said to have a subject and a predicate.

Logic using objects and predicates in place of propositions is called the **predicate calculus**. It introduces two new connectives:

$\forall$ represents "for all." So $\forall x Dx$ means "For all objects x, x is a terrible driver."

$\exists$ represents 'There exists." So $\exists x Dx$ means "There is some x, such that x is a terrible driver."

This allows us to write $\forall x Dx \rightarrow Df$ and $Df \rightarrow \exists x Dx$, which mean respectively: "If everyone is a bad driver, Fred is a bad driver" and "If Fred is a bad driver, someone is a bad driver."

So far it looks like all we've done is create a way to abbreviate sentences and certain sentence constructions into things that look less like language and more like mathematics. But does this really have any benefits?

It does in the same way that creating short forms of functions does. It allows one to pick up and examine the behavior of entire categories of sentences and statements, and to look into what's going on inside them. There are, however, a number of drawbacks which we'll get to.

## Logic in Math

The predicate calculus allows for a formalization of all mathematical statements. Every mathematical definition, statement, and proof can be written down using this logic. For example, suppose we have the predicate E which means "is an equivalence relation," then we can write:

$$E \equiv \Leftrightarrow (\forall a(a \equiv a) \ \& \forall a \forall b((a \equiv b) \rightarrow (b \equiv a)) \ \& \forall a \forall b \forall c((a \equiv b) \ \& \ (b \equiv c) \rightarrow (a \equiv c)))$$

This looks like a messy formula but let's break it down:

$E \equiv$ means $\equiv$ has the property of being an equivalence relation

$\Leftrightarrow$ if and only if

$\forall a(a \equiv a)$ for every a a has the relation $\equiv$ with itself (Reflexive property)

& and

$\forall a \forall b((a \equiv b) \rightarrow (b \equiv a))$ for every a and b if a has the relation $\equiv$ with b, then b has the relation $\equiv$ with a (Symmetric property)

& and

$\forall a \forall b \forall c((a \equiv b) \ \& \ (b \equiv c) \rightarrow (a \equiv c))$ for every a, b, and c if a has the relation $\equiv$ with b and b has the relation $\equiv$ with c, then a has the relation $\equiv$ with c (Transitive property).

Writing down mathematical statements and proofs in this form, it is possible to find patterns and abstract out the way forms of proof can work to prove propositions one does not even know in the same way that abstracting operations leads to an ability to prove things about operations without knowing what the operations are.

In this sense logic is a branch of algebra, abstracting beyond the components of a proof to the form of proof itself.

Furthermore, there is a practical equivalence between the predicate calculus and set theory.

Consider the predicate D above, which carries the meaning "is a bad driver." We can create a set **D** which is the set of all objects x for which Dx is true. That is, **D** is the set of all bad drivers. x is a member of **D** if and only if Dx is true (written $x \in \mathbf{D} \Leftrightarrow Dx$). For all practical purposes, D and **D** are the same thing; the predicate and the set of objects that make it true are one and the same.

Well, maybe. The truth of the matter is that things are not that simple. Indeed, it was the blending of the predicate calculus and set theory that caused a number of problems in the late 19th and early 20th centuries, some of them quite startling in their results.

Let's back up a bit and deal with an insignificant word that we've bandied about: Truth! and it's evil twin: Falsehood!

A lot of people have the naive idea that every statement is either true or false. But logicians discovered early on that there are some statements that aren't either one or the other. Some of these statements are fuzzy in meaning. "The sky is blue" is kind of true and kind of not true. Mostly it's just not exact enough for mathematics.

But there's another kind of statement to which the characteristics of truth or falsehood cannot be ascribed. This kind of statement is called a **paradox**. The most basic paradox is: "This statement is false."

Is this statement true or false? If it's true, then it's false. If it's false, then it's true. There are a number of books dealing with paradoxes as a source of amusement.

Symbolic logicians found paradoxes irritating, so they came up with the concept of **well-formed formula** (or **wff**). A statement that is a wff will not refer to itself. Symbolic logic says that only wffs are logical statements.

*So go away, you silly paradoxes!*

It didn't work out as well as they wanted because some paradoxes caused problems beyond this wffiness.

Set theory as originally formulated presumed that one could make up any set one wanted by putting things together. Set theory did not care if the things it put together were objects or other sets. Indeed, there was no problem in having sets as members of other sets (members, not subsets). So one could have {{a, b}, a, b}. This set has three members: the set {a, b} and the elements a and b.

Bertrand Russell, an English scholar who was born in 1872 and lived nearly a century during which he blew up many ideas that people were confident about in math, logic, philosophy, and social theory created what has been called **Russell's paradox**, which exploded this concept. Consider the predicate S, defined as "does not contain itself as a member." A set satisfies this predicate if it does not contain itself as a member.

*That's straightforward. Most sets don't contain themselves as members.*

True, but the equivalence of set and predicate means we can create a set **S** consisting of every set that satisfies the predicate S. In other words, a set is a member of **S** if it is not a member of itself.

*Uh-oh.*

**Yup, paradox city.** Is **S** a member of **S**? If it is, then it isn't. If it isn't, then it is.

Russell's paradox wrecked set theory as it was first conceived of. Since that time, the original theory of sets has been called **naive set theory**.

There have been a number of more formal set theories that prevent Russell's paradox and related problems. Russell himself provided one such. He called it the **theory of types**.

It works, but it's really messy both in theory and in practice and we'd like to avoid it for now. We've got bigger fallacies to fry.

Attempts to shut paradox out have failed in even more extreme ways. What follows is a very short version of a pretty complicated proof. We're leaving out the necessary ugly stuff to get to the meat of the matter.

Suppose we take predicate calculus statements where the only objects are whole numbers and the only predicates are properties of numbers. It is possible in such a system to create a code that takes any logical statement made using such objects and properties (and the usual logical connectives) and from it produce a unique number. The actual proof involves the arithmetic procedures necessary to produce those numbers.

It is possible in this formulation to create a predicate G that is stated in terms of properties of numbers, but for which Gn means "n is a number representing a statement which cannot be proved."

*Here we go again.*

**Yup.** And now, a little diagonalization. We can make an infinite list of statements of the form

G0

G1

G2

etc. With a little finesse, we can arrange for there to be a number a which represents the encoding of Ga. In other words, we've found a numerical way to say "This statement is false!"

The creator of this proof, Kurt Gödel, an Austrian mathematician, logician, and philosopher (not an uncommon combination at the time), was born at the beginning of the 20th century, and is credited or blamed with having destroyed the foundations of mathematics. He didn't. He, like Russell, pointed out problems in the attempts to abstract the processes by which mathematicians do mathematics.

*It's more than that.*

**All right. It is.** Gödel introduced the concept of **undecidables**. These are statements that cannot be proven or disproven. The undecidable above is just a parlor trick, but there are some that have meaning. One undecidable is the **continuum hypothesis**.

*Anything to do with the Power of the Continuum?*

**Everything to do with it.** The continuum hypothesis says that the number of real numbers is the second infinite number. That is, that $c = \aleph_1$. This cannot be proven one way or another. This is kind of disturbing because we're talking about not being able to tell if there are any numbers between $\aleph_0$ and c. We assume that if there were some number between, we could find a set that had that many elements (presumably a subset of the real numbers). But what the undecidability of the continuum hypothesis seems to mean is that we can't find such a set and we can't rule out its existence either.

These logic excursions show that some of the basic ideas of math (including sets and numbers) are not so simple as we like to think. And perhaps we do not understand what they mean as well as we thought.

This has a value in creating an increased level of mistrust and a worry about what is being done in mathematical proofs. But unlike the other ways of mathematical thinking, logical thinking does not change much in the way mathematicians do math or the way others employ math. The fringes of mistrust can be interesting, but they don't impact the day-to-day work of mathematicians the way a real advance in algebra, geometry, or analysis would.

## Logic and Meaning

Logic, as described above, works reasonably well for symbolically formulating mathematical statements, but logicians also want to extend logic to cover all of language and for some of them, all of thought. There have been numerous expansions of logic, such as the **modal logics**, which seek to add concepts such as necessity, possibility, moral obligation, and moral acceptance; **relevance logics**, which seek to make sure that statements only connect when they actually have something to do with each other; **temporal logics**, which seek to add time to the eternity of logical statements, and so on.

This desire to extend logic so that it covers all the concepts and connections of concepts that people use in life and language is part of the philosophical aspect of logic, the desire to go beyond the abstraction of mathematical proof. The logical mindset can be seen in this.

*Note*: A logical mindset does not mean a logical mind. It means a mind amenable to seeing things the way logicians see things. The logical mindset breaks down statements into components and sees how those components fit together. It thinks that the meaning of a statement can be found and preserved by such a process.

Richard is of two minds about logic. He was once enamored of it, held to the logical mindset strongly and felt it could be applied throughout life. Later on when he became a writer, he had to disown this mindset. His first efforts at fiction (mercifully unpublished and now nonexistent in any form) read like mathematical proofs. (*I'm not so sure about the "nonexistent" part. I may have the odd unpublished manuscript of yours lying around.* **What!! Burn all of those immediately!**).

Richard discovered that the meaning of a statement in everyday life and in story and poetry does not lie only in its pieces, but in their arrangements, their associations and connections. He learned that you can no more find the meaning of a piece of writing by breaking it down than you can find the experience of living in discerning the placement of the atoms that make up the molecules of the human brain.

This may sound like he's disillusioned, but it's not so. He's learned to see the place where logic is useful and where it isn't. Logic is a powerful tool in analyzing the structure of arguments and proofs to make sure they are sound. But it is a tool that is easy to misapply. It's like a powerful antibiotic that can kill many diseases but which can also damage living people if misused.

## Digging up the Roots

Looking back on these forms of thought, we can see that the apparent alienness of mathematicians lies not in their thinking and acting unlike everyone else, but in their taking of everyday processes of thought and extending and sharpening them.

Their care with sticks and stones and the leaps they make with them are age-old actions that everyone does unknowingly, but mathematicians do them with awareness, memory, and a care that some might find reverent (if they didn't tell so many bad jokes).

The carefulness of their abstraction, the intensity of their mistrust, and the depth of their laziness combine to make their substitution of the theoretical for the real and the real again for the theoretical not an unknown action but a more focused, more useful version of normal thought, a version that as we said at the beginning makes the relationship between world and mind closer, more intimate, and more workable.

The ways of algebra, geometry, analysis, and logic are nothing more than extended and controlled versions of everyday aspects of thinking.

There is nothing alien in this transformation of a human faculty into a refined skill any more than there is something unearthly in a practiced singer having the ability to produce effects with the voice that are startling to the untrained. And just as one can learn to sing late in life, so one can learn and hone any of the faculties that make possible the mathematical mind. Each of the ways of thinking has stumbling blocks in the learning, but each can be overcome by going slowly through the learning and taking care with understanding.

See variables as pronouns to learn algebra.

See shapes in terms of definition and drawing to learn geometry.

See processes as machines and then machines as functions to learn analysis.

See statements in terms of their structures and their interrelations to learn logic.

Most of the stumbling blocks come from mathematical shorthand. If one slows down the learning and examines each piece of notation and definition, one can get past the confusion that comes from people writing their explanations for people who already think like them.

Enough of the roots. It's time to climb the tree and look at some branches of mathematics.

# Part II

# Theory in Practice

Time to look at what math has wrought that makes it worth the time and effort to reclaim. It is not possible to treat the entire range of mathematics in the space of one, ten, or a hundred books. So the next few chapters, rather than being remotely comprehensive, will examine topics that have one way or another been entwined with human life and understanding. Some of the topics are vital to physics, some to other sciences, one to gambling, one to modern entertainment, and all to how math has come to be used and taught.

Many of the topics covered here have been discussed in other popularizations and articles written for those intrigued and interested by aspects of science. But most of them have shied away from the math itself. As a result, what is usually presented is the abstraction and application with only a cursory description of the theory. This is unfortunate, because mathematics is the language in which these theories are written.

This is rather like traveling in a foreign country whose language one does not speak accompanied only by a phrase book. You will hear a lot of words, some of which you might recognize; you will speak a lot of words, but you won't know what you're saying or if it makes any sense. You might be able to find a translator, but you won't know what kinds of nuances are lost in the translation.

If, however, you know at least some of the language, you will be able to understand better the places you are visiting, glean some of the history and culture, find out which tourist traps to avoid, and you'll be able to find the really good places to eat.

# 5

# *Analytic Geometry*

Geometry sees shape; analysis sees process. René Descartes, 17th-century French mathematician and philosopher, saw that processes have shapes, and he made a tool to reveal those shapes so that the mindsets of analysis and geometry could simultaneously be brought to bear on problems. He created **analytic geometry**. In addition to his contributions to mathematics, Descartes is regarded as one of the founders of modern philosophy. We do not know if the chain of reasoning below occurred to Descartes in any form, but it is one of the easiest ways to come to understand this branch of math.

Analytic geometry is another one of those mathematical objects made by taking something that already existed and transforming how it is thought of in order to create a new mental object. The fundamental tool from which analytic geometry was created is something that mathematicians had been staring at and using for millennia: the ruler.

We've looked at rulers as measuring sticks, but another way to see them is as lines marked with numbers. And by lines I mean geometric lines, those abstract straight infinite in both directions things geometers had been drawing using rulers. If one were to draw such a line and then put numbers next to it, each number at its corresponding place on the ruler, one would have something like that given in Figure 5.1.

The shift in thinking that was necessary to marry analysis and geometry is to transform the idea "the line is marked with numbers, each number corresponding to a point" to "the line is itself made up of the numbers used to mark it." In other words, the points on this line, those never really defined, infinitesimally small individual geometric objects, are actually numbers. They are lined up by ordering them so that 0 is to the left of 1 and between them is 0.5 and between 0 and 0.5 is 0.25 and so on, until we put every single one of the **power of the continuum** numbers into their place. What we end up with is that the line is the set of all real numbers put in their proper order.

Nowadays this concept is taught early and invisibly. Children are introduced to the number line without anyone pointing out what a startling idea it is. But before analytic geometry these ideas were separate. Points are meant to be locations, but real numbers were abstractions of amount.

How does one reconcile these two ideas into one?

By the convenience of 0. Yes, that lovely little nothing allows for one simple way to connect numbers and points. The number associated with a point is the distance from the point on the line marked by the number to the point marked by 0 (Figure 5.2).

On the line, the location of a point is uniquely determined by its distance and direction from the 0 point, also called the **origin**. On the line there are only two directions, right and left, so for any given distance there are only two points that have that distance from the origin, the one that lies that distance to the right and the one that lies that distance to the left.

Let's say we're talking about a distance of 5.26. If we say the point to the right is the positive number (5.26), and the point to left the negative (–5.26), we have one point for each real number and one real number for each point. We have created a one-to-one correspondence between the real numbers and the points on the line. This correspondence

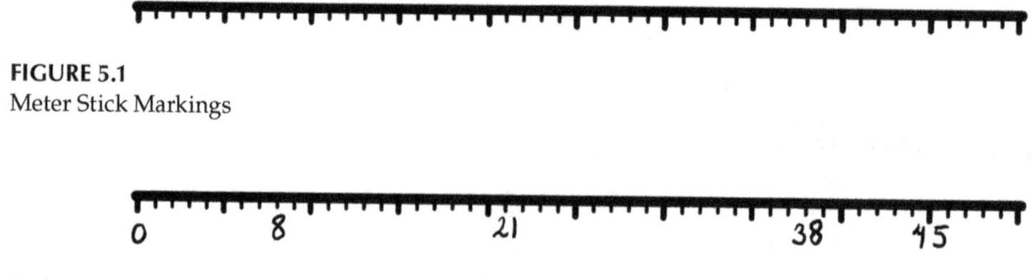


**FIGURE 5.1**
Meter Stick Markings

**FIGURE 5.2**
Number Line

preserves the ordering in that a point to the left of another point corresponds to a number that is less than the number corresponding to the other point. The distance between any two points will correspond to the difference between the numbers. (The distance between the point 3.5 and the point 0.5 is 3, which is also the difference between 3.5 and 0.5.) Thus, our correspondence takes analytic concepts (like "<" and "difference") and connects them to geometric concepts (like "left of" and "distance").

So not only have we created a correspondence of sets, but the correspondence translates the ordered structure of numbers (if you have two different numbers, one will be greater than the other) into the geometric ordering of points on a line (if you have two points, one of them will be to the right of the other). Furthermore, the crucial geometric concept of distance (what, after all, would the measurement of Earth be without measurement?) translates directly into the numerical action of taking the difference.

In technical modern math terms, if we carried this out with proper formalism, we could show that the number line is a model of the line. (A model of a theoretical object is another object, real or theoretical, that satisfies all the axioms of the theoretical object.)

The number line on its own is only so useful. For some people it helps in learning addition and subtraction, since one can see the process as moving around the line (to add 5 to 7, we go to 7 on the line and then go 5 more to the right, and we end up at 12).

In this little bit we can see a hint of one of the most intriguing aspects of analytic geometry. Most of the people who are helped by the number line are those with a geometric mindset. For them, the geometry of the number line acts as a support and a boost to the analysis. They see the shape of the operations they are doing, the shape of the numbers, and they come to understand numbers better.

There are others to whom the analysis acts as a prop to their weaker geometric thinking. For them, the idea of line becomes more vivid if they can see it as numbers. They get a better sense of points made into a line if they can think about individual numbers and ordering.

In short, analytic geometry is built in such a way as to have the two ways of thought that make it up support each other and create something greater than either. It is a marriage of ways of thought (a good and prolific marriage, the kind with mutual help and, as we will see later, a batch of kids, some not so well-behaved).

Not all fields that join ways of thinking work this way. Some require people who can think in both ways at once in order to work (linguistic translation, for example). A couple of mental fusions that do work this well are songwriting/singing (where music and poetry work together) and cartooning (where visual and written art blend together).

These examples point out what kinds of blends work like this: those where the corresponding concepts illuminate and assist each other by their correspondences. In songwriting, the rhythm of the music and the rhythm of the poetry provide clues to help in their mutual creation. In the number line, point and number reveal each other, relative position on line and in < ordering reveal each other, and distance as length of segment and distance as difference illuminate each other. Compare this to translation, which largely consists of frustrating searches for words or phrases that correspond to each other.

Of course, we're making pretty grand comparisons for a number line. And it is pretty thin (*very droll*). Analytic geometry really gets going when you go from a line to a plane.

## Planes and Space, Numbers by Numbers by Numbers

Take two lines. Set them up at right angles to each other. This makes a plane ("two intersecting or parallel lines determine a plane" is as old as Euclidean geometry). Now take any point in the plane and draw a perpendicular line segment from that point to each of the two lines. The line to the horizontal line is a vertical line segment, and the one to the vertical line is a horizontal segment (Figure 5.3).

How long are those two segments? Notice that combined with two segments on the two lines we started with, they form a rectangle. The two segments we drew are the same lengths as the segments parallel to them on the two lines, since opposite sides of a rectangle are the same length (Figure 5.4).

Now let's turn the two lines we started with into number lines. Since we've got two such lines now, we need to be able to distinguish which line's numbers we're talking about when we talk about position on the line. We'll call the line that goes from left to right the **x-axis**. The one from top to bottom we'll call the **y-axis.** Since the y-axis has no left or right sides, we'll say that the positive numbers are the ones above the 0 point, the negative numbers are below. We number the lines in such a way that the 0 points of both lines are at the point where the lines cross. We'll call this double 0 point the **origin** (Figure 5.5).

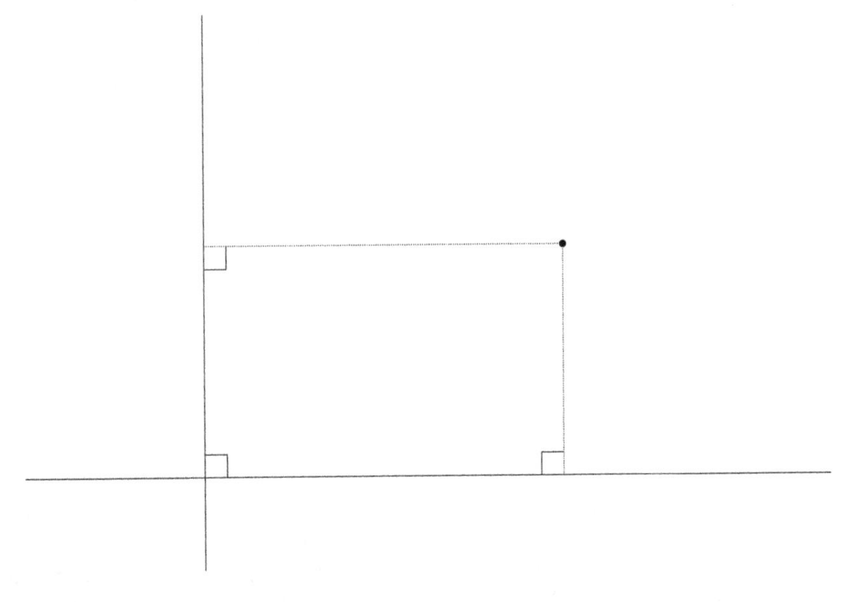

**FIGURE 5.3**
Horizontal and Vertical Rectangle

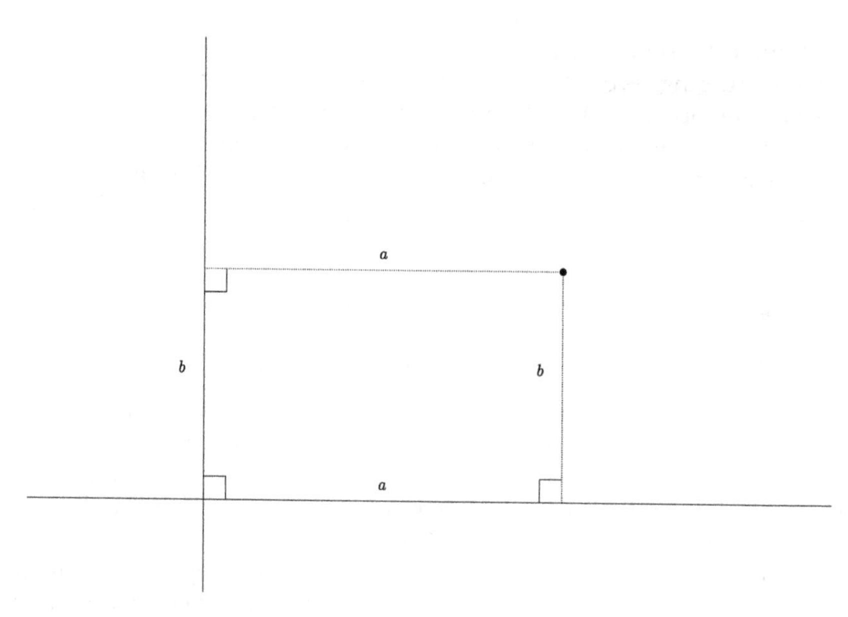


**FIGURE 5.4**
Lengths of Sides

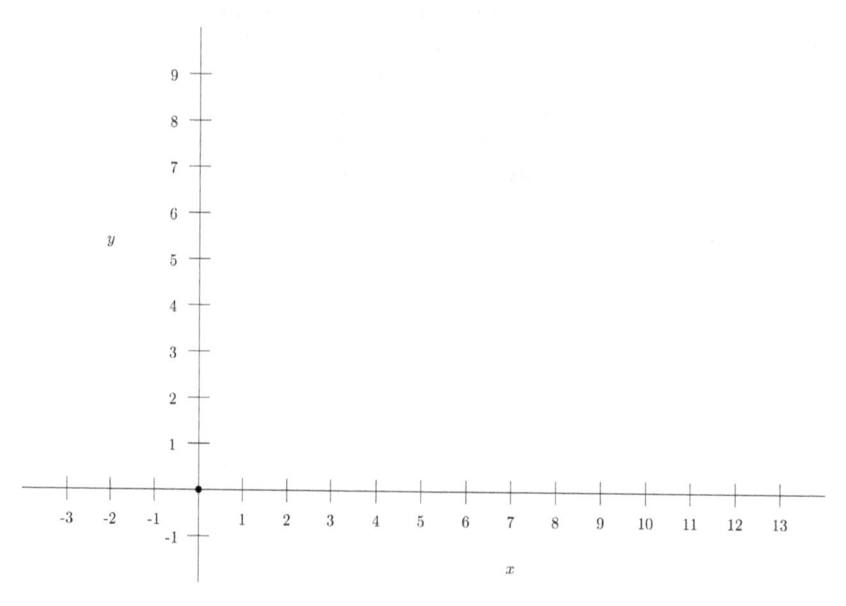


**FIGURE 5.5**
Cartesian Plane

Back to our rectangle. The two distances we figured out before are the points/numbers on the two lines where the line segments intersect them.

Now let's go a little slowly. Using the rectangle formed from the origin and the other point we started with, we've found two numbers (the numbers on the x- and y-axes, respectively). Can any other point in the plane give us those same two numbers?

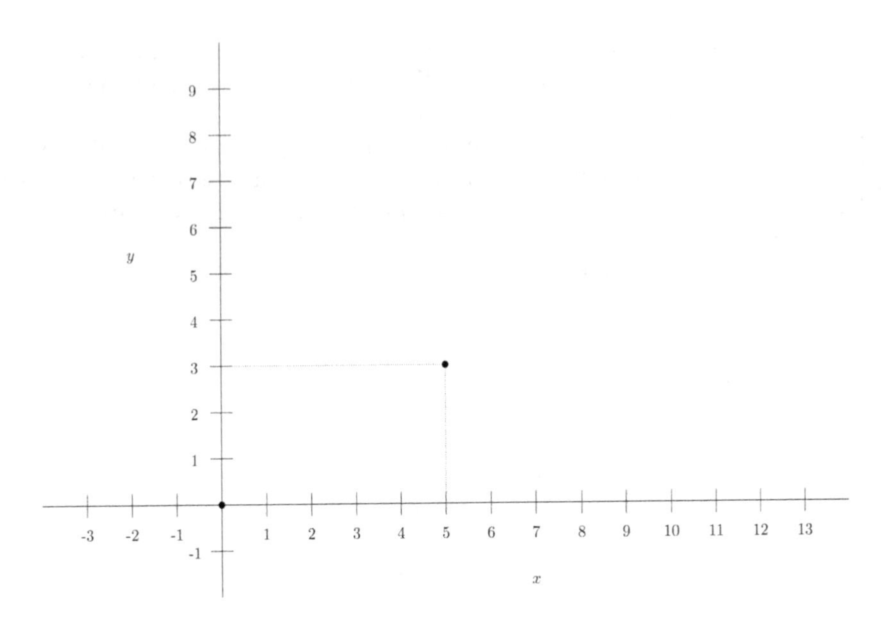


**FIGURE 5.6**
Cartesian Coordinates

No, because the rectangle that would have that point as its corner would be a different rectangle sharing the same two sides on the axes as this rectangle, which is geometrically impossible (any two of its sides determine a rectangle). So each point in the plane gives us a unique pair of numbers.

All right, suppose we start instead with two numbers. We can find a point in the plane by making a rectangle with the x and y line segments determined by those numbers (Figure 5.6).

This process clearly yields a unique point in the plane, and since we've created the same rectangle we had before, that point is the point that would give us the x-axis and y-axis values we just used.

In other words, we've come up with a one-to-one correspondence between the points in the plane and pairs of numbers. This is not that impressive. It just means they have the same number of elements.

But we've done it in such a way that distance and position are preserved when we move from points to pairs of numbers and vice versa. Two points and their corresponding pairs of numbers will have the same relative positions. We've identified points and number pairs in a way that preserves the characteristics we care about in geometry: distance and position.

With work one can prove that this set of pairs of numbers satisfies Euclid's axioms. We're not going to go through that. But this brings up an element of mathematical thinking that we've talked about but not beaten far enough into the ground.

For mathematics, anything that has the important characteristics of something can be treated as that thing. We did that implicitly in making sticks, where any characteristic that acted in a number-like fashion could be treated as a unit for numbers. Once we know that the numbered plane above acts like the Euclidean plane, we can treat it like the Euclidean plane.

This may not sound that impressive. But the mutual illumination between points and numbers that we saw in the number line is expanded greatly in the plane. As we will see, we can use numeric shortcuts to prove geometric results, and geometric awareness to learn more about the numerical representation. We'll be able to slip in and out of geometry and learn more about shape, thanks to the numbers.

Before we get to that, we need a little more terminology. We've identified each point in the plane with a pair of numbers, one number for the x-axis and one number for the y-axis. These two numbers are called the **x coordinate** and the **y coordinate** for that point.

We can specify a point by saying

x coordinate = (position on x-axis)
y coordinate = (position on y-axis)

To find the point geometrically, we go along the x-axis to the x coordinate. We draw a line at this point parallel to the y-axis. We go up or down this line an amount equal to the y coordinate (up if it's positive, down if it's negative). The point we reach is the point with the specified x and y coordinates.

As an example, the point where the x position is 5 and y position is 3 would be

x coordinate = 5
y coordinate = 3 (Figure 5.7)

The "coordinate =" notation above is cumbersome to write. We might shorten it to x = 5, y = 3. But we can do better than that by writing the ordered pair (5, 3) and remembering that the first value is the x coordinate and the second is the y coordinate (Figure 5.8).

The plane seen as pairs of numbers is called the **Cartesian plane,** named after Descartes, unsurprisingly.

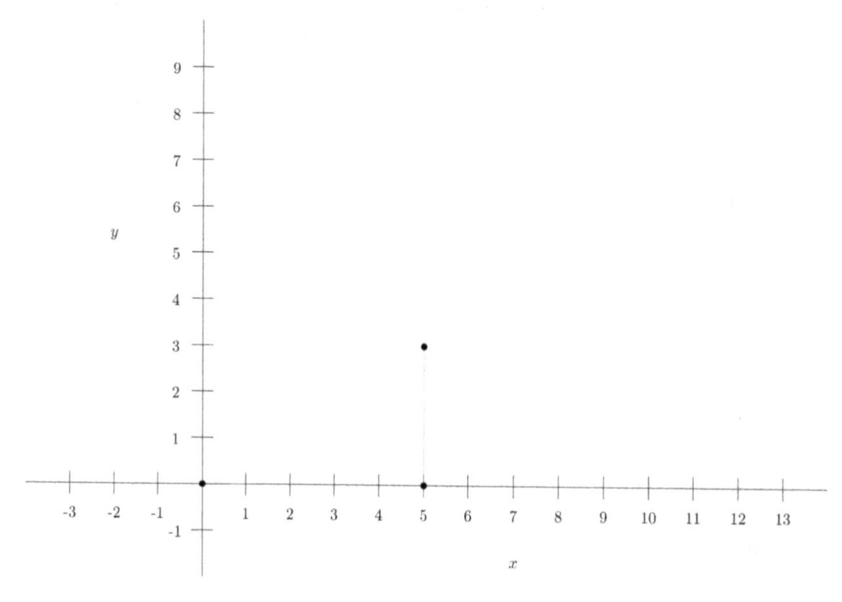


**FIGURE 5.7**
Finding a Point by Coordinates

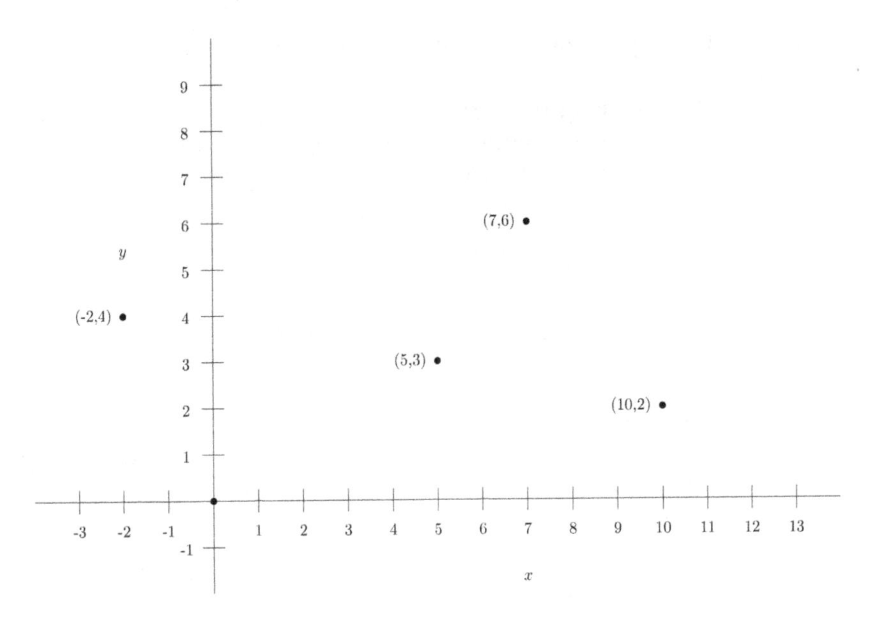


**FIGURE 5.8**
Points in Cartesian Plane

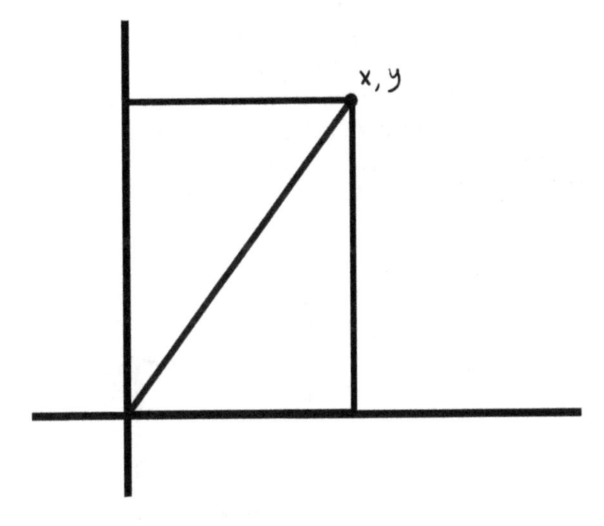


**FIGURE 5.9**
Point in Plane Connected to Origin

There is a lot of useful information in this simple way of describing a point as a pair of numbers. Consider a point (x, y) and draw the rectangle we showed above. Now draw the line connecting the point to the origin and have a look (Figure 5.9).

We have our old friend/nemesis a right triangle. The hypotenuse of this triangle is the line connecting the point to the origin, so the length of the hypotenuse is the distance of this point to the origin. But Pythagoras tells us:

> *"Don't eat the beans! What's this coffee you're drinking? This chocolate you're eating?"*

**Cocoa beans are not really beans. And anyway, that's not the saying of Pythagoras I was talking about.**

Pythagoras tells us that the length of the hypotenuse is the square root of the sum of the squares of the lengths of the sides. Those lengths are given by the x and y values (if we ignore the signs). That means the distance to the origin for a point (x, y) is

$$\sqrt{\left(x^2 + y^2\right)}$$

Our coordinate system has given us a fast way to determine distances to the origin. But we can do better than that. Take two points (x1, y1) and (x2, y2). Draw the rectangle that has these two as corners. Notice that the lengths of the sides are found by taking x2 – x1 and y2 – y1, ignoring the signs (Figure 5.10).

Now draw the diagonal between the points, and we have another right triangle with a hypotenuse representing the distance between the points. That distance is

$$\sqrt{\left((x2 - x1)^2 + (y2 - y1)^2\right)}$$

So given two points on the Cartesian plane, we can find their distance from each other by the above formula. No line drawing, no measuring, just coordinate values.

Take a look at that formula for a moment. It may be hard to see how phenomenal an advancement it alone represents. So imagine a numberless plane where points are created by complex constructions. Consider that points were so made by architects, astronomers, geographers, and everyone else who used geometry, and every one of those people at one time or another needed to find the distances between such points, whether for house building, star mapping or knowing when a ship was going to crash into an island.

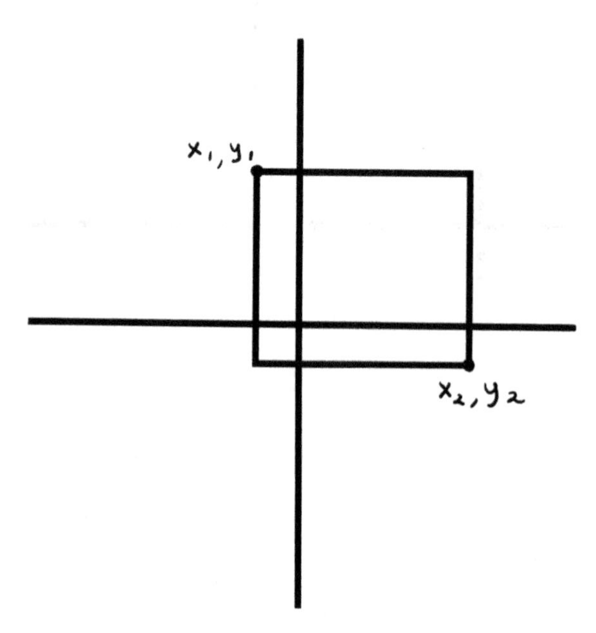


**FIGURE 5.10**
Points in Plane Making Rectangle

Thanks to Descartes' elaboration of Pythagoras, those problems were all reduced to finding the coordinates of the points and then doing the calculation above.

And we can handle even more annoying calculations than that. Let's go back to a single point and the right triangle it forms. Notice the angle between the hypotenuse and the x-axis. Call that angle θ (Figure 5.11).

Those trigonometric functions we had so much trouble figuring out, what are they for this angle?

sin(θ) = length of opposite side/length of hypotenuse.

But the opposite side is just the y value, and the hypotenuse is $\sqrt{(x^2 + y^2)}$,

so

$$\sin(\theta) = y \big/ \sqrt{(x^2 + y^2)}$$

Similarly, cosine is adjacent side (x) over hypotenuse:

$$\cos(\theta) = x \big/ \sqrt{(x^2 + y^2)}$$

And

$$\tan(\theta) = \text{opposite side}/\text{adjacent side} = y/x$$

All our messy trigonometry has just become easy to figure out.

*If you don't mind extracting square roots. Lots and lots of square roots.*

Furthermore, we are not confined to the plane in what we've done. Just as we can pinpoint any point on the line with one number and any point in the plane with two, we can get any point in space with three numbers.

We need to add a third axis. What to call it? Hmm, we did x and y. What comes next? (Figure 5.12)

Each point is here an ordered triple (x, y, z) (Figure 5.13).

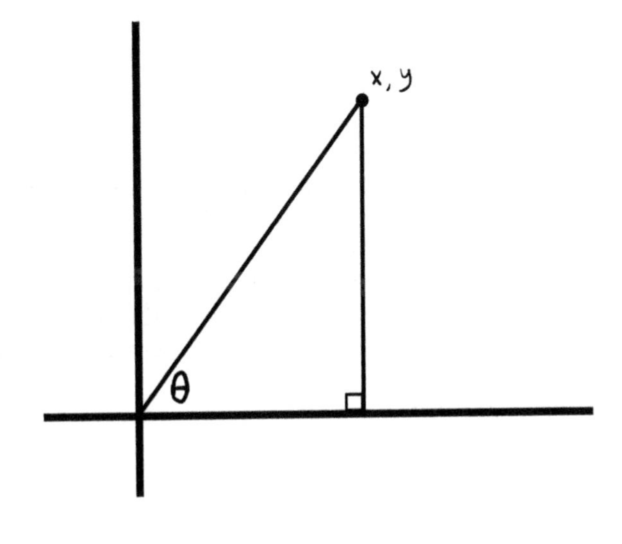


**FIGURE 5.11**
Right Triangle

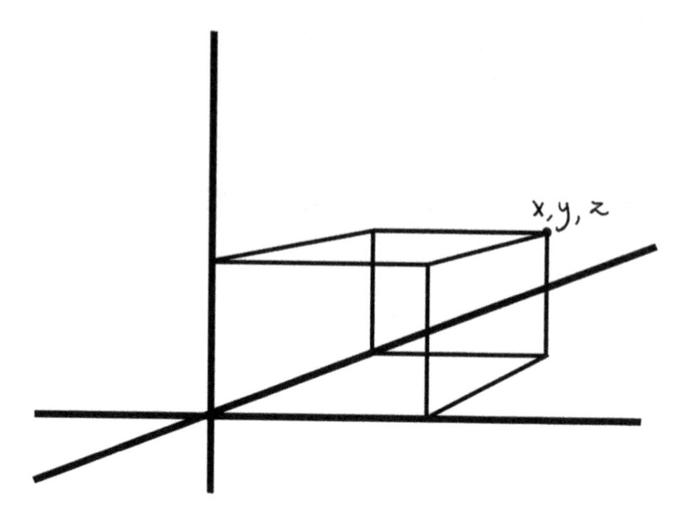


**FIGURE 5.12**
Cartesian Space

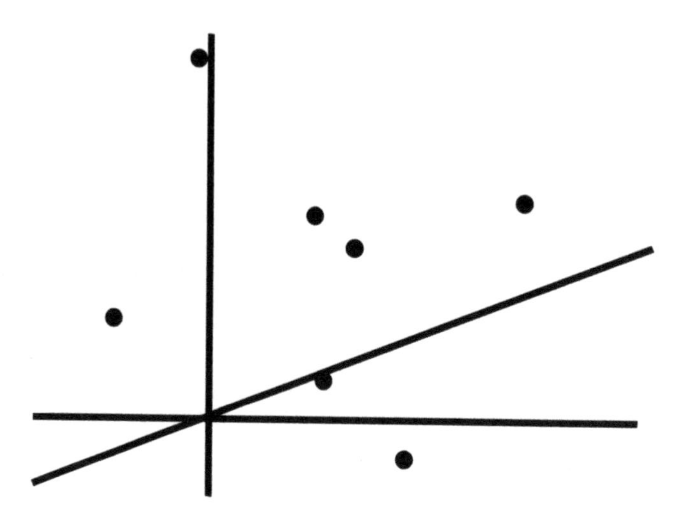

**FIGURE 5.13**
Points in Cartesian Space

Most of what we did before we can duplicate, including finding the distance between points. **We'll do the easier-to-draw distance to the origin if you don't mind.**

*Go ahead. Be as lazy as you want* (Figure 5.14).

Notice the right triangle made from the hypotenuse. One side is the z coordinate. The other is the hypotenuse of the right triangle we used to find the distance in the plane. We've built one right triangle on another, using the hypotenuse of the old one as one leg of the new one.

The length of our new hypotenuse is:

$$\sqrt{\left(\sqrt{\left(x^2+y^2\right)}\right)^2+z^2}$$

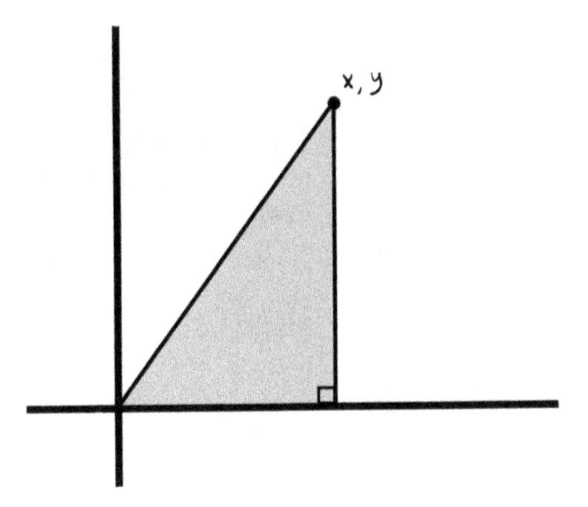


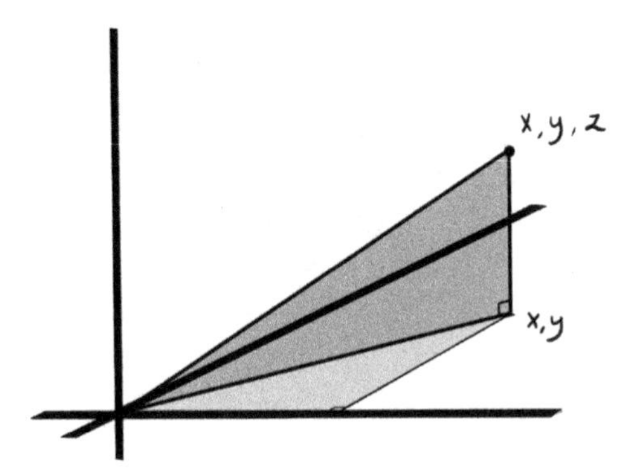


**FIGURE 5.14**
Right Triangles to Calculate Distance

But

$$\left(\sqrt{\left(x^2+y^2\right)}\right)^2 = x^2+y^2$$

So the length is

$$\sqrt{\left(x^2+y^2+z^2\right)}$$

Doing the same bit we did before, we find the distance from the point (x1, y1, z1) to (x2, y2, z2) to be

$$\sqrt{\left((x2-x1)^2+(y2-y1)^2+(z2-z1)^2\right)}$$

We'll do more stuff in Cartesian space later, but for now let's return to the plane. Something about what we've done may be troubling. We promised analytic geometry, but have we delivered? We identified numbers with points, lines, and planes, but isn't that connecting algebra with geometry? Where's the process? Where are the functions? Where's the analysis in analytic geometry?

## Function is Shape

Take a function, say $f(x) = x^2$. Up until this point we've been able to plug individual numbers in for x and get f(x) from them, but we haven't been able to take a look at the whole function in any way. It's as if we had been trying to see a beach by looking at each grain of sand in turn. Analytic geometry gives us a tool for seeing the whole. That tool is called **graphing**.

The graph of a function f(x) in the Cartesian plane is the set of all points (x, y), where $y = f(x)$.

So looking at our example above, $f(x) = x^2$, the graph would contain the points (0, 0), (1, 1), (–1, 1), (2, 4), (–2, 4), (1.5, 2.25), and so on – all the points where the y coordinate is the square of the x coordinate.

We can draw this graph and see the whole of the function as one shape (Figure 5.15).

Now we have something we can see and analyze. Geometry has opened our eyes to the shapes of functions (Figure 5.16).

Let's stop for a second and think about graphing. One of the subtle powers of geometry is its ability to present us with holistic views of things. This is embodied in the old phrase "a picture is worth a thousand words." Our ability to process and recognize characteristics of images is pretty phenomenal. We can recognize individual objects and people, spot little inconsistencies and changes in things we know, and so on.

Analysis on its own can suffer from the lack of an ability to see functions as a whole, to be able to apply our visual senses to their overly compact representations. Consider

$$f(x) = 3x^3 - 6x^2 + 7x - 20$$

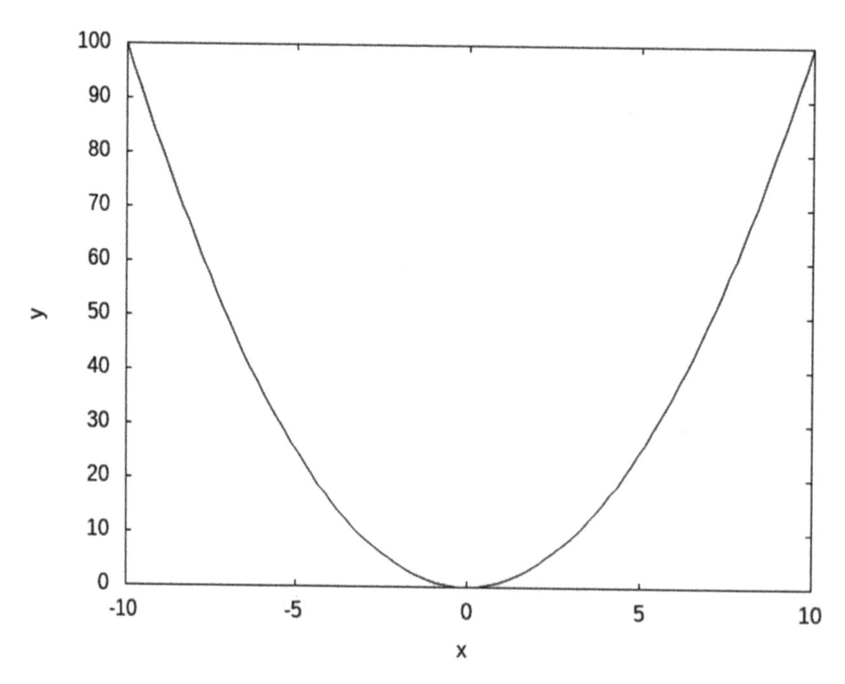


**FIGURE 5.15**
Parabola

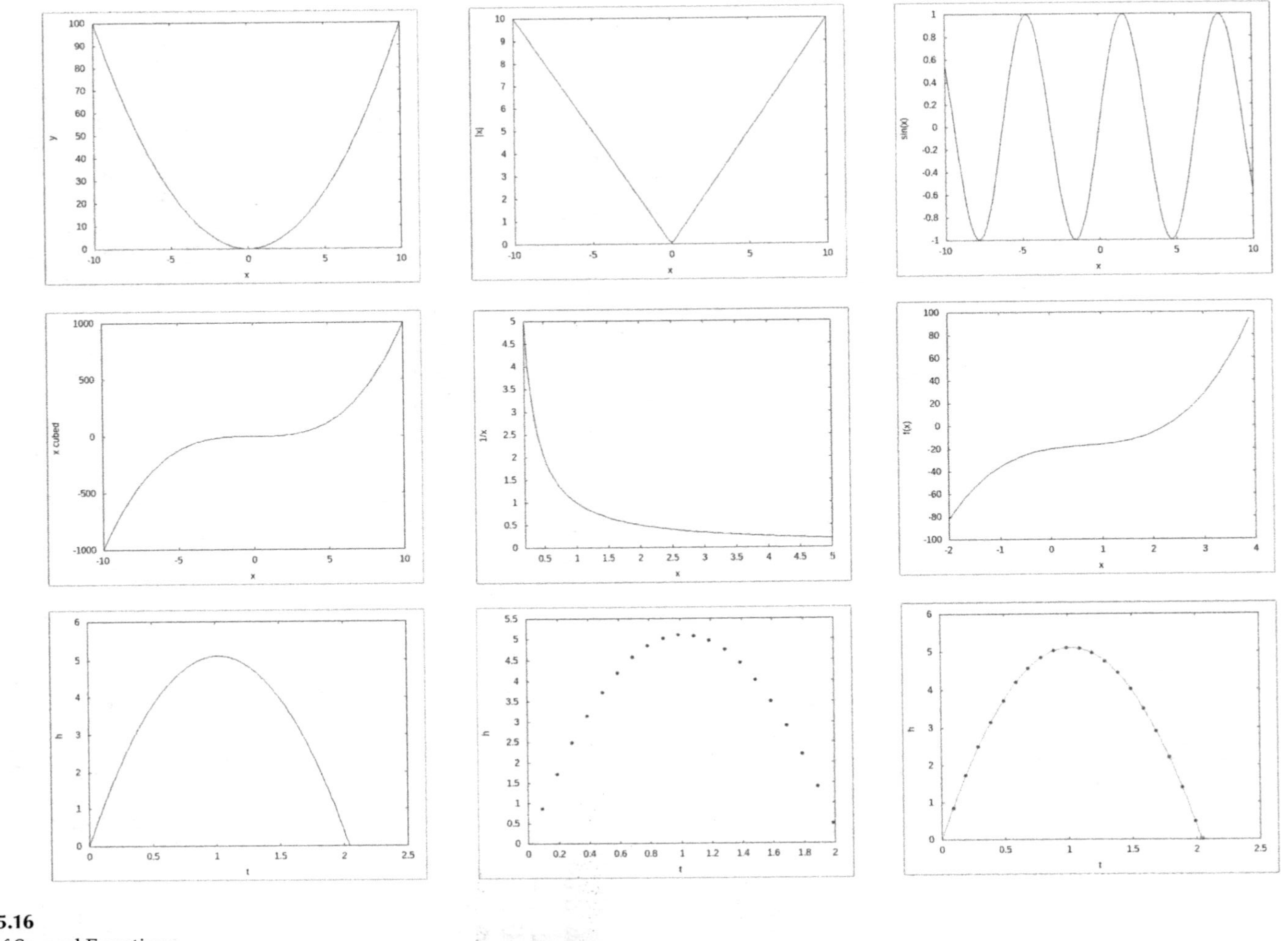


**FIGURE 5.16**
Graphs of Several Functions

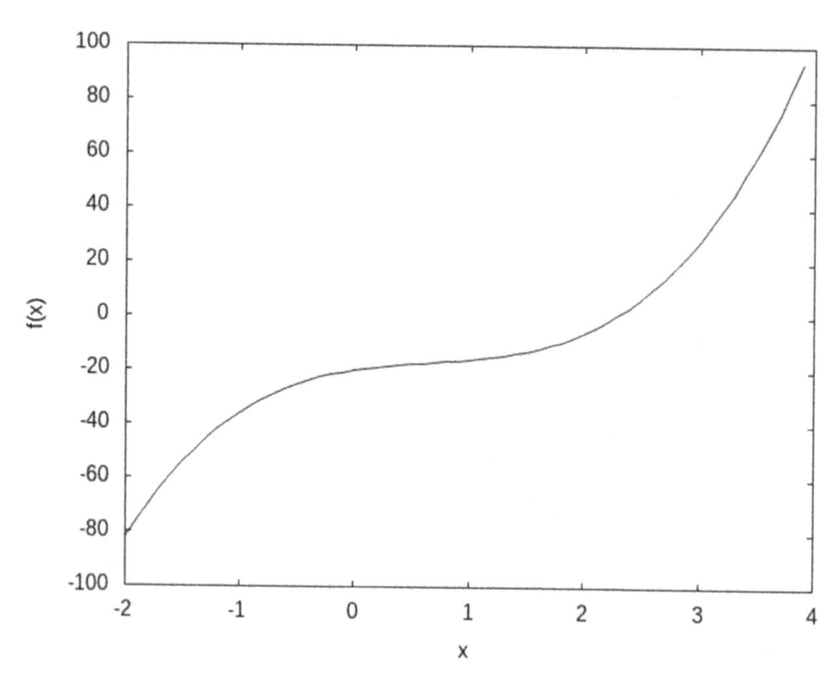


**FIGURE 5.17**
Graph of Cubic Function

It's not that hard to calculate individual values for this function, but it makes little impact on the mind. Whereas if the function is paired with its graph, (Figure 5.17)

the two mutually support each other. The function lets us find values; the graph reveals how those values relate to each other. We can look at the graph and know roughly what the value of the function should be for a given value of x. We can see where the function will be positive, where negative. If we know one value, we can see what nearby values should be.

The function creates the graph. The graph reveals the function.

If that were all it did, analytic geometry would be a boon only to mathematicians. But it was not created just for them. Just as we can use the function to make the graph, scientists can use graphs of detected data to discern the function that lies behind the data.

Scientists seek to find the underlying relationships between the quantities we measure in the real world. These relationships are often easiest to express in the language of mathematics, but more than that, their meaning is mathematical.

This may sound strange. After all, we've been talking about math as a process of leaving the real world with some measurements, doing some calculations and going back to the real world. How then can this intermediate object, the theoretical concept we are working with to do the calculations, be the meaning of something?

Here's the thing. All meaning is theoretical. Meaning is something the mind gives to things to comprehend what we think they are. "Tuna sandwich" is a meaning given to an agglomeration of a number of real-world objects in a given structure. It is a theoretical abstraction of a class of real-world objects.

Mentally the difference between "tuna sandwich" and "circle" isn't that the first is meaningful and the second isn't; it's that the meaning of the first is easiest to express in words or pictures (or tastes), but that of the second is easiest to express mathematically.

The laws of the universe seem to be easiest to express mathematically, therefore we would say their meaning is mathematical. Here the loss of heritage is much more serious. Those who have a difficult time with mathematical meaning are kept from knowing the meanings of universal conditions and actions.

Analytic geometry is a powerful tool to give these meanings to anyone willing to delve into it.

Before we can see how this works, we need to address the meaning of the number lines which we set at right angles to each other in order to construct the Cartesian plane and Cartesian space.

Left alone, x and y in the plane and x, y, and z in space represent positions relative to marker lines, how far along a point is in two or three directions. They allow positions to be numbered, and therefore they appear to be measurements of distance.

But they don't have to be. The same coordinate plane can have any measurable units assigned to an axis. The most common such an assignment is to make the x-axis time and the y-axis a quantity one wishes to measure over time. For example, one might have x be time and y be height and from this create a graph of the height of a particular person as he or she grows year by year. We might also change the labels on the axis, having a t-axis (for time) and an h-axis for (height) (Figure 5.18).

One might do the same with time and luminescence and show the brightness of a star as it goes through its life cycle, using a t-axis and an l-axis (Figure 5.19).

One might graph the height of a cannonball as it flies from the mouth of a cannon to the place where it lands (Figure 5.20).

Hey, that shape looks familiar. Indeed, this specific shape from this very problem is one of the reasons analytic geometry was created and was one of the things that led to Newton's laws of gravitation and the invention of differential calculus. We'll go over this in more depth in the next chapter. Our analysis now is simple. We look at this shape and we see what geometers call a **parabola**. (Our father likes to joke that he was so bad at math that

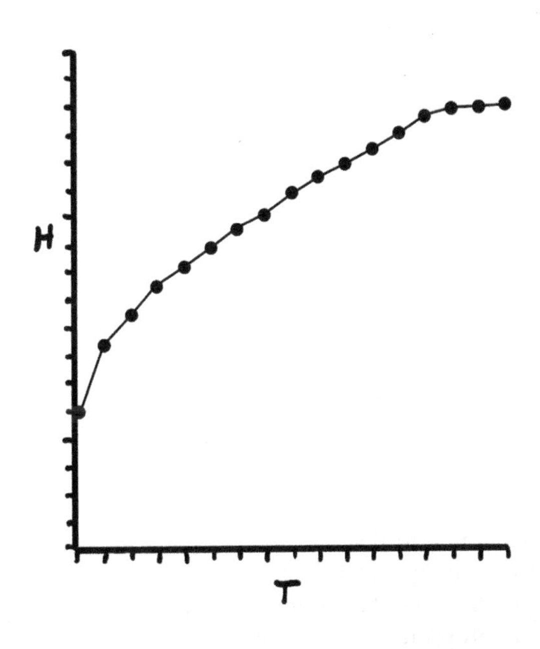


**FIGURE 5.18**
Height over Time

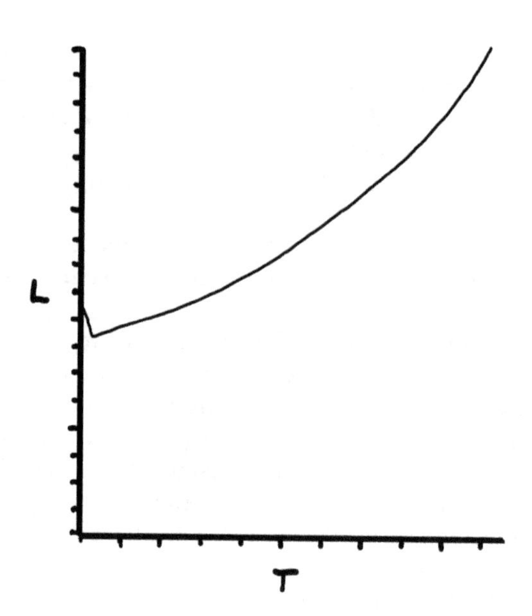


**FIGURE 5.19**
Luminescence over Time

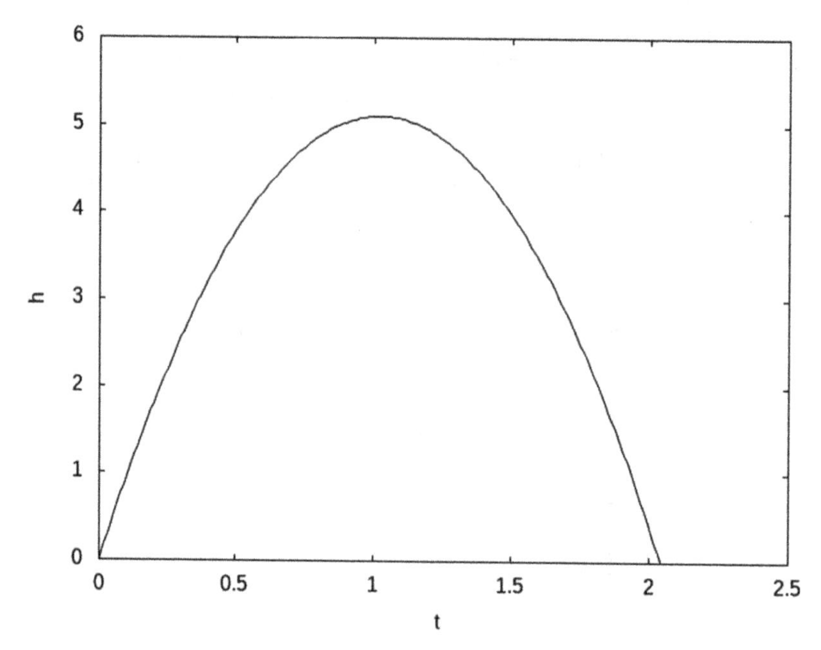


**FIGURE 5.20**
Cannonball Flight

whenever he had to take a math test, he would just guess on each question that the answer was "parabola." In this case he would be right) Parabolas were shapes that ancient geometers were familiar with. They even had a geometric definition analogous to that of circle.

A parabola is the set of all points equidistant (that is, having the same distance) from a given point (called the focus) and a given line (called the directrix) (Figure 5.21).

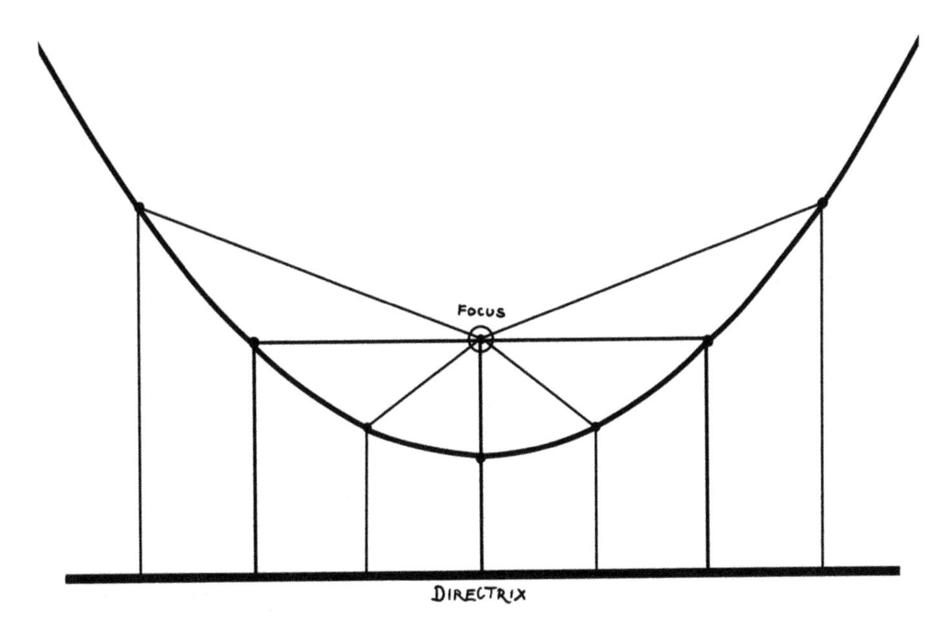


**FIGURE 5.21**
Classical Parabola

In analytic geometry, parabolas show up whenever one graphs a function of the form

$$y = ax^2 + bx + c$$

We see geometrically that the height of the cannonball graphed over time has the shape of a parabola. Therefore, we suspect that there is a law of physics which would express the height as a function of time, and that function would be of the form $h(t) = at^2 + bt + c$, since for appropriate values of a, b, and c, this function would have a graph like the above.

*Hold up a second! We're going way too fast on a really important point.*

**You're right. We need to come to a screeching halt, because we just casually passed for the second time in this chapter over one of the fundamental connections between math and physics, that river from which analysts and physicists both fish. That connection is the idea that laws of nature can be expressed as functions and equations.**

*On the face of it, this seems like a really outrageous claim, making about as much sense as claiming that laws of nature can be expressed as poetry or painting or as interpretive dance. But we do it for the best of all possible reasons. Because it works. Because when we create laws of physics in mathematical terms – if we do it right – those laws are accurate predictors of future behavior, a major goal of science. If we do it really well, those laws just need a few values plugged into them to calculate their predictions.*

**In other words, they are lazy solutions, a major goal of math.**

Back to the cannonball. One can graph an individual cannon shot by essentially tracking the height as it flies and recording the values. Then one makes a graph of the points as given in Figure 5.22.

Connecting the dots reveals (within the accuracy of the measurements) an approximate parabola.

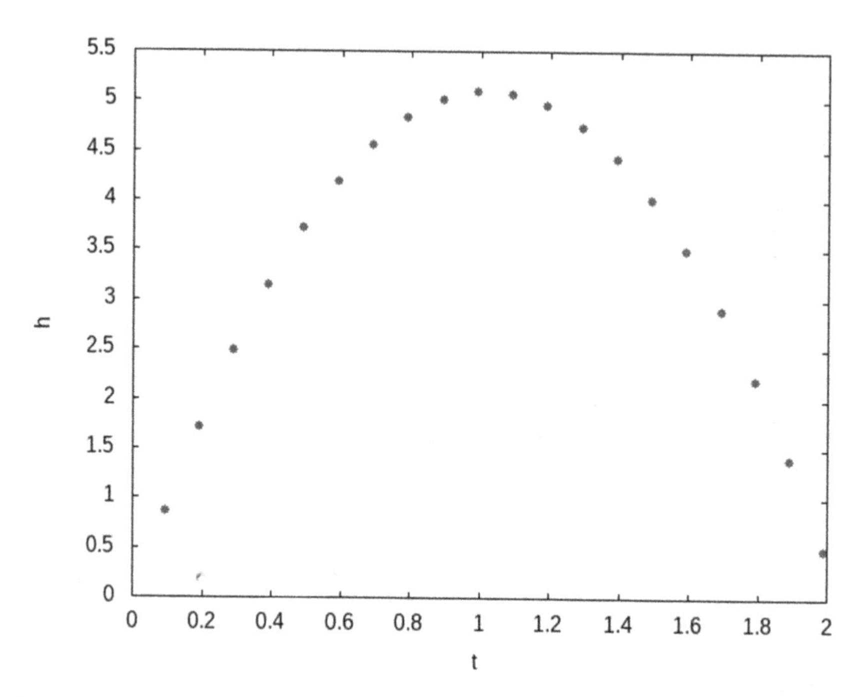


**FIGURE 5.22**
Cannonball Flight Redux

Now let's consider what we have. We have a lot of points, each with a t and an h value, and we have a function

$$h(t) = at^2 + bt + c$$

If we take, say, three points on the graph (t1, h1), (t2, h2), and (t3, h3) and plug them into this function, we end up with three equations.

$$h1 = at1^2 + bt1 + c$$

$$h2 = at2^2 + bt2 + c$$

$$h3 = at3^2 + bt3 + c$$

Now, this looks like it has a lot of variables, but remember that (t1, h1), (t2, h2), and (t3, h3) are all specific points. We really have only three unknown values, a, b, and c. We have three equations in three unknowns. We should be able to solve them and find the values for a, b, and c that will give us our function.

Alternatively, we can just look to see which values of a, b, and c give the best fit with all the data in the table. This is the approach used with more complicated functions and we will look at it in more detail in the chapter on probability and statistics.

We've glossed over parts of this, which we'll come back to later. The thing to be aware of here is that enough analysis of enough shots allows us to figure out what the a, b, and c are for each different shot and to associate those values with abstractions of real-world quantities.

$$h(t) = -4.9(m/sec^2)t^2 + v_0t + h_0$$

where $v_0$ is the upward muzzle velocity of the cannon (the upward speed the cannonball has when it leaves the cannon) and $h_0$ is the initial height from which it is fired. Notice that the number multiplied by $t^2$ does not change with the characteristics of the particular cannon shot. This number (−4.9) represents ½ the acceleration caused by the force that is pulling the cannonball down, the force of gravity. We'll wait until the next chapter to see why this is so.

The important thing here is the efficacy of getting experimental data, graphing the data, and then figuring out what kind of function fits that data, if any does. This process – gather, graph, and analyze – shows why analytic geometry is so powerful and useful. Let's break it down step by step.

1. Gather. Scientists working in the real world create experiments that measure quantities. This is (from our rarefied theoretical point of view) the grubby nasty world of experimental science which can involve things like ramming thermometers into the Antarctic permafrost or the backsides of various creatures, or painstakingly scanning the skies and marking down positions of the planets while your own backside feels like permafrost and so on. We're glad someone else is doing this work. (Note that David spent some time as a graduate student doing experimental particle physics. This experience convinced him that he was much better suited to being a theorist.)
2. Graph. By creating appropriate axes, one can plot the values determined in the data gathering in Cartesian planes and spaces. The graphs so done create shapes.
3. Looking at the shapes, we can figure out what kinds of functions are likely to produce such data. These functions are then used as hypotheses in later experiments that will confirm or deny the hypotheses.

This three-step process is so common in the sciences that the idea of working without it would be mind-boggling to many. And it is a beautiful process in its own way. It moves explicitly from the real world to the theoretical and then employs the sense and awareness of shape of the geometric mindset to generate an analytical possibility for later testing. It's a perfect scientific-mathematical cycle.

Unless, of course, there is no such function.

We could in theory graph any two quantities against each other, but if those two have nothing to do with each other (such as the average surface temperature of Venus on one axis, average pie consumption rates in North America on the other), then any function we might guess from the graph would fail as a predictor and therefore would not be expressing a real law of nature. The question of whether or not there is such a function is a vital one in science and we will look at that in depth in a later chapter.

From this, part of the relationship between math and science can be seen. Theory is not enough. Theory must be tested thoroughly before it can be accepted. And the functions derived from analysis of graphs are theories.

There's one more thing we can take away from our cannonball graph, a manner of translating problems between algebra, analysis, and geometry so that the same problem can be retrofitted to any particular mindset.

Let's start with the analytic question: when will the cannonball be on the ground?

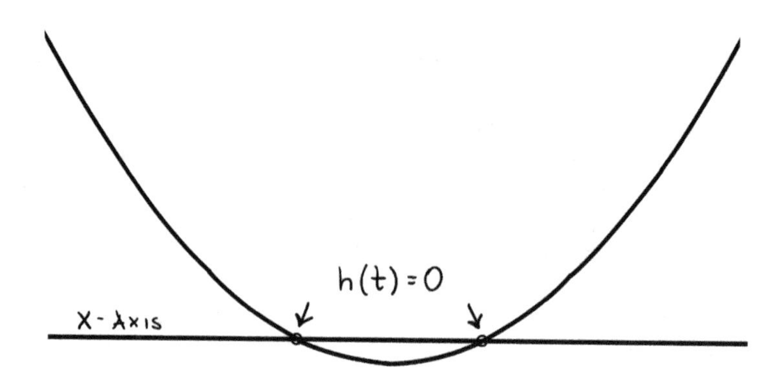


**FIGURE 5.23**
Intersecting the x-Axis

If we examine the graph, we see that this happens when h(t) = 0. Geometrically this occurs when the parabola we graphed above intersects the x-axis (Figure 5.23).

Algebraically, therefore, this occurs when

$$0 = -4.9\left(\text{m}/\text{sec}^2\right)t^2 + v_0 t + h_0$$

But this is a quadratic equation. Its solutions are

$$t = \left(-v_0 \pm \sqrt{\left(v_0^2 - 4(-4.9)h_0\right)}\right)\Big/2(-4.9) = \left(v_0 \pm \sqrt{\left(v_0^2 + 19.6h_0\right)}\right)\Big/9.8$$

The time value $(v_0 + \sqrt{(v_0^2 + 19.6h_0)})/9.8$ is when the cannonball lands. The time value $(v_0 - \sqrt{(v_0^2 + 19.6h_0)})/9.8$ occurs before the cannon is fired, so this solution, while correct for the equation, does not actually have any physical meaning. This shows the importance of remembering the real-world circumstances from which one is abstracted so that when the possible solutions are brought back to the world, one can discern which of them are real and which spurious.

Notice that knowing which of these answers is possible is not obvious from algebra alone. Algebra doesn't cross over into reality. Geometry does. Consulting the picture makes it clear what the reality of these times is.

This problem shows the advantages of the different ways of thinking supporting each other. Shape reveals function, which reveals equation, which reveals solution, which is then checked against shape to see if it makes sense. Furthermore, none of these is a particularly heavy use of any of the ways of thinking.

Analytic geometry is like an island in the middle of mathematics and physics, where it all comes together and becomes easier if you can learn to get there. It's worth the swimming or the boating or taking the bridge or any other metaphor because once you reach it, everything becomes available.

The graphed world, the world where data are laid down against each other in analytic geometry, is one of those parts of math that has permeated our everyday culture. Graphs are used all over the place to justify political positions, to convince people to buy things, to track disease progress, and to warn against dangers. But to be able to interpret those graphs, one must understand enough analytic geometry to be able to figure out whether they make any sense and whether they show what they purport to show.

## Parametric Graphing

There's another kind of graphing that plays at least as important a role in analytic geometry as the graphs shown above. This kind of graphing arises from a much older idea than the Cartesian plane: the map.

A Cartesian plane can be put like a transparent sheet on top of a map of some terrain to obtain a gridded map, one that uses coordinates to determine an exact position (Figure 5.24).

On such a map, it is a common and useful practice to draw a route, the path along which one might journey to traverse the wooded expanses from one place to another (Figure 5.25).

Let's take away the map and look at the route in the Cartesian plane (Figure 5.26).

This curve does not represent y as a function of x. Instead, it shows us position on the map dependent on time (Figure 5.27).

But time is neither the x- nor the y-axis on this graph. The coordinates are being employed to present position on the ground. Time is a variable not shown in this picture but clearly present, somehow. Since the positions depend on time, we can argue that this is a graph of some function that depends on time where time is not seen in the graph.

That may sound kind of silly, but the route laid down above is a useful intuitive idea, a depiction outside of time of a route laid down in time. This idea is a completely natural extension of the process of drawing. A line drawn is actually a track left behind, marking each place where the pen or pencil was as it progressed through the process of drawing. We can see the course of the pen point in the drawing it leaves behind, time captured in space.

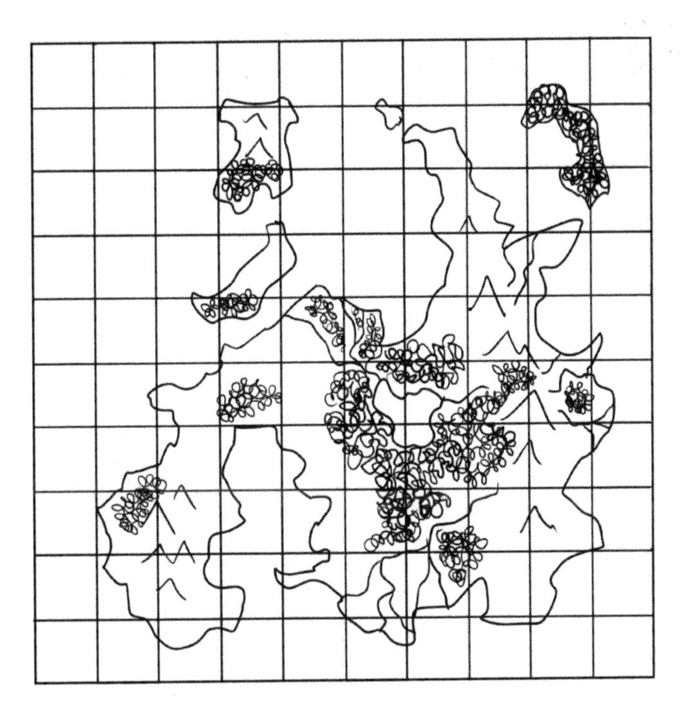

**FIGURE 5.24**
Gridded Map

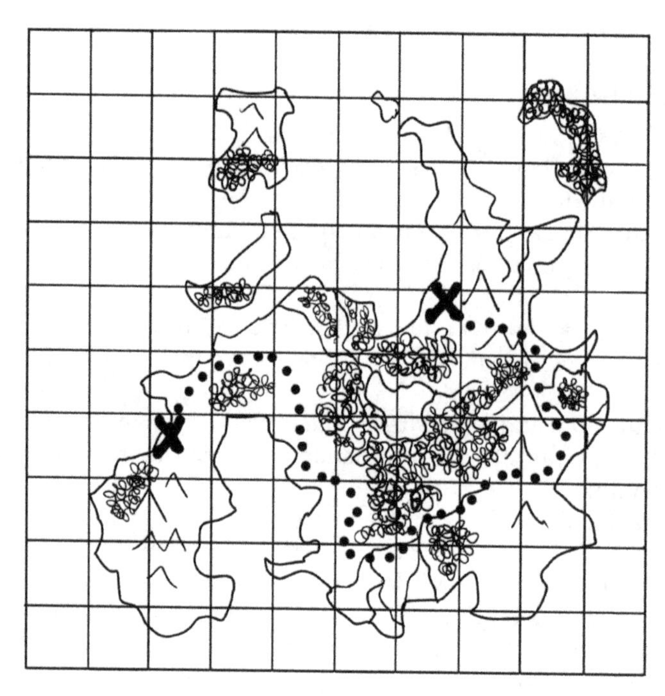

**FIGURE 5.25**
Gridded Map with Path

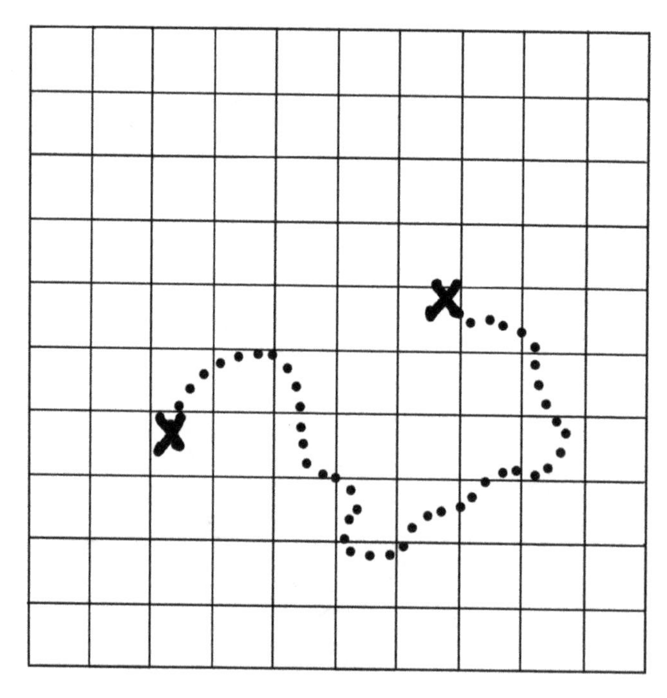

**FIGURE 5.26**
Path on Cartesian Plane

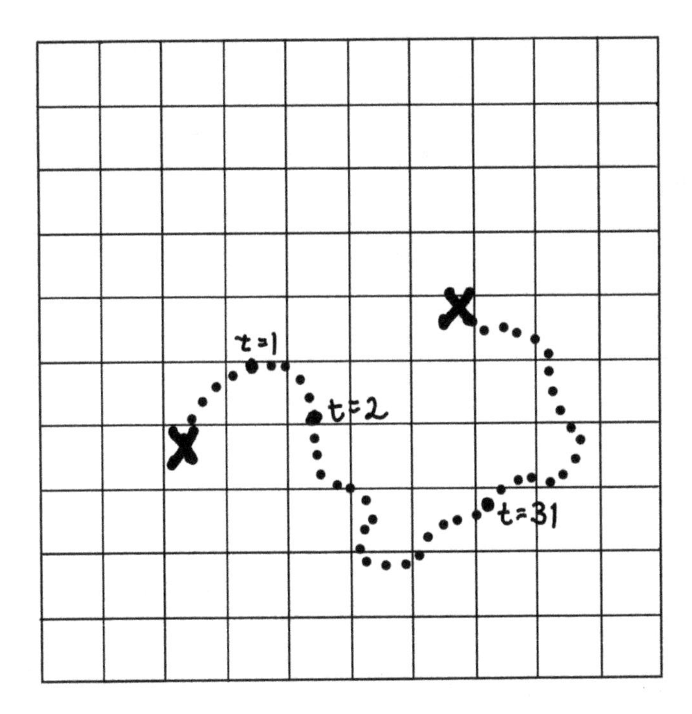


**FIGURE 5.27**
Path with Times Marked

The way to represent this analytically is to not treat y as a function of x, but to treat x and y each as functions of some other variable which we do not assign to an axis. We would write this as (f(t), g(t)) and say that we were graphing all points (x, y) in the plane where there is some value of t for which x = f(t) and y = g(t). So if we had, for example $(t^2 + 4, 2t - 5)$, then we would create the graph by selecting values of t, plugging them into the functions and plotting the results.

A few instances:

| t | point |
|---|---|
| 0 | (4, −5) |
| 1 | (5, −3) |
| 2 | (8, −1) |
| 10 | (104, 15) |
| −2 | (8, −9) |

The resulting graph would be as shown in Figure 5.28.

This kind of graph is called a **parametric graph** (a parametric curve in this case) because it depends upon an invisible parameter (t in this case), a parameter that is not attached in meaning to any of the axes involved. And yet, because it is not given any axial meaning, the meaning comes forth in the graph. The curve drawn is the route followed only because the curve does not show time. If one of the axes were time, then the graph would be in an abstract spacetime, not a graph across the real-world map we started with. (*Although later, when we talk about relativity, we will see that spacetime is less abstract and more real than one might think.*)

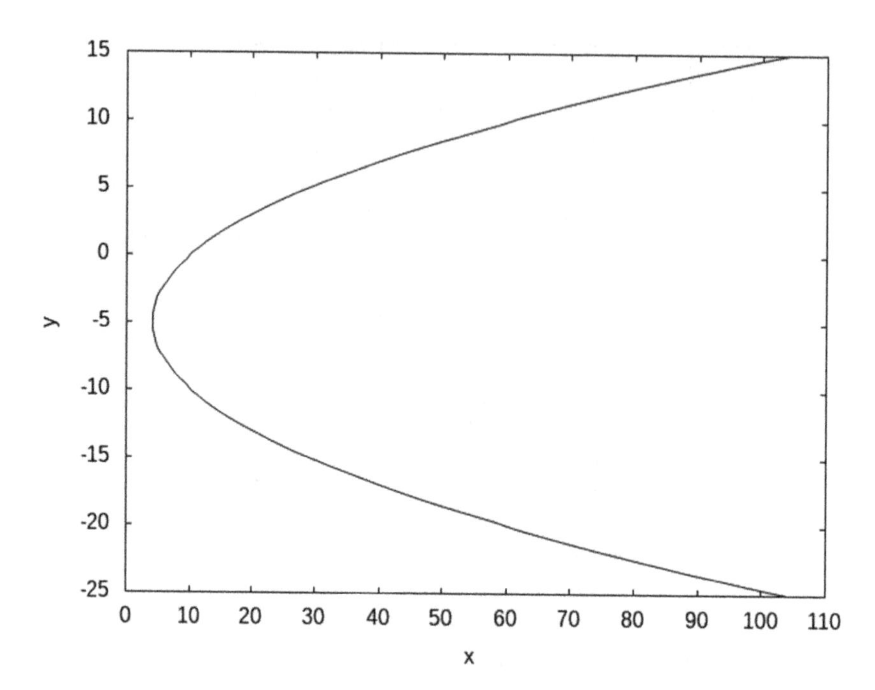


**FIGURE 5.28**
Graph using Instance Points

The parametric curve idea points up one of the aspects of analytic geometry that is closer to an art than anything else. Where do you put which meaning? We can measure and calculate a huge number of interrelated real-world quantities (position, time, speed, kinetic energy, etc.). Many of them are interrelated and many of those relationships could be revealed in properly chosen graphs. But when doing such a graph, you need to figure out what you are trying to see the shape of. Should a given measurement be an axis, a parameter, a constant value changed in each graphing, or omitted entirely? There are a multitude of different shapes that can lie inside a given set of data, and different awarenesses can come from each such shape.

Truth be told, what we're doing right now, graphing position and using time as a parameter, is one of the closest to our sense of reality. The paths we map above are like the tracks upon the ground of our journeys.

We can return to our earlier cannon example and graph the actual path of the cannon ball in flight.

In this case, the y function is exactly the same as before:

$$y = -4.9\left(m/sec^2\right)t^2 + v_0 t + h_0$$

The x function represents horizontal travel, how far the cannonball flies. This is

$$x = v_1 t$$

where $v_1$ is the horizontal muzzle velocity. $v_0$ and $v_1$ have a simple relationship with each other that depends on the angle the cannon is tilted ($\theta$) and the actual muzzle velocity (v):

$$v_0 = v \sin(\theta)$$

and

$$v_1 = v\cos(\theta)$$

Why this is so we'll explain in the next chapter when we discuss vectors.

Here are a few different cannon shot graphs, with the same cannon at different angles (Figure 5.29).

Two points of clarification before we move on. First, every graph done nonparametrically can also be done parametrically. If we want to graph the function y = f(x), we get the set of points (x, f(x)). But we can get the same set of points from (s, f(s)). That is, we have a parametric graph in which the function that determines the x coordinates is just x = s.

The other thing to make note of is that the same graph can represent more than one pair of functions. (s, f(s)) will give the same set of points as (2s, f(2s)), as well as (5s, f(5s)), (s/7, f(s/7)), and so on.

We can convert our parametric cannon graph into a nonparametric one by substituting $t = x/v_1$ into

$$y = -4.9\left(m/\sec^2\right)t^2 + v_0 t + h_0$$

to get

$$y = -4.9\left(m/\sec^2\right)\left(x/v_1\right)^2 + v_0 x/v_1 + h_0$$

While this has a parabolic graph, it does not look the same as our earlier function for a very simple reason. This is y as a function of x, not as a function of t. This graph tells you how

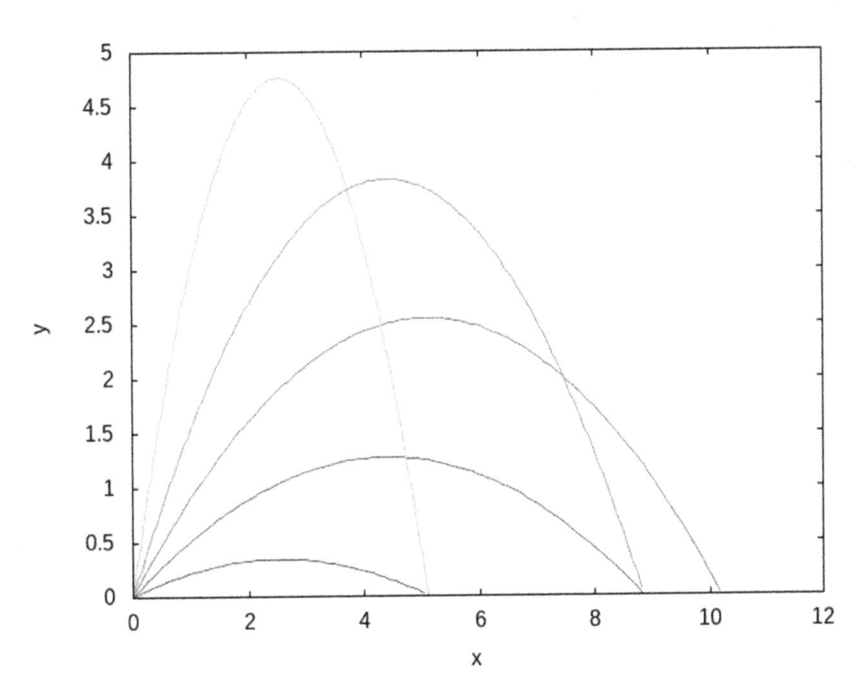


**FIGURE 5.29**
Multiple Cannon Shots

high the cannonball was when it traveled a horizontal distance of x. Time has completely disappeared from this graph.

This shows that there is more than one way to get a set of points, which leads neatly into our next subject.

## The Fault Is Not in Our Stars

The routes that interested mathematicians and physicists, the paths they wanted to map and wanted functions for, were not travel routes between cities. They wanted the route shown in Figure 5.30.

But what function draws these circles? (Actually they're ellipses, but we'll start with circles.) Our only definition of a circle is as a set of points. No functions seem to be involved.

*A circle is the set of all points a given distance (the radius) from a given point, the center.*

Let's translate this into analytic geometry terms. We need to find all points (x, y) which are a distance r, the radius, from a particular point $(x_c, y_c)$, the center.

The distance between (x, y) and $(x_c, y_c)$ is

$$\sqrt{\left((x-x_c)^2+(y-y_c)^2\right)}$$

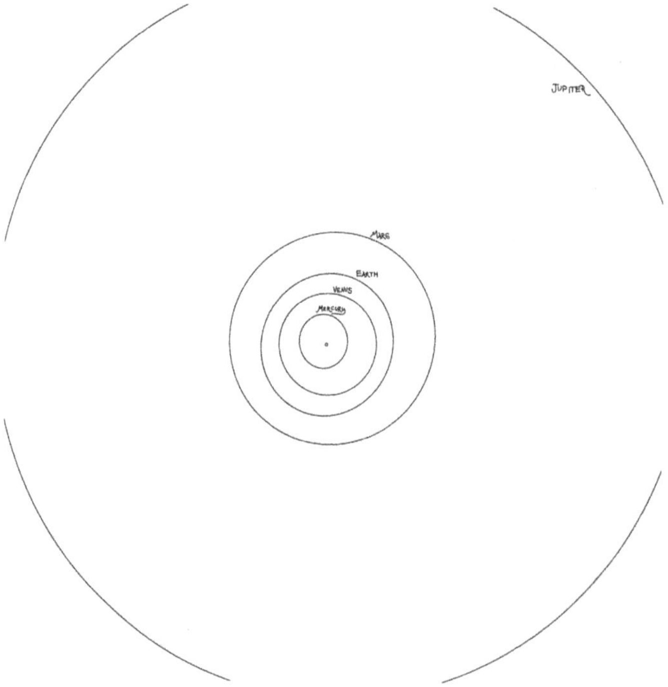

**FIGURE 5.30**
Some Orbits in the Solar System

So the requirement for a point (x, y) to belong to the circle is that

$$\sqrt{\left((x-x_c)^2+(y-y_c)^2\right)}=r$$

Squaring both sides gives us the easier-to-calculate formula:

$$(x-x_c)^2+(y-y_c)^2=r^2$$

We can sort of produce a function that gives us y in terms of x as follows:
We subtract $(x - x_c)^2$ from both sides to get

$$(y-y_c)^2=r^2-(x-x_c)^2$$

Taking the square root of both sides gives us

$$y-y_c=\pm\sqrt{\left(r^2-(x-x_c)^2\right)}$$

which yields

$$y-y_c=\pm\sqrt{\left(r^2-(x-x_c)^2\right)}$$

We'd like to say that the points of the circle are therefore

$$\left(x,\ y_c\pm\sqrt{\left(r^2-(x-x_c)^2\right)}\right)$$

There are two problems with this. The first is that the ± sign means that we have two y values for each x value. We can see this in Figure 5.31.

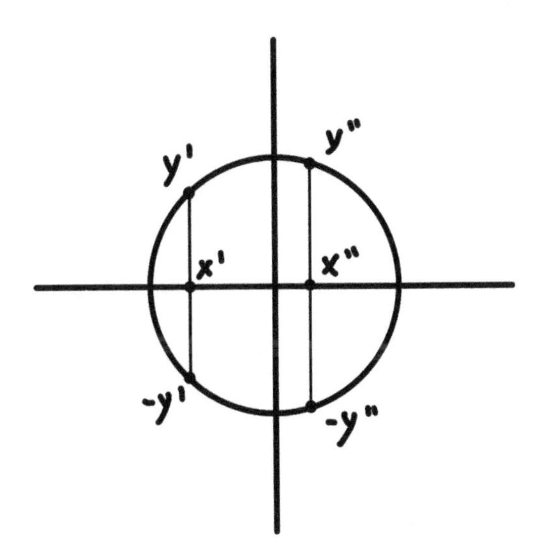


**FIGURE 5.31**
Circle Function Problems

This means that we don't really have a function, since a function gives one value for each particular set of parameter values. We can weasel this by saying that the square root is a **branched function**, that it's something that isn't a function but we want to treat as a function because it has enough function-like characteristics.

The second problem is that there are an infinite number of x values that the function is not defined for. Every x value that lies to the right of the rightmost point of the circle or to the left of the leftmost point would give us a negative value under the square root sign, and we cannot (at least, not yet) take the square root of negative numbers.

So yes, we can use the above to draw the circle, but we're a little queasy about it. Is there some other way, maybe a parametric way? We'd like such a thing, since an orbit is a graph of position over time.

Let's look again at a point on the circle. This time we'll draw the radius that touches that point (the line segment connecting the point to the center). To save wear and tear, we'll say the center is the origin. We'll deal with other centers later (Figure 5.32).

Where did that right triangle come from?

It came from the requirement that the point be a specific distance r from the center of the circle (the origin in this case). Let's take a look at the angle θ, and just for kicks (and because we're leading you down the garden, path toward a vista we already know is there) let's figure out what the sine and cosine of the angle are.

$$\sin(\theta) = \text{opposite side}/\ \text{hypotenuse} = y/r$$

$$\cos(\theta) = \text{adjacent side}/\text{hypotenuse} = x/r$$

$$\sin(\theta) = y/r$$

$$\cos(\theta) = x/r$$

Let's take these two equations and multiply both sides of both equations by r.

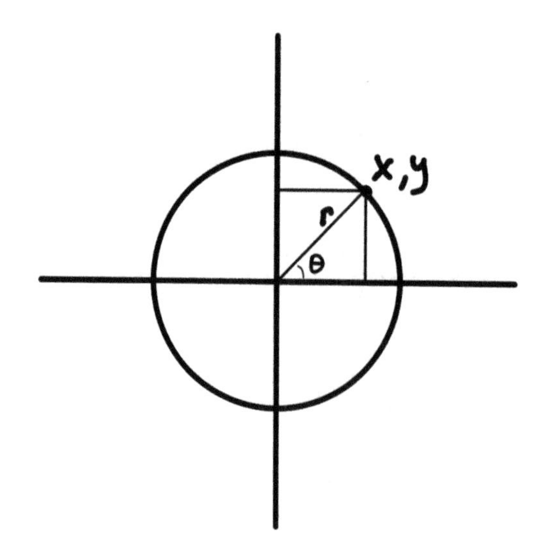


**FIGURE 5.32**
Circle with Radius

$$r\sin(\theta) = y$$

$$r\cos(\theta) = x$$

In other words, every point on the circle is (r cos(θ),r sin(θ)). This looks an awful lot like a parametric curve with θ as a parameter, and indeed it is. We can even take away the meaning of the angle, replacing θ with another variable, say t for time, and get (r cos(t),r sin(t)). This is a parametric curve which will trace out to a circle.

Furthermore, we can have any center we want simply by adding the coordinates of the center ($x_c$, $y_c$) to the point above to get ($x_c$ + r cos(t),$y_c$ + r sin(t)) (Figure 5.33).

*Hmm ... a circle made from triangles.*

**Yup**. Two shapes which look wholly unlike each other are found by analytic geometry to be intimately connected on the level of functions. Functions created from triangles are used to create circles. The reason the triangle and the circle come together is because the circle is defined in terms of distance ["a given distance (the radius)"], and the triangle defines distance (thanks to the Pythagorean theorem). There is a nonanalytic connection between the shapes.

Let's look at that a little more slowly. The fact that triangles define distance and circles are defined in terms of distance means that somewhere inside the geometric meaning "circle" lies the geometric meaning "triangle," even though the two shapes do not in any way resemble each other.

If planets orbited in circles, we'd be done. However, Johannes Kepler, the 16th- and 17th-century German astronomer, proved by examining mounds of data that planets orbit in ellipses. So we have a little more work to do, most of which we're going to omit because it's messy. We'll just hit the highlights.

An ellipse is defined as the set of all points whose distances from two points (called the **foci** of the ellipse) add up to a given value.

*That doesn't fall off the tongue or the pen as easily as the circle does.*

No, but one can see this geometric meaning by making a physical tool to draw ellipses. The tool is trickier than a compass but it can be constructed using household objects: two pushpins, a piece of twine tied into a loop, and a pencil. Also a piece of paper.

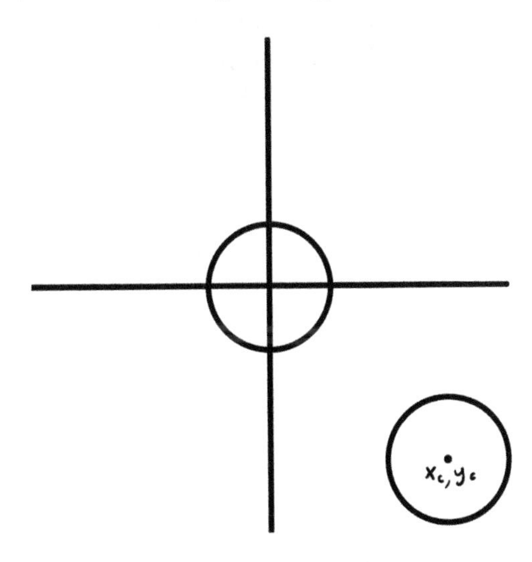


**FIGURE 5.33**
Two Circles

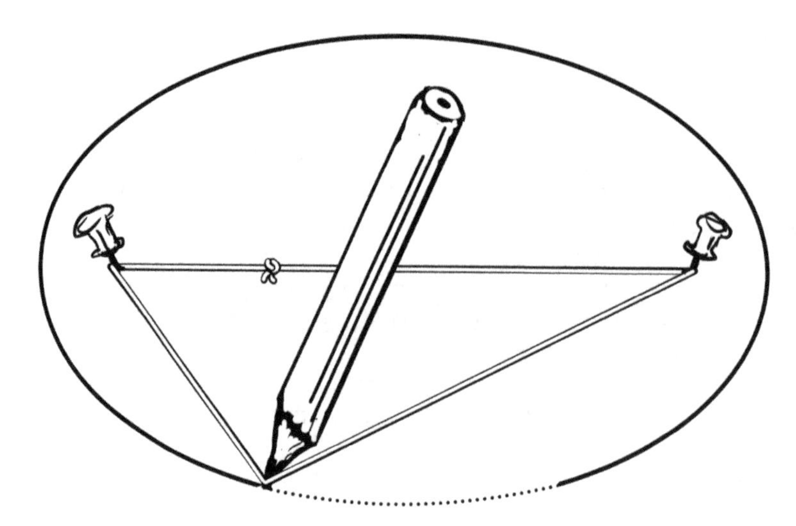

**FIGURE 5.34**
How to Draw and Ellipse

Stick the pins in at the points you have chosen for the foci, loop the twine around the pins, and tuck the pencil into the loop as well. Pull tight so the pencil is as far from the pins in a given direction as the loop will allow (*but not so tight as to yank out the pins).* Now draw, pulling the pencil around while maintaining the tautness of the twine loop. When you've come back to where you started, you've drawn an ellipse (Figure 5.34).

Notice that when the two foci are the same point we've made a crude form of compass and drawn a circle.

Since we're dealing with a sum of distances (which involves adding up two square roots), the calculations necessary to turn the equations into a parametric curve are a bit messy. Oddly enough, the calculation is easier if instead of choosing the origin at the center of the ellipse, we choose the center to be at one of the foci. Then the equation for the ellipse is $r = a(1 - e^2)/(1 + e\cos(\theta))$ and with this formula for r as a function of θ, the parametric form is $(r \cos(\theta), r \sin(\theta))$. Here a is half of the size of the long axis of the ellipse and e has to do with the ratio of the short to long axis of the ellipse. For the case where e = 0, the ellipse is a circle. For the more general case, where a focus is at the point (xc, yc) and the ellipse is rotated by an angle δ, the parametric form is

$$\left(xc + r\cos(\theta+\delta), yc + r\sin(\theta+\delta)\right)$$

So that's the description of the orbits of the planets (*other than the influence of the gravitational field of each planet on the orbit of the other planets, and the corrections of general relativity*).

## Curves and Surfaces

**Even neglecting the gravitational influence of planets on each other, there's a problem with this simple two-dimensional model of planetary orbits.**

*What's that?*

**Each planet orbits in a plane, but why should everything orbit in the same plane? For example, if the moon and earth were to orbit in exactly the same plane, we would have solar and lunar eclipses every month.**

*Yes, that's right. Astronomers call the plane of the Earth's orbit the ecliptic. Most planets orbit in planes that are fairly close to the ecliptic, but not exactly the same. This is because the planets all formed from the same disk of dust, ice, and gas in the early solar system. (Pluto's orbit is not all that close to the ecliptic, but now it's not considered a planet anymore.)*

To model this important bit of astronomy we need to plot these orbits in space rather than on a plane. This is not a major problem. We just need to add a parametric function for z and plot accordingly.

**Yes, now our plots will be fully three-dimensional.**

*What about our characters?*

**Flat, dull, easy for any Hollywood studio to cast.**

Curves in space work just like curves in the plane except they have more directions. They can pull, bend, and twist (Figure 5.35).

We can use these curves to represent the paths of objects through space, which is useful when animating air-to-air combat, (Figure 5.36)

or when charting the path of a particle bouncing around in moving air (Figure 5.37).

The path of the moon is an ellipse that has been rotated at an angle in space.

The graphing of curves in space is largely employed for determining paths and orbits.

This is all well and good, but none of these curves are the shapes we perceive in the world. What if we want to create analytic geometry objects for what we're actually looking at?

Consider the plane figure given in Figure 5.38.

We can express this easily in terms of inequalities. This figure is the set of all points (x, y), where

$$-5 \le x \le 12 \text{ and } 0 \le y \le 20$$

**FIGURE 5.35**
Curves in Space

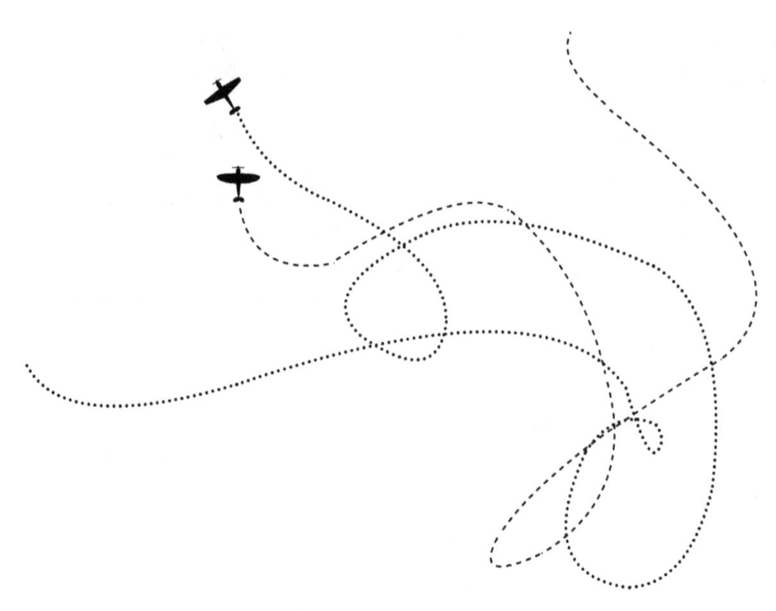

FIGURE 5.36
Dogfighting Airplanes

FIGURE 5.37
Brownian Motion

But there is also a way to express this parametrically as the surface (s, t), where s varies from −5 to 12 and t varies from 0 to 20. This may not look like much of a difference, and for the rectangle it isn't. But consider the following: The set of all points (t cos(s),t sin(s)), where s can be any number and t varies from 10 to 20.

This is like drawing an infinite set of circles, one after another, starting with one of radius 10 and ending with one of radius 20. In this figure, called an **annulus**, t and s both act as parameters (Figure 5.39).

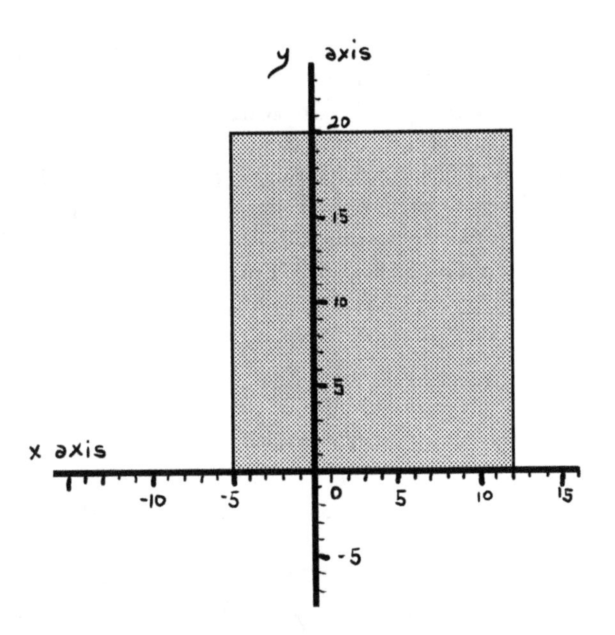


**FIGURE 5.38**
Solid Rectangle

**FIGURE 5.39**
Annulus

This is a two-parameter shape, commonly called a **surface** (as opposed to a one-parameter shape, which is a **curve**, as you might guess from the fact that we've been calling them that).

Surfaces in the plane are kind of boring. They tend to be the regions bounded by two planar curves. *They're flat, dull and uninteresting, but easy to cast – I mean, graph.*

In space, however, things get different and more like our perceptions.

Consider the set of points in space (t cos(s),t sin(s), t). When t is 0, we would have the point (0, 0, 0) regardless of what s is. This would draw a circle of radius 0 at the origin.

When t is 1, we have all the points (cos(s), sin(s), 1), which is a circle of radius 1 at a height of z = 1. There would be a circle of radius 3 when z = 3. In other words, the circles get bigger as you get higher. This shape is called the **cone** (Figure 5.40).

Or consider this surface: (10 cos(s),10 sin(s), t)

This draws a circle of radius 10 at each height, producing a **cylinder** (Figure 5.41).

A sphere, as noted in the previous chapter, has the same definition as a circle, but the points are not confined to a plane. **A sphere is the set of all points in space at a given distance from the radius from a given point, the center.** It can be represented as the set of all points (x, y, z), where

$$(x - x_c)^2 + (y - y_c)^2 + (z - z_c)^2 = r^2$$

Parametrically, this is messier than the circle, with

$$(r\cos(s)\sin(t),\ r\sin(s)\sin(t),\ r\cos(t))$$

In this case, s is the angle from the x-axis and t is the angle from the z-axis (Figure 5.42).

Surfaces can be used as graphs in the sciences in order to express relationships between three quantities at once. For example, the volume of a gas will depend on temperature and pressure. We can therefore graph a surface with x = temperature, y = pressure, z = volume that will show how these quantities relate to each other (Figure 5.43).

Surfaces as graphs can also be used for the same reverse process as curves to take data and discern functions from them. This requires knowing what kinds of functions produce what kinds of surfaces and determining which parameters for the surfaces give a best fit with the data. For sufficiently complex data distributions, this can involve a lot of calculation. Fortunately, we have computers.

We can also expand our earlier idea of plotting maps in Cartesian systems. Before, we used flat maps in the plane. Surfaces can make maps that show terrain, where the z-axis represents the height of each point on the map. These are called **contour maps** (Figure 5.44).

Surfaces can also be used, thanks to computers, to create art (Figure 5.45).

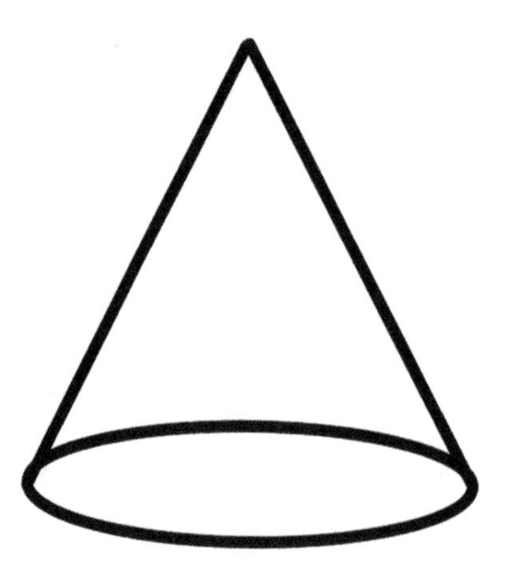

**FIGURE 5.40**
Cone

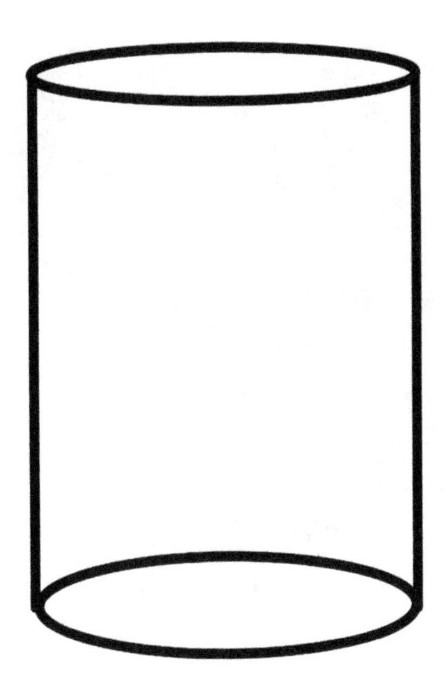

**FIGURE 5.41**
Cylinder

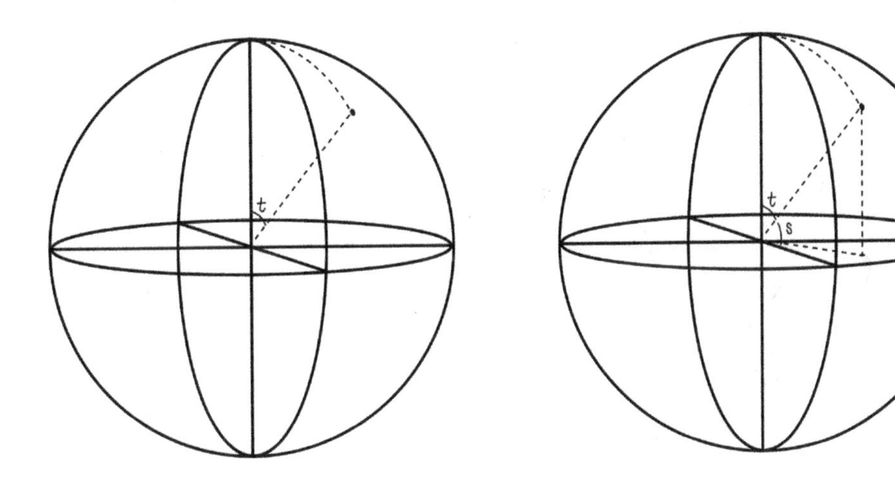


**FIGURE 5.42**
Parametric Sphere

To most people curves look and feel mathematical, but surfaces tend to look more real, because, of course, what we see of opaque objects in the real world are their outside surfaces. There is a feeling of actuality about these shapes that makes computer art possible, particularly animation. But grasping hold of them, taking the shapes into our minds, and being able to see and manipulate them – that, until analytic geometry, was the sole province of the geometric mind. With analytic geometry we can pick up shapes and carry them in the forms of functions. We can transport entire sophisticated objects in single packages and unwrap them on computers. More about this later.

So far we've been looking at space as if there were only one point of view possible. But we know that's not true. How does analytic geometry handle this concept?

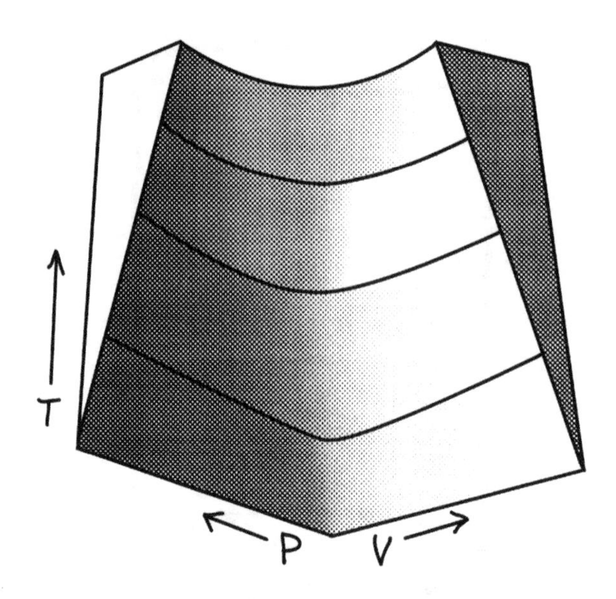

**FIGURE 5.43**
Ideal Gas Law Graph

**FIGURE 5.44**
Contour Map

**FIGURE 5.45**
Computer Model Butterfly

## Coordinate Systems

Imagine a person in a room. Call him Fred. There is a table 1 meter in front of Fred and 2 meters to his right. We can make a Cartesian plane with Fred at the origin, the x-axis as his left-right directions, and the y-axis as his front-back. The table, therefore, is at the point (2, 1) (2 meters along the x-axis to the right and 1 meter along the y-axis forward).

*Wait a minute. A table isn't a point. It's a three-dimensional shape made of real stuff, wood maybe, or plastic or whatever.*

**True, but we aren't going to be concerned with any of those aspects of the table.**

We only care about its position relative to a couple of people, so we can say that the table is at a particular point the same way we'd say the Earth is at a particular point in its orbit.

Now, in the same room there is another person. Call her Prunella. Prunella is on the far side of the table from Fred. It is 2.5 meters behind her and 1.25 meters to her left. We can also make a Cartesian plane with Prunella at the origin, the x-axis as her left-right directions, and the y-axis as her front-back. In this Cartesian plane, the table is at (–1.25, –2.5) (Figure 5.46).

What then are the coordinates of the table? (2, 1) or (–1.25, –2.5)?

Each is correct in its own frame of reference and wrong in the others. *A **frame of reference** is what a personalized, monogrammed, genuine Cartesian plane or space all your own is called.*

It is possible to shift from one frame of reference to another so that knowing where something is in one frame, one can figure out where it is in the other.

Let's examine Fred and Prunella's frames of reference. There are two things of importance to note. First, they are differently centered, one on Fred and the other on Prunella; and second, their axes point in different directions, Fred's along his front-back and right-left axis and Prunella's on hers.

There are two mathematical transformations implicit in these differences. If we do both of these transformations to a particular point (say the table) in the right order, we can transform Fred's coordinates for the table into Prunella's.

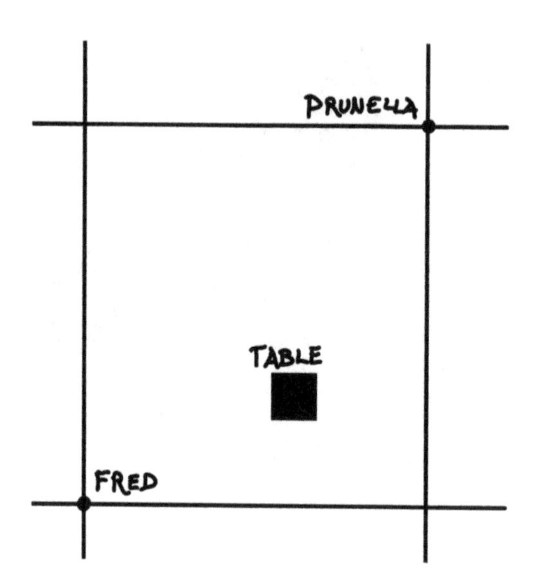


**FIGURE 5.46**
Fred and Prunella

These transformations are called **rotation** and **translation**.

Rotation is the turning of axis. Translation is the shift in origin. Let's examine these two separately.

Rotation involves spinning points around the origin. Rotation is perfectly easy to simulate in real life. Stand up, stay in one place, and turn a bit to the left. Or if you're sitting in a swivel chair, don't bother to stand up. Every object in your sight has now apparently shifted. You turned the entire universe with your heel.

The only quantity necessary to determine the rotation is the angle you made between your original position and your new position (Figure 5.47).

The effect of rotation can be calculated, but we have to be careful what quantities we're talking about when we do the calculation. When the frame of reference rotates counterclockwise an angle of $\theta$, it looks as if every object rotates clockwise an angle of $\theta$ (or counterclockwise an angle of $-\theta$). Through some fiddly trigonometric calculations it can be determined that if the object's old coordinates are (x, y), then the new coordinates will be $(x \cos(\theta) + y \sin(\theta), y \cos(\theta) - x \sin(\theta))$.

So if one had an object at (1, 0) that is directly to one's right and one spun oneself 90° counterclockwise, the object would be at $(1 \cos(90) + 0 \sin(90), 0 \cos(90) - 1 \sin(90))$. The sine of 90° is 1 and the cosine is 0, so the point is (0,–1), that is, directly behind one.

Rotation is a little messy to calculate. Translation is relatively easy.

Let's look at Fred and Prunella's frames of reference again. The origin of Fred's frame has coordinate values in Prunella's frame. To find the translation of the coordinates from Fred's to Prunella's frame we have only to subtract the coordinates of Fred's origin in Prunella's frame. In other words, we only have to see how far away Fred is and in what direction from Prunella in order to determine what the translation is.

To go from Fred to Prunella, we must then do the following: rotate so the axes match, then translate so the origins match. Using this we can go from any frame of reference to any other.

*Menacing music hinting at darker secrets to come when we come to discuss the mathematics of relativity.*

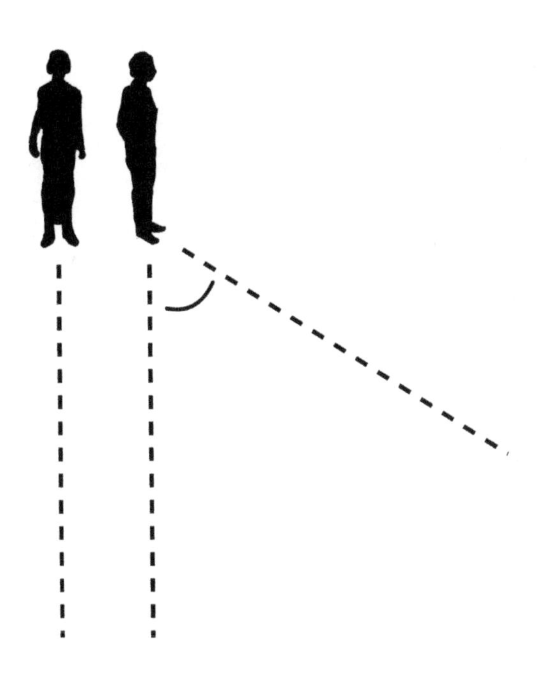

**FIGURE 5.47**
Person Turning Left

**That's some specialized menacing music you've got there.**

*Thanks. Would you like to hear my concerto for disposable characters in the opening scenes of horror movies?*

**Maybe later.**

Note that we should rotate first. If we translated first, we would not be moving the origin of Fred's system to the origin of Prunella's because we rotate in Fred's frame of reference. This harkens back to the discussion of composition of functions. The order matters. We can create an ordering of functions that will work for us that involves translation first and then rotating but it involves translating into Prunella's frame and then rotating in Prunella's frame.

The same effects can be done in space, but rotation is a more complicated process there and can produce wildly different views of the same thing (Figure 5.48).

Let's pause for a second and consider the psychological effects of this bit of mathematics. One of the most difficult things for people to do is see things from other people's perspectives. Indeed, many moral precepts are based precisely on this difficult requirement (such as "Do unto others as you would have them do unto you."). While the math of translation and rotation does not handle the moral dimension, it does take care of some of the purely spatial differences which can affect the moral (for example, the moral difference between a war going on halfway around the world and one going on in your backyard). It's startling how easy it is to see the differences in what matters in people's lives just by moving your around perspective. Physically, one would have to do this by changing position. In a virtual environment or a virtual model of a real environment, it is the math outlined above that takes care of changing views between frames of reference – two calculations to see the world in someone else's view.

But there's more than one way to change perspective. There are differences not just in how things are perceived but in how they are created and analyzed. These differences can

**FIGURE 5.48**
Aphrodite of Melos

make simple things complicated and complicated simple. To a great extent this is what analytic geometry is all about, creating different perspectives on functions and shapes in order to make them easier to work with and understand.

Different frames of reference are not the only way that the same points can be given with different numbers (and different importance). Let's go back to the way we came to the x, y coordinates.

We drew two axes and made a rectangle for each point. The lengths of the sides of that rectangle gave us the coordinates. These are called **Cartesian coordinates** because of Descartes.

In determining distance in the plane, we made a right triangle from the origin, the x-axis and the point (x, y). Let's look closely at the hypotenuse of that triangle (Figure 5.49).

Take a close look at that line segment. It has a distance from the origin of r, and it makes an angle θ with the x-axis. Is there any other point in the plane that has that same distance from the origin and makes that same angle?

No. For proof, consider the circle or radius r. Our point lies on that circle, as does every other point with distance r from the origin. Every point that lies at an angle θ to the x-axis lies on the ray going out from the origin at that angle (Figure 5.50).

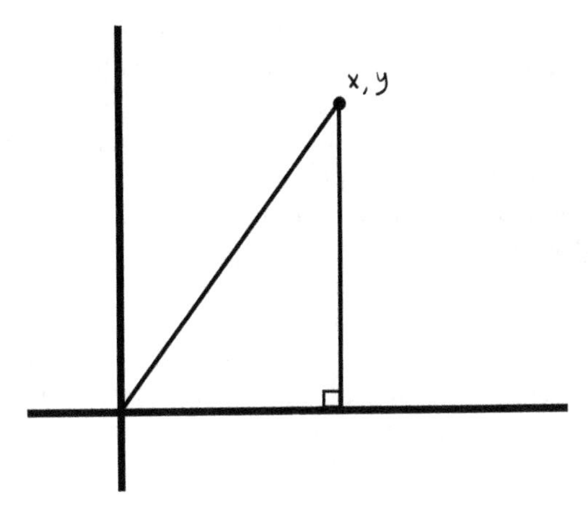


**FIGURE 5.49**
Right Triangle

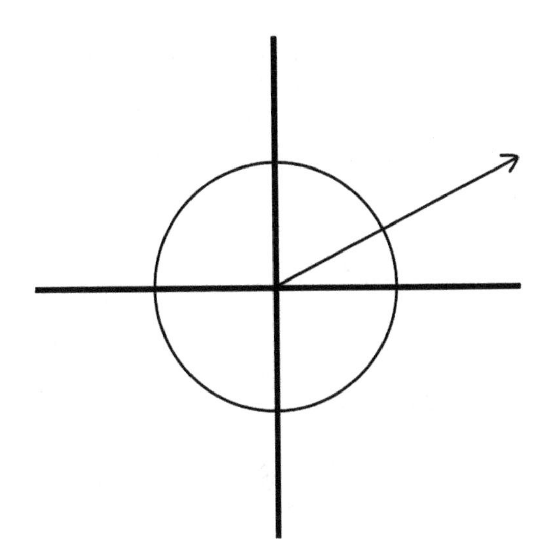

**FIGURE 5.50**
Circle and Ray

The ray and the circle intersect only at that one point.

This means that every point is uniquely determined by giving r and θ, unless r = 0, in which case the angle can be anything. We determined that we could use x and y coordinates for points precisely because a given (x, y) pair corresponded to a unique point. But any other way of generating a unique correspondence between numbers and points could also be used.

*Hold up. It wasn't just the correspondence, it was the geometric character of that correspondence.*

**Yes, and here with r and θ, we have another pair of correspondents with its own geometric character. Distance and angle are fundamental geometric concepts and here we are using them.**

Rather than expressing a point by x, y values, we can use r and θ. A point given by its distance from the origin and angle from the x-axis is said to be in **polar coordinates**. This is usually expressed as [r, θ].

Just as we can create graphs in rectangular coordinates by using functions that generate x and y values, we can do the same in polar coordinates with functions that generate r and θ. Figure 5.51 shows some examples of this.

Notice that certain shapes that take a lot of effort in rectangular coordinates, such as circles and spirals, are easy to generate in polar coordinates.

Furthermore, notice the ease of rotation in polar coordinates. To rotate a point [r, θ] by an angle $\theta_1$, just add $\theta_1$ to θ, giving the point $[r, \theta + \theta_1]$.

Translation, however, is not easy in polar coordinates.

It is possible to calculate polar coordinates from rectangular coordinates and vice versa, but we're not going to worry about that. Polar coordinates are often used in navigational plotting because rather than the rectangular process of stating the point one is starting at and the point one is going to, polar navigation simply states direction and distance and heads off that way, which is much more like real navigation.

Just as we expanded (x, y) in the plane to (x, y, z) in space, so we can expand polar coordinates into space. But rather than one coordinate system, we get two: **cylindrical coordinates** and **spherical coordinates**.

Cylindrical coordinates take polar coordinates and then use a rectangular z for how far up or down from the plane the point lies. This (r, θ, z) positioning has definite uses when navigating and flying planes, since one still navigates as if in a ship by using planar direction and distance, but one also deals with altitude (the z value) (Figure 5.52).

Shapes can also be generated using functions for these three coordinates (Figure 5.53).

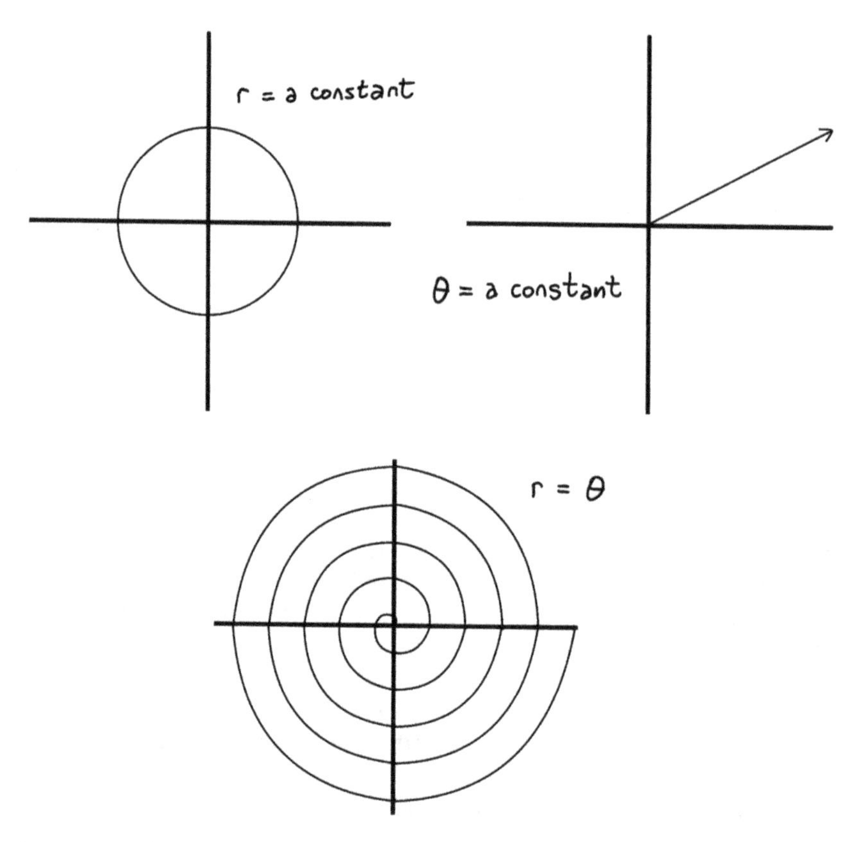


**FIGURE 5.51**
Circle, Ray, and Spiral

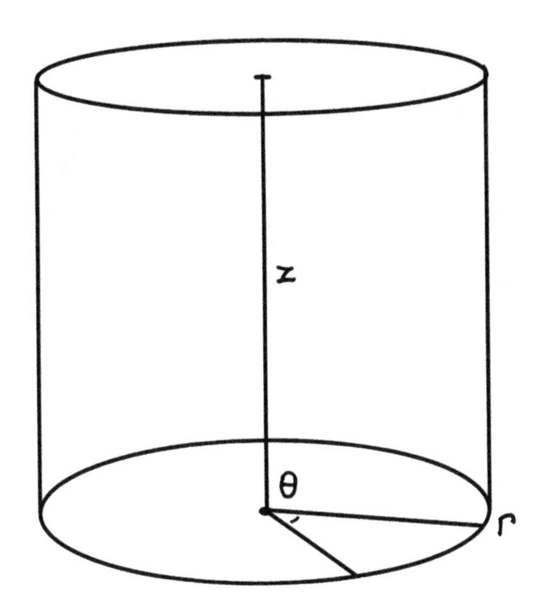


**FIGURE 5.52**
Cylindrical Coordinates

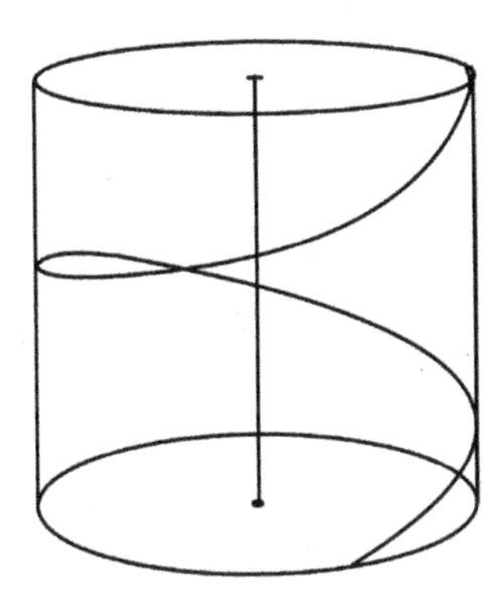

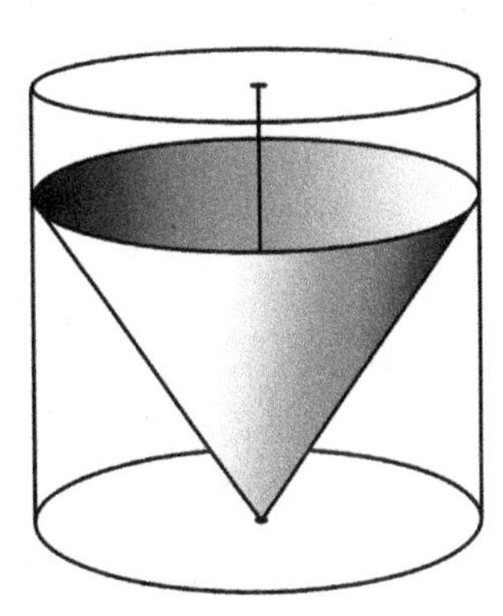

**FIGURE 5.53**
Shapes in Cylindrical Coordinates

Notice that the r in the cylindrical coordinates is the planar distance from the origin, not the distance in space.

Spherical coordinates, by contrast, use the distance in space from the origin (also called r, to confuse things) and two angles – θ, which is the angle from the z-axis, and angle φ (phi), which is the angle from the x-axis – when the point is pushed into the x-y plane. Spherical coordinates are useful in marking places in the sky by noting the angles of their position on the celestial sphere and the distance from the Earth (Figure 5.54).

And, of course, we can make shapes using these coordinates as well (Figure 5.55).

There are other kinds of coordinate systems that can be created, some of which we'll look at later. The important thing to note about all of them is that each system makes certain actions and shapes easier to depict and analyze, and others harder. The choice of the coordinate system is usually a question of convenience: which way of looking will bring the thing under observation into clearest focus or make the calculations simplest.

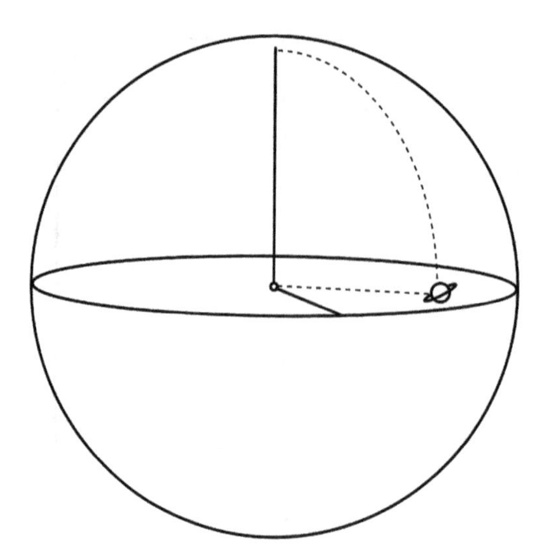

**FIGURE 5.54**
Spherical Coordinates

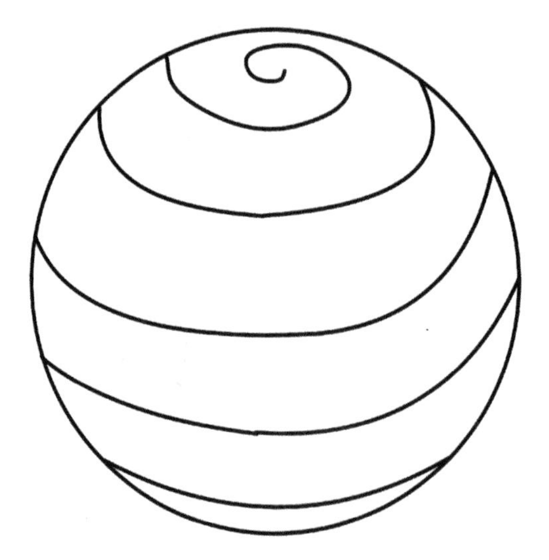

**FIGURE 5.55**
Spherical Shape

## More Dimensions than You Can Shake a Stick at

There's a word we've used very sparingly in the last chapter because we wanted to save it up for the end. That word is *dimension*. It's a word that has gotten a lot of sloppy use in various venues including ...

*Come on, fess up.*

... **including (mumble mumble science fiction writing mumble mumble).**

The word is easily misused. Until fairly recently it was a simple concept (for the recent complexity, see the chapter on fractals). All a **dimension** is is an independent quantity that you are using as an axis in a graph.

*Oh, is that all?*

**Look. Here's a graph of jelly donut consumption versus time with no facts in it whatsoever.**

*Fact-free donuts?*

**Quiet** (Figure 5.56).

In the graph in Figure 5.56, there are two dimensions: Jelly donut consumption and time. It is graphed in a two-dimensional plane.

If we had jelly donut consumption, time, and temperature, we would have a graph with three dimensions, which is graphed in three-dimensional space, and no, we're not going to waste time making up another fake graph and giving it three-dimensional shading and all that stuff.

While it's true that dimensions in a graph can be any stick-like quantities, when we graph in planes and spaces meant to represent space itself, the dimensions are positional, x, y, and z in Cartesian coordinates representing three dimensions of location (sometimes called **spatial** dimensions). Because we use Cartesian space as an image of the real world, we have a sense that these three dimensions are more real than the others mentioned above, which are mere conveniences used for graph generation.

To add to the confusion of the word, every shape drawn in analytic geometry has a dimension equal to the number of different parametric variables it takes to create a graph for that shape. That's not a precise definition, but it is a practical way of looking at it (at least until we get to fractals). A circle is graphed with one parameter, so it is one-dimensional; a sphere is graphed with two parameters, so it is two-dimensional. All the things we've been calling curves are one-dimensional and the surfaces are two-dimensional.

Here's where the language gets messy. A circle is a one-dimensional object that exists in a two-dimensional space (that is, a plane). A helix is a one-dimensional object that exists in a three-dimensional space (that is, space). Both of them are curves. Both have an **inherent dimension** of 1. But the lowest dimensional space that can hold them is not the same. The circle can be "embedded" in a two-dimensional space, but the helix needs a minimum of three dimensions (Figure 5.57).

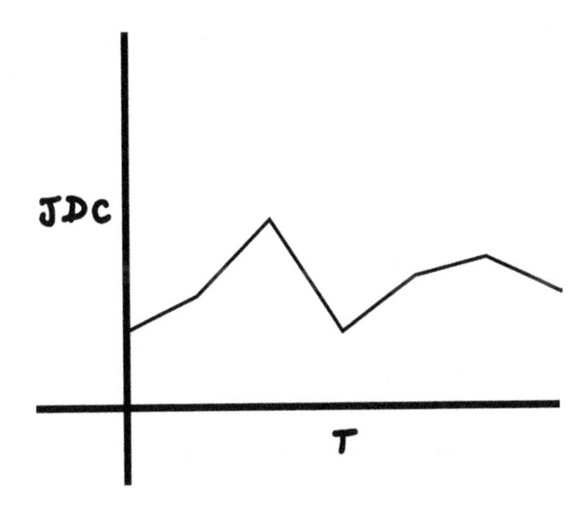


**FIGURE 5.56**
Bogus Jelly Donut Graph

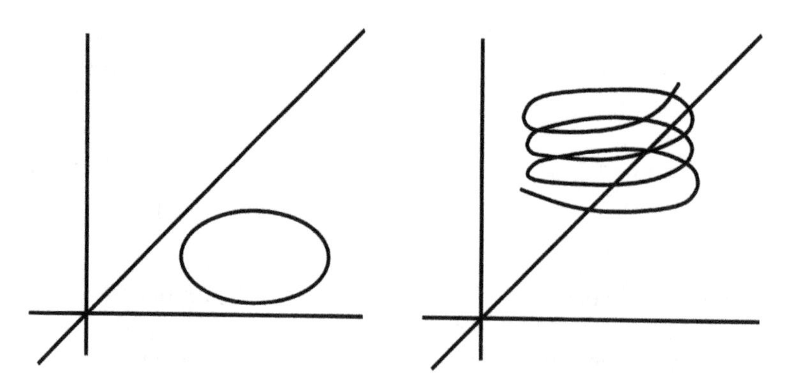

**FIGURE 5.57**
Circle and Helix

For those of you with geometric minds, hold on to your analysis. It's about to get brain melting.

Before analytic geometry, it was obvious that there were only three dimensions, because any point anyone could point to had to be in space.

But here's what analytic geometry says a point is:

In one dimension, a point is a number, x.

In two dimensions, a point is an ordered pair of numbers, (x, y).

In three dimensions, a point is an ordered triple of numbers, (x, y, z).

There's no reason we can't jump the dimensional shark and say:

In four dimensions, a point is an ordered quadruple of numbers, (x, y, z, a).

In five dimensions, a point is an ordered quintuple of numbers, (x, y, z, a, b).

In n dimensions, a point is an ordered n-tuple of numbers, $(x_1, x_2, \ldots, x_n)$

Once we say that dimensions don't have to correspond to our spatial reality, we can have as many of them as we want. Furthermore, we can have shapes in these n-dimensional spaces. That is a weird concept. We're talking about shapes that exist in nonexistent spaces that cannot be perceived.

We're about to do one of those bits of mathematical twisting around that happens often. Let's go through it slowly. The process works like this:

1. Take a concept and define it mathematically.
2. Abstract the characteristics of the resulting mathematical object that delineate it, so that everything of this kind of object has these characteristics.
3. Generalize these characteristics to create a class of mathematical objects that resemble the initial concept in such a way as to make the initial concept an instance of the more general.
4. Now treat the more general object as the basic kind of object you wish to work with.

We saw this in algebra, where we abstracted some of the properties of addition and multiplication and from them created the idea of an abelian group.

In the present circumstance, we took the idea of shape, and using analytic geometry created the idea of a shape as a parametric graph. But a parametric graph, analytically, is a list of functions, each assigned to produce the values for a coordinate. If we then shift the meaning of shape to such a parametric graph, we can generalize shape to higher dimensions using this analytic definition of shape. We can make a curve in any number of dimensions just by specifying a list of functions of one parameter, one function for each dimension. We can make a surface in any number of dimensions by specifying a list of functions of two parameters – one function for each dimension. We can make a hypersurface in –

*OK, we're going to lose them with that word "hypersurface." They're going to think we've gone all science fiction with hyperspace stuff with all the stars going whoosh?*
**Well, it is sort of like that. Let's start with a four-dimensional space.**
*But not one that you can get to.*
**Right. A four-dimensional space is a mathematical construct.**
In a four-dimensional space, a **hypersurface** is a shape specified by a function of three parameters.

*Got an example?*
**Sure.**
A hypersphere is the set of all points in four-dimensional space a given distance r from a given point called the center (x, y, z, a) $x^2 + y^2 + z^2 + a^2 = r^2$. We can make this parametric with angles the same way we did with circles and spheres.

*You're just sticking hyper- in front of everything to make it sound cool.*

**Sort of.**
**Hyper-** is a Greek prefix meaning "over." Its Latin equivalent is **super-**.

*Supersphere?*
**Yes.**

But instead of fighting for Truth, Justice, and the American Way (three dimensions), Supersphere fights for Truth, Justice, the American Way, and the use of subtler coloration in comic books, because that four-color process is just harsh on four-dimensional figures.

Along with Super-, I mean the hypersphere, we have a hypercube, which is made up of three-dimensional faces, each of which is a cube the same way a cube is made up of two-dimensional faces, each of which is a square and a square is made up of one-dimensional edges, each of which is a line segment (Figure 5.58).

A shorter term for hyperspace is four-space. This is also an easier term to go up the dimensions in. We can have 5-space rather than hyper-hyperspace and 100-space rather than hyper-hyper-(95 more hypers)-space.

Another term used for these spaces is $R^n$, where n is the number of dimensions. So the line is $R^1$, the plane $R^2$, etc.

*That R looks an awful lot like the symbol for the real numbers.*

**So it is.** $R^n$ is not a geometric idea, it's from set theory. It means the set of ordered n tuples of real numbers (**n-tuple** just means a sequence of n somethings, so a three tuple is a triplet). This is an attempt to take the idea of dimension and push it into a more general set theoretic idea with no geometry in it whatsoever. The idea is called the **Cartesian product**. If you have two sets, A and B, the Cartesian product $A \times B$ is the set of all

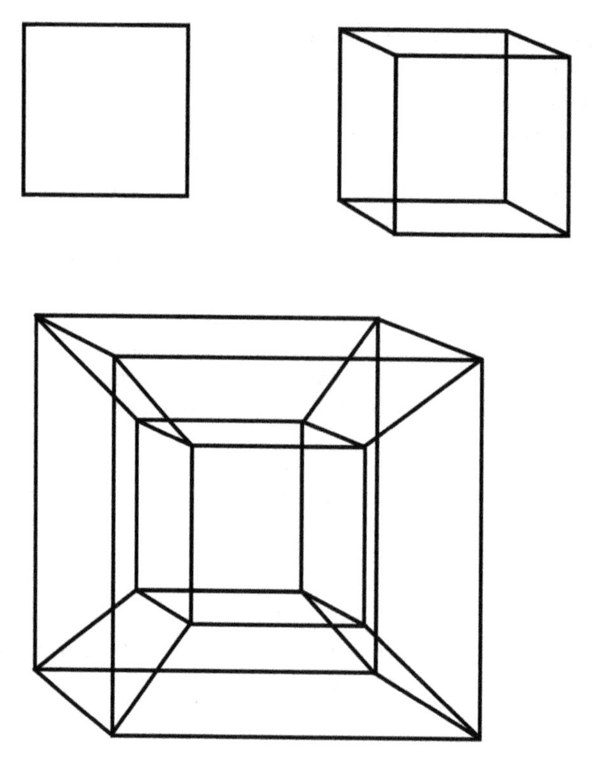

**FIGURE 5.58**
Square, Cube, and Hypercube

ordered pairs (a, b), where a is a member of A and b is a member of B. For example, the Cartesian product of {1, 2} and {a, b, c} is {(1, a), (2, a), (1, b), (2, b), (1, c), (2, c)}. You may notice that the number of elements in A × B is the number of elements in A times the number of elements in B. $R^n$ is the Cartesian product of R with itself n times.

Notice that this process abstracts the idea of dimension completely away from the concept of a space that holds shapes and makes it only a set with coordinates taken from various other sets. But $R^n$ obtains a geometric structure from the fact that distance between two points is given by the n-dimensional version of the Pythagorean theorem.

*I suppose you're now going to claim that "alternate dimensions" comes from* $R^n$.
**No. That's not a math term.**
*Aha. This is another of those –*

**Yes, the phrase "alternate dimensions" is the fault of science fiction writers.**

Imagine a cube in space, and imagine it sliced up infinitely thinly in two-dimensional layers (Figure 5.59).

Now imagine that one of those planes is really our three-dimensional universe and that all the other planes are universes like ours. Then we would say that each one is an "alternate universe," which is a perfectly acceptable term; so is "alternate plane of existence." But someone decided, for whatever reason or confusion, that "alternate dimension" sounded cool, and the term stuck even though it has confused things even more.

So what's the point of all this? Is this multidimensional stuff anything more than a way to abstract spaces beyond the space we can see?

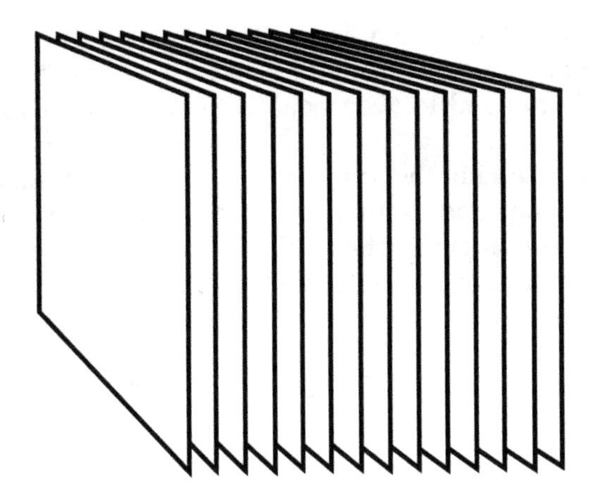

**FIGURE 5.59**
Sliced Cube

In part it is. The idea of space has become very abstract indeed in modern mathematics, but really what's going on here is that sometimes you need more than three dimensions to express a relationship between quantities. Sometimes the functions and shapes you need to graph simply have more values than three-dimensional space can contain. So we simply add more dimensions as we need them. We're not obligated to confine our mental objects within our physical limitations.

The multidimensional approach has liberated analytic geometry and made it possible to prove theorems about shapes that are beyond the ability of the mind to visualize. Paradoxically, this has removed the geometric advantage of the whole field. It was the ability to see function as shape that made analytic geometry so valuable, to allow the geometric and analytic mindsets to work together. But even skilled geometric minds have difficulty in imagining shapes in four or five dimensions, and even more difficulty bringing geometric awareness to bear on a theorem that works in n dimensions.

The further paradox of all this is that it was this very liberation that led to the branch of mathematics that permits modern physicists to try to discern the properties of the space we live in. We do not inhabit a neat, orderly, rectangular Euclidean three-space with (x, y, z) coordinates all laid out in a row. Our universe, as Einstein discerned, ineluctably combines time with space into a four-dimensional spacetime, and not a Euclidean one either, but one whose curvature is a manifestation of gravity.

Geometric thinking launched the mindset that leads to relativity, and it still helps in imagining what the universe is like around us, but it can no longer hold that universe in its thinking and see it clear and entire as it once could.

The regrettable part of this is that the classical modes of geometric thinking have, for all these reasons, been deemed superfluous by many modern mathematicians and physicists. But looking back over this chapter and ahead to the next one, one can see that even the most abstracted concepts of shape and dimension that lie beyond the ability of the mind's eye to conceive still rest on a foundation of geometric imagination.

Those with geometric mindset have the easiest time beginning the practice of analytic geometry. But they can get lost in the transitions that abstract away from the visual character of shape to its analytic character. It seems to us that the hurdles necessary for them to shift from *geometry* to *analytic* geometry are not insurmountable. One helpful approach is to treat the functions of analytic geometry as if they were instructions in how to draw the shapes. The sense of a shape being drawn in a space by the behavior of functions of parameters can be extended to higher and more abstract dimensions in such a way as to allow the geometrically minded to keep hold of their thinking while multidimensions turn around them.

Analytic geometry is arguably the foundation of all modern mathematical physics. It brings together all the mathematical mindsets and grows shapes from the sticks we first measured with. If there is a single most useful, most accessible treasure of the lost inheritance, it lies here. The great thing about it is that anyone can get to it. There is road inward from any direction of mathematical thinking and roads out that lead to every aspect of modern math.

Unfortunately, one of those roads has an unnecessary "Beware" sign over it, as we will see in the next chapter.

# 6

# *Calculus: Motion and Size*

**The dark-cloaked figure sidles across the hall. His steps echo doom for all that look upon him. He comes before the terrified student body, flinging back his cape, clasped with a d/d and embroidered with long, snakelike s's. His voice rises to a shriek as of failing grades and he cries a single word over and over into the night.**

**"Calculus! Calculus! Calculus!"**

**All but a few flee, and only they learn what the creature is or why it is cloaked in confusion.**

Calculus is the last major falling-off point that separates those who will be mathematically educated from those who won't, the final sink or swim that will separate the people who say "I never really got math" from those who treat math as a part of their mental toolkit.

But what is it? What is this bugaboo that culls people away from math and physics at the moment of entering college? What is the calculus?

*For that matter, our readers might wonder which calculus we're talking about. We've already shown them propositional and predicate calculuses. Calculi? Whatever the plural is. Which calculus is this?*

**Fair enough, let's tell them.**

A calculus is simply a method of reasoning. The calculuses (that's the plural) we are talking about are the two branches of **infinitesimal calculus**. Infinitesimal simply means infinitely small. Why this is the name will become clearer as we go along.

Infinitesimal calculus is a strange plant. It grows in the wilds of analytic thinking and like kudzu tends to take over the land around it. It has two branches, **differential calculus** and **integral calculus**, seemingly unrelated, grown from different roots, but intertwined at every curl and turning. It declares itself, correctly, to be the right tool to solve a startling variety of problems, from rocketry to wine barrels to the shape of can that holds the most volume for the least amount of metal. (*So calculus is used by cheapskates.*)

So useful is it that physicists reach for it automatically. It colors the modern mathematical view of the world. Like Euclidean geometry before it, it makes us see certain shapes as normal and others as strange. And for physics it is bar none the most important treasure in the inheritance of mathematics. Understand calculus and you can understand most of physics.

For the rest of this chapter and most of the book the word calculus now means differential and integral calculus.

*So we're going to only use the word to mean something we haven't defined yet.*

**We're getting to it.**

As processes (since calculus is a branch of analysis, it naturally concerns processes), each branch of calculus was created using a trick that turns a single unsolvable problem into an infinite group of solvable ones and then finds a way to cheat to make a new single solvable problem from that infinite list of solvable problems. Solving the single solvable problem thus created solves the single unsolvable problem that one started with. Thus is the unsolvable dissolved into solution.

*OK, that sounded a bit overly dramatic and more than a bit unclear. Why not tell them that we're going to find an approximate solution to the problem, and a way of making the approximation better and better, until finally by looking at all these approximations we find the solution?*

**Boy you physicists sure know how to take all the fun out of things.**

Anyway, we've seen the method of turning a problem we can't solve into one we can by transforming the problem or our perspective on the problem. We started it with sticks. By making a stick into a unit, we turned measurement into counting so that our stone counting could now be applied beyond its origins. In the process we opened up the vistas of the real numbers, of the creation of units and dimensional analysis and so on. The trick of reducing something to a previously solved problem can also make available whole new areas of the theoretical universe.

In calculus we pull a variation on this stunt. Rather than create a single solvable problem, we create an infinite number of solvable problems, none of which are quite the problem we need to solve, but then we find a way to look at and calculate solutions using that infinite number of problems.

One of the major steps we're going to do is to find a different way of looking at the infinite number of problems that contains more information than the infinite number of problems do.

We've seen this before, sort of. Remember that we turned the infinite number of statements

$0 < 1$

$1 < 2$

$2 < 3$

...

into the single statement

that for any number a, $a < a + 1$.

This statement contains more information than just the infinite number of statements before because it can be easily substituted into proofs. Whenever we have a situation where in a proof we have some number a, we can assert $a < a + 1$, just because we have a number and because we know the universal truth embodied in the single statement.

In calculus, we will do a similar transformation from the infinite number of problems into a new form that contains more usable information than the infinite set of problems.

But first let's find the unsolvable problems that lead to the infinite sets.

*Let's start with physics!*

**OK, what part of physics?**

*The most basic part: $F = ma$.*

Newton's second law of motion $F = ma$ is about how forces affect the motion of objects. Here F is the force, m is the mass, and a is the acceleration. Here acceleration is how fast velocity is changing and velocity is how fast the object is moving.

So we start with a basic question of physics: how fast is that thing going?

We have a stick for this. Speed = distance/time. We start our stopwatch at the beginning of the race and stop it when the runner we are timing crosses the finish line. The distance is the length of the racecourse, and the time is the time elapsed on the stopwatch. The runner's speed is that distance divided by that time. But this is only an average speed for the whole race. The runner could have kept up that same steady pace for the whole race, or they could have started out fast and then tired and slowed down. Or, they could have put on an extra burst of speed right at the end to win the race. None of this nuance is captured in the average speed measured by the average speed measurement. It also won't help in calculating the acceleration, since that is all about how rapidly the speed changes. We need to be able to measure how fast the runner is going at every moment of time. How do we do that?

## The Derivative

Two people came up with the answer, which is differential calculus: Newton and Leibniz. Isaac Newton, the greatest physicist of all time and one of the greatest mathematicians, needs no introduction. Gottfried Wilhelm von Leibniz was a 17th- and 18th-century German mathematician and philosopher, as prominent for his contributions to the philosophical school of rationalism as he is for his contributions to mathematics. There has been an intermittent argument as to whether or not each of them came up with this branch of calculus independently or whether someone was swiping stuff from someone else. The general consensus these days is to give them both credit and not get bogged down in an intellectual property dispute that old.

*Boy, wouldn't that have been a fun patent infringement case? Newton v. Leibniz. Everyone would be wearing wigs and no one would understand any of the arguments and – You know, it's not that different from today. (In the UK, they even have the wigs.)*

The beginning of the way of thinking that leads to differential calculus is fairly simple provided one has taken in and is comfortable with the math that comes before it. While many of the difficulties people have learning calculus involve the concepts and their presentation, there is another problem that arises because to do any calculus requires the ability to use high school algebra with ease and facility. It also helps to be solidly grounded in analytic geometry. The difficulty is that, as we'll show, the ideas of calculus can be acquired without these skills, but the use of calculus demands an ability to employ the tools that were taught before.

This is true of all math learning. Each step of math relies on earlier understanding. We've seen this already in the abstraction of sticks and stones into numbers. A person needs to be at ease making numbers out of things before they can handle the operations of addition, subtraction, multiplication, and division of whole numbers. Facility with these is needful in order to learn to handle fractions and decimals. These in turn are needful in doing algebra since one must be able to multiply 32.5 by 16.7 before one can do $(32.5a + 16.7b)^2$.

This facility in turn must be developed before going on to calculus. The problem is that a student who can handle the concepts but cannot handle the problems cannot actually learn to use the mathematical tools. One must have both in a math class. We'll talk more about this much later. For now, back to the concepts.

We can't directly figure out how fast something is moving, but we can make approximations. That is, we can come up with numbers that will be reasonably close to what we want. Let's suppose we have a curve in the x-t plane representing the distance an object has traveled at each point in time, done as $x = f(t)$, and we want to know how fast it's moving at a certain point (t1, f(t1)). We can make an approximation by taking a time near to t1 – say, t2 – and looking at whatever point on the curve has this t-value. That would be (t2, f(t2)). Draw the line between these two points (this is called a **secant line**) and figure out the slope of this line. This slope will be an approximation to the speed (Figure 6.1).

The slope is $(f(t2) - f(t1))/(t2 - t1)$

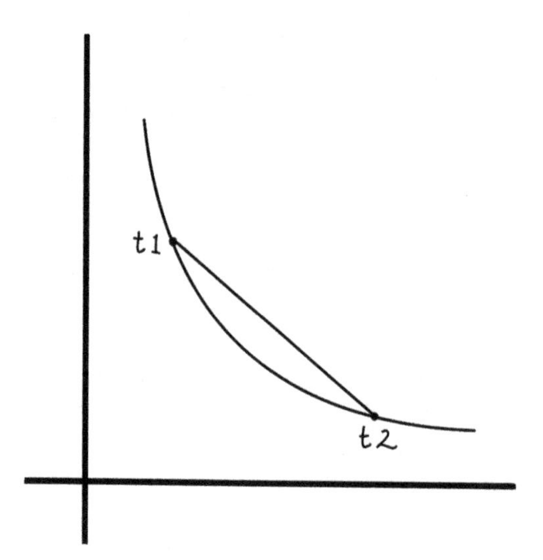


**FIGURE 6.1**
Secant Line

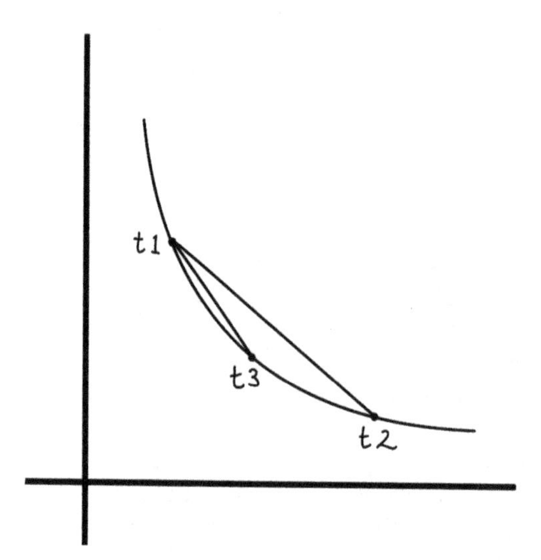


**FIGURE 6.2**
Secant Lines

This process is used all the time in the real world when measuring the speed at which people run. Two measurements of distance at two different times are taken and then the ratio of differences is calculated. We then say that the person was traveling at that speed, even though there would be variation in the speeds during that time. (People can speed up or slow down a little even in small stretches of travel.)

Now suppose we pick another moment in time closer to t1 than t2. Call this t3. We can draw another secant line from (t1, f(t1)) to (t3, f(t3)). This line's slope, (f(t3) – f(t1))/(t3 – t1), should be even closer to the speed we are after than our first approximation. We can continue this process with points that are closer and closer and get a sequence of secant lines and a sequence of slopes (Figure 6.2).

It appears that as we get closer and closer to our initial time value t1, our secant lines get closer and closer to a line that goes through our initial point (t1, f(t1)), the slope of which should be our desired speed. This line we long to reach but cannot is called the **tangent line** at the point (t1, f(t1)).

This is the part of the process where we have taken an insoluble problem (finding the slope of the tangent) and replaced it with an infinite number of solvable problems (finding the slopes of all the secants).

This hardly sounds practical unless we can find a way to turn that infinite number of solvable problems into one solvable problem.

We can do that, but first we need to examine a mathematical idea that causes people to drop calculus courses and swear off math forever. (Calculus has a number of these scary moments. This is just the first.) The concept is that of **limits**.

*Should we have a dramatic crash of thunder and a nearby bolt of lightning? Oh, wait. I'd better check the surge protectors first.*

---

## Limits

In everyday usage, the word "limit" means a boundary or an edge of something. Math plays on this idea to mean something that you can't reach directly but can sneak up on by taking an infinite number of steps. Mathematicians study limits in order to be able to skip those steps, rather like a reader flipping to the last page in a murder mystery.

*I don't care about the plot twists or the characters, I just want to know who done it.*

**From a writing perspective I abhor this idea. In math I applaud it.**

*Passing swiftly over your hypocrisy ...*

**Thank you.**

Let's take a closer look at what a limit means. We'll start with the simplest kind of limit, called the **limit of an infinite sequence.**

An infinite sequence is a countably infinite list of numbers, often written ($a_1$, $a_2$, $a_3$, ...)

Suppose you have an infinite list of numbers such as

1, ½, 1/3, ¼, 1/5, 1/6, and so on. In this sequence $a_n$ (the nth number in the sequence) is equal to 1/n. As we go along this sequence, the numbers get closer and closer to 0 but don't reach it. 0 is not a number in the sequence, yet we have a sense that if there were a last element of the sequence, an $\aleph_0$-th element as it were, it would be 0. Mathematical limits take this intuitive sense of where this sequence is going and say that the sequence **converges** to 0.

Not every sequence converges to a number. Most of them don't. A sequence is said to diverge if it does not converge. Here's a simple example:

1, 2, 3, 4, 5, etc.

This sequence **diverges** to infinity. There's more than one way to diverge.

1, −1, 1, −1, 1, −1, etc.

diverges because if it had a last value, that value could be 1 or −1, but there's no real way to pick one over the other.

Intuitively, a sequence converges if in the long run it gets arbitrarily close to a single number. We'll give an exact definition of this in a moment (*Get your running shoes on*).

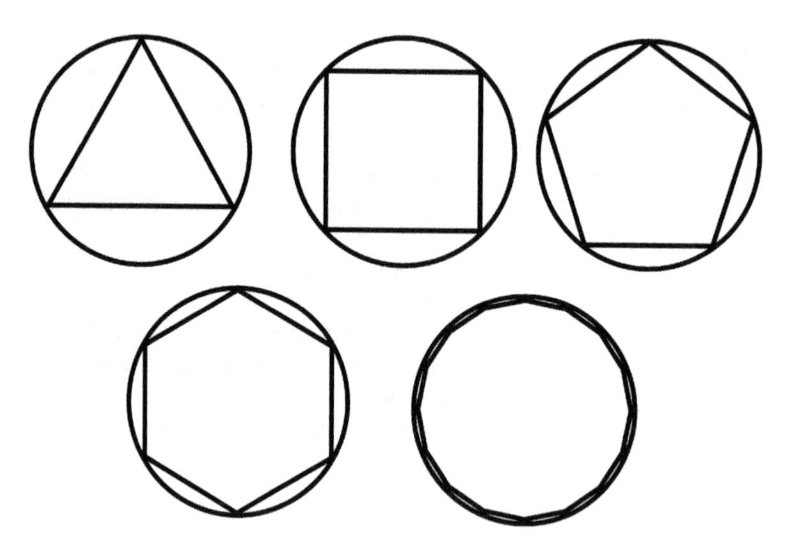

**FIGURE 6.3**
Polygons Inscribed in Circle

First we want to make it clear that mathematicians were using limits long before they had a precise definition of the process. One of the oldest examples of limit use is the problem of finding the circumference of a circle (that is, the distance one travels if one walks once around a circle). The Greeks reasoned thus as follows: take a circle. Inscribe an equilateral triangle (that is, a triangle whose sides are the same length) so that its vertices (*the pointy bits*) touch the circle. Figure out the perimeter of the triangle (that is, the distance traveled if one walked around the triangle). Now do the same with a square, a pentagon, a hexagon, etc (Figure 6.3).

Each polygon has one more side than the previous one. We can see that as the number of sides increases, the polygons become closer and closer approximations of the circle, so the perimeters of the polygons become better and better approximations to the circumference of the circle. Without having a definition of limit, the people who carried out this process eventually came to a formula for the circumference: $2\pi r$, where r is the radius of the circle. They also used the same process to determine that the area of the circle is $\pi r^2$. They had only an approximation for the value of $\pi$. Indeed, this method was the first way that such an approximation was determined.

The point of all this is that limits were being used intuitively for centuries before being put on a solid, if bewildering, footing.

Let's pause for a second and consider what's about to happen and why it's going to happen. Convergence as defined above is not a difficult idea. The concept of getting really, really close to something and not touching it is something fairly easily graspable (*ask any pool player who's seen a ball get just to the edge of the pocket and not fall in).* But in order to make the concept rigorous enough for mathematical mistrust, the idea is going to have to be put in pretty weird language that does not sound at all like the sense we have of convergence.

Here's the first of the limit definitions:

> A sequence $a_1, a_2, a_3, \ldots$ converges to a value v if for every real number $\varepsilon$ ("epsilon") > 0, there exists a positive integer n such that for every $m > n$ $|a_m - v| < \varepsilon$. This is often written

$$\lim(\text{n goes to infinity})a_n = v$$

We'll dissect this definition in a moment to see why it fits the intuitive idea. Before we do, we'd like to express some frustration that this is used not as a piece of mathematics but as a sink-or-swim test in teaching math.

It's not uncommon for calculus texts and classes to use this definition for that purpose. They present the definition of limit without teaching the underlying idea, nor explaining why this definition is a mistrusting translation of that idea. Those students who can take this definition and from it gather its meaning and usage are seen as fit to go on to the intricacies of calculus. Those who don't tend to drop the class.

Yet if approached from the right direction, this definition, while it may be persnickety, is understandable.

Let's go over again what we mean intuitively by a sequence converging to a particular value.

We mean that if we go far enough along on the sequence, we get arbitrarily close to the value. But what do we mean by arbitrarily close?

It essentially means this: give me a positive number. If we go far enough along on the sequence, then every number in the sequence will be closer than that positive number to the value we say it converges to. In other words, I don't care what measuring stick you give me, the sequence will eventually be closer to the value than that stick can measure.

Translated, this says: pick a positive number. Call it $\varepsilon$ (epsilon). If you go far enough along the sequence (that is, past some particular entry, say the nth entry), every number in the sequence beyond that point (say, the mth entry) is closer to the value v than the positive number $\varepsilon$ you picked to check with. This means that $|a_m - v|$ has to be less than $\varepsilon$. If we go through one more layer of persnicketiness, we'll get back to the definition above.

What happens in the calculus class after this definition is introduced? Students are given piles of problems that involve proving which sequences converge and what values they converge to.

This sounds like reasonable practice before moving on to figuring out derivatives. But in many such classes, the derivative and the motion problems that motivated the teaching of limits have not at this point been mentioned. Limits and sequences and convergence are dropped in students' laps without cause or explanation. Those who catch on keep going. The rest swap down to an easier math class or drop out entirely.

We've talked a lot about math being a lost inheritance. In most cases, mathematicians are not culpable in that loss. They don't like the idea that they can't explain what they're doing to most people and they hate math phobia. But in this particular case, they are at least partly to blame.

Why do they teach this way? One way to understand their motivation is to consider the following analogy between calculus class and college classes in foreign languages: the main use of learning a foreign language is to be able to communicate with people who don't speak your language. However, there is a secondary use: to be able to read the literature written in that language and thus to acquire a deeper understanding of that literature through not losing those things that are inevitably lost in translation.

College language professors are scholars in the literature of the language they teach, so their inclination is to emphasize the literary aspects of their courses. Similarly, college mathematics professors are mathematics researchers. The main goal of mathematics research is to produce proofs of mathematical theorems, even though for all the rest of us the main use of mathematics is to do calculations. Thus, the inclination of the mathematicians is to emphasize those aspects of their courses that involve proofs and the techniques used to prove theorems. The rigorous definition of limit and the use of epsilon is the first such proof technique encountered in college courses, and is the foundation for the techniques of mathematical analysis. (We will return in subsequent chapters to the difference in point of view between mathematicians and the users of mathematics.)

The next limit definition is trickier but still can be handled if you start with intuition and move on to definition. This limit concerns the behavior of a function. Up until this point, we have looked at functions as simply processes to turn one number into another. We've graphed functions and done some studying of their behavior. But we haven't tried to say whether a function is in some sense "well-behaved." Take a look at the following function (Figure 6.4):

This function has what's called a **discontinuity** at 0. We look at it and somehow feel that it should have f(0) = 0, not f(0) = 38. Our geometric sense of aesthetics points us toward this idea, which is called **continuity**. A function is continuous at a given point if the values of the function near that point get arbitrarily near to the value of the function at that point.

In order to test if a function is continuous at a point, we need to know how it behaves near that point. This leads us to an ugly definition of limit of a function at a value:

The limit of a function f(x) as x tends to a is v if for every $\varepsilon > 0$, there exists a $\delta$ ("delta") > 0, such that if $|x - a| < \delta$, then $|f(x) - v| < \varepsilon$.

In other words, if as x gets close to a, f(x) gets close to v, then the limit of f(x) at a is v or

$$\lim[x \to a]f(x) = v$$

If the limit of the function f(x) at a is equal to f(a), then f is said to be continuous at a. So, if the value of the function is the value our eyes tell us it should be when we look at its

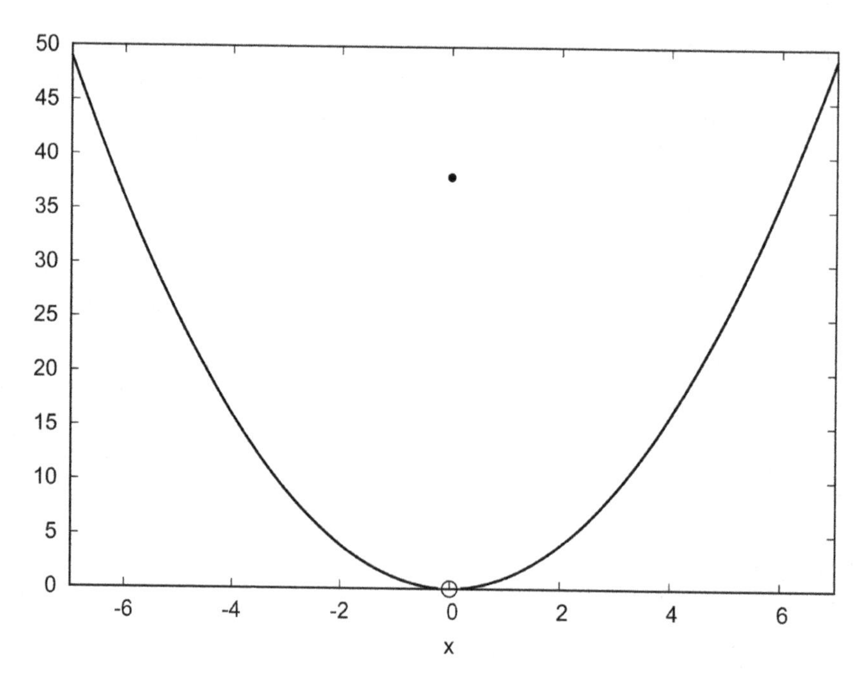


**FIGURE 6.4**
Discontinuity

graph, then the function is continuous. Continuity is a geometric idea, but it is expressed in analytic terms in the definition above. When students are given this definition, they are then given the job of doing what are called εδ proofs (epsilon-delta proofs), which consist of proving continuity or discontinuity of various functions at various values. This does the next batch of culling for those who came through the sequence proofs.

Only those students who pass through this go on to actually face the derivative.

Why is calculus done this way? Here we will make use of another analogy, this one to a part of the brilliant historical spoof *1066 and All That* where the authors describe the two sides of the English Civil War as "Right but Repulsive" and "Wrong but Wromantic." Newton and Leibniz did not use epsilons and deltas. Instead, they came up with the notion of infinitesimals, numbers that were larger than zero but smaller than any regular number. They made their tangent lines from secant lines by choosing t2 to be t1 plus an infinitesimal.

Newton and Leibniz knew that the idea of infinitesimals was highly questionable at best, but the calculations worked and end results were sensible, so they kept at it. The epsilons and deltas are a later invention of Bernhard Riemann, who proposed them to finally put the infinitesimal calculus of Newton and Leibniz on a firm mathematical footing. (Riemann, a 19th-century German mathematician, also pioneered the treatment of curved spaces, which later became the mathematical foundation for Einstein's theory of general relativity.) The infinitesimals are wrong, but intuitive. The epsilons and deltas are a right but highly nonintuitive way of doing the same thing.

## Derivative Redux (Reduced, That Is)

When last we left our intrepid concept for instantaneous speed, it had given us an infinite number of problems. Now we have a theoretical idea of how to turn that into a single problem. We make a limit of our secant lines which will hopefully converge to the tangent line. The slopes of these lines are the measured speeds which should converge to an instantaneous speed. We will do so in the following form: rather than take two times, t1 and t2, we will take a time t and a time a little later than it, t + h, and we will take the limit as h tends to 0 of the slopes of the secant lines (Figure 6.5).

We want to take the limit as h tends to 0 of the slope, which we calculate as (f(t + h) − f(t))/(t + h − t).

Since t + h − t = h

this becomes

limit as h tends to 0 of (f(t + h) − f(t))/h

This is called the **Newton quotient.**

**Leibniz: Curse him for all eternity, that credit-stealing British guy.**

*Will you quit putting words in the mouths of people who aren't here to defend themselves? What do you think this is, a historical novel?*

There is a single basic trick to doing most limit problems, which we will now use over and over again to prove things with Newton's quotient.

Suppose you're trying to find the limit of some continuous function g(x) as x goes to a. If you can simply calculate g(a), you're done. If not, you transform g(x) into some equivalent function that you can calculate at a and then calculate it.

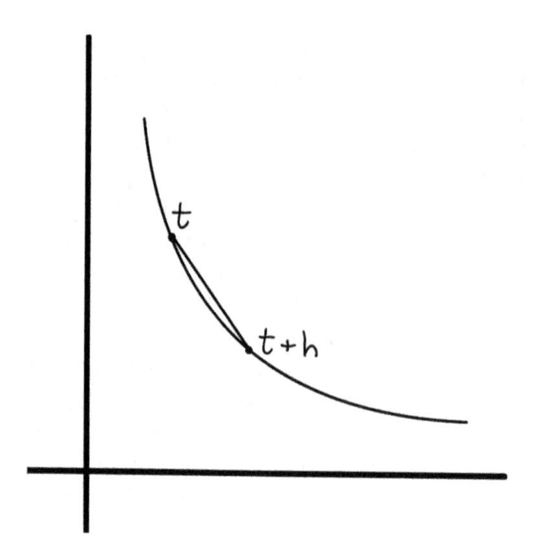


**FIGURE 6.5**
Secant Line at t and t + h

Here's a simple example.

Suppose g(x) = x/x and we want the lim as x goes to 0 of g(x). We can't just substitute 0 for x since that would be 0/0, which has no definite answer. But h(x) = 1 is an equivalent function to g(x) since x/x = 1 if (x is not 0). Since in our limit we never have x = 0, we can make this substitution, so we can substitute 1 for x/x.

This gives us the limit as x goes to 0 of 1, which equals 1.

Newton's quotient always starts with this same difficulty. If we just made h equal to 0, we would have (f(t) – f(t))/0 which would be 0/0.

So the trick is to mess around in legal algebraic fashions with (f(t + h) – f(t))/h until we can turn it into some form that is not 0/0.

Let's try this with something not quite as simple but still not so bad. Suppose f(t) = vt, where v is a constant. In this example, we have our old friend constant velocity, so if the derivative gives us speed at time t, we expect that we should get the answer v. Let's try it out:

(f(t + h) – f(t))/h

=

(v(t + h) – vt)/h. Note that v(t + h) is v times (t + h), not a function called v working on t + h as its parameter value. *Great. More annoying notation problems.*

=

(vt + vh – vt)/h

=

vh/h

=

v

Now substitute 0 for h and we still have v. So our idea for calculating speed works for things we already know the speed of. That's good. If it didn't, we'd have to start all over again.

This is one of those basic tests of mistrust that every theory must pass. Does it work for the things we already know? It's amazing how many theories outside the sciences don't even try to do this.

Let's try our cannonball shot formula. We're going to change the notation a bit. Rather than say $f(t) = at^2 + bt + c$, we're going to put in different symbols for the numbers a, b, and c. We'll replace a with a/2, b with $v_0$, and c with $s_0$. We're doing this because these names will be given meanings once we're done. (This is the standard sort of cheating where we already know how the trick turns out, and dress things up so they'll look right when it's done.) We'll leave out the = between lines. Each line below is equal to the one above it unless we say otherwise.

*Hear that, functions and equations? You do what we say.*

f(t): Only if you tell the truth.

*We are going to tell the truth, right?*

**Yes.**

$f(t) = at^2/2 + v_0t + s_0$. Now let's stick this into Newton's quotient.

$$\left(f(t+h)-f(t)\right)/h$$

$$\left(a(t+h)^2/2+v_0(t+h)+s_0-\left(at^2/2+v_0t+s_0\right)\right)/h$$

$$\left(a\left(t^2+2th+h^2\right)/2+v_0t+v_0h+s_0-at^2/2-v_0t-s_0\right)/h$$

Notice that we are adding $v_0t$ and $-v_0t$ as well as $s_0$ and $-s_0$. Those add up to 0, so we can remove them, giving us:

$$\left(a\left(t^2+2th+h^2\right)/2+v_0h-at^2/2\right)/h$$

We now apply the distributive law to multiply out $a(t^2 + 2th + h^2)/2$

$$\left(at^2/2+ath+ah^2/2+v_0h-at^2/2\right)/h$$

Now we have $at^2/2$ added to $-at^2/2$. These add to 0, giving us:

$$\left(ath+ah^2/2+v_0h\right)/h$$

Notice that now everything in the numerator has an h as a factor, which means we can actually carry out the division by h and get:

$$at+ah/2+v_0$$

Now we can take the limit by substituting 0 for h. We get

$$at+a0/2+v_0$$

but $a0/2 = 0$.

So we get the formula for the speed, which is

$$at+v_0$$

Let's examine this formula from the standpoint of dimensional analysis. It's supposed to yield speed, which has units of distance/time. Since $v_0$ is a constant, it must have the same units as our answer, so it must represent a speed of some kind. But what are the units of a? Since at is some kind of speed and t has units of time, a must have units of (distance/time)/time or distance/time$^2$. Therefore, a must represent an acceleration (that is, a change of speed over time ... *but we knew that*). Furthermore, a is a constant acceleration.

By the way, it was in order to get a in the formula above that we use the form $f(t) = at^2/2 + v_0t + s_0$. In order for the graph of cannonball shot height to have the shape it does, there must be a constant acceleration a, an initial upward speed $v_0$, and an initial height $s_0$. That's the meaning of the three constants.

Before Newton did this calculation, Galileo had, by experiment, determined that objects falling to Earth undergo a constant acceleration, which is now designated g: g = 9.8m/sec$^2$. g is a positive number even though gravity pulls downward. It's a peculiar convention of this formula. If you want to orient the direction of motion normally, put a minus sign in front of the g. If you look back in the previous chapters at earlier forms of the cannonball shot formula, we had a term, $-4.9t^2$. $-4.9 = -9.8/2$, which is where that term came from.

Before we go on, we need a little clarification and a little notation. First of all, the form of the Newton quotient given above, $(f(t + h) - f(t))/h$, is called **the derivative of f with respect to t.** f is the function we are trying to find the rate of change of, and t is the variable we are changing in order to find the rate. We do this by bumping t up by h before dropping h to nothing.

We are not confined to taking derivatives with respect to time. If we had a function f(x), we could take the **derivative of f with respect to x** at a value $x_0$, which would be the limit as h goes to 0 of $(f(x_0 + h) - f(x_0)/h$. This represents the rate of change of f as x changes, not the rate of change of f as time changes, unless x represents time. For example, if we had a totally bogus made-up graph of demand for water versus salted popcorn consumption, (Figure 6.6) we could figure out the rate at which demand for water goes up when people eat 6 kilograms of popcorn in a day. This would be the derivative of w (water demand) with respect to p (popcorn consumption).

Next, notice that while we calculate the derivative at a given value, we actually end up not just with specific values, but with a whole new function. If, as we had before, height = $f(t) = at^2/2 + v_0t + s_0$, we end up with a function for speed that is at $+v_0$. Notice that this function is also a function of t, the variable we took the derivative with respect to. We could give that function a new name like g(t), but it would be better if we had some way of writing it down that expresses the fact that g(t) is the derivative of f(t). It needs to be simpler than

$$g(t) = \text{lim as h goes to 0 of}\big(f(t+h) - f(t)\big)/h$$

There are three common forms of notation, and each of them has drawbacks. Newton's notation was to put a dot over the thing he was taking the derivative of, so it would be $\dot{f}(t)$. There are a batch of problems with this notation. First of all, Newton meant this notation to be derivative with respect to time. There is no way in the dot notation to make note of the variable involved. Also, dots over something get lost in the reading.

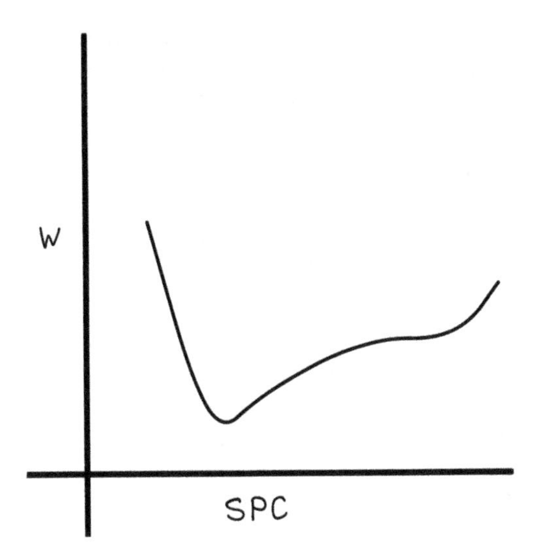


**FIGURE 6.6**
Bogus Water and Popcorn Graph

There is the more comprehensive df/dt, which is read as "dee eff dee tee" or "dee eff by dee tee," which gives us all the information we need but is kind of cumbersome. Also, it looks like instructions to multiply f and d and then divide by the product of d and t. It also looks like it should cancel out to f/t, which doesn't make any sense at all. There is a reason for this notation, and were this a calculus text, we'd have already explained it. But we're going to save it for later for a cheap parlor trick. We'll use this notation when necessary, but in general will use the third one, which is to put a ′ mark between the function and its argument like this:

f′(x), which is read "f prime of x"

So if $f(t) = at^2/2 + v_0t + s_0$

then

$f'(t) = at + v_0$

To add to the confusion of notation, if we have y = f(x), we might write dy/dx for df/dx, or y′ for f′(x).

Each of these notations has its uses. Sometimes you want a quick way of writing the fact that you are taking the derivative, in which case you might use the ′ notation. Sometimes you concern yourself with the variable you are taking the derivative with respect to, in which case the df/dx notation is useful.

In terms of units, the df/dx notation is revelatory since the units of the derivative of f with respect to x are the units of f divided by the units of x. We can see this in the Newton quotient. The numerator of Newton's quotient, f(x + h) − f(x), is a difference of things with the units of f. The denominator h has to be in the units of x since we are adding h to x in the numerator. So the units of the derivative are the units of f/the units of x. If f is the distance and x is the time, we have df/dx in units of distance/time or speed.

Once one has the derivative of a function f, one has created a new function f′. We can then take the derivative of that new function.

*And that gets us back to physics! Because that's exactly what we need to do to get the acceleration starting from position.*

Let's go back to our classic example:

$$f(t) = at^2/2 + v_0 t + s_0$$

$$f'(t) = at + v_0$$

To represent the **second derivative** of a function f with respect to t, we have the same three variants of notation. We could use Newton's notation and put a second dot over the f.

We could use the df/dt notation, which looks like this d(df/dt)/dt and is usually shortened to $d^2f/dt^2$ (read "d squared f d t squared"). Or we can use the ′ notation, in which case the second derivative is f″(t).

$$f''(t) = \text{limit as h goes to 0 of} \left(f'(t+h) - f'(t)\right)/h$$

which is

$$\left(a(t+h) + v_0 - (at + v_0)\right)/h$$

$$\left(at + ah + v_0 - at - v_0\right)/h$$

$$ah/h$$

$$a$$

Taking the limit, we get

$$f''(t) = a$$

Or the rate at which the speed changes is the acceleration. That sounds about right.

By the way, another way to write f″(x) is $f^{(2)}(x)$. Similarly, f‴(x) = $f^{(3)}(x)$, etc.

So far our prescription for how to take the derivative is to form the Newton quotient and then take the limit. If we want to use a computer to take the derivative, it's even easier than that. Computers calculate approximate answers, and the Newton quotient is an approximation to the derivative. So all we have to do is pick a small h and let the computer calculate the Newton quotient.

But if we don't want to use the computer method, you might worry that for any function that we want to differentiate, we would have to first form its Newton quotient and then perform the sort of manipulations that were done above to find the limit as h goes to zero. But recall the power of mathematical laziness: find the answer to a general problem that can then be applied to many specific cases.

We're going to list a batch of these general rules without proving any of them. Most of them rely on sticking the function listed in the left-hand column below into Newton's quotient and calculating things out before taking the limit. Some of them are very general (such as if f(x) = g(x) + h(x), then f′(x) = g′(x) + h′(x)) ). Others are more specific.

| Function | Derivative |
|---|---|
| 1. $f(x) = c$ (a constant) | $f'(x) = 0$ |
| 2. $f(x) = ag(x)$ (a is a constant) | $f'(x) = ag'(x)$ |
| 3. $f(x) = x^n$ | $f'(x) = nx^{n-1}$ (power law rule) |
| 4. $f(x) = g(x) + h(x)$ | $f'(x) = g'(x) + h'(x)$ (sum rule) |
| 5. $f(x) = g(x)h(x)$ | $f'(x) = g'(x)h(x) + g(x)h'(x)$ (product rule) |
| 6. $f(x) = g(x)/h(x)$ (quotient rule) | $f'(x) = (g'(x)h(x) - g(x)h'(x))/(h(x)^2)$ |
| 7. $f(x) = g(h(x))$ | $f'(x) = g'(h(x))h'(x)$ (chain rule) |
| 8. $f(x) = \sin(x)$ | $f'(x) = \cos(x)$ |
| 9. $f(x) = \cos(x)$ | $f'(x) = -\sin(x)$ |

That's plenty. The process of figuring out the function that is the derivative of another function involves careful and sometimes repetitive applications of these rules. Let's go back to

$$f(t) = at^2/2 + v_0t + s_0$$

and find its derivative using the rules above.

This function is the sum of three other functions, $at^2/2$, $v_0t$, and $s_0$. Rule 4 (called the summation rule) tells us that the derivative of the sum is the sum of the derivatives. So we need to find the derivatives of $at^2/2$, $v_0t$, and $s_0$ and then add them up.

$at^2/2$ is the constant $(a/2)$ times $t^2$, so using rule 2, the derivative of $at^2/2$ is going to be $(a/2)$ times the derivative of $t^2$. Rule 3 tells us the derivative of $t^2$ is $2t^{2-1}$

$2 - 1 = 1$ and $t^1$ is just t. So the derivative of $at^2/2$ is $(a/2)2t = at$.

The same rules will tell us the derivative of $v_0t$ is $v_0\,(1t^{1-1})$

$1 - 1 = 0$ and $t^0 = 1$ (anything to the 0 power is 1). So the derivative of $v_0t$ is $v_0\,1 = v_0$

Finally, the derivative of $s_0$ is 0 since $s_0$ is a constant and rule 1 tells us that the derivative of a constant is 0.

Adding these up, we get:

$$f'(t) = at + v_0$$

which is the answer we got before with the Newton quotient.

**Well, that's enough physics for now. It's time to turn to the application of calculus for cheapskates.**

*Can we at least start the cheapskate discussion with some physics?*

**Oh, all right.**

Differential calculus is not limited to determining rate of change. It can reveal even more information that lies waiting in functions. Let's take our cannonball shot one more time and ask how high the ball will fly. The height is given by

$$f(t) = -gt^2/2 + v_0t + s_0$$

where we have used the fact that the acceleration is $-g$. If we draw in the tangent line at the place where the ball is highest, we can see that the line is flat, or in other words that the derivative is zero (Figure 6.7).

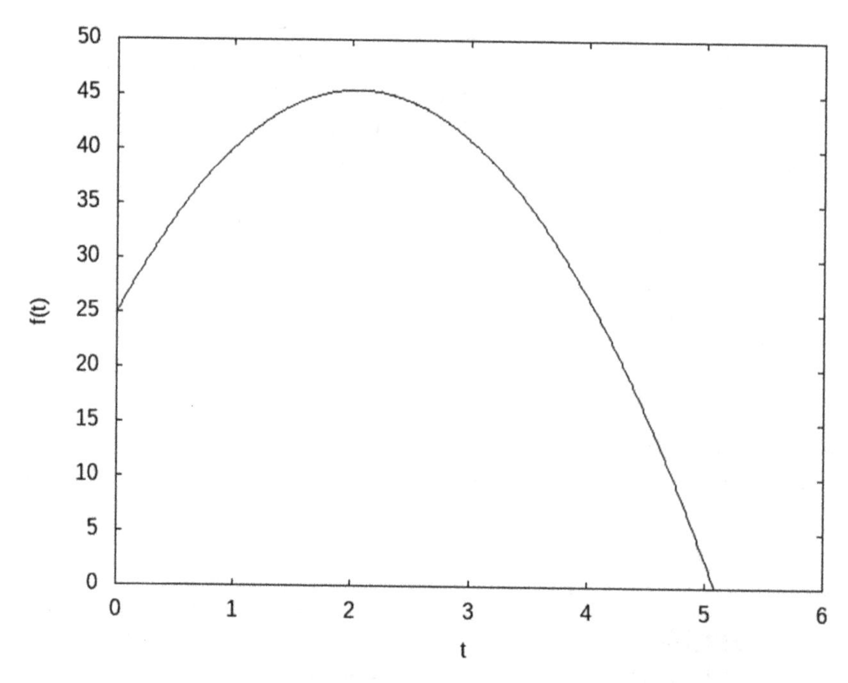


**FIGURE 6.7**
Parabola

Intuitively, we can see why this is so by recalling that f′(t) is the velocity of the ball. As long as f′(t) is positive, the ball is going up and hasn't reached its maximum. When f′(t) is negative, the ball has passed its maximum and is going down. It is just when f′(t) is zero that the ball has stopped going up, but hasn't yet started going down, that it has reached its maximum height.

In order to find the maximum height, we need to know at what time the ball is at that maximum. Once we have the time t, we can substitute it into the original equation to find the height. To find the time, we need to know at what time the slope of the tangent line is 0.

The slope of the tangent to the curve is the derivative of the function that graphs the curve. That is, the slope is 0 when

$$f'(t) = -gt + v_0 = 0$$

$$-gt + v_0 = 0$$

$$gt = v_0$$

$$t = v_0/g$$

Plugging this time value into f(t) gives

$$f(v_0/g) = -g(v_0/g)^2/2 + v_0(v_0/g) + s_0$$

$$-gv_0^2/2g^2 + v_0^2/g + s_0$$

$$-v_0^2/2g + v_0^2/g + s_0$$

$$v_0^2/2g + s_0$$

That's the maximum height of the shot.

More generally, we find the maximum of a function by finding the place where its derivative is zero. Since cheapskates want to get as much as they can for a certain amount of money, this means that calculus should be the favorite mathematics of cheapskates. (Or put another way, since the economics textbooks assure us that companies make decisions in such a way as to maximize profits, calculus should be a tool of wide application in business.)

By way of illustration, we will pose and solve one such cheapskate problem: given an amount of sheet metal of fixed area A to be made into a can, what should the proportions of the can be so that it can hold the maximum amount? Suppose that the radius of the can is r and the height of the can is h. Then the can is a cylinder of radius r and height h and therefore has a volume

$$V = \pi r^2 h$$

However, r and h cannot be varied independently since the total area of the metal is A. The top and bottom of the can are each circles of area $\pi r^2$, while the side of the can is made from a rectangle of metal bent into a circular shape. One side of the rectangle is the height of the can, while the other side is the circumference of the circle, so its area is $2\pi rh$. Thus, we have

$$A = 2\pi r^2 + 2\pi r h = 2\pi\left(r^2 + rh\right)$$

We think of r and h being functions of some parameter s, where the functions are chosen so that A doesn't change as s varies. From the power rule and chain rule, we have $d(r^2)/ds = 2r(dr/ds)$. From the product rule and the chain rule, we have $d(rh)/ds = r(dh/ds) + h(dr/ds)$. Using these two results, we then find

$$0 = dA/ds = 2\pi\left[(2r + h)(dr/ds) + r(dh/ds)\right]$$

We therefore find that

$$r(dh/ds) = -(2r + h)(dr/ds)$$

Now taking the derivative of the formula for V, and similarly using power rule, product rule, and chain rule, as well as the above result, we have

$$dV/ds = \pi r^2 (dh/ds) + 2\pi r h(dr/ds)$$

$$= \pi r\left[r(dh/ds) + 2h(dr/ds)\right]$$

$$= \pi r\left[[-(2r + h)(dr/ds) + 2h(dr/ds)\right]$$

$$= \pi r\left[h - 2r\right](dr/ds)$$

To make $dV/ds = 0$, we should choose $h = 2r$. Or to put it another way, the cheapskate's favorite can should be as tall as it is wide.

Two cautions about this method are in order. One is that exactly the same reasoning that says that the derivative is zero at a maximum (highest point) also says that the derivative is zero at a minimum (lowest point). So having found a place where the derivative is zero, we should try to check whether it is a maximum or a minimum.

One way to do this is that a place with zero derivative and positive second derivative is a minimum, while a place with zero derivative and negative second derivative is a maximum.

The second caution is that a maximum found by this method is a local maximum. That is, it is larger than any nearby points, but may not be the largest of all. Thus, one would need to check all possible local maxima (as well as the endpoints of the interval if the function is only defined on a finite interval) to see which is the largest of all.

**And speaking of cautions ... .**

*Oh yes, you're right. We left out a big one.*

**Well let's tell them about it now.**

## When Can We Differentiate?

We cut to the chase before in order to introduce the derivative, but in doing so left proper mathematical caution behind. We did not answer a fundamental question. Does the process we used to make the Newton quotient and find its limiting value always make sense?

The answer is no, not at all. We can only do this for functions that are sufficiently "well-behaved," where, as any sufficiently cynical parent knows, "well-behaved" means "acting the way I want them to, and doing what I want them to when I want them to."

It's easy to come up with functions where Newton's quotient doesn't have a limit. For example, we could do this with breaks, or corners.

A function with a break in it – that is, a place where the values jump around – is called **discontinuous**. For example, the **signum function** is –1 at every negative value, 0 at 0, and 1 at every positive value (Figure 6.8).

This function is well-behaved except at 0, where it jumps in literally no time from –1 to 0 and then to 1. It has, as it were, an infinite speed at 0. We can't make sense of the Newton quotient at 0 because if h is positive, then $(\text{signum}(0 + h) - \text{signum}(0))/h = (1 - 0)/h = 1/h$, which diverges when we substitute 0 for h. Similarly, if h is negative, we get $-1/h$, which also diverges when we substitute 0 for h.

For a function with corners, let's look at the **absolute value function** (Figure 6.9).

On the negative side of 0, this function has a derivative of –1 because that half is a line with slope –1. On the positive side, it has a derivative of 1 because that half is a line with slope 1. What's the derivative at 0? Is it 1 or –1?

The problem is that at a corner, the derivative calculated from the left and the derivative calculated from the right are not equal, so there can't be a derivative at that value.

So how does calculus deal with this issue that not all functions have derivatives? In just the formal, abstract way that you might expect: a function is called differentiable if the Newton quotient has a limit (the same limit from both sides). Otherwise, it is not differentiable. This approach is not quite as unsatisfying as you might think, because the calculus rules given in the table above are also rules for making differentiable functions.

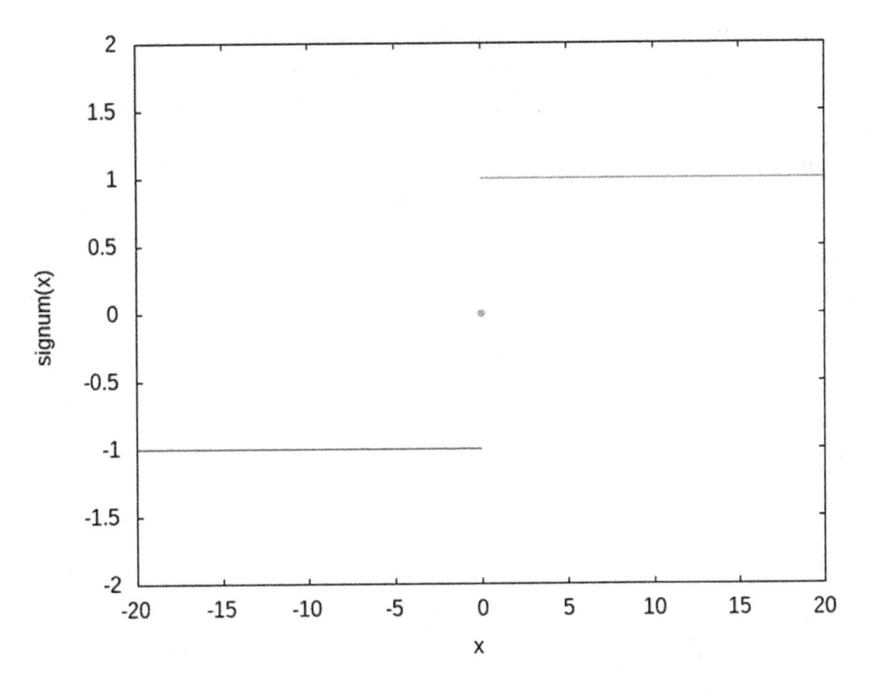


**FIGURE 6.8**
Signum

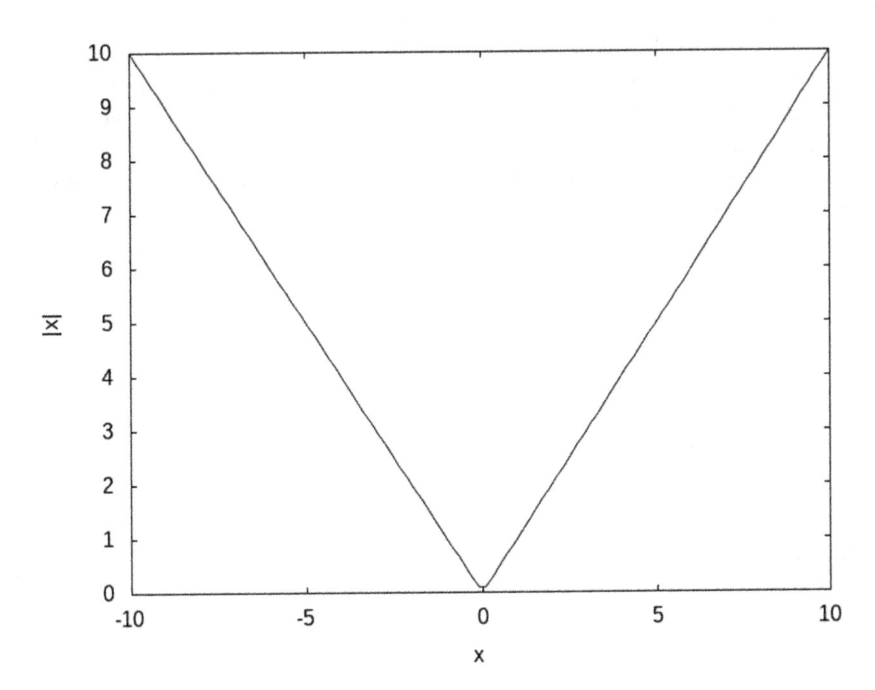


**FIGURE 6.9**
Absolute Value

So, for example, the sum rule says that the sum of differentiable functions is a differentiable function. The product rule says that the product of two differentiable functions is a differentiable function. The quotient rule says that the quotient of two differentiable functions is a differentiable function (except where the function in the denominator is zero. Even in calculus, you're not allowed to divide by zero.) The chain rule says that a differentiable function of a differentiable function is a differentiable function. In this way, we can build up huge families of differentiable functions.

When we take the derivative of a differentiable function $f(x)$, the new function $f'(x)$ may or may not be differentiable. If it is, then the original function $f(x)$ is called twice differentiable. Similarly, functions can be three times differentiable, four times differentiable, and so on. Then there are functions that are infinitely differentiable. No matter how many times you take the derivative, you still get a differentiable function. Why would we care about that many derivatives? Well, it turns out to be important for actually calculating functions.

## Function Approximation and Taylor Series

When we talked about analysis, we passed without noticing over a practical question: what kinds of functions are easy to calculate? It's not much use to say that we have all these functions representing physical quantities and actions if we have no practical ways to determine their values. So what mathematical operations are simple? What can we do that isn't too much work for us or for our computers?

Addition, subtraction, multiplication, and division. That's it. Everything else is hard work. Whatever functions we can reduce to these four bits of sentimental arithmetic are going to be far easier to work with than those we can't.

Expanding this, we can see that the next easiest functions to calculate are those derived from these four. We can do powers, since that's just multiplication, and we can add numbers derived by calculating different powers.

There's a couple of terms we need to introduce. A product of numbers and variables raised to powers is called a **monomial.** Here's a few:

$3$

$45x^7$

$16xyz^7b^{19}$

Calculating monomials is just a matter of multiplication. If we add a batch of monomials together, we get a **polynomial**. We've used several of these already.

$mx + b$

$ax^2 + bx + c$

$x^2 + y^2$

$x^2 - 2xy + y^2$

Calculating polynomials is just a matter of adding up the results of calculating monomials. We can calculate polynomials pretty easily. Certainly, our computers can. But so many of our functions (sines, for example) are not polynomials. Is there some way we can go from those mean functions to the nice ones we can compute?

Let's look again at functions and tangent lines in Figure 6.10.

What is the formula for the tangent line at $(x_0, f(x_0))$?

We know its slope, $f'(x_0)$. And we know it contains the point $(x_0, f(x_0))$. So let's try to find its formula in the form $y = mx + b$.

We know $m = f'(x_0)$. To find b, we only have to substitute the x and y values for a point we know belongs to the line and solve for b. Oh, there's a point: $(x_0, f(x_0))$

$$f(x_0) = f'(x_0)x_0 + b$$

so

$$b = f(x_0) - f'(x_0)x_0$$

Which means that the formula for the line is

$$y = f'(x_0)x + f(x_0) - f'(x_0)x_0$$

We're going to rewrite this in a slightly odd form:

$$y = f(x_0) + f'(x_0)(x - x_0)$$

Now let's look a little closer at the part of the curve around the point $(x_0, f(x_0))$. Notice that the tangent line is pretty close to the curve there. In fact, this line is the best line approximating the curve. If the curve does not bend much near $(x_0, f(x_0))$, then we can use this line to get approximate values for the function f(x) if we are close to $x_0$. That can be pretty useful. We can calculate one value of the function and one value for the derivative and use

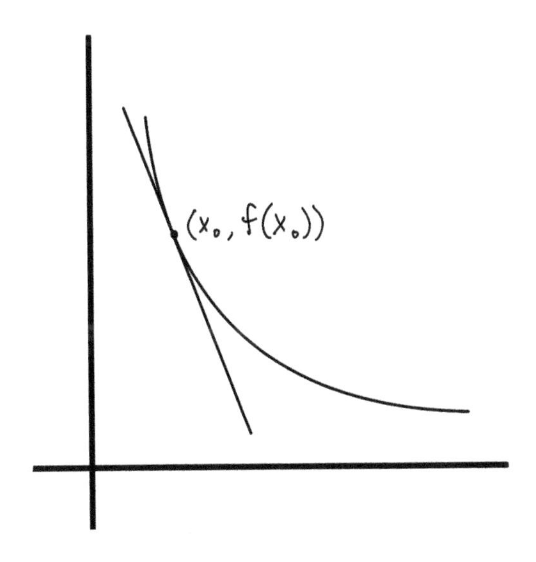


**FIGURE 6.10**
Function with Tangent Line

that as an approximation for the function itself. Once we've done those two messy calculations (one function value, one derivative value), everything else is just multiplication and addition.

Still, it's not so great. It is a line with no curvature and we don't have to go very far from our initial point for the approximation to be too far from the real value. Can we do better in approximation?

We can. We can use the second derivative to combine instantaneous acceleration with instantaneous speed to get something that looks like our old friend $at^2/2 + v_0t + s_0$. After some work, which we're not going to show, we end up with this:

$$y = f(x_0) + f'(x_0)(x - x_0) + f''(x_0)(x - x_0)^2/2$$

Better. We now have some curvature and a curve that's a closer approximation. There's more work, since we need to calculate a second derivative at a single value, but since its curvature is closer to that of the original curve, we can use it as an approximation in a broader area (Figure 6.11).

Brook Taylor, an 18th-century English Mathematician, and Colin Maclaurin, an 18th-century Scottish mathematician, took this idea and expanded it to produce what is called the **Taylor series**. It takes a function f and creates an arbitrarily close approximation to that function by continuing this process out to some number n of terms.

$$y = f(x_0) + f'(x_0)(x - x_0) + f''(x_0)(x - x_0)^2/2!$$
$$+ f^{(3)}(x_0)(x - x_0)^3/3! + \cdots + f^{(n)}(x_0)(x - x_0)^n/n!$$

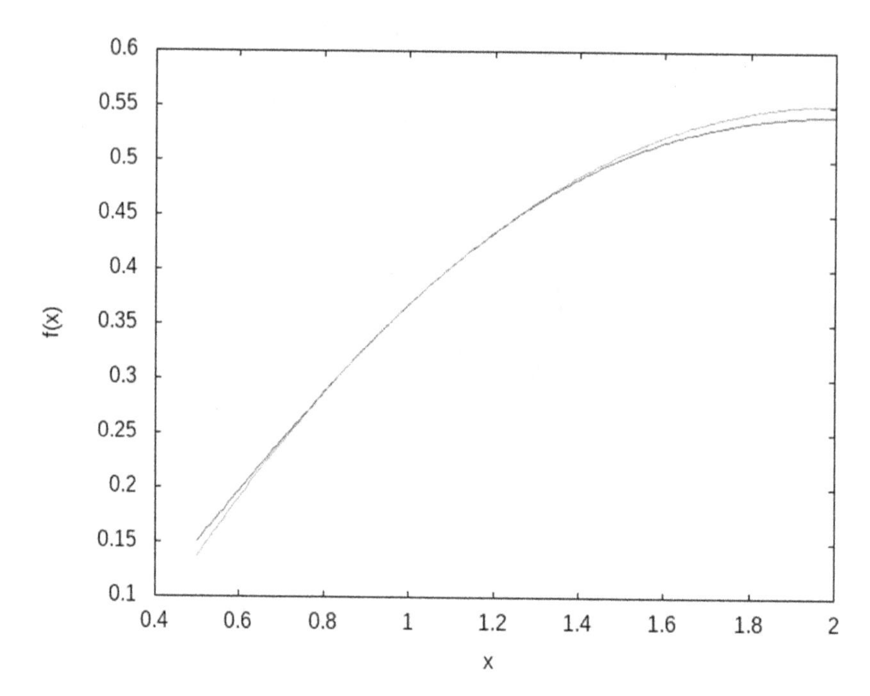


**FIGURE 6.11**
Curve with approximate Parabola

The ! above is not for emphasis. It's the factorial operation. n! = n(n – 1)(n – 2) … 1.

So 1! = 1

2! = 2 × 1 = 2

3! = 3 × 2 × 1 = 6

etc.

If we look at each of these Taylor polynomials as a function, we can see that we have a sequence of functions:

$$f_0(x) = f(x_0)$$

$$f_1(x) = f(x_0) + f'(x_0)(x - x_0)$$

$$f_2(x) = f(x_0) + f'(x_0)(x - x_0) + f''(x_0)(x - x_0)^2/2$$

$$f_3(x) = f(x_0) + f'(x_0)(x - x_0) + f''(x_0)(x - x_0)^2/2 + f^{(3)}(x_0)(x - x_0)^3/3!$$

$$f_n(x) = f(x_0) + f'(x_0)(x - x_0) + f''(x_0)(x - x_0)^2/2$$
$$+ f^{(3)}(x_0)(x - x_0)^3/3! + \cdots + f^{(n)}(x_0)(x - x_0)^n/n!$$

We can keep going and get an infinite sequence of functions.

Getting back to ideas of convergence, a sequence of functions $f_n$ converges to a function f in an interval (a, b), if at every value v in that interval, the sequence

$f_0(v), f_1(v), f_2(v), f_3(v), \ldots$

converges to the value f(v).

Proofs of function convergence can get messy, and we're not going to bother with them. The relevance is that if the function f is sufficiently well-behaved, then the sequence of Taylor polynomials created from f converges to f when the parameter value is close to the $x_0$ you used to form the polynomial.

This is how computers calculate functions like the sine and cosine. They don't use the functions themselves. They use Taylor polynomials. Computers can only add and multiply (actually they do even less than that, as we'll see later), so building a function that only does these operations but approximates other functions is a major boon. Before this, sines and cosines were calculated painstakingly and written up in big books; people looked them up when they needed to know the answers.

On the face of it, the Taylor series method doesn't seem all that helpful. We have a function we can't directly calculate, but we have to somehow find the function and all of its derivatives at some point $x_0$ to even get started. To illustrate how this works, we'll look at the case of sine and cosine. We're going to pick $x_0 = 0$, so the first thing that we need is sin(0) and cos(0). In a right triangle, as the angle goes to zero, the length of the opposite side goes to zero and the length of the adjacent sides becomes the length of the hypotenuse. It immediately follows that sin(0) = 0 and cos(0) = 1. What about the derivatives? In our table of rules, we have that the derivative of sine is cosine and the derivative of cosine is minus sine.

This means that if we take two derivatives of either sine or cosine, we just get back what we started with, but multiplied by –1. This fact is going to be very important in the next chapter, but for now note that it means that for sine, any even number of derivatives is zero at x = 0 and any odd number of derivative is plus or minus 1 at x = 0 with the plus and minus signs alternating. Similarly, for cosine, any odd number of derivatives is zero at x = 0, and any even number of derivatives is plus or minus 1 with the plus and minus signs alternating. Putting these facts into the Taylor series formula, we find

$$\sin(x) = x - x^3/3! + x^5/5! - x^7/7! + \cdots$$

$$\cos(x) = 1 - x^2/2! + x^4/4! - x^6/6! + \cdots$$

Figure 6.12 shows sin(x) and cos(x) along with the first four terms in their Taylor series approximation.

Here the angle is measured in radians, not degrees, where a whole circle (360°) is 2π radians. We use radians because that is the angle measure needed for the derivative formulas we gave above for sine and cosine to work.

**You know, from a writer's point of view, there's a major problem with this chapter so far.**

*What's that?*

**So far, we've been talking about single functions f(t). This means that all our characters are one-dimensional. It's almost as if we were writing a clichéd action novel.**

*Yes, that's a major problem from the point of view of physics too. So far we have an excellent description of objects moving along a line. But real objects move in space. It's time to get three-dimensional.*

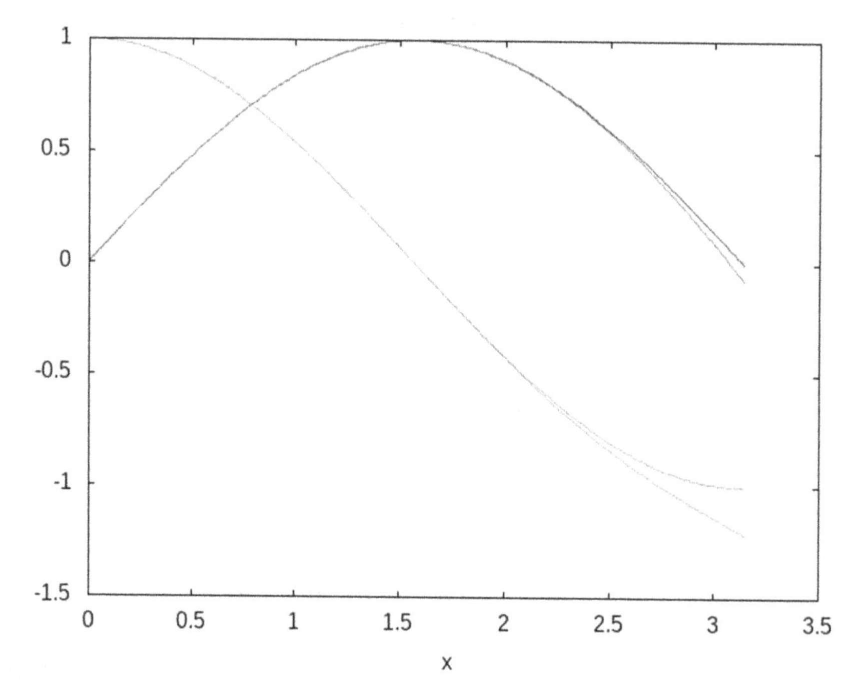


**FIGURE 6.12**
Sine and Cosine

**OK, but we're going to lose the lucrative action novel market. Our sales will plummet.**

We already have a way to describe motion in space: the parameterized curve introduced in the last chapter, where the parameter is time. That is, motion in space is described by a triple of functions (x(t), y(t), z(t)), where x(t) is the x coordinate of the moving object given as a function of time, and correspondingly for y(t) and z(t). The velocity is then also a triple of functions ($v_x$, $v_y$, $v_z$) where $v_x = x'(t)$, $v_y = y'(t)$, $v_z = z'(t)$. Similarly, the acceleration is a triple of functions ($a_x$, $a_y$, $a_z$) where $a_x = v'_x(t)$, etc. Since F = ma, this means that force is also a triple of functions ($F_x$, $F_y$, $F_z$). It's cumbersome to keep writing triples of functions, so the textbook notation is to use a boldface letter to stand for the triple. That is **r** is an abbreviation for (x, y, z), and **v** is an abbreviation for ($v_x$, $v_y$, $v_z$), and **a** is an abbreviation for ($a_x$, $a_y$, $a_z$), and so on.

Of course, thinking of the triple (x, y, z) as a single entity **r** gives the algebraic mindset a whole new thing to abstract and generalize, which leads to the mathematical subject of vectors.

## Vectors

One consequence of thinking of **r** as a single entity called a vector, rather than a collection of three numbers, is a simple technique for visualization. Draw an arrow, starting at the origin (0, 0, 0) and ending at the point **r** = (x, y, z). As we saw from spherical and polar coordinates, the length of the vector and the direction that it points are information equivalent to the information in the numbers x, y, and z. In particular, the length of the arrow is $\sqrt{x^2 + y^2 + z^2}$. But we can do that same trick with any vector. For example, for the vector v = ($v_x$, $v_y$, $v_z$), we can think of a space where the ($v_x$, $v_y$, $v_z$) are the Cartesian coordinates and draw an arrow from the origin to that point. The length is that arrow is $\sqrt{v_x^2 + v_y^2 + v_z^2}$ which is the speed of the object, and the direction of the arrow is the direction in which that object is moving at that time.

Thinking of **r** as a single entity makes it tempting to form the Newton quotient and its limit directly with **r**, rather than separately with its components x, y, and z. That is, we would like to have **v** as the limit as h goes to zero of (**r**(t + h) − **r**(t))/h. In order to do that, we would need a way to add and subtract vectors and to multiply and divide them by numbers.

It turns out that there is a simple way to do this: add vectors by adding their components. Multiply a vector by a number by multiplying each of its components by that number.

For simplicity, we will start with vectors with only two components, representing objects situated in a plane.

Let's add the vector (3, 2) to the vector (4, 7). The rule that says add the components gives (3, 2) + (4, 7) = (3 + 4, 2 + 7) = (7, 9).

In the plane, the vector (x1, y1) + the vector (x2, y2) gives the vector (x1 + x2, y1 + y2).

In space, the vector (x1, y1, z1) + the vector (x2, y2, z2) gives the vector (x1 + x2, y1 + y2, z1 + z2).

Similarly, for a constant c, we have that c multiplied by the vector (x, y) is the vector (cx, cy) and c multiplied by the vector (x, y, z) is the vector (cx, cy, cz).

That's enough information to form the Newton quotient for vectors and take its limit as h goes to zero. And not surprisingly, taking the derivative in this vector way gives exactly the same answer as taking the derivative of each component and then combining those component derivatives to form a vector.

For the purposes of using calculus to describe motion in three dimensions, we don't need to say much more about vectors. However, from the point of view of the algebraic mindset, vectors with their rule for addition and their rule for multiplication by numbers are a concept just begging to be abstracted and generalized.

To begin with, for the algebraic mindset, it is confining to restrict to two or three components. How about four components or five components? How about n components, where n can be any whole number that we like? In four-space vector, addition is done by adding each of the four coordinates of two points, and so on.

The rules for vector addition comply with the axioms introduced above for abelian groups. Now, we can prove the group axioms for vectors. We'll do it in the plane, but the same principles can be used in any number of dimensions.

First, closure. If we have two vectors (x1, y1) and (x2, y2), their vector sum is (x1 + x2, y1 + y2), which is clearly a vector.

Next, associativity. If we have three vectors (x1, y1), (x2, y2), and (x3, y3), then

$$\left((x1, y1)+(x2, y2)\right)+(x3, y3)=\left((x1+x2)+x3, (y1+y2)+y3\right)$$
$$=\left(x1+(x2+x3), y1+(y2+y3)\right)=(x1, y1)+\left((x2, y2)+(x3, y3)\right)$$

This proves associativity.

Commutativity is even easier:

$$(x1, y1)+(x2, y2)=(x1+x2, y1+y2)=(x2+x1, y2+y1)=(x2, y2)+(x1, y1)$$

There is a 0 vector (0, 0). To prove this acts as an identity, we add

$$(x, y)+(0, 0)=(x+0, y+0)=(x, y).$$

Finally, given a vector (x, y), there is an inverse, the vector (−x, −y) since

$$(x, y)+(-x, -y)=(x+-x, y+-y)=(0, 0)$$

Let's pause for a second, because this illustrates two common methods for extending mathematical ideas that often work together.

First, we're essentially saying that we want to treat vector addition as an addition-like operation. To do that, we lay down a set of conditions, the axioms, that an addition-like object must satisfy, and then we prove those conditions. Once we've done so, we can use all the theorems that apply to addition-like objects. We can do this because the premises in those theorems are the very axioms we just proved. The theorems say that if an object satisfies these axioms, then this conclusion will be true. Proving the axioms proves the premises, which by modus ponens proves the conclusions. In math, this method is used constantly to show that an object fits an abstract structure and therefore can exploit what has already been proven about that structure.

The second trick is to take something we already know how to prove and use it to prove something we did not. Let's look at the proof of commutativity above and go a little more slowly.

(x1, y1) + (x2, y2) = (x1 + x2, y1 + y2). This is the definition of vector addition.

(x1 + x2, y1 + y2) = (x2 + x1, y2 + y1) because x1 + x2 = x2 + x1, and y1 + y2 = y2 + y1. In other words, this step is true because plain old number addition is commutative.

(x2 + x1, y2 + y1) = (x2, y2) + (x1, y1). Again, this is the definition of vector addition.

The middle step relies on something we already know to be true (numerical addition is commutative), which we've embedded in the proof of something built on it (vector addition is commutative).

These two processes, building up by proving axioms and reaching down to what we already know, are critical applications of laziness. They minimize the work needed to gain access to truths already shown.

So vector addition is addition-like enough for our purposes.

Abstract vector spaces also have axioms for multiplication by numbers. They are the following: for any vector $\mathbf{x}$, we have $0\mathbf{x} = \mathbf{0}$ and $1\mathbf{x} = \mathbf{x}$. For any vector $\mathbf{x}$ and numbers c and d, we have $(cd)\mathbf{x} = c(d\mathbf{x})$ and $(c + d)\mathbf{x} = c\mathbf{x} + d\mathbf{x}$. For any vectors $\mathbf{x}$ and $\mathbf{y}$ and any number c, we have $c(\mathbf{x} + \mathbf{y}) = c\mathbf{x} + c\mathbf{y}$.

Our vectors as pairs of numbers with the component rule for multiplication by numbers satisfy these multiplication axioms. This can be checked using essentially the same techniques we used to show that the vector addition axioms are satisfied. Multiplying components by zero gives the vector with all zero components, which is the zero vector. Multiplying all components by 1 leaves the components unchanged, which leaves the vector unchanged. The other axioms are essentially vector versions of the associative and distributive laws and are true for our vectors by virtue of the fact that they are true for ordinary numbers. In classic annoying math book fashion, we will not carry out any of these proofs and instead leave them as "exercises for the reader."

This abstraction of the notion of vectors doesn't help much in describing motion in three dimensions. However, as we will see later, it plays a major role in quantum mechanics.

One interesting consequence of motion in more than one dimension is that we can have acceleration without changing speed. The easiest way to understand this is to think of a vector in terms of magnitude and direction. The speed of the object is the magnitude of the velocity vector. The acceleration is the rate of change of the velocity vector. If the velocity vector has constant magnitude, but changes its direction, then there will be acceleration without changing speed.

We will illustrate this with the case of an object going around a circle at constant speed. This motion is described by the parameterized version of the circle given in the last chapter, but with time as the parameter. We will write this in a slightly different way:

$$(x, y) = \left(r\cos(2\pi t/T), r\sin(2\pi t/T)\right)$$

Here r is the radius of the circle, t is the time, and T (called the period) is the amount of time that it takes the object to go once around the circle. We are measuring angle in radians, so when t changes by an amount T (one trip around the circle), the quantity $2\pi t/T$ changes by $2\pi$ (the number of radians in a circle).

To find the acceleration, we take two derivatives of the position with respect to time. From the rules for derivative of sine and cosine, we found that if we take two derivatives of either sine or cosine, we get back what we started with multiplied by −1. However, because in the formula the sine and cosine are functions of $2\pi t/T$, the chain rule says that each time we take a derivative, we get an extra factor of $2\pi/T$. The acceleration vector is then given by

$$(a_x, a_y) = -\left(4\pi^2/T^2\right)\left(r\cos(2\pi t/T), r\sin(2\pi t/T)\right)$$

Or in vector notation $\mathbf{a} = -(4\pi^2/T^2)\,\mathbf{r}$. In terms of magnitude and direction, this means that the magnitude of the acceleration is $a = 4\pi^2 r/T^2$ and the direction is toward the center of the circle. Since **F** = m**a**, this means that whatever force is keeping the object in circular motion must also point toward the center of the circle.

There is a silly myth that Newton was inspired to develop his theory of gravity by being hit on the head with an apple. The truth is much cooler. People had been hit on the head with apples for longer than recorded history. And people had been observing the moon for longer than recorded history. But only Newton thought to make a quantitative comparison between the motion of the apple and the motion of the moon. Apples (and everything near the surface of the Earth) fall with acceleration g.

But according to Newton's laws of motion, with no force on it, the moon would travel in a straight line, not a circular orbit. So the moon is also falling, even though it never gets closer to the Earth: the Earth's gravity holds the moon in its circular orbit. So which falls faster, the apple or the moon? And by how much? Newton knew the distance to the moon as well as the time (a little less than a month) that it takes the moon to orbit the Earth. Using the formula $a = 4\pi^2r/T^2$, he found the moon's acceleration. The moon is falling more slowly than the apple, but not just by any old amount. It's smaller by a factor of $R_E{}^2/r^2$, where $R_E$ is the radius of the Earth (and thus the distance of the apple from the center of the Earth) and r is the distance of the moon from the Earth. Thus, gravity gets weaker the farther you get from an object. The $1/r^2$ means that when you get twice as far away, gravity gets four times weaker. Newton also applied the formula $a = 4\pi^2r/T^2$ to Kepler's measurements of the motion of the planets and found that the Sun's gravity also has the same $1/r^2$ behavior. Thus, by reading the motion of the moon and planets using the language of calculus, Newton found his law of gravity.

Differential calculus has taken us from the infinitesimal through can design to simplified function calculations to the interplanetary.

What will we find when we look down the other branch of the tree of calculus?

## Integral Calculus

Differential calculus uses the basic subtraction and division process used to calculate the slopes of lines to allow us to do the more complex work of finding the slopes of curves.

Integral calculus allows us to use the simple multiplication that determines the areas of rectangles and from them find the areas of practically any two-dimensional shape.

Integral calculus attained its present completely mathematically correct form in the work of Riemann. But it had a number of earlier forms and precursor problems, not the least of which was finding out how much wine a barrel could hold without first filling it up and emptying it out.

Before we pour drinks, a little notation.

Let us introduce the Greek letter Δ, capital Delta, delta to its friends and relations, relative of our own local letter "d," which we used when dealing with derivatives. d and Δ both get a lot of use in calculus. We met lowercase delta δ earlier. It mostly does scut work in limit proofs.

Capital Delta is often used in math to represent the difference in some variable when its value changes. For example, if x changes from 3 to 7, $\Delta x$ would equal $7 - 3 = 4$.

Let's go back to the Newton quotient. Instead of describing it in terms of a function f and a limit of h, we can also look at two points on the curve, (x1, y1) and (x2, y2). The slope of the secant line between these two points is

$$(y2-y1)/(x2-x1)$$

which we could also write as $\Delta y/\Delta x$ – that is, the change in y divided by the change in x. As we go through the process of making secant lines that are closer and closer to the tangent lines, we are essentially shrinking the $\Delta x$ value and determining the corresponding $\Delta y$ value. This turns the definition of the derivative into this:

$$\text{lim as } \Delta x \text{ goes to 0 of } \Delta y/\Delta x = dy/dx$$

This is just a rephrasing of the stuff we've already been over. It uses different symbols for the same meaning. The earlier version with f(x + h) – f(x) is easier to understand at the beginning. We're swapping symbols now because they matter in the way integration is done.

*Can we get back to the wine?*

**In a little bit.**

We'll start with something simpler: the area under a curve.

Take a graph of y = f(x) (Figure 6.13).

Suppose you want to find the shaded area.

*Hmm ... our readers will probably wonder why you would want to do that?*

You probably wouldn't, but it forms the easiest introduction to the technique that leads to the wine barrels. In any case, you might. Let's say you had a piece of ground with one sharp border (like a property line) and one ragged border (like a river).

We can't directly find the area with what we already know how to do, but we can make an approximation.

*Cue the music for the strange sense of déjà vu. We make better and better approximations and then take the limit.*

**That's right.**

The approximation we are going to make is called a **Riemann sum**, named after guess who.

*Lobachevsky?*

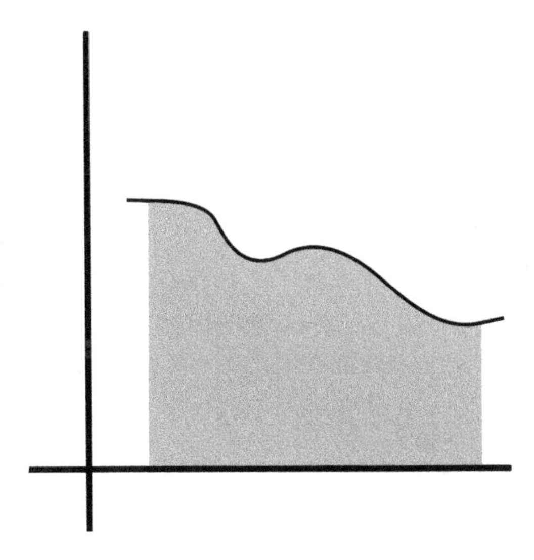

**FIGURE 6.13**
Area under a Curve

**No, we'll get to him in a later chapter.**

The way we find this area is to take the interval of the x-axis that represents the straight border and divide it up into a number of pieces, each of the same length. We'll call that length $\Delta x$. Then at the endpoints of each of these pieces, we'll draw vertical lines up to the corresponding point on the curve. Then from each of those points, we'll draw a horizontal line across to the next vertical line (Figure 6.14).

We've now got a bunch of rectangles. If we calculate the areas of each of those rectangles and add them up, we'll have an approximation of the area between the x-axis and the curve for this stretch of the x-axis.

Each rectangle has the same width ($\Delta x$), but each one has its own height which is equal to $f(x)$, where the x is the value on the x-axis where we start the rectangle.

The area of each rectangle is then $f(x)\Delta x$.

The sum of the areas is written like this $\Sigma f(x)\Delta x$. The letter $\Sigma$ (capital Sigma) corresponds to the English S.

*So why does everyone use it as E in pseudo-Greek fonts?*

**I don't know. Maybe because it looks like an E. It is rather annoying.**

If we take the limit of this Riemann sum as $\Delta x$ goes to 0, we will end up with a value for the area between the curve and x-axis. This is called the **integral** of $f(x)$ and is written $\int f(x)dx$ (read as "the integral of f(x)dx" or "the integral of f(x) with respect to x").

More precisely, if we want the area between an x value of a and an x value of b, we have $\int_a^b f(x)dx$ (read as the integral from a to b of f(x)dx). This is called a **definite integral**.

Definite integrals show again the basic technique of taking something we can do for simple things (like find the area of rectangles) and by a twist of thought turn it into something applicable elsewhere. This kind of innovation, the slippery, twisty, lazy kind, is much valued in mathematics. But it's also useful elsewhere. A lot of inventions, both physical and mental, employ this kind of chicanery. Learning to think in the way that leads to this kind of stunt is also a part of the legacy of math.

*Oh, come on. People have been sneaky longer than they've been piling rocks up. After all, piling the rocks is one of these sneaky bits.*

**FIGURE 6.14**
Riemann Sum

**That's true, but the formalized form of sneakiness is part of mathematical character.** It is precisely the action of substituting what you can do for what you can't (mathematical abstraction), then working out a way to put together the can do's into a way of doing the can't do (laziness) and then making sure it works (mistrust). It's the twisting around to find substitutes that work out (lots of rectangles for one big irregular shape) that shows the mathematical mind in action.

From definite integrals, it is possible to create a somewhat more sophisticated object, the **indefinite integral**, but it's a little trickier.

If you have a function f(x), then a function F(x) is said to be an indefinite integral of f(x) if for any values of a and b

$$\int_a^b f(x)dx = F(b) - F(a)$$

First of all, notice that a function will have more than one indefinite integral, since if F(x) is an indefinite integral of f(x), then G(x) = F(x) + 4 is also an indefinite integral. The proof is as follows:

$$G(b) - G(a) = F(b) + 4 - (F(a) + 4) = F(b) - F(a) = \int_a^b f(x)dx$$

And of course that added bit doesn't have to be 4. If F(x) is an indefinite integral of f(x), then F(x) + c is also an indefinite integral of f(x) if c is a constant. So any function with one indefinite integral actually has a number of indefinite integrals equal to –

*Can I say it?*

**All right. Go ahead.**

*The power of the continuum!*

One more point to make note of: while we said that the definite integral gives the area under the curve, that's not quite right. Notice that if the curve lies below the x-axis, the integral will be negative (since all the y values in those rectangles will be negative) (Figure 6.15).

Furthermore, if the curve has part under and part over the x-axis, the integral will be the area of the part above minus the area of the part below (Figure 6.16).

Just as with the derivative, there are two ways of doing integrals, the computer, or numerical, way and the human, or brain-hurting, way. The computer way involves choosing a value for $\Delta x$ and calculating the Riemann sum. However, unlike the derivative, this is not a simple one-off calculation. If you look at the picture of the Riemann sum, you'll see that a computer doing this calculation needs to first figure out f(x) for each and every value on the x-axis that has a vertical line at it, and then multiply that by $\Delta x$ and add up the results. For a small-enough value of $\Delta x$, this can be an incredibly long calculation. On the other hand, it doesn't tax modern computers to calculate definite integrals to any practical degree of accuracy.

So let's look at other uses for integration.

*Hey, aren't we going to tell them about the human method?*

**Later. It's a surprise.**

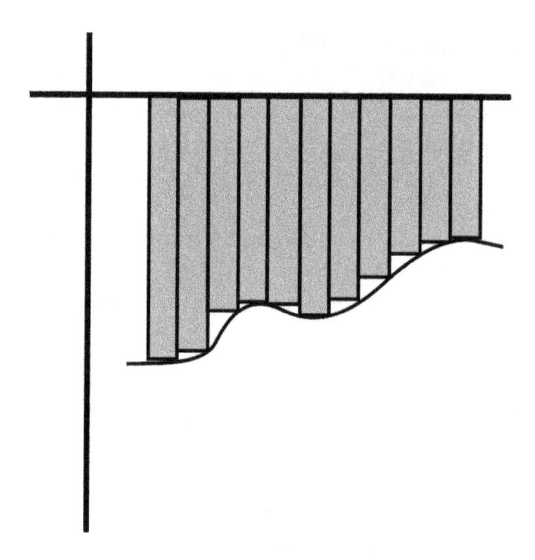

**FIGURE 6.15**
Curve Below x-Axis

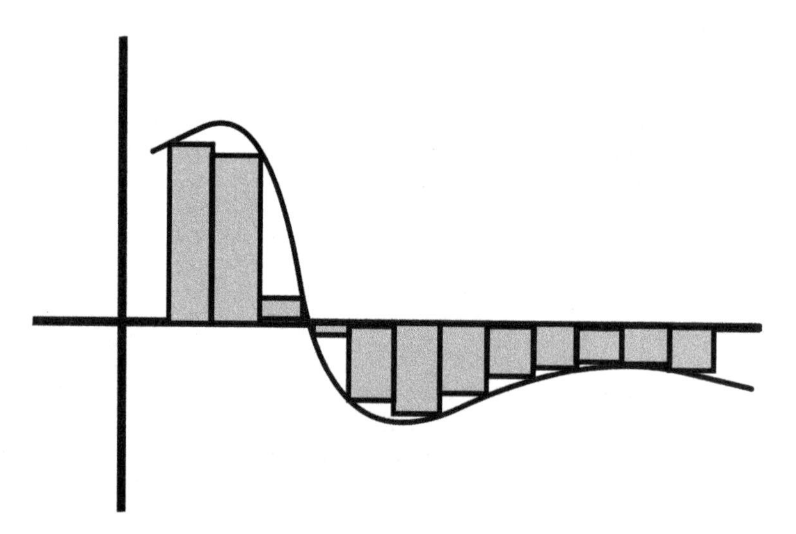

**FIGURE 6.16**
Curve Both Over and Under

Let's look at the units that result from taking the integral of a function that represents a real-world quantity. Suppose we have a function f(x) representing distance, and x also represents a distance. We'd have this situation if we were finding the area under a curve that is supposed to represent a physical boundary. Then the pieces of the Riemann sum (the $f(x)\Delta x$) have units of distance times distance, which is area. Since the integral is the limit of Riemann sums, its units are the same as those of the sum, so the integral also has units of area.

To continue the hint, let's suppose that f(t) is a function of time representing speed (i.e., distance/time) and t is the time. Then the units of the sum $(f(t)\Delta t)$ are speed times time, which is (distance/time) times time, or distance. So the integral of speed with respect to time is distance.

Now, let's look at other applications of the integral.

*After which we'll show them the human method?*

**Yes.**

*All right. When can we get to the wine?*

**Right now.**

The trick in turning a real-world problem into an integral is to find a way to slice up something you can't calculate into pieces you can calculate that take the form of lots of little $f(x)\Delta x$, and then add them up and take the limit.

So here's a wine barrel. We want to know its volume. We'll slice it up into a lot of little disks (Figure 6.17).

Each disk has a volume of $\pi r^2$ times the height of the disk. That height is $\Delta x$, so each disk is $\pi r^2 \Delta x$. All we need to do is make a Riemann sum from this and turn it into an integral that will give us the volume. Look at this cross section of the wine barrel. The radius depends on where along the barrel you are. So r is actually r(x) (Figure 6.18).

So the integral is

$$\int_a^b \pi r(x)^2 \, dx$$

We could do the same thing to find the volume of a pyramid (Figure 6.19).

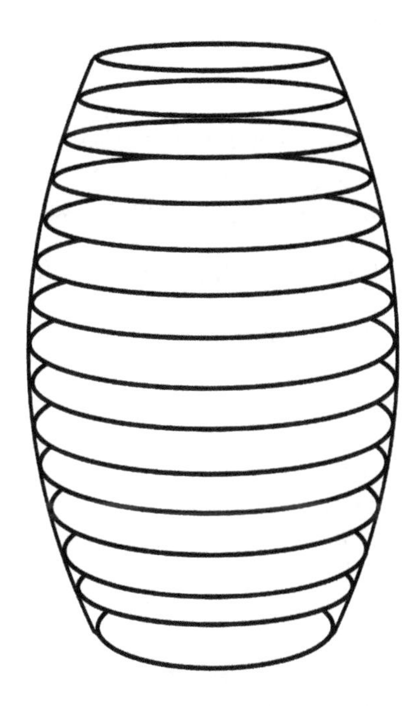

**FIGURE 6.17**
Sliced Wine Barrel

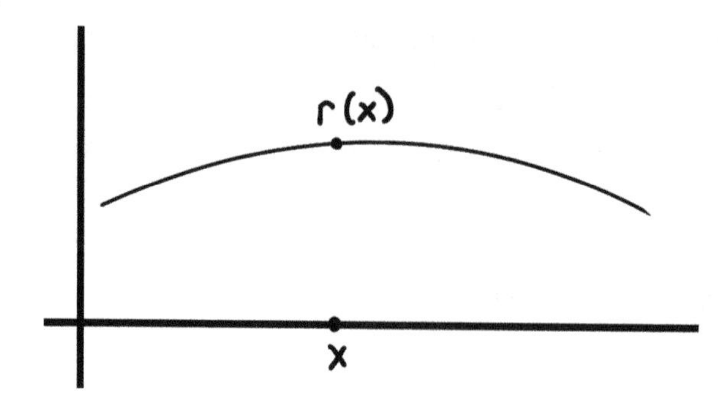


**FIGURE 6.18**
Wine Barrel Cross Section

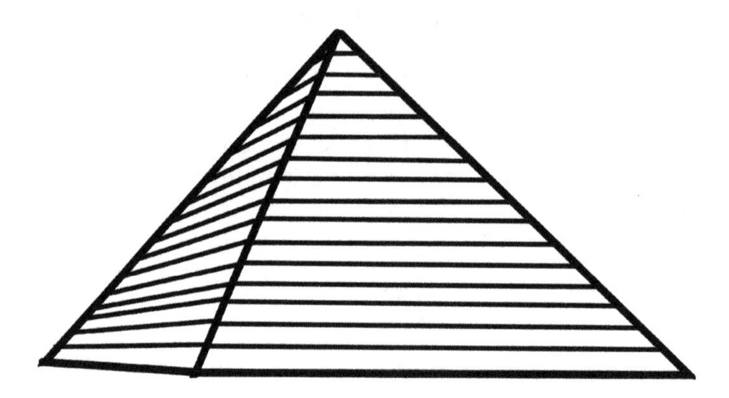

**FIGURE 6.19**
Sliced Pyramid

We could vary the trick to find the volume of a sphere by slicing it up into concentric shells, each of which has an approximate volume of $4\pi r^2\Delta r$. In this case, we end up integrating with respect to r, that is, going from a radius of 0 to that of the radius of the sphere (Figure 6.20).

We can do a similar stunt to find the area of a circle by slicing it into concentric circular pieces and summing $2\pi r\Delta r$.

These uses alone show the practicality of integral calculus. They can turn measures of curves into measures of surfaces, and measures of surfaces into measures of volume. Think how many irregular shapes there are we might need to know the size of. After all, most plots of land aren't really rectangles and most containers aren't cubes.

We've reached another of those disconnects between real life and math. People fill things up all the time and wonder how much something can hold. Can I get more in there? Will it burst? Does my neighbor have a bigger gazebo than me? But the mathematics that answers these questions is walled off behind the bugaboo of calculus. Because of that, the very people who need the questions answered have no reason to think math is of any use to them at all.

Wait, there's more. You know that annoying question of whether a function is differentiable. The integral doesn't have any of that: if a function is continuous, then it is integrable.

*But our readers want to know how to integrate. Let's tell them!*

**Since you ask so politely, I give you:**

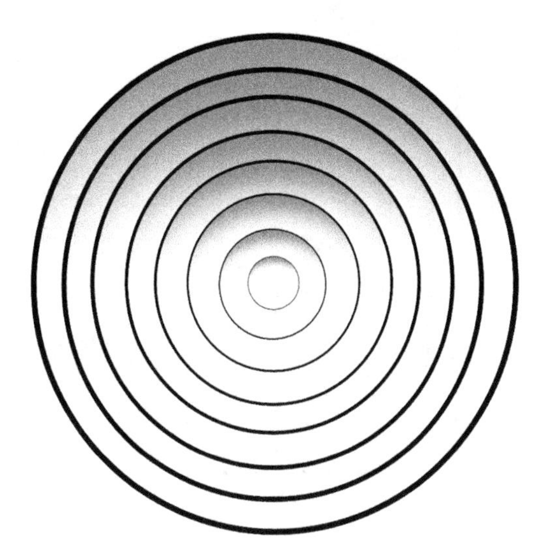

**FIGURE 6.20**
Spherical Shells

## The Fundamental Theorem of Calculus

*Finally! – though that is a pompous name. Are we going to show them a proof of the theorem?*

**Not exactly. Instead, we'll give them a demonstration of why the theorem is true.**

A proof has to fit the highest standard of mistrust. But sometimes such proofs demonstrate without edifying. The process by which they show that something is true does not illuminate what causes it to be true. What follows does the opposite, showing why without proving it.

Let's take a function F(x) which is an indefinite integral of f(x).

What is dF/dx? What, in short, is the derivative of an integral?

Since F(x) is an indefinite integral of f, it satisfies $\int_a^b f(x)dx = F(b) - F(a)$.

Instead of a and b, let's put in x and x + h. (For the mathematically accurate, we know that we shouldn't have the x under the integral sign as well as in the limits of the integral, but it looks better.) And let's divide both sides by h, so we have a Newton quotient on the left-hand side.

So $\left(F(x+h)-F(x)\right)/h = \left(\int_x^{x+h} f(x)dx\right)/h$

Now let's convert the right-hand side into a Riemann sum:

$$\sum f(x)\,\Delta x$$

We have on the left-hand side a value h that we picked to be small and on the right-hand side a similar value Δx that we picked to be small. We could make these values the same arbitrary small value h = Δx.

What does that leave us? Let's look at a graph (Figure 6.21).

The whole area above is F(x + h), which is the integral from 0 to x + h of f(x).

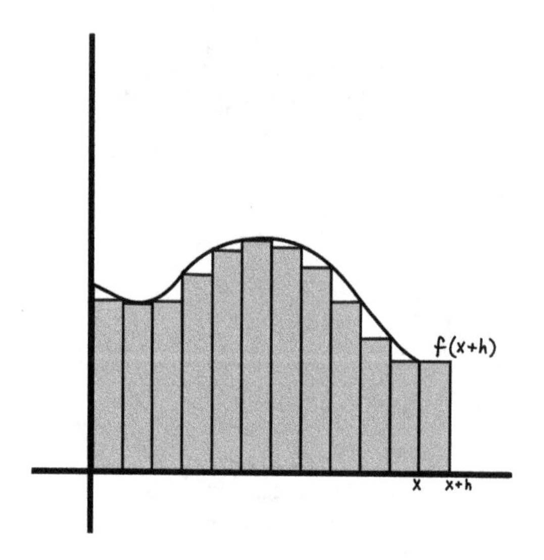


**FIGURE 6.21**
Riemann Sum + h

The shaded area on the left is F(x), which is the integral from 0 to x of f(x). Then there's that one little rectangle that is the difference between these two, which is F(x + h) – F(x) or the integral from x to x + h of f(x). Notice that the definite integral is approximated by this one little rectangle, the area of which is f(x + h)h.

Let's go back to our Newton quotient. What we've said is that

$(F(x + h) - F(x))/h = (\int_x^{x+h} f(x)dx)/h$ is approximately equal to $f(x + h)h/h$ $= f(x + h)$.

What happens when h goes to 0?

We get

dF/dx = f(x). Or, the derivative of an integral of f is f.

What about $\int_a^b f'(x)\,dx$?

First let's turn this into a Riemann sum.

$$\sum f'(x)\,\Delta x$$

Let's pull a reversal of the same stunt. We'll make a Riemann sum of the Newton quotient that computes f′(x) from f(x) using Δx for both.

We get $\Sigma\ ((f(x + \Delta x) - f(x))/\Delta x\ )\Delta x$

Then the divided by Δx and the multiplied by Δx cancel out, giving us

$$\sum f(x+\Delta x)-f(x)$$

Let's look at the definition of a Riemann sum (Figure 6.22).

Notice that each point on the x-axis that we sum over is exactly Δx away from the previous one. The first point is f(a), the second is f(a + Δx), the third is f(a + 2Δx), and so on up to f(b).

So let's look at what we are summing when we do $\Sigma f(x + \Delta x) - f(x)$.

The first entry is

$f(a + \Delta x) - f(a)$

to which we add

$f(a + 2\Delta x) - f(a + \Delta x)$

and then we add

$f(a + 3\Delta x) - f(a + 2\Delta x)$

and so on until we reach the last two which are

$f(a + n\Delta x) - f(a + (n - 1)\Delta x)$

and

$f(b) - f(a + n\Delta x)$

Notice that we are adding and subtracting each entry except for f(b) and f(a). So when we sum everything, those two entries disappear, leaving us with

$f(b) - f(a)$

So

$\int_a^b f'(x)\, dx = f(b) - f(a)$

So one of the indefinite integrals of the derivative of f(x) is f.

The fundamental theorem of calculus tells us that integration and differentiation are what are called **inverse operations**. Do one and then the other and you end up back where you started, more or less.

In other words, to find integrals we **anti-differentiate**. To see what this means, let's go back to our table of derivatives and swap things around.

| Integral | Function |
|---|---|
| 1. $\int f(x) = c$ (a constant) | $f(x) = 0$ |
| 2. $\int f(x) = a\int g(x)$ (a is a constant) | $f(x) = ag(x)$ |
| 3. $\int f(x) = x^{n+1}/(n + 1)$ | $f(x) = x^n$ |
| 4. $\int f(x) = \int g(x) + \int h(x)$ | $f(x) = g(x) + h(x)$ |
| 8. $\int f(x) = \sin(x)$ | $f(x) = \cos(x)$ |
| 9. $\int f(x) = -\cos(x)$ | $f(x) = \sin(x)$ |

Always remember that you can generate more indefinite integrals for the same function by adding an arbitrary constant c.

**So it's as simple to integrate as to differentiate.**

*Liar, liar, pants on fire.*

**All right. You caught me.**

Actually, integration is a pain in the neck. See, if you have a function in closed form, its derivative will also be a function in closed form. But that isn't true of integrals. Some functions that look simple, like the bell curve (which we'll cover in the probability chapter) (Figure 6.23) are totally nonintegrable by any normal technique.

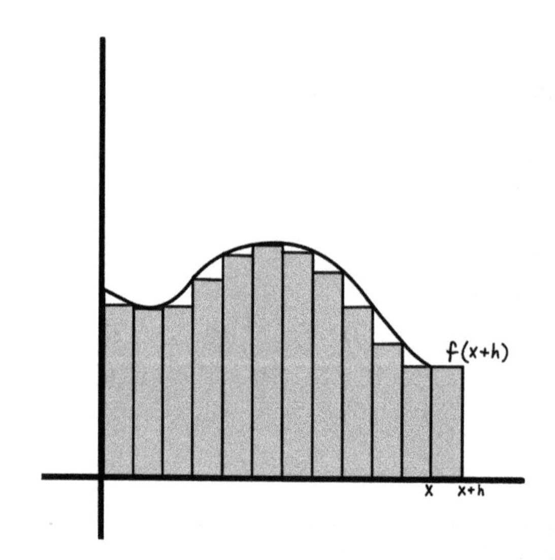


**FIGURE 6.22**
Riemann Sum + h

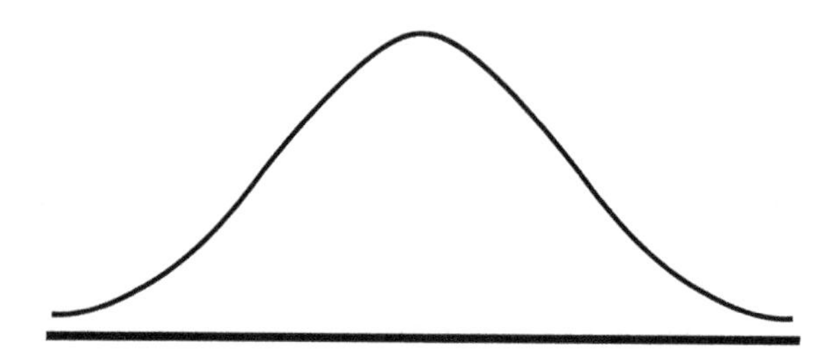

**FIGURE 6.23**
Bell Curve

Integration involves tricky guesswork and requires a problem solver to figure out how a batch of rules can be twisted to apply to a particular problem. Some of those rules are derived from trying to undo the product, quotient, and chain rules. Most of the tricks involve clever applications of substitutions that can be fun to figure out if your mind works the right way.

We now illustrate one of those sneaky tricks called "integration by parts." We start with the product rule $(fg)' = fg' + gf'$. Integrating both sides and then rearranging terms, we find

$$\int fg'dx = fg - \int gf'dx$$

The idea for how to use this equation is the following: suppose that you have some expression that you don't know how to integrate, but at least you recognize it as a product of two terms f and g', where g' is something that you know how to integrate to get g. Then you could try using the above expression in the hope that maybe gf' is easier to integrate than the original fg'.

Let's use this trick to integrate x sin(x). Here we will choose $f = x$ and $g' = \sin(x)$. Since $g' = \sin(x)$, that means g must be the integral of sin(x), which is $-\cos(x)$. Since $f = x$, it follows that $f' = 1$, so the integration by parts formula gives us

$$\int x\sin(x)dx = -x\cos(x) + \int \cos(x)dx = \sin(x) - x\cos(x)$$

Note that we can check that this really is the answer by taking its derivative and seeing that we get x sin(x).

A lot of students who have survived through differential calculus find themselves stumped when they reach the techniques of integration sections of their books. This is often the last culling point in calculus. Those who make it through this have sufficiently analytic minds to go on to more advanced mathematics or to dig into physics. Those who don't usually decide math is not for them.

This is a fallacy because the analytic and substitution tricks are really only a single kind of mental puzzle, and there is a lot of math that does not involve that kind of puzzle solving. Indeed, many of the college math classes that come after calculus do not require any such analytic ability at all.

Besides, there is a cheat around this problem. That cheat only works if the functions involved are sufficiently well-behaved – and yes, that's the same well-behaved we needed for Taylor series, because the cheat involves Taylor series.

We're going to write the series a little differently. If you look at the terms of the series, each is of this form: $f^{(n)}(x_0)(x - x_0)^n/n!$

We can write the series as a sum using the same Σ for sum that we used with Riemann sums as follows:

f(x) is approximately $\Sigma\, f^{(n)}(x_0)(x - x_0)^n/n!$

where n starts at 0 and goes up as high as you are doing the series (It can even go to infinity, but we'll worry about that later)

Suppose we want to find the derivative of this Taylor series. Since it's a sum, we simply need to find the derivatives of each term and then add them up.

So let's look at each term $f^{(n)}(x_0)(x - x_0)^n/n!$

If $n = 0$ (that is, if we are at the first term), then this term is a constant which will have a derivative of 0.

Every term where $n > 0$ is easy to differentiate.

$f^{(n)}(x_0)\ /n!$ is a constant, so the derivative of the term will be $f^{(n)}(x_0)\ /n!$ times $d(x - x_0)^n/dx$. This last is just $n(x - x_0)^{n-1}$.

So the derivative of the term will be

$(f^{(n)}(x_0)\ /n!)\ n(x - x_0)^{n-1}$

We can simplify this a little since $n! = n(n - 1)(n - 2) \ldots 1$

$n/n! = 1/(n - 1)(n - 2) \ldots 1 = 1/(n - 1)!$

So the term is

$f^{(n)}(x_0)\ (x - x_0)^{n-1}/(n - 1)!$

The derivative of the Taylor series is therefore

$\Sigma\, f^{(n)}(x_0)\ (x - x_0)^{n-1}/(n - 1)!$

We can pull a similar stunt using integrals and the rule that

$\int x^n = x^{n+1}/(n + 1)$

If we do this to the Taylor series, we have that

$\int\Sigma f^{(n)}(x_0)(x - x_0)^n/n! = \Sigma f^{(n)}(x_0)(x - x_0)^{n+1}/(n + 1)! + c$ (c some constant)

In other words, for a well-enough behaved function, we can use a Taylor series to find its integral. Since Taylor series are not so bad to compute, we can use this as a sometimes easier numerical method for integration.

The fundamental theorem of calculus is not just a means to calculation. It also points toward some interesting relationships between characteristics of geometric shapes. Unfortunately, these are often neglected in teaching calculus, which tends to focus on the analytic side of things. This is too bad, because in many respects they are the meaning of the fundamental theorem, which is, despite its analytic character and use, at heart a geometric relationship.

Let's take our old friend the circle. We can find the area of a circle using a Riemann sum that amounts to slicing the disk interior of the circle into successive circular strips (Figure 6.24).

Each of these strips has an area approximately equal to $2\pi r\Delta r$. If we sum and take the limit, we end up with an integral $\int_0^r 2\pi r dr$. As $2\pi$ is a constant, so this becomes $2\pi\int_0^r r dr$. The integral of rdr would be $r^2/2$. So this integral is $2\pi r^2/2$ which is $\pi r^2$. The integral of the circumference of a circle is the area of that circle. The derivative of the area is the circumference.

As something a bit more complicated, we will now use calculus to find the formula for the volume of a cone. Suppose that we have a cone of radius R and height h. We draw the axis of the cone (the line from the vertex to the center of the base) and denote by z the distance along the axis from the vertex. We're going to slice the cone into a bunch of thin disks with position z and thickness dz (Figure 6.25).

If the radius of one such disk is r, then the volume of the disk is $\pi r^2 dz$. However, r increases from 0 to R while z goes from 0 to h. So we have $r = (R/h)\,z$. We then find that the volume V of the cone is

$$V = \int_0^h dz\pi(R^2/h^2)z^2$$

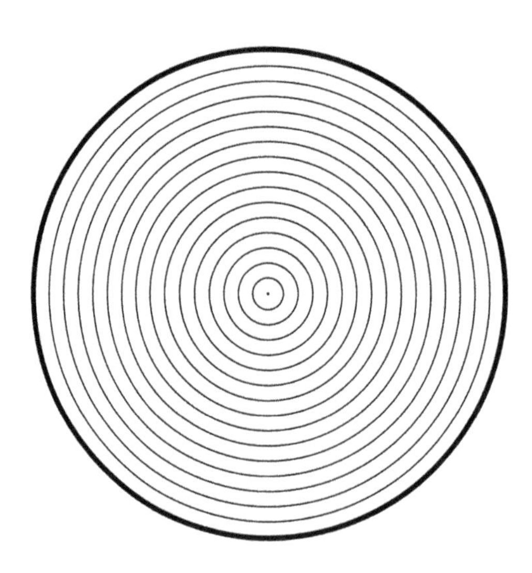

**FIGURE 6.24**
Sliced Disk

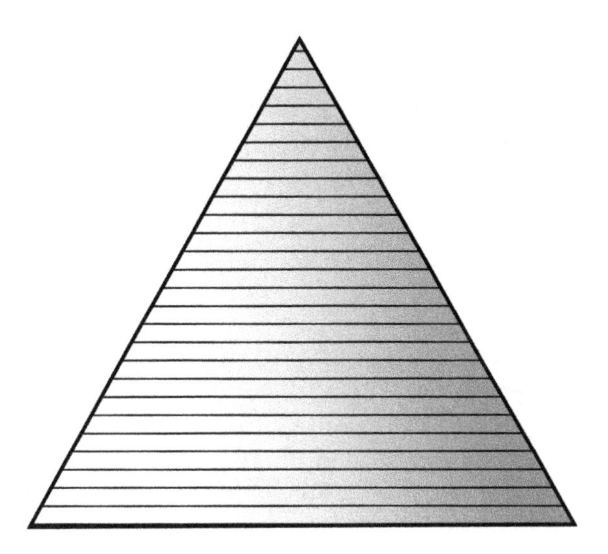

**FIGURE 6.25**
Sliced Cone

The integral of $z^2$ is $z^3/3$. So we find

$$V = \pi\left(R^2/h^2\right)\left(h^3/3\right) = \left(\pi/3\right)R^2h$$

Similarly, the volume of a sphere of radius r can be calculated by chopping it up into disks. The integral is more complicated, and we're not going to show it, but will just quote the well-known end result $V= 4\pi r^3/3$. Once one has this formula for the volume of the sphere, we can use it to derive the formula for the surface area of the sphere. This is done by chopping up the sphere in a different way, as a collection of concentric spherical shells of radius r and thickness dr. Adding the last spherical shell gives a volume of Adr, where A is the area of the sphere. So we find $A = dV/dr = 4\pi r^2$. The volume is the integral of the surface area, and the surface area is the derivative of the volume.

This relationship is embodied in some complicated theorems (**Green's theorem**, **Stokes' theorem**, and **Gauss' theorem**), but what they amount to is that when you have a shape with a boundary, there is an equivalence between the size of the shape (size being area, volume, hypervolume, whatever) and an integral taken using the boundary (circumference, area, volume, etc.). There is also an equivalence between a derivative taken on the whole shape and the boundary itself. We're being deliberately fuzzy here because it would be a lot of work to define terms more exactly. While the true utility of these theorems only comes from fully knowing and understanding them, they are all extensions of the basic relationship between a circle's circumference and its area.

The relationship here is simple geometrically. If you imagine a disk as made up of an infinite number of concentric circles, then if you increase the radius just a little you are adding a new circle to the outside and the amount of stuff in the circumference of that circle is added to the area. So the circumference of the circle is the rate of change of the area. This applies to any shape that you can in effect grow by layering on shapes of a lower dimension. This is the underlying geometry of the relationship between derivative and integral. Each new layer's size is the rate at which the object itself is growing. The layer therefore is the rate of change (derivative) of the object and the object is the sum (integral) of its layers.

Any mathematician reading this will probably want to whack us upside the head, and with good cause. We're being extraordinarily sloppy in this description. But we're not trying to show how to use these theorems, just trying to show their roots, which can be understood even without the elaborate machinery of calculus.

## Logarithm and Exponential

There's a cliché that great discoveries in science are not accompanied by the word "Eureka!" but rather by the phrase, "that's funny … ." We're now going to look at one of the "that's funny" pieces of calculus. Recall that one of the derivative formulas was that the derivative of $x^n$ is $nx^{n-1}$. From this and the fundamental theorem of calculus, it follows that the integral of $x^n$ is $x^{n+1}/(n + 1)$. But wait a minute, even calculus doesn't allow us to divide by zero, so we better not use that formula when $n + 1 = 0$, or in other words when $n = -1$. That's funny, every power of x has a nice simple formula for its integral, except $1/x$.

Nonetheless, $1/x$ is a perfectly good integrable function. It has an integral, just not one that we have a simple formula for. As we will see, this integral has some very interesting properties. Let's define ln(x) to be (a particular version of) this integral:

$$\ln(x) = \int_1^x du/u$$

Now let's compute ln(ab) where a and b are numbers:

$$\ln(ab) = \int_1^{ab} du/u$$

Now we're going to divide the interval from 1 to ab into two pieces, from 1 to a, and from a to ab:

$$\ln(a\,b) = \int_1^a \frac{du}{u} + \int_a^{ab} \frac{du}{u}$$

The first integral on the right-hand side is ln(a). But what about the second integral? If we define w to be u/a, then u = aw, so du/u = a dw/(aw) = dw/w. Also, when u goes from a to ab, w goes from 1 to b. So we find

$$\int_a^{ab} \frac{du}{u} = \int_1^b \frac{dw}{w} = \ln(b)$$

Putting it all together we find

$$\ln(ab) = \ln(a) + \ln(b)$$

This should look oddly familiar. Recall the isomorphism between addition and multiplication $10^{a+b} = 10^a 10^b$, where 10 to the power x turns addition into multiplication. The function ln acts in the opposite direction, turning multiplication into addition.

This suggests that the inverse of ln should be some sort of power. So let's define $e^x$ to be the inverse of ln(x). That is, $y = e^x$ if $x = \ln(y)$. We will also sometimes write this function as exp(x). Now let's take the equation $x = \ln(y)$ and take the derivative of both sides with respect to x. The left-hand side is easy: derivative of x with respect to x of x is 1. On the right-hand side, since ln(y) is the integral of 1/y, it follows that the derivative of ln(y) with respect to y is 1/y. Applying the chain rule, we have

$$1 = (1/y)(dy/dx)$$

From this it follows that

$$dy/dx = y$$

So $e^x$ is that function that is equal to its own derivative.

Then no matter how many times we take the derivative of this function, we always just get the same function back. In particular, since $e^0 = 1$, it follows that all derivatives of $e^x$ are equal to 1 at $x = 0$. This immediately gives us a Taylor series for $e^x$.

$$e^x = 1 + x + x^2/2! + x^3/3! + \ldots$$

and since $e^1 = e$, the Taylor series gives us an expression for the number e:

$$e = 1 + 1 + 1/2! + 1/3! + \ldots$$

from which we find that e is approximately equal to 2.71828.

Now with a slight change of notation and an assist from the chain rule, we find that the function $N(t) = N_0 \exp(\lambda\, t)$, where $N_0$ and $\lambda$ are constants, satisfies the equation

$$dN/dt = \lambda N$$

This innocent-looking equation is the most frightening equation in all of mathematics, as well as giving rise to the most widely misused mathematical adjective, "exponentially."

What's so frightening? Well let's consider the case where $\lambda$ is positive and define the quantity $\tau$ to be $\ln(2)/\lambda$. Note that since $\lambda$ times t has no dimensions, and t has dimensions of time, it follows that $\lambda$ has dimensions of 1/time and therefore that $\tau$ has dimensions of time. $\tau$ is called the doubling time, because from its definition it follows that $\exp(\lambda\tau) = 2$, so after a time $\tau$ has elapsed, N is twice as large as when it started. Then after another amount of time $\tau$, N is twice as large as that, or four times as large as it started.

This is an old mathematical game: start with something small and double it a number of times to see how large it gets. This amounts to listing the powers of 2: 1, 2, 4, 8, 16, 32, 64, 128, 256, 512, 1024, 2048, 4096, 8192, 16384, 32768, 65536, 131072, 262144, 524288, 1048576, 2097152, etc. Not only does exponential growth lead to large numbers from small beginnings, but it does so at an ever-increasing rate: after all the time it takes to go from one million to two million is the same time that it takes to go from 1 to 2.

Because of this startling growth encapsulated in the exponential function, the word "exponentially" has become a favorite piece of pseudo-mathematical hyperbole. Anytime someone wants to call attention the fact that something has grown, they say that it is "exponentially" bigger.

What about the case where $\lambda$ is negative? That's not frightening at all. In this case, we define the time $t_{1/2}$ (called the half-life) by $t_{1/2} = \ln(2)/|\lambda|$ and we find from the definition that $\exp(\lambda t_{1/2}) = ½$. In each half-life, the original amount goes down by half: from $N_0$ to $N_0/2$, and then to $N_0/4$, and then $N_0/8$, and so on. After a certain number of half-lives (depending on the original amount), there's hardly any left at all.

Despite its startling properties, the exponential function is only frightening if it has some application in the real world. (Imaginary monsters are much less frightening than real monsters.) So is there any real-world application? Boy is there ever! And as this book is being written, we are living though one such application. Recall the equation that leads to exponential growth

$$dN/dt = \lambda N$$

says that the rate at which something grows is proportional to how much of it there is. So what acts like that? An infectious disease does! Each person infected with a virus can pass that virus on to other people. So the rate at which the number of infected people grows is proportional to the number of infected people. It is then a consequence of calculus that the number of infected people really grows exponentially. This is a literal not melodramatic use of the term. From a single original case, that exponential growth gives rise to frighteningly large numbers with the potential to create worldwide devastation.

As we write these words, the numbers are already frightening, and each of your authors lives in a US state under a variety of "social distancing" and "shelter in place" orders. But what is the point of these orders? How can you fight an exponential? It turns out that you can, because the monster in the equation goes away when $\lambda$ turns from positive to negative. Each person infected with the virus will either fight it off and recover, or lose their battle with the virus and die.

It matters very much which of these outcomes happens, but in terms of the number of infected people, either outcome results in that person no longer being infected. This means that the number of infected people also tends to *decrease* at a rate proportional to the number of infected people. Thus, whether the overall $\lambda$ is positive or negative depends on which of these two rates, the rate of increase or the rate of decrease, is larger. Without social distancing, the overall $\lambda$ is positive leading to exponential increase. With enough social distancing, $\lambda$ can be turned negative, leading to exponential decrease.

Oddly enough, the same exponential reasoning that we apply to the virus can also be applied to us. Each human population has a birth rate and a death rate. If the birth rate is larger than the death rate, then the number of people increases in proportion to the number of people already there, leading to exponential growth. And indeed the overall number of people has grown dramatically in the past 200 years, which coupled with the growing industrialization of that same time period has led to great pressure on the environment, including climate change and the widespread loss of species.

There's another monster lurking here. Both exponential growth and exponential decay have applications in nuclear physics: growth to the explosion of an atomic bomb, and decay to radioactive dating. In an atomic bomb made with uranium, the uranium nuclei are unstable, occasionally breaking apart into two smaller nuclei and also releasing two neutrons. This process is very slow in the sense that a uranium nucleus has a very long half-life. However, a neutron released in the fission of one uranium nucleus can hit another uranium nucleus and cause it in turn to split apart (a process called nuclear fission), releasing more neutrons, which split more uranium nuclei, and so on.

In a small piece of uranium, each neutron is more likely to escape the sample than to split another nucleus. However, for a certain sample size, called critical mass, on average one of the two neutrons produced in a fission will cause another fission. Any mass larger than critical mass corresponds to a positive $\lambda$ and thus exponential growth of the number of fissioning nuclei. The doubling time is so short that the whole mass of uranium explodes in a very short time. The bomb is made by taking two masses of uranium, each smaller than critical mass but that have more than critical mass when put together. To set the bomb off, the two masses are rapidly assembled into one, and then exponential growth of the number of fissioning nuclei produces the explosion.

A more benign use of this process occurs in the archeological process of radioactive dating. Radioactive dating is done using the fact that certain unstable nuclei decay into other nuclei with a measured half-life. In particular, suppose that you have a purported historical document, but are worried that it might be a forgery. One way that you might check it is to see if the paper is as old as it is claimed to be. Paper is made from trees which, like all plants, make their material using carbon dioxide from the air. Most of the carbon in carbon dioxide is carbon 12, that is, its nuclei have 6 protons and 6 neutrons. However, a small proportion of the carbon in carbon dioxide is carbon 14, with 6 protons and 8 neutrons.

Carbon 14 is made by cosmic rays in the upper atmosphere, but is eventually mixed throughout the atmosphere, where some of it can be incorporated into plants. Unlike carbon 12, which is stable, carbon 14 is unstable and has a half-life of 5730 years. Once the tree is made into paper, the original number $N_0$ of carbon 14 atoms gets smaller by the equation $N = N_0 \exp(\lambda t)$, where now $\lambda$ is a negative number related to the half-life by $t_{1/2} = \ln(2)/|\lambda|$. Now divide the equation by $N_0$ and take ln of both sides leading to

$$\lambda t = \ln(N/N_0)$$

Since $\lambda$ is negative, we have $\lambda = -|\lambda|$, and by the properties of ln we have $\ln(N/N_0) = -\ln(N_0/N)$. Putting all that together with the expression for half-life, we find

$$t = t_{1/2} \ln(N_0/N)/\ln(2)$$

So in other words, if we know $t_{1/2}$, $N_0$, and N, then we can find out how old our sample is. The half-life is well known by having been measured in the lab by counting radioactive decays of a sample of carbon 14. N for a small sample of our paper is measured using a mass spectrometer. $N_0$ for that same sample is inferred from the (previously measured) proportion of carbon 14 in living plants.

*Thus, calculus and physics make life easier for historians and archeologists.*

**And harder for forgers.**

## The Absence and Presence of Calculus

What then does a person separated from the understanding of calculus lack that someone who understands it possesses?

An awareness of the character of motion and of size.

Think about that. Size is the thing we made sticks for. Motion is what physics has been about from the beginning, from thrown sticks to moving stars. Calculus is the language of these. If analytic geometry is one of the most fertile grounds in the theoretical universe, calculus is its most vital crop. And there's no need for anyone not to be able to grow and harvest this crop. Calculus can be difficult and complex to practice, requiring finesse, twisty thinking, and a lot of computing time, but understanding how it works and what it does is not that hard. It is regrettable that it is set as a dividing marker between the mathematically minded and everyone else. Much of what is taught after calculus is specialized practice distant from the world, but calculus itself is intimately bound up in how things are the way they are and move the way they move. There's no need and no good in denying that understanding to anyone.

Fortunately, with a little work and thought it can be reclaimed. And what comes to the mind that reclaims it? A world where motion is comprehensible in its controlled and uncontrolled ways; a world where the strangest shapes can be modeled and molded, their sizes discerned; a world where motion, shape, and transformation are all interconnected, where processes reveal their outcomes without being seen through to the end, and where the mysterious changes in things can be anticipated, grasped, and acted upon; a world where each process can be made to give up its secrets with study and analysis; a world where best, worst, least, and greatest can be found; a world where you know how fast the wine flows, and how many glasses the barrel will fill.

# 7

## *The Language of Motion*

An object in motion tends to remain in motion at the same speed and in the same direction unless acted upon by a force.

That's Newton's first law of motion. We've all heard it so many times that it's become rote. From our perspective, it's hard to see how revolutionary a concept it was.

Our everyday experience of motion is that moving things eventually stop. Throw a ball and it will hit the ground. Run and you get tired and eventually need to stop running. Cars run out of gas and stall on the highways.

From the human perspective, things don't keep going. Newton's first law says that any time you see things stopping, there's a force acting on them. That without those forces, the ball would keep flying, you would keep going without having to pump your legs, and you'd never need to gas up your car unless you wished to turn or speed up.

Accepting Newton's first law requires a change of perspective that alters the universe we perceive. All well and good, but on its own, the law doesn't tell us how a force will change the way an object is moving.

The second law takes care of that. Elaborated slightly, the law says that when a force F acts on an object of mass m, that object experiences an acceleration a, which is equal to F/m. This is usually shortened to

$$F = ma$$

This equation looks like a tame piece of high school algebra, one of those give me any two of the terms and I'll find you the third kinds of things. However, for motion, we want to know position as a function of time. And when we ask this question, the equation above leaves the domain of algebra and enters that of calculus.

*Calculus!! Calculus!! Bwah. Hah. Hah.*

**Now cut that out!**

*All right. Sheesh. I don't even get to provoke a bit of mathphobia.*

To find position from $F = ma$, we need to use two facts of calculus:

$a = dv/dt$, i.e., acceleration is the derivative of velocity with respect to time, and

$v = dx/dt$, i.e., velocity is the derivative of position with respect to time.

If we put these two together, we get:

$a = d^2x/dt^2$, i.e., acceleration is the second derivative of position with respect to time.

This lets us restate Newton's second law as

$$d^2x/dt^2 = F/m$$

This is a differential equation, or at least it becomes a differential equation once we specify how F depends on x. In a differential equation, the unknown function x(t) occurs on both the right-hand side and the left-hand side of the equation.

Nearly every aspect of physics can be mathematically described using differential equations. That sounds great; after all, we've got one branch of mathematics that we can use to solve everything in physics. But first we have to figure out how that branch works. Let's start with an example we've seen before: we take an object and throw it straight up. We want to find its height y as a function of time. The force of gravity on the object points downward and has a magnitude mg, where $g = 9.80\ m/sec^2$. Since y increases in the upward direction, we have $F = -mg$ and therefore our differential equation is

$$d^2y/dt^2 = -g$$

But this is something we know how to solve: we recall that the integral undoes the derivative and so we take the integral of both sides to get

$$dy/dt = -gt + c_1$$

where $c_1$ is some unknown constant. Now we integrate again and find

$$y = -(1/2)gt^2 + c_1 t + c_2$$

where $c_2$ is some other unknown constant. But what are these unknown constants doing in our formulas? And what good is a formula with an unknown constant? How are we supposed to use it to calculate anything? To answer this, we think about what the object is doing at time zero, the time when we throw it. Let $v_0$ be its velocity at time zero. Then, from the first of the above two equations evaluated at $t = 0$, we find $v_0 = c_1$. Let $y_0$ be the height of the object at time zero. Then, from the second equation we find $y_0 = c_2$. Our solution for y then becomes

$$y = -(1/2)gt^2 + v_0 t + y_0$$

This is a general feature of the equations of motion of physics. We get to pick the initial position and velocity, and then the solution of the equation of motion tells us where the object will be at any subsequent time.

This first example of a differential equation is perhaps too simple: recall that what makes a differential equation difficult is that the unknown quantity x(t) occurs on both sides. But for a falling object, the force is constant and so we could solve the equation by just doing integrals. Now, let's consider a more complicated example. We put a mass on a spring. As long as the spring is neither stretched nor compressed, it exerts no force.

But if we stretch the spring a little amount, it exerts a force pulling back, and the more we stretch, the more it pulls. In mathematical terms, the force is given by $F = -kx$, where x is the amount the spring is stretched, and k is a constant called the force constant of the spring. The negative sign tells us that the force is in the opposite direction to the position.

It works the same if the spring is compressed instead of stretched, the more we compress the spring, the harder it pushes back. We don't even need a separate equation: in $F = -kx$, a compression corresponds to a negative value of x, and the formula gives the amount of the push. Now our differential equation $F = ma$ becomes

$$d^2x/dt^2 = -(k/m)x$$

Our unknown quantity x(t) occurs on both sides of the equation. How are we supposed to solve this? We are basically looking for a function where when we take its derivative twice, we get back the same function but multiplied by a negative constant. Do we know any functions like that? It turns out that we do. We encountered two of them in the previous chapter. Recall that the derivative of sine is cosine, and the derivative of cosine is minus sine. This means that if we take two derivatives of sine, we get minus sine. And if we take two derivatives of cosine, we get minus cosine. But how do we get the factor of k/m? Suppose we try $x(t) = \sin(\omega t)$, where $\omega$ is a constant. Then, taking one derivative we find $dx/dt = \omega\cos(\omega t)$ and taking derivative of that we find $d^2x/dt^2 = -\omega^2 \sin(\omega t)$. That is, we have found that $d^2x/dt^2 = -\omega^2 x$. This is a solution of our differential equation provided that $\omega^2 = k/m$. That is, we must choose the constant $\omega$ to be the square root of k/m. What works for sine also works for cosine and for any combination of sine and cosine. So the general solution is

$$x(t) = c_1 \cos(\omega t) + c_2 \sin(\omega t)$$

where $c_1$ and $c_2$ are arbitrary constants. As before, we find the values of those constants using initial conditions. Let $x_0$ be the position and $v_0$ the velocity of the object at $t = 0$. Since $\sin(0) = 0$ and $\cos(0) = 1$, we find that $x_0 = c_1$. Taking the derivative of the general solution we have

$$dx/dt = -\omega c_1 \sin(\omega t) + \omega c_2 \cos(\omega t)$$

which evaluated at $t = 0$ yields $v_0 = \omega c_2$. Our general solution then becomes

$$x(t) = x_0 \cos(\omega t) + (v_0 / \omega) \sin(\omega t)$$

## Linear Digression

Let's back out of calculus for a bit because we just casually used a tool of math that comes from another field entirely. That field is called linear algebra. In the previous chapter we introduced vectors as a way to represent positions, velocities, accelerations, forces, and other multidimensional objects.

We also gave properties of vectors and the axioms that govern them. More broadly, anything that obeys those axioms is called a vector space. Without too much elaboration, a vector space consists of objects that can be added together and that can be multiplied by real numbers. So if v and w are vectors and a is a number, then v + w is a vector and av is a vector.

The equation of motion for a mass on a spring is called a linear differential equation. It has the property that if you add any two solutions of the equation, you get a solution, and if you multiply any solution by a number, you get a solution. This means that (in a completely abstract mathematical sense) the solutions of the equation are vectors.

Whenever mathematicians create a kind of algebraic object, they immediately ask about functions from one of that kind of object to another such that preserves the algebraic structure. What this means is that if you use the algebra to combine objects that are parameters for the function, then apply the function and what you thus get would be the same as if you applied the function to the objects individually and then combined them using the algebra.

That's a bit abstract, so let's apply the idea to a vector space. A function T from one vector space to another (or to itself) is called a linear transformation if

$$T(v+w)=T(v)+T(w)$$

and $T(av) = aT(v)$

Let's see what effects restricting our attention to linear transformations has. We'll go back to plain old boring three-dimensional Euclidean space. And we'll consider a linear transformation from this space to itself. We'll write a vector in this space as $xi + yj + zk$, where

$$i=(1,0,0),\quad j=(0,1,0),\quad \text{and}\, k=(0,0,1)$$

Let's see what happens if we apply some linear transformation T to this vector.

$T(xi + yj + zk) = T(xi) + T(yj) + T(zk)$ because of the first property of linear transformations. And

$T(xi) + T(yj) + T(zk) = xT(i) + yT(j) + zT(k)$ because of the second property.

So,

$$T(xi+yj+zk)=xT(i)+yT(j)+zT(k)$$

In other words, to find out what this transformation does to any vector, we only need to know what it does to the i, j, k vectors. But this is a transformation from $R^3$ to itself. So, the resulting vectors can also be written as linear combinations of i, j, and k.

In other words,

T(i), T(j), and T(k) are themselves sums of numbers multiplying i, j, and k. For reasons that will become clear in a bit, we're not going to write these numbers as x, y, z. Instead, we'll say

$$T(i)=a_{11}i+a_{12}j+a_{13}k$$

$$T(j)=a_{21}i+a_{22}j+a_{23}k$$

$$T(k)=a_{31}i+a_{32}j+a_{33}k$$

So, with nine numbers we can completely specify the linear transformation. And indeed (warning: eye-blurring boring calculations ahead) that means we can calculate $T(xi + yj + zk)$ as

$$\begin{aligned}&xT(i)+yT(j)+zT(k)\\&=x(a_{11}i+a_{12}j+a_{13}k)+y(a_{21}i+a_{22}j+a_{23}k)+z(a_{31}i+a_{32}j+a_{33}k)\\&=(a_{11}x+a_{21}y+a_{31}z)i+(a_{12}x+a_{22}y+a_{32}z)j+(a_{13}x+a_{23}y+a_{33s}z)k\end{aligned}$$

So, in order to calculate this operation, we only need to do nine multiplications and six additions. Isn't that simple?

Ah, hmm. Maybe, simple but annoying. These days, of course, computers can do this in much less time than it takes to twitch at the idea of doing the calculations by hand. But linear algebra started in the 17th century and had its heyday in the 18th and 19th centuries. Many mathematicians were confronting this kind of fussy calculation by hand. Naturally, they tried to find lazy solutions.

We're not going to dive into everything they did, because a lot of it is fussy and didn't yield much beyond techniques for simplification. But one method they came up with yielded results so deep that they became one of the bases of quantum mechanics. Would these ideas have been found if they had the computers we have now? It's difficult to say.

What about more complicated forces? Is there always some trick that results in a formula for x(t)? Sadly, no. But if you are willing to make do with an approximate solution, there is a general computer method. Let's start by writing F = ma as two equations:

$$dx/dt = v$$

$$dv/dt = F/m$$

Now let's get sloppy and forget all that stuff about limits and write these two equations as

$$\big(x(t+dt)-x(t)\big)/dt = v(t)$$

$$\big(v(t+dt)-v(t)\big)/dt = F\big(x(t)\big)/m$$

Here dt is some small number, and the equal sign really means "approximately equal with the approximation getting better when we pick a smaller dt." Now, rearranging the terms in the above equations we have

$$x(t+dt) = x(t) + dt\,v(t)$$

$$v(t+dt) = v(t) + dt\,F\big(x(t)\big)/m$$

Remember that we are supposed to pick $x_0$ and $v_0$, and then the equations of motion are supposed to tell us the rest. Using $x_0$ and $v_0$ on the right-hand side of our equations gives us x(dt) and v(dt). But then using those on the right-hand side gives us x(2dt) and v(2dt), and then using those … (you get the idea). On a piece of graph paper, we plot the points (0, $x_0$), (dt, x(dt)), (2dt, x(2dt)), etc., and then play the old child's game "connect the dots" and we have the curve x(t) (or at least a good approximation of it).

Why do we call this a computer method? It certainly didn't start out that way. It is called Euler's method and was invented by the great Swiss mathematician Leonhard Euler, who was also a physicist, and engineer, who lived from 1707 to 1783: long before the invention of the modern electronic computer.

However, for a fixed time interval, the smaller you make dt, the more computations you need to do to get to the end of the time interval. Computers excel at doing a very large number of computations at high precision, and in a very short amount of time. (And unlike humans, computers don't make computational errors and don't complain about the boredom involved in doing the same damn task over and over.)

So is this picture of objects, forces, motion, and differential equations for $x(t)$ the last word in physics? Nope. (You knew things couldn't be that simple, right?) One issue, called "action at a distance," bothered Newton's contemporaries to no end and eventually led to a new kind of differential equation called partial differential equations (or PDE to the people who work in this area).

Let's start with Newton's law of gravity, which states that two masses M and m a distance r apart exert a force on each other of magnitude

$F = GMm/r^2$, where the constant G is called Newton's gravitational constant. Now let's consider the gravitational force that the Sun exerts on the Earth. When the Sun moves, its distance r from the Earth changes, and therefore (according to Newton) the Earth instantly feels a change in the force acting on it.

How, wondered Newton's contemporaries, could the information that the Sun had moved travel the enormous distance between Earth and Sun in no time at all? Newton's reply to these concerns was "hypotheses non fingo," a Latin phrase that is usually translated as "I frame no hypotheses." But if one wanted to get at what Newton meant by that, it might be more like the following: "Look man, I don't tell Nature what to do. I just take what Nature actually does and write that down as a bunch of equations."

What was merely disquieting to Newton's contemporaries became a fatal flaw when Einstein formulated his theory of special relativity. One consequence of Einstein's theory is that nothing can travel faster than light. Thus, "action at a distance," in which changes are instantly felt by distant objects, is not allowed. Instead, all physics theories must be field theories: any cause must have immediate effects only in its immediate vicinity, only gradually propagating outward to have later effects in more distant regions.

Ten years after producing special relativity, Einstein produced his theory of general relativity, an exotic theory in which gravity involves the warping of space and time, but which nonetheless reduces to Newtonian gravity when objects move slowly compared to light and when the gravitational field is weak. In general relativity, all changes in gravity travel at the speed of light: when the Sun moves, the Earth feels it eight minutes later.

What does the mathematics of a field theory look like? Let's *not* start with general relativity, which is a very complicated theory. Instead, let's start with something simpler: a guitar. The sound of a guitar is made by vibrating strings, each one under tension, with each string tension individually adjustable by a key. How does such a guitar string move? Let x be distance along the guitar string and y the height of that part of the guitar string above the flat level that the string would have if it weren't vibrating. Thus, to describe the motion of the string, we need a function of two variables $y(x, t)$: the height of the piece of string at position x and time t.

To find the motion of the string, we are going to use $F = ma$, but we are going to apply it separately to each piece of the string. Let T be the tension in the string and $\mu$ be the mass per unit length of the string (the mass of the guitar string divided by its length).

Now let's pick a small dx and apply $F = ma$ to the tiny piece of string between x and $x + dx$. The mass of the piece of string is $m = \mu dx$. The piece of string is pulled by force T by the piece of string to the right of it and also pulled by force T by the piece of string to the left of it. These forces point in almost exactly opposite directions and therefore they almost cancel out. The reason that they don't completely cancel out is that the string has a curved shape, and so the direction along the string at $x + dx$ is slightly different from the direction along the string at x. Calculating the net force on the tiny piece of string and applying $F = ma$ to it yields the following equation:

$$(1/v^2)\partial^2 y/\partial t^2 = \partial^2 y/\partial x^2$$

This equation is called the wave equation. Here v is a constant equal to the square root of $T/\mu$, and the $\partial$ symbol denotes a particular kind of derivative called a partial derivative. Recall that y is a function of two variables, x and t. So how do we take the derivative of y? $\partial y/\partial t$ means keep x constant and take the derivative with respect to t. In other words, sit at a single position x on the string and think of y at that position as a function of t, and take its derivative. $\partial y/\partial x$ means keep t constant and take the derivative with respect to x. In other words, take a snapshot of the string at a single instant of time t, and think of the y in that snapshot as a function of x, and take its derivative.

How are we supposed to solve the wave equation? In the next chapter, we will look at a method having to do with the properties of sine and cosine. However, for now note the following remarkably simple solution method: let f(s) be any function, and try $y(x, t) = f(x - vt)$. Then the calculus rules of the previous chapter along with the rule for how to take a partial derivative yield the following:

$$\partial^2 y/\partial x^2 = f''(x - vt)$$

$$\partial^2 y/\partial t^2 = v^2 f''(x - vt)$$

From this it immediately follows that y(x, t) satisfies the wave equation. Here f(x) is the shape of the wave at time zero. At any subsequent time t, the wave has the same shape, but just moved over to the right (i.e., in the positive x direction) by an amount vt. In other words, the wave keeps its shape, but moves to the right at speed v. Exactly the same considerations apply to any function $g(x + vt)$, which is a solution of the wave equation representing a wave moving to the left at speed v. The general solution of the wave equation is

$$y(x,t) = f(x - vt) + g(x + vt)$$

That is, a combination of a right moving wave and a left moving wave.

Remarkably, the wave equation isn't just for guitar strings. It also describes ripples in the water on the surface of a pond, sound waves traveling through the air, radio waves, microwaves, or light waves traveling through the vacuum of space, or through any other transparent medium. It even describes the gravitational waves of general relativity when they are far enough away from their source to be weak.

Of course, there's more to physics than the wave equation. There's the heat equation, which describes how heat flows from hotter objects to colder objects. This same equation (but now called the diffusion equation) describes how a blob of cream in a cup of coffee spreads out to eventually make a uniform coffee and cream mixture. There's Poisson's equation, which describes how electric charges make electric fields, and also how masses make gravitational fields. There's Schrodinger's equation, which describes the behavior of the wave function in quantum mechanics. We will encounter these equations, and ways to solve them, in the next chapter.

# 8

# *Sound, Notes, and Harmonics*

Mathematics evolves to help model the things in the universe that matter to us as humans. Some of those are fundamental to the way things work, like motion. Some are purely human inventions, like economics. Some seem like human inventions and turn out to be fundamental, such as probability.

And some are so deeply important to humans personally that we'll go to any mathematical lengths to model them, so we can get better at those things. And then we discover that while the thing we've modeled isn't fundamental, the model itself is.

For that, let us give an ear to music.

Music is something that every human culture of any time period has developed. Early flutes have been found that are more than 40,000 years old.

Music is both pan-human and deeply personal. People have strong attitudes toward what music sounds good, what music inspires them, what sets their feet tapping and hands clapping. What they sing when they're sad, and dance to when they're happy. It all matters to us as individuals, and it matters often to societies what music is popular, what is religious, what is praised, and what is deemed unsuitable.

The awareness that there is some underlying mathematics to music goes way back. The Pythagoreans felt that there was some deep relationship. Indeed, they claimed that the planets had associated notes that acted in harmony (watch this word for later developments). This is where we get the phrase "music of the spheres" from.

It's perhaps not too surprising that mathematics can be used to help understand music. Far more surprising is that music can return the favor by turning calculus into algebra and providing a powerful method for solving partial differential equations.

This method was discovered by Joseph Fourier, a French mathematician who lived from 1768 to 1830. Fourier's essential insight is that any sound can be written as a sum of pure notes. A sound wave is a disturbance in the normal air pressure, with the pressure alternating between slightly higher than normal and slightly lower than normal. If one were to graph this departure from normal pressure as a function of time, it would look like a sine wave for a pure note. This sine wave is characterized by its frequency $f$, the number of times in a second that the pressure goes from high to low and back again to high. The higher the frequency, the higher the pitch of the note. A general sound is a combination of sine waves of all different frequencies. Here one should not think of the different notes as being played one after the other as they are in music. Rather one should think of all the notes being played at once in one giant chord (or perhaps discord?) to produce the sound.

Oddly enough, Fourier wasn't studying sound: he was studying heat. He knew that heat flows from hotter objects to colder objects, and that when heat flows into an object, it raises the object's temperature. So if a bar of metal were to start with a nonuniform temperature, the flow of heat would gradually even things out until the temperature became uniform.

What would be a mathematical description of this process? Suppose that at time zero the temperature $T$ as a function of position $x$ along the bar is $T_0(x)$. What is the temperature at all later times? That is, what is $T(t, x)$?

To start with, Fourier needed a partial differential equation describing the temperature. Since the rate of heat flow depends on how steeply temperature changes from position to position, and since the rate of change of temperature depends on the net heat flow (heat flow in minus heat flow out), Fourier deduced that the equation for temperature is

$$\partial T/\partial t = C\partial^2 T/\partial x^2$$

Here C is a constant that depends on both how well the metal conducts heat and how much heat it takes to raise its temperature. This partial differential equation is called the heat equation.

But how to solve the heat equation? Here's where Fourier's insight comes in: just as any sound is a combination of notes, so any $T_0(x)$ is a combination of sin(kx) and cos(kx) for all possible values of the constant k. So, let's ask how a single one of these "Temperature Notes" depends on time and then add them all up to get T(t, x).

That is, we seek a solution of the heat equation of the form

T(t, x) = A(t) cos(kx). From the rules for how to take partial derivatives we have

$$\partial T/\partial t = (dA/dt)\cos(kx)$$

$$\partial^2 T/\partial x^2 = -k^2 A \cos(kx)$$

Note that this last equation means that the complicated calculus process of taking the second partial derivative with respect to x is replaced by the simple algebraic process of multiplying by $-k^2$.

Now, using our partial derivative calculation in the heat equation, we find

$$(dA/dt)\cos(kx) = -Ck^2 A \cos(kx)$$

which simplifies to

$$dA/dt = -Ck^2 A$$

But this is an equation that we have seen before: we are looking for a function where when we take the derivative, we get back the same function multiplied by a constant. This is the exponential function. So we find

$$A = \exp\left[-Ck^2 t\right]$$

So the behavior of a single "note" of heat is

$$T(t,x) = \cos(kx)\exp\left[-Ck^2 t\right]$$

And if we want to know what happens to our particular $T_0(x)$, we simply decompose it into notes, use the formula above for the time development of each note, and add up the contribution of all the notes to get T(t, x).

Furthermore, we can tell qualitatively from the formula for a single note what is going to happen: we remember that for any positive constant b, exp[bt] gets very large, once t is somewhat large compared to 1/b. But exp[–bt ] = 1/exp[bt], so it gets extremely small once t is somewhat large compared to 1/b. So, all the unevenness in temperature eventually goes away, with the most uneven parts going away first.

Iron bars are one-dimensional, so we could describe the temperature by a function that depends on only one spatial variable x. But suppose we instead want to know how temperature behaves in a three-dimensional object, like a cube of iron, a golden egg, or the Earth. Then, we would need to find a function T(t, x, y, z) satisfying some more complicated version of the heat equation. However, this more complicated version of the heat equation isn't that much more complicated, and remarkably Fourier's trick still works, and in almost exactly the same way. For three-dimensional objects, the heat equation becomes

$$\partial T/\partial t = C\nabla^2 T$$

where $\nabla^2$ is an abbreviation for $\partial^2/\partial x^2 + \partial^2/\partial y^2 + \partial^2/\partial z^2$. This is mathematical shorthand. It means that to apply $\nabla^2$ to a function, we find its second derivative with respect to x, add that to its second derivative with respect to y, and then add that to its second derivative with respect to z.

As with a lot of mathematical shorthand, a lot of calculations are concealed, ahem, compacted into a few symbols. This saves a lot of space on blackboards and in books, until you have to actually apply it.

This may look like one is only being lazy with the symbols. But in a lot of cases, as we are seeing in these examples, it is possible to do the work once and then figure out how it applies in very different cases.

Now consider a single "note" of the form

$$f(x, y, z) = \sin(k_x x)\sin(k_y y)\sin(k_z z)$$

where $k_x$, $k_y$, and $k_z$ are numbers. Then, the rules for partial derivatives give

$\nabla^2 f = -k^2 f$, where now $k^2$ is an abbreviation for $(k_x)^2 + (k_y)^2 + (k_z)^2$. And just as before, this leads to a time dependence of the form $\exp[-Ck^2 t]$. So the temperature evens out, and the more jagged it was to begin with, the faster it evens out. Even in the three-dimensional case, calculus becomes algebra and the complicated second partial derivatives of $\nabla^2$ just become multiplying by $-k^2$.

Now let's move on from iron cubes to coffee with cream. When we pour cream in our coffee, we stir it to get the same amount of cream in each part of the coffee. But if we weren't so impatient, we could just pour in the cream and leave it alone, and eventually the cream would spread out to a uniform density in each part of the coffee. This happens through a process called diffusion: the molecules in the cream bounce around at random. This means that just through random chance, a region of high concentration of cream, since it has more cream to begin with, will tend to lose cream to its surroundings. Similarly, a region of low concentration is likely through random chance to gain cream from its surroundings. In other words, cream in coffee behaves like heat in a block of iron.

But this is more than a slight qualitative similarity. If we call the concentration of cream in the coffee cup X, then the differential equation it obeys is

$$\partial X/\partial t = D\nabla^2 X$$

This is called the diffusion equation, and the constant D is called the diffusion constant. But note that despite the change of name, and the change of the situation being described, the diffusion equation is really just the heat equation. This often happens in mathematics: the same equation can be used to describe two entirely different physical situations. Thus, we can apply "mathematical laziness" and assert that we have already solved this problem: for any solution of the heat equation, just change the T to an X (and the C to a D) and voila! we have a solution of the diffusion equation, good for describing cream in coffee.

Now let's apply Fourier's ideas to the guitar string treated in the last chapter. For a single note, we should have

$$y(t,x) = A(t)\sin[kx] + B(t)\cos[kx]$$

and we should use the wave equation to find $A(t)$ and $B(t)$. Let $L$ be the length of the guitar string, and let's choose our $x$ coordinate to be zero at the left-hand end of the string: that is, the string goes from $x = 0$ to $x = L$. Guitar strings are held down at either end, which means that no matter what time it is, $y(t, 0) = 0$ and $y(t, L) = 0$. Since $\sin(0) = 0$ and $\cos(0) = 1$, the statement that $y(t, 0) = 0$ becomes $B(t) = 0$. So even though we started with both sines and cosines, the fact that the string is tacked down on the left end means that we will only use sines.

The statement that the right-hand end of the string is tacked down becomes $\sin[kL] = 0$. This means that most values of k aren't allowed at all! The guitar string can only have particular values of k. What are these allowed values of k? Well, to start with, we are looking for all possible angles whose sine is zero, and then the allowed values of k are those for which kL is equal to one of these special angles. What are the angles whose sine is zero? In degrees, it is 0, 180, 360, 540, 720, etc.

However, we are measuring angles in radians, where a whole circle (360 degrees) comes to $2\pi$ radians. (We are using these units so that we have the convenient result that derivative of sine is cosine and derivative of cosine is minus sine.) So in these units, the angles with zero sine are $0, \pi, 2\pi, 3\pi$, etc. Picking $k = 0$ would just give us the flat guitar string, not making any sound, so we summarize the allowed values of k as $k = n\pi/L$, where $n = 1, 2, 3, \ldots$.

Now let's consider the wave equation

$$\left(1/v^2\right)\partial^2 y/\partial t^2 = \partial^2 y/\partial x^2$$

and apply it to the function $y(t, x) = A(t) \sin(kx)$. We remember Fourier's trick that for such functions, taking the second derivative with respect to x is just the same as multiplying by $-k^2$. So we find

$$\left(1/v^2\right)d^2 A/dt^2 = -k^2 A$$

But we know how to solve that equation! It's the one we've been seeing over and over in the last two chapters. So we find

$$A(t) = a\cos(\omega t) + b\sin(\omega t)$$

where a and b are constants and $\omega = vk$. So the guitar string motion is given by

$$y(t,x) = [a\cos(\omega t) + b\sin(\omega t)]\sin(kx)$$

That is, the guitar string vibrates up and down with each point on the string moving just like a mass on a spring.

What is the pitch of the note played by the guitar string? To answer this, we need to know how many times the string vibrates up and down in a second. Since sine and cosine repeat themselves when their angle changes by $2\pi$, in a time t, the string goes through $\omega t/(2\pi)$ cycles. This means that the frequency f of the note played by the guitar string is $f = \omega/(2\pi)$. Furthermore, since $\omega = vk$ and $k = n\pi/L$, we find that $f = nv/(2L)$. So short strings make higher notes than long strings, and strings with a higher wave velocity make higher sounds. When we tune a guitar, putting the string under more tension leads to higher wave speed, which leads to a higher note. Putting the string under less tension leads to lower wave speed, which leads to a lower note. We fiddle with the string tension until we get just the right note.

But wait! The number n can be 1 or 2 or 3 or any whole number. So rather than a single note, the guitar string is playing several notes at once. That's right. The $n = 1$ note is the one that we think of as the note of the string. The $n = 2, 3, 4$, etc. notes are called higher harmonics, and it is their presence that gives each instrument its characteristic sound.

Over the hundreds of thousands of years since humans first made music, instrument makers garnered a great deal of practical knowledge about what materials in what shapes would produce what kinds of sounds. Nowadays, they can make use of the underlying mathematics to help them.

But more importantly, because of the ability to turn music into mathematics and then mathematics into music, it has become possible to create virtual instruments and digital music. In the last two generations, the math of music has made it possible to create and affect sounds in ways that would be nearly impossible without the mathematical manipulation.

*Richard: My younger child took two years of electronic composition classes in high school. For their final project they created virtual instruments that transformed ordinary kitchen sounds into musical harmonies.*

Having had success describing guitars, let's now turn to drums. A drum head consists of a stretched membrane held to a circular edge. But that description seems like a two-dimensional version of the description of the guitar string. And sure enough, the drumhead satisfies the two-dimensional version of the wave equation. Let x and y be coordinates for points on the drumhead, and let h be the height of a point of the drumhead above its resting position. Then, $h(t, x, y)$ satisfies the two-dimensional wave equation

$$\left(1/v^2\right)\partial^2 h/\partial t^2 = \nabla^2 h$$

where in this case $\nabla^2$ is an abbreviation for $\partial^2/\partial x^2 + \partial^2/\partial y^2$.

So now you might think we just look for solutions that are a function of t multiplied by $\sin[k_x x]\sin[k_y y]$ and then $\nabla^2$ just becomes multiplication by $-k^2$. But unfortunately, drums are not quite so simple. We can get perfectly good solutions of the wave equation this way, but they won't describe a drum. The reason is that h must be zero at the edge of the drum, and the solution above doesn't do that.

Nonetheless, part of the old strategy survives: we are going to look for a solution of the form $h(t, x, y) = \cos[\omega t]\, f(x, y)$, where $f(x, y)$ satisfies

$$\nabla^2 f = -k^2 f$$

for some value of k, and where f is zero on the edge of the drum. That is, we are simultaneously looking for the function f and the constant k. In the jargon of mathematics, $\nabla^2$ is called an "operator," and this process of finding the k and f is called "finding the spectrum of the operator" or "finding the eigenfunctions and eigenvalues of the operator."

As we saw earlier, we can take the innocent eigenvalues and eigenvectors of finite-dimensional vector spaces and make them creep out into vast function spaces. The methods we use here to solve the drum equations will be used again later in the chapter when we talk about quantum physics.

So how do we find the f and k for the drum? To find the right solution, the x and y coordinates are not the most convenient ones. Instead we use polar coordinates r and θ, given by $x = r \cos \theta$, $y = r \sin \theta$. That is, r is the distance from the center of the drum, and θ is the angle with respect to the x-axis. We look for a solution of the form $f = R(r) \cos(mq)$, where m is a whole number 0, 1, 2, etc. Then, the equation for f becomes the following equation for R

$$d^2R/dr^2 + (1/r)dR/dr + \left(k^2 - m^2/r^2\right)R = 0$$

The allowed values of k are those for which there is a solution of the above equation where $R = 0$ at the radius of the drum. For each allowed value of k, the frequency of the sound made by the drum is then $vk/(2\pi)$.

Do we know how to solve the equation for R? Well, not really. That is none of the conventional functions we have to work with (polynomials, trigonometric functions, exponentials, and logarithms) will actually provide such solutions.

So, we solve this by not really solving it. We will follow a longstanding mathematical tradition: both the differential equation that we don't know how to solve and its solutions are named after whichever 19th-century mathematician first studied them. The equation for R is called Bessel's equation, and its solutions are called Bessel functions, named for the 19th-century German mathematician, astronomer, and physicist Friedrich Wilhelm Bessel, who is also the first person to make accurate measurements of the distances to the stars.

While that may sound unsatisfying, both the computer method and the power series method can be used on Bessel's equation to produce graphs of Bessel functions to whatever accuracy we desire. (When David was a postdoctoral researcher, one of his supervisors tried this method on him: "Steve. Do you know how to solve this differential equation?" "Well, David since you are working on this problem, I hereby name this differential equation the Garfinkle equation, and so its solutions are the Garfinkle functions. There ya go. You're welcome." Despite this cheap trick, Steve was actually extremely good at all sorts of calculations involving differential equations. So much so that his wife Sandy joked that the only reason Steve could remember her name is that it could be written $\int_a^n dy$).

Unlike the guitar strings, the notes of the drum are not evenly spaced. That is, there is a note of lowest frequency $f_0$ and higher notes $f_1$, $f_2$, $f_3$, etc. However, the higher notes are not simple multiples of the lowest note.

We've been jumping around in this chapter from guitars to heated iron bars to drums, but all of those are human scale sorts of things: stuff you can pick up in your hands. Okay, you can pick up the hot iron bars, but it's not a good idea.

But again, mathematical laziness doesn't confine our solutions to the problems we started with. Once we solve something, we can look at other things in the universe that have processes that work in corresponding ways. If the ways processes work correspond, then the mathematics that describes and predicts those processes will likely have similar forms.

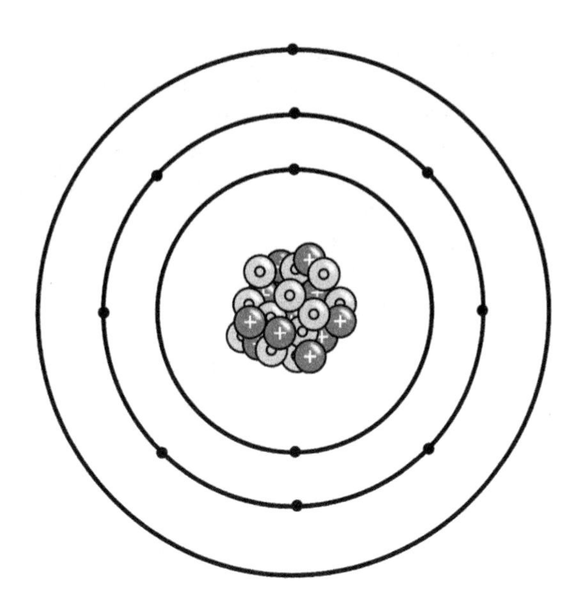

**FIGURE 8.1**
Sodium Atom

So without further delay –

Drum-roll please

We give you, the musical foundations of quantum mechanics.

Cast your eyes upon these atoms. Okay, they're hard to see individually, so here's an illustration (Figure 8.1).

An atom consists of a nucleus in the center surrounded by electrons. The nucleus has a positive electric charge, while electrons have a negative charge. The fact that opposite charges attract means that the electrons are attracted to the nucleus, which is what holds the atom together.

From this description, one would think that an atom is like a miniature solar system, with electrons orbiting the nucleus like planets going around the Sun. And indeed, this is often how atoms are depicted. But this picture is very misleading.

At any given time, we don't actually know the precise positions of the electrons. The best that we can do is to calculate the probability that the electron is in any given region by using a function called the wave function. This wave function in turn is a solution of a differential equation called Schrodinger's equation.

We are not going to write down Schrodinger's equation, but will instead make do with a qualitative description emphasizing analogies with musical instruments. Schrodinger's equation has solutions that are a combination of $\cos(\omega t)$ and $\sin(\omega t)$ multiplied by some function $\psi(x, y, z)$. As in the case of the drum, $\psi$ is a solution of some operator equation, only now the operator is not $\nabla^2$ but is instead $\nabla^2$ plus an extra piece that has to do with the forces between the electron and the nucleus.

Unlike the drumhead, space has no edge. But the fact that the electron is confined to the atom means that the probability of finding it in any given place must get very small when we get very far away from the nucleus. This translates to the mathematical condition that $\psi$ approaches zero as x, y, or z gets very large. This sort of "$\psi$ goes to zero at spatial infinity" condition plays the same role as the "h goes to zero at the edge of the drumhead" condition and similarly yields the condition that only certain values of $\omega$ are allowed. Let's pause for a moment to reflect on what this means. Schrodinger's equation tells us that for atoms, neither position nor energy is as simple as we thought it was in Newtonian mechanics. Position and energy are quantities that we are used to thinking of as stick measurements. That is, we think they are continuous values and that an object can occupy any position and have any energy. But Schrodinger's equation turns this around in two ways.

First, it formulates position in terms of likelihood rather than as an absolute. We'll return to this subject near the end of the next chapter. For the moment let's look at energy. There is a direct relationship between energy and frequency. There is a universal constant called Planck's constant and denoted by the letter h, such that for a solution of Schrodinger's equation with frequency f, the energy of the particle is hf. Note that since $f = \omega/2\pi$, it also follows that the energy E is given by $E = (h/2\pi)\omega$.

But as we saw from the properties of Schrodinger's equation, frequency can only be one of the eigenvalues of an operator. So, our stick quantity, energy, has become a stone quantity. Quantum physics shows that many of the fundamental physical values are not continuous; they are discrete and can only take on specific values.

This confuses people a lot, because these are values that from our direct experience seem to be continuous. Mass and energy look like they're stickish. How can they be stony?

The answer to that comes from the fact that the stones represent very small quantities. Atoms are very small, and the universal constant h is a very small quantity. As an analogy, think about rice. Rice comes in individual grains, so in principle we should measure it in a stony way by talking about the number of grains of rice. But in practice, the amount of rice that we use for a recipe contains so many grains, and we are indifferent to the exact number, so we say one cup of rice (a stickish measure) rather than worry about the number of grains. Similarly, there are so many atoms in any ordinary size object, and the energy of each atom is so tiny compared to the total energy of the object that we can treat mass and energy as though they could take on any value.

On the other hand, chemistry as we know it, and therefore life as we know it, depends on this quantization. The structures that make us up would not exist if the universe did not work this way. To understand that, let's resume looking at sound and atoms.

For musical instruments, we want to know the lowest allowed $\omega$ because that corresponds to the fundamental tone, what we think of as "the" note played by the instrument (or for the case of a guitar string by that part of the instrument). In quantum mechanics, lowest $\omega$ corresponds to lowest energy of the system. Since a system that can give up energy to its surroundings will settle down to its lowest allowed energy state, this means that the lowest $\omega$ and its corresponding $\psi$ provides a description of what the quantum system settles down to.

This "lowest note" property of quantum mechanics also provides an explanation for the basic chemistry fact that atoms bind together to form molecules. It is a property of Schrodinger's equation that in certain cases a lower note can be reached if an electron goes from one kind of atom to another (a phenomenon that chemists call an ionic bond), or if an electron is shared between two atoms (covalent bond) or between multiple atoms (metallic bond).

Therefore, because bonded atoms can be of lower energy than nonbonded atoms, molecules can be stable, continuing to exist until and unless they are affected by appropriate energy. This stability of chemistry leads to stability of biochemistry and to the evolving stability of life.

The energy we're talking about has perceivable consequences. The relationship between energy and frequency also applies to light. And we perceive the different energies of light as colors. Looking at a standard spectrum, red is the lowest energy and violet the highest in the range of visible light. But violet is only about twice the energy ergo twice the frequency of red. In music, a doubling of frequency is equivalent to a single octave (say a jump from middle C to treble C). Quadrupling of frequency is a jump of two octaves (middle C to high C).

Because of the narrowness of our visual range, we cannot see real harmonics in light, the way we can hear them in sound. But that doesn't mean all that invisible light has no effect; infrared, radio waves, and microwaves are all too low frequency for us to perceive, and ultraviolet, x-rays, and gamma rays are all too high frequency.

When a particle of light (a photon) impinges on an atom, one of the electrons of that atom might absorb that photon and jump up to a higher energy state. That can only happen if the energy of that photon would raise the energy of the electron to one of its higher allowed energies. Similarly, if an atom starts out in one of its higher allowed frequencies, it can jump to a lower frequency, emitting a photon whose frequency is the difference in the two frequencies of the atom.

This process means that if you shine a light on something, the light that comes out of that something will depend on the chemical composition of that something. In short, our color vision shows us objects because of the quantum harmonic structure that came out of the eigenvalues of the operator in Schrodinger's equation.

So in the end, the Pythagoreans were right about the "music of the spheres," though not nearly in the way they imagined: planets don't make music, but atoms do. And it's not some universal harmony, but rather a sort of "quest for the lowest note" that binds everything together and lets us see a one octave slice of the universe around us.

# 9

# *Probability and Statistics*

There is a part of the inheritance of mathematics that looks like it is not lost. News sources blast out statistical claims every day, and odds are quoted for every sport in the world. It would seem that probability and statistics are everyday tools for most people.

Unfortunately, ignorance of probability and statistics is as widespread as ignorance of calculus. Even people who know they don't know calculus think they understand probability and are bombarded with statistics that most people misinterpret. This branch of math is lost in plain sight. It's as if there were a common household object that everyone thought they knew how to use, but in fact did not. (That's computers. More about that later.)

Mathematical probability may be the oddest branch of math. Where other branches abstract shapes or number or process, probability abstracts the fact that "stuff happens."

Plants grow or die, rain falls or it doesn't, rocks tumble from mountains or stay up so a person traveling beneath is hurt or safe. Things happen beyond human control but affect human life.

In many ways probability hearkens back to nonmathematical abstractions, to the creation of the idea of an event occurring or not occurring. The abstraction of this truth, that stuff happens, was not originally mathematical. From the way this idea fits in human thought, it appears that it was storytellers who took happenings and from them created events. This was not necessarily the formal art of storytelling practiced by professional tale-tellers so much as everyday person-to-person gossip and boasting.

> "That trout was as long as my arm, I tell you. I nearly had it, but it got away."
>
> "There were ten of them chasing me. I managed to take one down before the others were on top me. I tell you, if I hadn't rolled into the dirt those rabbits would have torn me limb from limb."
>
> "We were lost in the desert, three days without water, mirages everywhere. Then on the last day, the three of us who survived out of the twenty who began the trek found an oasis where no one had ever been before."

Such stories have been around longer than recorded history. They have excitement, drama, risk (*except for the one with the rabbits*). People feel tension when they hear such tales, longing for things to come out one way, hoping that they will, fearing they won't.

Underlying these hopes and fears is one of the basic truths of human existence. Sometimes the world goes as you need, sometimes it goes against it, and sometimes stuff just happens.

A large number of scientific, philosophical, and religious explanations have been crafted for the why of these events because people rarely leave "stuff happens" alone. They want to know more, to look deeper into cause and effect, to banish the uncontrollability of that happening, to make events graspable as delineated tales, tales within human control. This may be why there was such a long gap between the first observation of stuff happens (long before history) and the mathematics that is used to study the happening of stuff (begun in the 17th century). The math requires acceptance that sometimes things are beyond anyone's power to change.

*That's just a theory of history created to grasp why this thing, the gap in creation of probability, happened.*

**True. I never claimed not to be susceptible to such ideas.**

There may be no explanation as to why there is such a gap in time, or it may simply be that no one did the work to formalize ideas that had been bubbling below the surface of thought. There is no way of knowing the why, only the what. But that's okay in this field of math.

In order to understand the mathematics of probability, it is necessary to turn a blind eye to how and why things happen and look only at what can happen.

Probability looks at an event as a sequence of occurrences:

1. An event is set up to occur.
2. Stuff happens.
3. Outcome of the event.

To see things from the point of view of probability, you must ignore the middle and look only at the setup and what the outcomes might be. In effect, to see events as probability sees them, you must ignore the story.

This may seem like a weird kind of abstraction. Wouldn't it be better to abstract the processes that make the event come about or not? In short, why have probability when we can use analysis.

That would make sense for events that have a small number of contributing factors whose contributions can be charted and graphed. Such events are susceptible to the techniques of analysis, because the functions involved have a small number of parameters. But if the event is too complicated for this, then we can only look at the beginning and the end and treat the process as a black box in which stuff happens.

This kind of event description where the story is omitted, sees Hamlet like this:

> Hamlet comes home to find his father dead and his uncle on the throne and married to his mother.
>
> – five acts of stuff happening –
>
> All the major characters are dead and this guy Fortinbras is on the throne.

If that's all you knew, you might ask, how did that happen? This could lead to reading or seeing the play.

Or you might say that the outcome doesn't seem very likely. Wasn't it more likely that either Hamlet or his uncle would die and the survivor end up on the throne?

Likeliness is another crucial idea that probability abstracts from real life and from storytelling. From the tales we are told and the events we witness, we develop a sense of what should be the outcome or the viable outcomes of a situation. Most people like to think that they develop that sense from their experience, but in truth it comes only partially from reality. It also comes from the human habit of telling stories and wanting them to cohere in a certain fashion, even though the real world does not end up like that most of the time.

Satisfying stories consist of interleaved meaningful actions that impact and are impacted by the lives of their characters. It's generally considered not a good idea to pepper one's tales with haphazard occurrences that have no meaningful connection to the characters. This is not a matter of story as a mirror of reality but story as acceptable to audience.

These acceptabilities are often codified as story-tropes and writers either work with or go against them. Tropes are not based on reality, but adhere to culturally taught expectations.

Our story sense of correct outcome is at odds with how outcome occurs in the real world. Inevitably, when such a disjuncture between world and thought occurs, someone figures out how to make money out of it. There are a lot of ways to do this, starting with convincing people that if they do what you tell them, the story will work out correctly.

We're not going to deal with that way to make money. We'll concern ourselves with people's sense that likelihood should bend in their favor. In other words, we're talking about gambling.

To be clear on the terminology, we need to distinguish gambling from gaming. Although there has been a recent attempt to make the words synonyms, they are not the same concept. **Gaming** is the playing of a game. **Gambling** is the laying of wagers. You can game without gambling, since a game need not have stakes, and you can gamble without gaming because wagers can be placed on anything. People have often gambled on things like who can run faster than whom, who can jump higher, who can beat up whom, and so on. We have no idea how old games are, nor how old gambling is, but both of them go back before recorded history.

The mathematical theory of probability started in an arena of human hope and fear where gambling and gaming overlap, in games of chance.

Games are generally divided by how much they rely on skill (that is, on human control) and how much there is an element of that which is beyond our control: chance, or fortune. There are games that are purely in one realm or the other. Chess is purely skill. Most dice games are purely luck. Most card games partake of both.

Probability theory was an attempt to examine the luck elements of games and use them to figure out which gambles in such games are good gambles. Among those who worked on these problems were 16th-century Italian mind-of-all-trades Gerolamo Cardano, 17th-century French mathematician Pierre De Fermat who confounded centuries of other mathematicians by claiming to have proven an incredibly hard theorem that was only finally proven in 1995, and Blaise Pascal who along with probability theory did enough work in physics to have the standard unit of pressure named after him.

A good gamble is one where the person making it has a reasonable expectation that he or she will come out ahead without having to have too many unlikely things happen in his or her favor. In other words, a good gamble is one where the gambler does not need to be "lucky," just not "unlucky."

In order to determine what a good gamble is, there needs to be a measure of the quality of the bet being offered.

> "I'll bet you $20.00 that you won't roll 12 on two dice."
>
> "That's silly."
>
> "All right. If you roll a 12, I'll give you $100.00, but if you don't, you give me $20.00. Do we have a bet?"

Most gambles are like that. There's the risk and the payoff. The question to be answered is: is the risk worth the payoff. But how to measure that?

*By how much the person offering the bet is smirking?*

**No, some of them are good at concealing the smirks. They don't call them poker faces for nothing.**

Probability theory starts by looking at the event, in this case rolling two dice. To be exact, this is the roll of two fair six-sided dice, each marked 1, 2, 3, 4, 5, and 6. The dice are rolled and the results added up. The event has a set of **possible outcomes**, in this case a total roll of {2, 3, 4, 5, 6, 7, 8, 9, 10, 11, or 12}.

Probability theory sees the event in terms of its possible outcomes. It does not examine the way the dice are thrown in the air, the wind speed, or the bounciness of the surface on which they will land. It removes all of those like it removes the five acts of Hamlet and sees the event solely as its possible outcomes.

Let's look at a simpler event, the throw of a single six-sided die. (For shorthand purposes, we will adopt the conventions of role-playing and war games and call this the roll of 1D6 – that is, the roll of one die with six sides.)

What are the possible outcomes? The die could come up 1, 2, 3, 4, 5, or 6. *Or something weird could happen like the die be snatched out of the air by a nearsighted eagle and swallowed. The die could roll down a sewer drain. Or the die could crack on impact and fall apart, scattering dice shrapnel.*

**While these weird outcomes are possible, in most games of dice if one of these did happen, that outcome would be ignored and the roller would be required to roll a new die in its place.** In effect, the only outcomes that are accepted in the roll of a die are coming up 1, 2, 3, 4, 5, or 6.

What can be done with this humble knuckle-boned beginning to help create a branch of mathematics? Here is where probability makes its first stick. It abstracts the concept fundamental to the way we look at the world that we mentioned earlier: **likeliness.**

Which of these two stories is more likely?

1. "I took the cookies, mommy."
2. "Tiny chocolate chip aliens came down and took back all their kin, saying they had to take them out of the cookie batter they were baked in."

*If you answered number 2 and you have children ...*

Likeliness is one of those ideas that does not reduce easily to other ideas, the same way size as an idea does not. It's one of those things we think we understand intuitively. But it's an idea that we feel has a "more" and a "less" to it.

More and less are our determiners of whether something should be a stick. It's a good idea to make that stick because people's sense of likeliness is very strongly biased in favor of how they think things should come out. In short people equate likelihood with trope-following. Most people suffer from the view that things are likelier to go as they think they should. It's as if in coming up with a measure for length everyone thought that things were longer than they were and needed a stick to help them remove their mismeasure.

In any case, we can make a stick for likeliness, but it's not a simple, cut-a-limb-off-a-tre e-and-put-it-up-against-something stick.

Probability uses a stick with the lowest value 0, representing the likeliness of an impossible event (the chocolate chip alien story is not impossible, but it's close to 0), and highest value 1, representing the likeliness of an event that is definitely going to happen or has already happened. Where an outcome lies on this short stick is called the **probability** of the outcome.

So to be clear, all probability values will be real numbers between 0 and 1.

*That's all well and good, but we have to tell them how to assign probability values to the outcomes of events: where to stick things on this stick.*

**I'm getting to that.**

Let's go back to our roll of 1D6. We said it was a fair die, which means that rolling each side should have the same probability value. We further said that the only possible outcomes should be rolling 1, 2, 3, 4, 5, or 6. So let's consider the outcome where we win if we roll any of 1, 2, 3, 4, 5, or 6. The probability of that outcome should be 1. In other words, it's guaranteed that we roll some face of the die. Can we use this to figure out what the probability of an individual roll is?

We can, but in order to do that we need to determine some mathematical characteristics of the probability function. P(E), where E is an event (actually the outcome of an event), and P(E) is a number between 0 and 1 representing the likelihood of E happening.

*But you don't have a function. How can you talk about it if you haven't defined it?*

We're going to construct the characteristics of the probability function P by examining what we mean by probability and abstracting that into a rough set of axioms that P will have to obey. If we can do that, we will have created a probability function.

First of all, let's look at what we mean by an **outcome** of an event.

We mean any combination of the ways that event might resolve. From probability's perspective, remember, an event is not what the process of happening is, but only the set of outcomes. We might look at the die roll and see these as events:

{1} rolling a 1

{2} rolling a 2

{1, 2} rolling a 1 or a 2

{1, 2, 3, 4, 5} rolling anything except a 6

{1, 2, 3, 4, 5, 6} rolling anything

So far we want to say the following:

P({1, 2, 3, 4, 5, 6}) = 1. That is, it is certain that we will roll a 1, 2, 3, 4, 5, or 6,

and

P({1}) = P({2}) = P({3}) = P({4}) = P({5}) = P({6}). That is, the probability of rolling any one face on 1D6 is the same as the probability of rolling any other face. We say this because we declared at the beginning that the die was fair. And as the crying child teaches, equality is the mathematical abstraction of that fairness.

We'd also like to say the following:

P({7}) = 0. That is, we can't roll a 7.

P({cheese, eggplant, Apollo 13}) = 0. The die cannot come up cheese, eggplant, or Apollo 13.

These last are only theoretically important when we consider all possibilities, not just the possible possibilities.

*Ah, you want to measure impossible possibilities. What's next, measuring green yellow things?*

**Sure, the amount of green in a yellow thing depends on how you make the yellow –**

*Never mind.*

So we have a few definite probability values, but we need to do more to determine the characteristics of the function. Let's consider

$$P(\{1, 2\})$$

What is the probability of rolling a 1 or 2?

We look at this and it seems to make sense that the likelihood of rolling a 1 or a 2 should be the likelihood of rolling 1 combined somehow with the likelihood of rolling 2. The combination that makes most sense seems to be addition since we are talking about an outcome that consists of either of two possible outcomes. It would make sense that the probability of rolling 1 or 2 is the sum of the probability of rolling 1 and the probability of rolling 2.

Let's stop for a second. That last paragraph contains something of the way math is created. The process of abstraction, of creating a new mathematical idea from the world, relies on the translation of real-world actions into appropriate mathematical operations. We are working from what we understand to create a theory of what we understand. We want our mathematical objects to mirror reality in theory. That's the only way to ensure that they will work as abstractions of that reality. In order to do that we need to consider how things work in the real world and find their theoretical mirrors. In this case, the either 1 or 2 outcome points toward:

$$P(\{1, 2\}) = P(\{1\}) + P(\{2\})$$

This in turns leads us down a path of finding out what the probability of each roll is, because from this idea we can write down the probability of rolling any outcome in terms of the probabilities of rolling each possible outcome:

$$P(\{1, 2, 3, 4, 5, 6\}) = P(\{1\}) + P(\{2\}) + P(\{3\}) + P(\{4\}) + P(\{5\}) + P(\{6\})$$

But we know that

$$P(\{1\}) = P(\{2\}) = P(\{3\}) = P(\{4\}) = P(\{5\}) = P(\{6\})$$

So

$$P(\{1, 2, 3, 4, 5, 6\}) = 6P(\{1\})$$

But we know

$$P(\{1, 2, 3, 4, 5, 6\}) = 1$$

Therefore,

$$1 = 6P(\{1\})$$

So

$$P(\{1\}) = 1/6$$

And since

$$P(\{1\}) = P(\{2\}) = P(\{3\}) = P(\{4\}) = P(\{5\}) = P(\{6\}), \text{ each of these is equal to } 1/6.$$

In other words, the probability of rolling any given side on 1D6 is equal to 1/6, or 1 out of 6. This certainly makes sense to our intuitive sense of probability.

Let's look at a few consequences of this.

$$P(\{1, 2, 3\}) = P(\{1\}) + P(\{2\}) + P(\{3\}) = 1/6 + 1/6 + 1/6 = 3/6 = ½$$

That looks good. Getting one result from half of the possible outcomes has a ½ probability.

$$P(\{4, 5, 6\}) = P(\{4\}) + P(\{5\}) + P(\{6\}) = 1/6 + 1/6 + 1/6 = 3/6 = ½$$

So is the probability of the other half of the outcomes.

$P(\{1, 2\}) = P(\{1\}) + P(\{2\}) = 1/3$. That also looks right.

Notice that $\{1,2\} = \{1\} \cup \{2\}$ (remember that if A and B are sets, $A \cup B$ ("the union of A and B") is the set that contains all the elements of A and all the elements of B).

So

$$P(\{1, 2\}) = P(\{1\} \cup \{2\}) = P(\{1\}) + P(\{2\})$$

and

$$\{1, 2, 3\} \cup \{4, 5, 6\} = \{1, 2, 3, 4, 5, 6\}$$

So

$$1 = P(\{1, 2, 3, 4, 5, 6\}) = P(\{1, 2, 3\} \cup \{4, 5, 6\})$$

and

$$P(\{1, 2, 3\}) + P(\{4, 5, 6\}) = ½ + ½ = 1$$

Hmm. I wonder if we can generalize this to

$$P(A \cup B) = P(A) + P(B)$$

In other words, we'd like to enshrine in our theory that the probability of either of two sets of outcomes is equal to the sum of the probabilities of each of those outcomes.

*Not so fast, Sunny Jim.*

**Sunny Jim?**

*Whatever. Be a little more careful.*

**OK, you caught me.**

Let's look at

$$P(\{1, 2\}) + P(\{2, 3\}) = 1/3 + 1/3 = 2/3$$

But

$$\{1, 2\} \cup \{2, 3\} = \{1, 2, 3\}$$

and

$$P(\{1, 2, 3\}) = ½$$

**Oh look, you counted the 2 event twice.**

*What do you mean I counted it? I'm the one who warned you.*

The point is we need to be careful that we don't double count when taking the union of sets of outcomes. We can do this by subtracting the probability of the intersection of those sets. (Remember that the intersection of two sets A and B, $A \cap B$ is the set of all elements that belong to both A and B.)

The actual general rule on adding probabilities is

$$P(A \cup B) = P(A) + P(B) - P(A \cap B)$$

Watch:

$$½ = P(\{1,2,3\}) = P(\{1,2\} \cup \{2, 3\}) = P(\{1, 2\}) + P(\{2, 3\}) - P(\{1, 2\} \cap \{2, 3\}) =$$
$$1/3 + 1/3 - P(\{2\}) = 2/3 - 1/6 = ½$$

That seems to make sense.

If we look at the 1D6 situation, we can offer a further generalization. If you have n **equiprobable** outcomes (equiprobable means equally likely) that cover all possibilities, and you have an event consisting of m such outcomes, then the probability of this event is m/n. If you have 6 outcomes on 1D6, then the probability of getting any one of them is 1/6. The probability of getting any of two of them (say, 1 or 5) is 2/6, and so on. The probability of getting no outcome is 0.

We've more or less mined what we can about 1D6.

But we are by no means done with dice. Let's expand our horizon dramatically, add to our understanding, and strive forth into a brave new world. In short, let's add another die. Let us see what we can learn from 2D6.

Oddly enough, we can learn quite a lot.

What's the probability of rolling a 9 on 2D6?

In truth you do not roll a 9. You roll two six-sided dice, a first die and a second die, and you add up the results.

A roll of 9 is not a single outcome but any member of this set of outcomes: {(3, 6), (4, 5), (5, 4), (6, 3)}, where the first number in each pair is the roll of the first die and the second number is the roll of the second die.

Let's try to make a list of outcomes of rolling 2D6. We are doing a two-step process, taking the pairs of rolls and then grouping together those that add up to the same number. In the example above we are treating the four outcomes {(3, 6), (4, 5), (5, 4), (6, 3)} as one outcome, since what we care about is not the rolls but the sum of the rolls.

Let's pause for a moment, because we're slipping past an important concept. We defined an event as a set of outcomes, but the meaning of the event is not the set of outcomes but what human event they comprise in the game we are playing. "Rolling a 9" is a human view (more or less) of the outcome because in many dice games it's the sum of the dice that matters, not the individual rolls. But {9} isn't really an outcome because the actual underlying probabilistic event is the rolling of two dice. So the "real" outcome list is {(3, 6), (4, 5), (5, 4), (6, 3)}.

In this case, to discern the probabilities we need to list the number of different real outcomes (that is, dice rolls) that produce the same outcome we care about (the sum of the rolls). We do this by counting the number of roll pairs that sum to the total we are interested in. *Note*: No summed outcome can be less than 2 or more than 12, since the lowest you can roll on both dice is (1, 1) and the highest is (6, 6).

| Total | Outcomes | Count |
|---|---|---|
| 2 | (1, 1) | 1 |
| 3 | (1, 2), (2, 1) | 2 |
| 4 | (1, 3), (2, 2), (3, 1) | 3 |
| 5 | (1, 4), (2, 3), (3, 2), (4, 1) | 4 |
| 6 | (1, 5), (2, 4), (3, 3), (4, 2), (5, 1) | 5 |
| 7 | (1, 6), (2, 5), (3, 4), (4, 3), (5, 2), (6, 1) | 6 |
| 8 | (2, 6), (3, 5), (4, 4), (5, 3), (6, 2) | 5 |
| 9 | (3, 6), (4, 5), (5, 4), (6, 3) | 4 |
| 10 | (4, 6), (5, 5), (6, 4) | 3 |
| 11 | (5, 6), (6, 5) | 2 |
| 12 | (6, 6) | 1 |

The Count column tells us how many outcomes match the total. By the way, keep your eyes on that belling-out-and-in shape in the Outcomes column. It's one of the most important things in the table above.

Assuming we have two fair dice, the outcomes of the underlying event are equiprobable even though the outcomes of the human event are not. How many possible underlying outcomes are there?

We can start by counting them. After making a pile of rocks, we get 36. We could also look at it in the following way, which will matter later:

> Look at the ordered pair we are trying to fill, (a, b). There are six possibilities for a and six possibilities for b. We could write it out in a square as follows

| | a | 1 | 2 | 3 | 4 | 5 | 6 |
|---|---|---|---|---|---|---|---|
| b | 1 | (1, 1) | (2, 1) | (3, 1) | (4, 1) | (5, 1) | (6, 1) |
| | 2 | (1, 2) | (2, 2) | (3, 2) | (4, 2) | (5, 2) | (6, 2) |
| | 3 | (1, 3) | (2, 3) | (3, 3) | (4, 3) | (5, 3) | (6, 3) |
| | 4 | (1, 4) | (2, 4) | (3, 4) | (4, 4) | (5, 4) | (6, 4) |
| | 5 | (1, 5) | (2, 5) | (3, 5) | (4, 5) | (5, 5) | (6, 5) |
| | 6 | (1, 6) | (2, 6) | (3, 6) | (4, 6) | (5, 6) | (6, 6) |

Notice that in order to create the paired outcomes we simply listed all possible pairs of outcomes, producing a number. So the number of possibilities for (a, b) is 6 (for a) times 6 (for b) = 36.

These 36 outcomes are all equiprobable since they consist of an equiprobable outcome for a and an equiprobable outcome for b. We can use the formula above, which says that the probability of an event consisting of a batch of equiprobable outcomes is equal to the number of outcomes in the event/number of total outcomes. The Count column of the table we made above for a roll of 2D6 gives the number of outcomes in each game event (such as rolling a 9), and we know that the number of all possible outcomes is 36. So we can finally compute the probabilities for rolls on 2D6.

| Total Roll | Count | Probability |
|---|---|---|
| 2 | 1 | 1/36 |
| 3 | 2 | 2/36 = 1/18 |
| 4 | 3 | 3/36 = 1/12 |
| 5 | 4 | 4/36 = 1/9 |
| 6 | 5 | 5/36 |
| 7 | 6 | 6/36 = 1/6 |
| 8 | 5 | 5/36 |
| 9 | 4 | 4/36 = 1/9 |
| 10 | 3 | 3/36 = 1/12 |
| 11 | 2 | 2/36 = 1/18 |
| 12 | 1 | 1/36 |

**We already have enough information in this table to form misleading questions.**

*Here's one: What is the most likely outcome in rolling 2D6?*

We look at the chart and see that rolling a 7 has the highest probability. So the answer is 7.

But is 7 a likely outcome?

No. There's only 1 chance in 6 that a roll of 2D6 will come up 7.

*Most likely* does not mean *likely*. People are often hoodwinked by that simple confusion.

**In gambling, there are a lot of words for people who think they know the odds and are wrong.**

*"Sucker" is the most well-known. Although I think in the stock market the term is "individual investor."*

We'll look more at this kind of confusion a bit later. But before that we've still got a ways to go before we can discern a good bet from a bad bet.

Let's start by adding a new concept to the probability function. We can start by asking this relatively innocuous question: what is the probability of rolling a 5 on 2D6 if the roll of the first die was a 1?

We might look at this and form the following false chain of reasoning: the only way to get a 5 if the first roll is a 1 is to get a (1, 4), so there's only one outcome, which means the probability is 1/36.

This is wrong because the set of possible outcomes is no longer all the (a, b) combinations. We said that we know that the first roll is a 1. Therefore, the possible outcomes are {(1, 1), (1, 2), (1, 3), (1, 4), (1, 5), (1, 6)}. In other words, we are down to only six possibilities and our desired outcome is {(1, 4)}, which means the probability is 1/6.

The general term for this kind of probability is **conditional probability**. It asks what is the probability of event A given that event B has already happened. This is written P(A | B) and read "The probability of A given B."

This allows us to more carefully parse the kinds of probability questions that show up in gambling. For example, what are my odds of getting blackjack if my first card is a 10?

We'll dig into cards in a little while. First, let's try to answer the basic gambling question of when to bet and when not to. We can figure out probability to some extent now, but how do we know if a gamble is a good one?

We need to know more than just the probability function to figure out if a gamble is a good one. We need to know what we can expect if we win or lose the bet. For that we need to introduce a new concept: expectation value.

Let's play a game. You roll 1D6. If you get a 1, 2, 3, or 4, you pay me the number rolled in dollars. If you roll a 5 or a 6, I'll pay you the number rolled in dollars.

Should you play this game?

In order to answer this question we have to shuck off nearly all the meanings of the word "should." If you are morally opposed to gambling or are a gambling addict, then you shouldn't, no matter what; the game is a bad idea for you.

When we try to use math to answer the question above, it becomes clear that we are dealing with a very narrow definition of the word "should," the definition that says you should do something if you are more likely to make money than lose money. This is the economist's definition of should. And this "should" should, morally and logically, be kept separate from all other definitions of should for reasons of morality and logic.

So, rephrasing:

> Should (economically) you play this game?

There is a way to put a stick to this question and so come up with a numerical rating for the economic shouldness of any gambling game. This stick is called **expectation value**, and it measures how much money you "expect" to get out of any round of playing this game.

Expectation value is not much harder to calculate than probability. For each outcome of the game you multiply the probability of that outcome by the amount of money that outcome will net you (this is called the **payoff** of the outcome). Then you add all of these values up and that is the expectation value.

This definition makes sense because it's not as if the payoff of an outcome is money in your pocket. You only get it if you win. We're in effect giving each payoff a weight based on the likelihood of getting it and basically saying that a 1 in 10 chance of getting \$20 is the same as \$2 you have now. Since the expectation of \$20 at a probability of 1/10 is equal to \$20/10 = \$2.

Expectation value is a mathematical formalization of the old hunter's maxim that "a bird in the hand is worth two in the bush." That is, a bird you've already taken is worth the same to you as two birds you're still hunting for.

Let's look at the game mentioned above.

| Die Roll | Payoff Gained | Probability | Payoff Times Probability |
|---|---|---|---|
| 1 | −1 | 1/6 | −1/6 |
| 2 | −2 | 1/6 | −2/6 |
| 3 | −3 | 1/6 | −3/6 |
| 4 | −4 | 1/6 | −4/6 |
| 5 | 5 | 1/6 | 5/6 |
| 6 | 6 | 1/6 | 6/6 |

Now we add up all the entries in the Payoff times probability column:

$$-1/6 + -2/6 + -3/6 + -4/6 + 5/6 + 6/6 = 1/6$$

So your expectation value for this game is 1/6th of a dollar. This means that the game is weighted in your favor and therefore is a good gamble for you. Economically you should play this game.

**Unless, of course, you can make more money doing something else – giving advice on economics, for example.**

*Or having other people pay you to teach them infallible gambling systems that you never seem to have time to use because you're so busy teaching.*

At this point it would be a good idea to introduce the concept of a **random variable.** This is a variable that ranges over a set of outcomes to an event. In other words, a random variable is the mathematical embodiment of the question "What's the outcome?" From the variables are pronouns perspective, a random variable is just a pronoun for outcome. We would describe the playing of a round of the game above as a random variable (call it X) and would say that the expectation value above is a function E(X) of the random variable. E(X) = 1/6 in this case. Notice that the random variable needs to know both probabilities and payoffs for each outcome.

*These numbers we've been throwing around for probabilities and expectations. We haven't said anything about their units: about what kind of sticks they are.*

If we look at the expectation value, we ended up with an expectation of 1/6 of a dollar, but that was also the units of the payoffs. So expectation value has the same units as payoff, usually some monetary unit.

Since expectation values are calculated by multiplying probabilities by payoffs and adding them up, the units here would be (units of probability) × (units of payoff). But since the result also has units of payoff, dimensional analysis yields:

(units of probability) × (units of payoff) = units of payoff

Dividing both sides by units of payoff gives

Units of probability = 1

This means that probability, while it is a stick, is not really a unit. Probabilities are what is called **dimensionless.** They have no units. You can multiply them by other units without affecting those units, but you can only add them to other probabilities.

Back to the random variables. There is a shorthand in expectation values that is sometimes used with random variables to get a sense of what the "average outcome" of an event that has numerical results is. The terminology gets a little sloppy. It's common to talk about rolling a die as a random variable in which the result of the dice roll is the same as the payoff (in other words, rolling a 7 on 2D6 has a payoff of 7, rolling a 1 on 1D6 has a payoff of 1, etc.), whereas the random variable is really the question of what the outcome is and by implication what the payoff is.

The expectation value of rolling 1D6, E(1D6), is found in the same way as before:

| Die Roll | Payoff Gained | Probability | Payoff Times Probability |
|---|---|---|---|
| 1 | 1 | 1/6 | 1/6 |
| 2 | 2 | 1/6 | 2/6 |
| 3 | 3 | 1/6 | 3/6 |
| 4 | 4 | 1/6 | 4/6 |
| 5 | 5 | 1/6 | 5/6 |
| 6 | 6 | 1/6 | 6/6 |

E(1D6) = 1/6 + 2/6 + 3/6 + 4/6 + 5/6 + 6/6 = 21/6 = 3.5

We can do something similar to find E(2D6).

| Die Roll | Payoff Gained | Probability | Payoff Times Probability |
|---|---|---|---|
| 2 | 2 | 1/36 | 2/36 |
| 3 | 3 | 2/36 | 6/36 |
| 4 | 4 | 3/36 | 12/36 |
| 5 | 5 | 4/36 | 20/36 |
| 6 | 6 | 5/36 | 30/36 |
| 7 | 7 | 6/36 | 42/36 |
| 8 | 8 | 5/36 | 40/36 |
| 9 | 9 | 4/36 | 36/36 |
| 10 | 10 | 3/36 | 30/36 |
| 11 | 11 | 2/36 | 22/36 |
| 12 | 12 | 1/36 | 12/36 |

E(2D6) = 252/36 = 7.

It is important to keep track of the meanings of words here. Expectation value does not mean that in playing one round of the game you expect to get that amount of payoff. It means something a little more subtle, which has led to one of the two most common delusions about probability. We'll go through this slowly because this delusion is so common.

First, a bit of stuff that belongs in the statistics part of this chapter (which has its own delusions).

If you have a set of n values, the **mean** of those values is equal to the sum of the values divided by n.

The mean of {4, 7, 2} = (4 + 7 + 2)/3 = 13/3 = 4 1/3

The mean of {6, 6, 6, 6, 6} = (6 + 6 + 6 + 6 + 6)/5 = 6

Most people use the word "average" for "mean," but in statistics there are several different kinds of averages and the mean isn't always the right one. We'll stick to mean for now (*we'll be nice later*).

Suppose you have a game, like rolling 1D6 and getting a payoff of the amount you rolled. Suppose you play this game over and over and over for, say, n times. Each round of this game is a separate random variable; call the variable for the ith round $X_i$. These variables are distinct, even though they all look alike, because each round of the game is a distinct event with its own outcome. These random variables are what are called **independent random variables**, so called because the outcome of one has no effect on the outcome of any of the others. One roll doesn't care what the earlier rolls were.

Now, here's something important about independent random variables. Suppose you have two such variables, called X and Y (say X is the roll of 1D6 and Y is the roll of another 1D6). Then the probability of a particular joint outcome P(X & Y) is equal to P(X)P(Y). For example, the probability that X will roll a 5 and Y will roll a 3 is P({X = 5, Y = 3}) = P({X = 5})P({Y = 3}) = (1/6)(1/6) = 1/36, which is the probability of any particular paired outcome on 2D6, which is what you would expect. We can see this just by going back to our chart of roll outcomes on 2D6.

Back to our set of multiple rolls of 1D6: $X_1, X_2, \ldots, X_n$

If we take the mean of the results of these n rounds of the games, we would get a measurement for how well one did on average for these rounds. Let's call this $A_n = (X_1 + X_2 + \cdots + X_n)/n$

Now, here's where things get a little weird. $A_n$ is a random variable. It's not independent of the $X_i$ – indeed, it depends completely on them – but it is itself a random variable, a variable that ranges over outcomes, a question of how well one did on average. This variable is dependent on the outcome of various conjoined events.

$A_n$ might be as little as 1 and as much as 6. But for $A_n$ to be 1, every $X_i$ from 1 to n would have to be 1, otherwise the mean would be more than 1. In other words, for $A_n$ to be 1, one would have to roll n 1s in a row. The probability of that is $(1/6)^n$. Each roll of 1 has a probability of 1/6. The probability of all the $X_i$ rolling 1 is the product of the probabilities of each $X_i$ rolling 1. Since there are n of them, this product is $(1/6)^n$.

Let's look at what this is for the first few values of n:

| Rolls | Probability |
|---|---|
| 1 | 1/6 |
| 2 | 1/36 |
| 3 | 1/216 |
| 4 | 1/1296 |
| 5 | 1/7776 |

These chances get really low, really fast.

By the way, note that the probability column above is not only the chance of getting five 1s in a row, it's also the chance of getting any specific sequence of rolls. The odds of rolling (1, 2, 3, 4, 5) are the same as of rolling (1, 1, 1, 1, 1), as are the odds of rolling (2, 2, 4, 5, 6). Any specific sequence of outcomes is as unlikely as any other such sequence. This is just an extension of the 2D6 chart above, where we listed all the outcomes in a square. Each of the ordered pairs was equiprobable. In the case of the five-element sequences above, each of the 5-tuples is equiprobable.

But although any given sequence is equally likely, the means of those sequences are not. To get a mean close to 1 you need to have a sequence that is mostly 1s. Similarly, to get a mean close to 6 you need to have a sequence that's mostly 6s. But to get a mean close to 3 you can have sequences all over the place. Here's a few ordered 4-tuples whose means are 3:

(1, 2, 4, 5)
(2, 2, 4, 4)
(3, 3, 1, 5)
(6, 4, 1, 1)

This list is by no means exhaustive. In other words, if we look not at the individual $X_n$ random variables, but at the random variables of the averages, the $A_n$, we find that the probabilities push toward the middle. It's more likely in the long run that the averages will be nearer the middle. If we look at the $A_n$ as n increases, the highest probabilities all tend toward a certain value. That value is the **expectation value** of X, where X is a generic random variable representing a single play of the game. In other words, in the long run, averages tend toward expectation values.

Now, here's the highly misinterpreted theorem. It's called the **law of large numbers**, or sometimes called the **law of averages**.

For every $\varepsilon > 0$, limit as n goes to infinity of $P(|A_n - E(X)| > \varepsilon) = 0$.

*Please, parse this one out. It looks as unreadable as those calculus limit theorems.*

**Okay.**

Suppose we keep playing the game an arbitrary number of times and we keep track of the averages ($A_1$, $A_2$, ...). This is an infinite sequence. It is not guaranteed to converge at all. And it might converge to any number in the range of possible expectation values (in the case above, any number between 1 and 6). But the probability that such a sequence does not converge to the expectation value of the game being played over and over does converge to 0. In other words, it's not that the averages have to eventually come out to be the expectation value, it's that the *likelihood* that they don't eventually come out to be the expectation value becomes itself vanishingly small.

The important thing to remember is that any specific sequence of results is itself unlikely. The longer the sequence, the less likely it is. Suppose you guess beforehand what the results of a sequence of 100 dice rolls will be. If you then roll that many dice and record the results, whatever list you get is unlikely to be identical to a result you predicted beforehand. The probability that the results you got were the results you got is 1 (since any known fact has a probability of 1), but the probability before you did the rolling that you would get that sequence is quite small.

The meaning of the law of large numbers is not that one is guaranteed to end up near the expectation value; it is that one is likely to end up near there. This is not a statement about the probability of events but about the probability of the probability of events.

It's more likely that what is likely to happen will happen more often than what is unlikely. That's a probability of probability. It is this second-order probability that messes people up. This is one of the critical tripping points in probability theory, and it has created considerable confusion.

People think that the law of averages means that if things have been going one way (say a string of 1s), they have to reverse somehow to make things average out. People think that an opposite result is "due." But you could continue to roll 1s from now until the universe dies. It's unlikely – but that's all it is. Unlikely is not impossible, and when you put together all the things that have happened at any given point in time, the probability that they would have happened is itself unlikely. Yet they did happen (more about this below).

Furthermore, remember, dice rolls are independent. The dice don't remember and the universe doesn't care. The universe does not fix things so that the expected happens. Expectation happens when it happens because it's more likely, not because someone's keeping score and cheating to make things even out.

The law of large numbers is one of those results in math and science that connects in an unfortunate manner with the way people like to think. It seems to be a statement about luck, that events have to turn around sometime, that streaks of luck, good or bad, have to end at some point.

**This fits a certain kind of story that was very popular as a morality story in the Middle Ages. These stories revolved around an image, the Wheel of Fortune.**

*Not the game show version of hangman, where they don't hang anyone.*

The Wheel of Fortune was based on the idea that success on Earth rose and fell and was not attached in any way to the merit of the person who received the fortune, good or ill. Owing to theological changes during the Reformation, Wheel of Fortune stories became less common, replaced by more deterministic ideas of earthly outcomes. In-depth analysis of this change of thinking belongs to a history of religious thinking rather than mathematical thinking.

What matters here is that the sense of the rise and fall of fortune and the misapplication of mathematics leads people to a feeling that they should continue to gamble when they've been losing because the odds "have to" turn around.

The thing is the outcomes don't have to even out and the gambler doesn't have to keep gambling. That's not what the math is saying.

A person who misunderstands this piece of mathematical legacy is using math as superstition. The one who understands it knows that the dice will fall according to the same probabilities each round, and that if you analyze the expectation values of any casino game, you'll find that in the long run the only one who wins is the house.

## Cards and Randomness

We've talked enough about how the dice don't go your way. What about cards?

Let's start simply.

What is the probability that the first card you pull from a standard deck of 52 playing cards will be an ace?

There are 52 possible outcomes, and 4 of these are the desired outcome. So the probability is $4/52 = 1/13$.

If you pull a card out, check it, put it back, shuffle the deck enough and then pull another card out, the probability that this card will be an ace is again $1/13$.

But suppose you pull a card out and it is an ace and you keep it. What is the probability that the next card you pull will be an ace?

Now there are 51 possible outcomes and the number of desired outcomes is 3. So the chance is $3/51 = 1/17$.

Suppose instead that one has pulled a card out, kept it, and it is not an ace. What then is the probability of drawing an ace? There are 51 possible outcomes and 4 desired, giving $4/51$.

In the last two cases, while each pull of the cards is a distinct random variable, they are not independent but **dependent random variables** because each previous outcome affects the probabilities of the ones that follow. The changing probabilities as cards turn up makes determination in card games more difficult than in dice games, although it can be done. One of the things that makes expert card players experts is the ability to refigure the odds based on the information they are presented as the game progresses.

*Don't forget their ability to figure out how their opponents are thinking and to bluff them into thinking that they're thinking something completely different from what they are thinking so that they think that they think that.*

**You lost your pronoun antecedents. Anyway, that applies to some but not all card games. Probability applies to all of them.**

The question of independence and dependence extends far beyond their sources in probability. Dice and cards and the ideas of random variables they spawn reveal complex questions on the nature of events that end up mattering a great deal in the atomic and subatomic structure of the universe **(not that we'll be talking about that in this chapter).**

*Darn. That's my favorite part of probability. What's the probability we'll talk about it later?*

**1. We already wrote the later chapter, so it's already happened. The probability of all events that have already happened is 1.**

There is an invisible underlying question in probability theory, the meaning of **randomness.** Random is a word we use to mean something where the outcome just happens and any of the possible outcomes could happen. In effect random is a removal of cause and effect, the excision as we said before of the middle of the story.

But are things really random? Say one looks at the toss of a die not as a random event but as a question in mechanics – that is, a study of motion. One would look at the forces on the die (initial throw, gravity, air motion, and resistance) and perhaps be able to formulate an unholy mess of a differential equation that if solved would predict which way the die would land. But this is impractical, and besides, the equation would change with each throw of the die because all those forces (except for gravity and air resistance) would change, as would the height from which the die was thrown, the initial velocity and so forth. It is not practical to create and solve that equation for each roll.

The mechanical systems used in games of chance – coins, cards, dice, roulette wheels, etc. – have a property often referred to as chaos, which means sensitive dependence on initial conditions. That is, small changes in the way the die is thrown, or the coin is tossed, or the roulette wheel is spun, can lead to large changes in the outcome. Thus prediction, while possible in principle, is totally impractical. We'll talk more about chaos in the next chapter.

This impracticality of calculation makes each throw of the die a black box, an unopened, unsolved means of determination that has so many different factors, susceptible to big effects for slight changes, that we might as well call it random.

What about cards? A deck of cards once shuffled has no randomness in it. The place of each card is determined. If someone cheats and looks at the cards, he or she can know exactly how things will be dealt, who will get what cards when.

*How is that random?*

**It isn't really.** But we can treat it as random. If no one looks, the deck is as much of a black box as the unthrown die. Properly shuffled by someone who is not cheating, any card can end up anywhere in the deck.

*But that's just concealing information. It's not real randomness.*

**The thing is, randomness does not have to be real any more than lines have to be real.**

Randomness is a mathematical concept that, in defined circumstances, decently models certain kinds of real-world events, just as lines, carefully delineated, decently model certain real-world shapes.

The funny thing is that even though randomness isn't real on the level of cards and dice, on the quantum level randomness does seem to be real.

*I thought you said we weren't going to talk about that until a later chapter.*

**We didn't. That was foreshadowing.**

## Combinatorics: Counting Without Counting

You may have noticed that there came a time above where we stopped counting actual outcomes and started calculating the number of possible outcomes. That was a simple application of a branch of math called **combinatorics**, which can be defined as figuring out what you would get if you did the actual counting.

We're not going to dig too far into this field, interesting though it can be. We just want to look at a few aspects of it relevant to probability and make clear a difficulty people often have with large numbers and the way they can be tricked by them.

Most basic combinatorics involves asking the question, How many possible somethings are there? For example, how many possible rolls of 5D6 are there? Combinatorics looks at this and sees that the possible outcomes here are really combinations of outcomes. Let's take our 5D6 and roll them one at a time in a predefined order: first die, second die, third die, fourth die, fifth die. We could write down an outcome like this:

First die = 3

Second die = 2

Third die = 5

Fourth die = 6

Fifth die = 2

But that's wasteful of space in our document and could be annoying if we had a lot of dice. But we've seen something like this before when we were writing out coordinate values, shorting x-coordinate = 3 and y-coordinate = 7 to the ordered pair (3, 7). We can similarly turn the above into an ordered 5-tuple

(3, 2, 5, 6, 2). That's much more compact.

More generally, if we have a sequence of random variables, $X_1, X_2, \ldots, X_n$, we can write out an outcome of these as an ordered n-tuple, sometimes called a finite sequence.

$$(a_1, a_2, \ldots, a_n)$$

where each $a_i$ is a possible outcome for the corresponding $X_i$.

Combinatorics looks at this and says that the number of such sequences of possible outcomes is the number of choices for $a_1$ times the number of choices for $a_2$ and so on down the line to the number of choices for $a_n$. Those choices are exactly the possible outcomes for the corresponding random variable. Therefore, the number of possible outcomes for the whole is equal to the product of the number of possible outcomes for each of the component variables.

Let's see where this gets us for our roll of 5D6. We have five random variables, each of which has six possible outcomes. The total possible number of outcomes therefore is $6 \times 6 \times 6 \times 6 \times 6 = 6^5 = 7776$.

More generally, if you have n independent random variables of the same type (like each is a single roll of a die), each of which has m possible outcomes, the number of total possibilities is $m^n$. And therefore the probability of any given such sequence occurring is $1/m^n$.

This is an aspect of probability that is somewhat counterintuitive. Think of a license plate that is a sequence of three letters and three numbers, say in the order of number, letter, letter, letter, number, number. There are 26 possibilities for each letter and 10 possibilities for each number. The total number of possible license plates is therefore $26^3$ times $10^3$ = 17,576,000. The probability of any given license plate is then 1/17,576,000, which is approximately 0.0000000569. And yet you look around and see cars with license plates, each of which is fantastically unlikely.

## Counting Cards

The calculation of outcomes above works for independent random variables. What about some that aren't? Let's try to figure out how many possible shufflings of a standard deck of cards there are.

Once again each possible outcome is a finite sequence. Top card, card below the top, card below that, etc. We can write each such sequence as a 52-tuple.

$$c_1, c_2, \ldots, c_{52}$$

But in this case, the number of choices for each $c_i$ changes depending on where we are in the sequence. We have 52 choices for $c_1$ but only 51 for $c_2$, because we can't choose the card we had for $c_1$. It's gone from our set of possible choices.

The number of choices for $c_3$ is 50, and so on down to $c_{52}$, where we have only one choice. The number of such sequences is the product of the number of choices possible for each $c_i$, which is

$$52 \times 51 \times 50 \times \cdots \times 1$$

We've seen this before. This is 52! (52 factorial). A brief bout on a calculator tells us 52! is approximately $8 \times 10^{67}$. Here's another giant number of possibilities, this one generated every time somebody shuffles a single deck of cards.

Suppose we want to count something simpler with cards, like the number of possible 5-card poker hands. We can do something similar to what we just did, but instead of using up all the cards, we'll just use five of them. We're making a 5-tuple, with 52 choices for the first card, 51 for the second, 50 for the third, 49 for the fourth and 48 for the fifth. So the number of such possible hands is

$$52 \times 51 \times 50 \times 49 \times 48 = 311{,}875{,}200.$$

By the way, this is also equal to

$$52!/(52 - 5)!$$

We don't recommend using this as a calculation as it can be quite inefficient.

*Misdeal!*

**What do you mean?**

*You miscounted.*

**How so?**

*You said poker hands not card sequences. Poker hands don't care about the order in which the cards occur they just care what the cards are.*

**Actually, that's poker players. But you're right.**

We need to group our raw outcomes into larger events to get poker hands, just as we needed to group our rolls of 2D6 to get the raw outcomes that lead to dice totals. What we're doing is making an equivalence relation between outcomes saying that they're the same outcome if they represent the same poker hand. Another way to look at this is trying to eliminate the ordering from the 5-tuple.

How do we do that?

We start by partitioning the set of outcomes into equivalence classes.

When are two hands the same hand? When you can rearrange one to get the other. With a little work we could prove that's an equivalence relation, but let's not bother.

Let's see if we can count how many equivalents each hand has.

Each hand has five cards. Each card can go in one of five slots. So we have five choices for the first slot, four for the second, three for the third, two for the fourth, and one for the fifth. That means we have 5! equivalent hands in each equivalence class. That's darned convenient. If the equivalence classes are the same sizes, then we can figure out how many equivalence classes there are by dividing the number of elements in the whole set by the number of elements in the equivalence class.

*Wait. What?*

Look, if we have 12 objects and we divide them up into sets of 3 each, we can figure out how many sets we have just by dividing 12 by 3. This is one of the oldest uses of division, fair distribution.

So we had

52!/(52 – 5)! and we divide that by 5! (which is 120), giving us

52!/(52 – 5)!5! = 311,875,200/120 = 2,598,960.

The first result we counted, the number of sequences of 5 cards from a standard deck of 52, is called the number of **permutations**. The second result, the number of actual hands, is called the number of **combinations**.

In general, if you have n things and you want to make sequences of m of those things without repetition, the number you calculate is called the permutations of n things taken m at a time, which is n!/(n – m)!.

The number of combinations of n things taken m at a time (that is the number of m element subsets of a set of n things) is n!/(n – m)!m!

Please note two things about this. Combinatorics is impressively lazy, since it involves counting without counting. And again note how easy it is to get large numbers.

Combinatorics is a field that can surprise us by its results because intuition about numbers fails when we start dealing with possibilities. It's valuable getting used to the way large numbers are generated by possibility so as not to be fooled by people playing games with them.

So much for stones. On to sticks.

## Probability Distributions: Stick the Chances

**I've got one single die.**

*Okay.*

**I'm going to toss it and you bet.**

*On what comes up?*

**No, on where it lands.**

*What?*

**Where it lands.**

*But it could land anywhere. There aren't specific delineated outcomes. How can I figure the probabilities? What does probability even mean if you don't have a list of outcomes?*

Cards and dice are like the stones of probability theory. The theories based on them work when outcomes can be listed. But where something lands? Where a cannonball hits, where a traffic accident might occur, when rain might start to fall, when someone is likely to pass out from drinking too much? These are the sticks of probability. A branch of probability theory deals with them.

Let's take the case of an archer shooting at a target. Suppose the archer is a good shot, pretty likely to hit near the bull's eye and less and less likely to do so the further you get from the center. We might represent this graphically as a shaded disk in which the darker the area, the more likely the arrow is to hit there (Figure 9.1).

Now let's turn this picture sideways, and instead of making the more likely areas darker, we'll make them higher (Figure 9.2).

We're assuming that this likelihood of hitting a point on the target depends only on the distance from the center r. So, just displaying it as a function of r, we get the curve shown in Figure 9.3.

We haven't said what this is a graph of, nor the meaning of the y-axis (the x-axis is the distance from the center where the arrow hits).

The y-axis does not represent probability as such. If it did, the chance of hitting the dead center would be the value of the function at r=0.

What do we mean by hitting the bull's eye? Do we mean that there is a single point, the origin, and if the arrow is not at (0, 0), it missed the bull's eye?

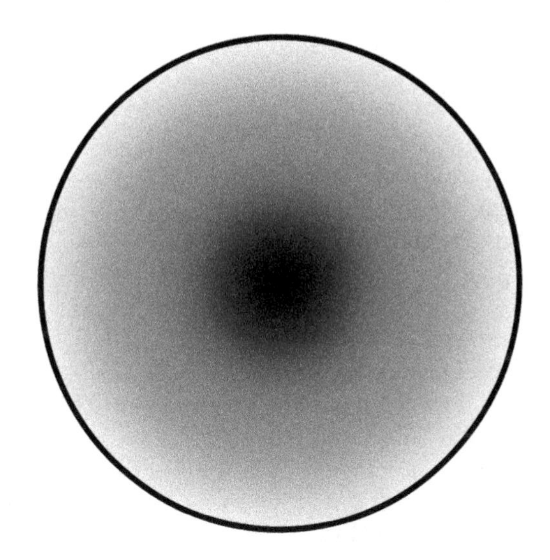

**FIGURE 9.1**
Probability Target

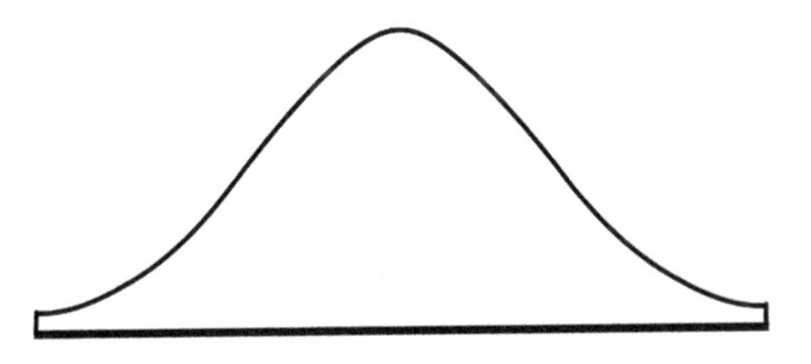

**FIGURE 9.2**
Sideways View of Target

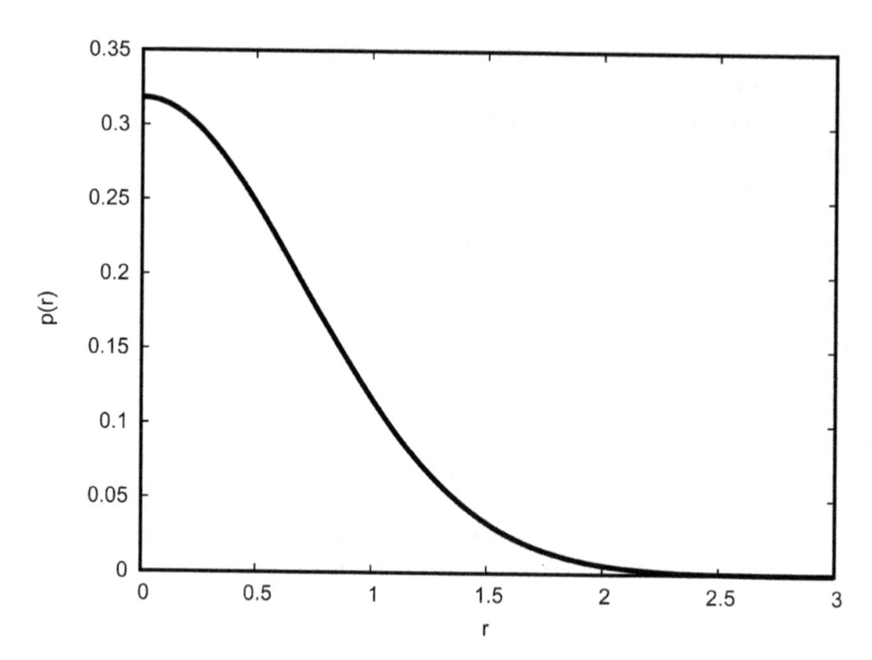


**FIGURE 9.3**
Positive Half of Bell Curve

No, we mean a region around the origin. On a line, it would be an interval. In a plane, it's a disk. If the arrow hits anywhere in that area, we would count it a bull's eye. So the question we are asking is, "what is the probability that the arrow hits in a given area?"

The function that we have graphed is a probability per unit area: to get the probability that the arrow lands in a given region of the target, we should multiply the probability per unit area by the area of that region. Or at least that is what we would do if the probability per unit area were constant. But we've seen this problem before and we know how to deal with it: chop the region up into tiny subregions where the function is approximately constant. In each subregion, multiply function times area and then sum up the results. In other words, the probability that the arrow hits the region is the integral of the probability per unit area over the region. In our case, we would multiply the probability per unit area p(r) by the area of a disk of radius r and width dr, that is, $dA=2\pi r\,dr$, and then integrate. So the probability that the arrow hits the target in the region between r=a and r=b is

$$P([a,b]) = \int_a^b p(r)2\pi r dr$$

Note that we can define from the probability per unit area p(r) a probability per unit radius $p_1(r) = 2\pi r\, p(r)$ so that the above integral becomes

$$P([a,b]) = \int_a^b dr p_1(r)$$

Note, however, that $p_1(r)$ has different properties than p(r). p(r) has a maximum at r=0, because our archer is a good shot and r=0 is what he is aiming for. However, $p_1(r)$ is zero at r=0, because circular strips of smaller radius have a smaller area, and however good a shot our archer is, a smaller area is harder to hit than a larger area (Figure 9.4).

For a variable x, the function p(x) giving the probability per unit length for finding an object in a small interval containing x is called a **probability distribution function**. Similarly, there are probability distribution functions p(x, y) for an object in a plane and p(x, y, z) for an object in space.

Notice we've had a sudden very serious jump in the mathematical skill and tools needed to approach this aspect of probability theory. Dice and cards, the stones of probability, can be done using nothing messier than division and multiplication. The sticks require integral calculus.

Rather than dig deep into the math here, we'd like to look at the difference in thinking caused by this stick/stone dichotomy.

Mathematicians who study the probability theory we started with use combinatorics. Combinatoric thinking is algebraic.

Mathematicians who study the theory of probability distribution functions are using calculus, relying on analytic thinking. This means that to shift from stones to sticks requires a change from algebraic to analytic thinking, so the people who are comfortable doing stone mathematics and those doing stick mathematics need not be the same people. If they are, they need to be able to move back and forth between these two easily.

Furthermore, while the probability function P(X) seems to have the same meaning as it did with cards and dice – the abstraction of likelihood – there is a big difference in meaning between a definite one-among-many answer and a somewhere-in-a-ballpark answer. The former is harder to fuzz than the latter. Anyone who has heard (or been) a child complaining that a 1-point difference in test score means a whole difference in grade knows the inherent fuzziness that comes from trying to make a definite meaning out of an indefinite likeliness.

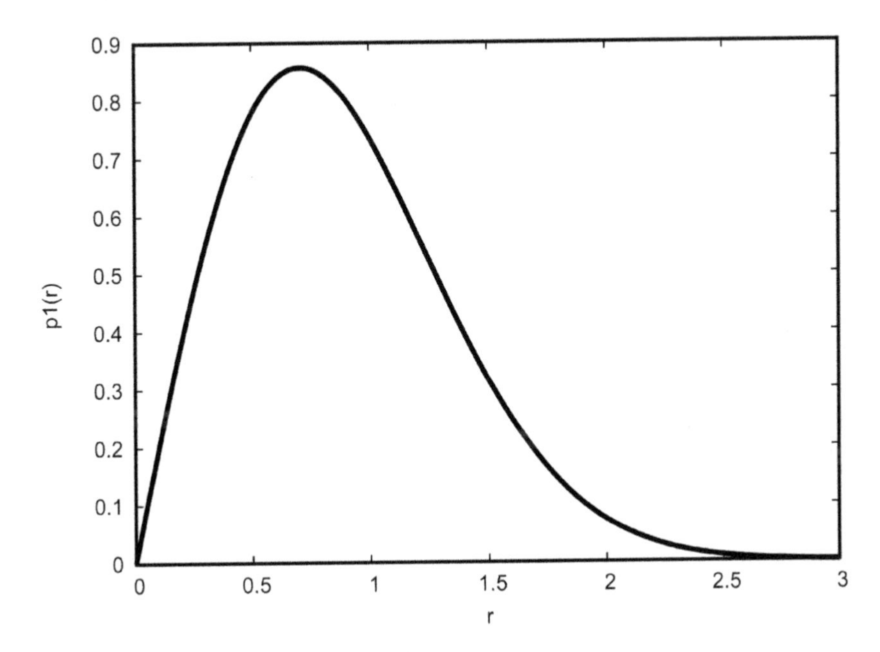


**FIGURE 9.4**
Probability Distribution Graph

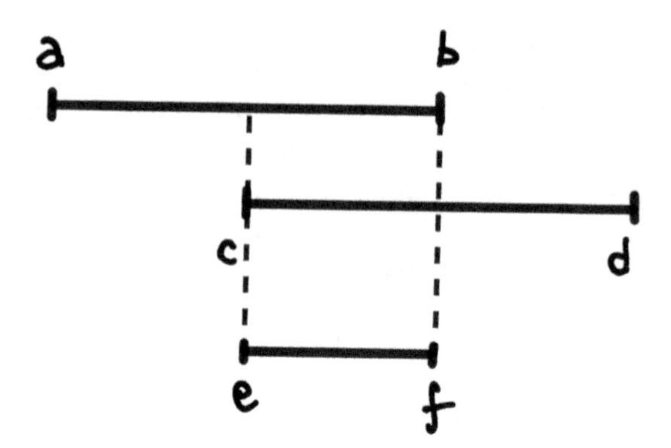


**FIGURE 9.5**
Overlapping Intervals

Despite all this difficulty, we can transplant all the processes we used in stones to sticks, including expectation value, the law of large numbers, and good and bad bets. But we have to do integrals instead of addition, so we're not going to do very many examples of this.

Here's one bit. Suppose we have two intervals [a, b] and [c, d] and we want to measure the probability that the arrow falls into either of them. Just as before, we might think that the probability is equal to the sum of the probabilities.

$$P([a,b] \cup [c,d]) = \int_a^b p(x)dx + \int_c^d p(x)dx$$

But again there is the problem of overlap of intervals and the fact that we are in effect counting an area twice. We need to subtract the integral over the interval of overlap if any (Figure 9.5).

So it's really

$$P([a,b] \cup [c,d]) = \int_a^b p(x)dx + \int_c^d p(x)dx - \int_e^f p(x)dx$$

This is an instance of

$$P(A \cup B) = P(A) + P(B) - P(A \cap B)$$

where in this case, A and B are sets of intervals in the line and one is taking a batch of integrals and adding up the results. Similar translations from stones to sticks can be done.

We would like to point out something about stick probability versus stone. Stone probability works based on very controlled random variables, very narrowly confined possible outcomes. Stick probability is closer to reality, where random events can vary widely in their results.

## Stochastic Processes

In both kinds of probability, there is an important field of study that concerns the long-term behavior of an object that is repeatedly being affected by some probabilistic phenomenon that can change its state. Such a process is called a **stochastic process** (stochastic means randomly determined).

Suppose you have an object on a number line starting out at 0, and suppose that every minute a coin is flipped. If the coin comes up heads, the object moves 1 to the right. If it comes up tails the object moves 1 to the left. This kind of process is called a **random walk**. Suppose you do this for a large number of coin flips. Where is the object most likely to end up?

If you guessed the middle, you would have made the most common guess, and one that's substantively wrong. In the long run, the object is likely to push away from the center.

To see why the object slides away, let's look at the first five steps and figure out what the probability is that the object will be at each value from –5 to 5.

At step 0, the object is at 0.

At step 1, the object will be at –1 (50% chance) or 1 (50%) chance.

At step 2, if the object was at –1 in step 1, it has a 50% chance of being at –2 and a 50% chance of being at 0. But these depend on an event with a 50% probability, so there is a 50% times 50% = 25% chance of being at –2 and a 25% chance of being at 0. But if the object was at 1, it has a 25% chance of going to 2 and a new 25% chance of going to 0. This means the chances are –2, 25%; 0, 50%; and 2, 25%. We can keep doing this process and see what happens for five steps:

| | 0 | 1 | 2 | 3 | 4 | 5 |
|---|---|---|---|---|---|---|
| –5 | 0 | 0 | 0 | 0 | 0 | 3.125% |
| –4 | 0 | 0 | 0 | 0 | 6.25% | 0 |
| –3 | 0 | 0 | 0 | 12.5% | 0 | 15.625% |
| –2 | 0 | 0 | 25% | 0 | 25% | 0 |
| –1 | 0 | 50% | 0 | 37.5% | 0 | 31.25% |
| 0 | 100% | 0 | 50% | 0 | 37.5% | 0 |
| 1 | 0 | 50% | 0 | 37.5% | 0 | 31.25% |
| 2 | 0 | 0 | 25% | 0 | 25% | 0 |
| 3 | 0 | 0 | 0 | 12.5% | 0 | 15.625% |
| 4 | 0 | 0 | 0 | 0 | 6.25% | 0 |
| 5 | 0 | 0 | 0 | 0 | 0 | 3.125% |

There are lumps forming in our probability graph, causing the results to spread out and causing there to be a reasonable chance that the object will end up far away from the center. The longer we carry out this process, the farther we are likely to end up.

Why is this happening? Our intuition says that because we are going back and forth with even probability, we should end up at the center. But that intuition is wrong.

Let's look not at the first step but at the second, where we are either at 1 or –1. Suppose that rather than the first step being the first step, the second step was our first step, that we began at 1 or –1. Then, our intuition would tell us that we should end up back where we started, at 1 or at –1, whichever we began at. At step 2, if we forgot how we got there, we would think that doing the random walk would bring us back to 2, 0, or –2, wherever we started.

At each step, our intuition tells us that we have reached a point of long-term equilibrium, and at each step our intuition is wrong.

Analysis of these long-term processes produces the kinds of counterintuitive results that are vital to figuring out what will happen. Random walk models are used to examine stock market behavior, usage patterns in downloading over the internet, and a variety of other things. Predicting the likely outcomes makes it easier to design a stock-picking system or an internet communication protocol. It also makes it easier to create models for some of the odder phenomena in nature, such as Brownian motion.

If you've ever seen a floating mote of dust suddenly seem to dart off in a direction unexplained by wind or gravity, you've seen Brownian motion. What's happening is that the dust is being continually smacked around by molecules in the air. The dust mote is small and light enough that it can be pushed around by things too small to see. We can look at the molecules in the same way we do the coin flip in the random walk, as random sources of redirection. Using a three-dimensional version of a random walk we can get a more accurate description of where the dust mote is likely to fall.

Enough of probability for now. There's another side to this whole field, one that is misused even more.

## Statistics

Ahh, statistics. What other branch of math is splashed daily across front pages and trotted forth with such glowing pride by its users and misusers? What other branch is immortalized in a quote attributed to Mark Twain, and by Mark Twain to Disraeli, "There are three kinds of lies: lies, damned lies, and statistics."

What other branch has been so thoroughly abused by so many? Yes, many branches of math are hated by those who had a hard time learning them. But statistics is abused by those who claim to love it, to employ it, to give it a home in their publications and their hearts.

> What other branch is so commonly heard but so little understood?
>
> What is a statistic, and what does this branch of math actually do?

A **statistic** is nothing more than a number derived from data. Any number derived from data is a statistic. It does not matter how it was derived or if the number has any meaning. Throw data into a function and you get a statistic.

The branch of mathematics called statistics is supposed to be concerned with the generation of and testing of meaningful statistics. Statistics as a field exists largely for the benefit of science. Experiments produce data. These data need to be analyzed to figure out what, if anything, the experiment shows.

We will emphasize experiments that count things, and start with a phenomenon we have treated before: radioactivity. In Chapter 6, we noted that the ages of certain objects could be determined using Carbon 14, which has a half-life of 5730 years. We confidently asserted that half-life could be measured in the lab but didn't say anything about how to do it. How would you measure the half-life of a radioactive substance?

Since the half-life is the time it takes for half of the original sample to disappear, we could imagine starting our stopwatch, then making a series of measurements of how much of the sample is left, and then when we get to half of the original amount, stopping the stopwatch and reading off that time as the half-life. However, 5730 years is a long time to do an experiment. Just to make the issue more stark, let's consider the half-life of Uranium 238, which is 4.5 billion years. How do we measure that?!

When a Uranium 238 nucleus decays, it emits an alpha particle (basically the nucleus of a helium atom) and turns into a Thorium 234 nucleus. Each alpha particle is emitted with such a large amount of energy that we can detect each one of them as it is emitted, and over a time period $\Delta t$, we can count the number of emitted alpha particles. We could count all the emitted alpha particles by surrounding the sample of Uranium 238 with detectors in every direction. But that's both impractical and unnecessary: if we have a single detector of area A at a distance r from the sample, then (since a sphere of radius r has area $4\pi r^2$), we know that our detector captures the fraction

$A/4\pi r^2$ of the emitted alpha particles.

How does detecting the alpha particles help us to measure the half-life of the Uranium 238? The equation for the number N of atoms of Uranium 238, written in terms of the half-life $t_{1/2}$, is

$$dN/dt = -\left(\ln(2)/t_{1/2}\right)N$$

But each lost Uranium 238 nucleus results in the creation of one alpha particle, so in a time $\Delta t$, the number of alpha particles produced should be

$$\Delta t\left(\ln(2)/t_{1/2}\right)N$$

And since our detector detects the fraction $A/4\pi r^2$ of them, we should expect our detector to detect n alpha particles given by the formula

$$n = (A/4\pi r^2)\Delta t\left(\ln(2)/t_{1/2}\right)N$$

Solving this equation for $t_{1/2}$, we find

$$t_{1/2} = (A/4\pi r^2)\Delta t \ln(2)(N/n)$$

So that's how the half-life is measured. We can measure an enormous half-life in a much tinier time $\Delta t$, because in the formula the tiny $\Delta t$ is multiplied by the huge number (N/n) (Figure 9.6).

There's one problem with the method given above. The number n is the expected number of alpha particles. But radioactive decay is a completely random process, so even if we expect n, there should be some probability that instead we get some other number m. And indeed there is. It is given by the Poisson probability function:

$$P(m;n) = n^m \exp[-n]/m!$$

This is the probability of getting m counts in a random process where n counts are expected. Sure enough, if one takes the expectation value of m, the result is n. But because a given experiment could get any number, this means that calculating the half-life based on the number m that we actually do count is likely to give an error in the half-life.

How large an error should we expect? The answer to this is given in a quantity called the standard deviation $\sigma$. We will first give a general formula for $\sigma$ for any probability function, then explain that formula, and finally specialize to $\sigma$ for the Poisson probability function.

$$\sigma = \sqrt{E\left(X - E(X)\right)^2}$$

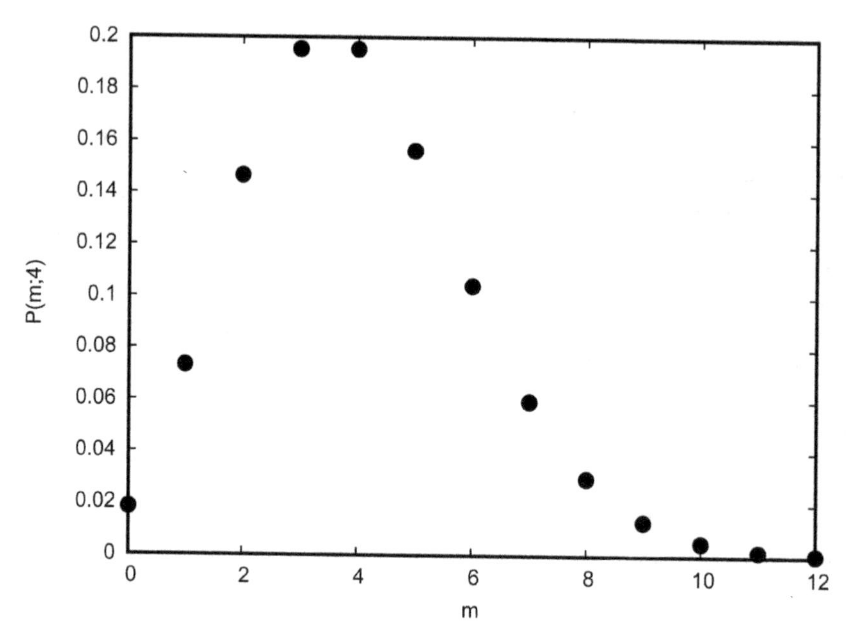


**FIGURE 9.6**
Values of Poisson Distribution

What this formula says is that for a random variable X, we take its expectation value E(X), which gives the average value of X. Then X-E(X) is how much X differs from its average value. If we were to average this quantity, then we would just get zero. Instead, we square, then average, and then take the square root. What results is essentially the average amount that X differs from its average. Hence the name, average difference = standard deviation.

For the Poisson probability function, the result is that $\sigma = \sqrt{n}$. But our formula for the half-life is proportional to n, this means that the fractional error in the half-life is $1/\sqrt{n}$. Or to put it another way, the percentage error is $100/\sqrt{n}$. So if I have counted 100 alpha particles, I have probably made about a 10% error in the half-life. If I have counted 1,000,000 alpha particles, I have probably made about a 0.1% error.

The fact that the more measurements we take, the more accurate our result is likely to be is the most basic practical application of the law of large numbers. It is also fundamental to the way science is done these days.

This is a general property of measurements made by counting: the percentage error is $100/\sqrt{n}$. The more you count, the more accurate your measurement, but to get twice as accurate a measurement, you have to count four times as many (of whatever it is that you are counting). To get a measurement that is ten times as accurate, you have to count 100 times as many.

What other measurements are made by counting? How about political polling? There are several difficulties with doing accurate polling. One is making sure that your population of survey respondents is actually a random representative sample of the population you want to poll (say likely voters in an upcoming election). Another problem is making your survey questions sufficiently neutral that they don't nudge respondents to give a particular answer. (Our father, who is an expert in market research, likes to joke that the only truly neutral survey question is the one word Yiddish question, "Nu?" whose rough translation in English is something like "Well?" or "So?"). However, in addition to all this, there is always the issue of Poisson statistics.

Political polls sometimes come with the notation "this poll has an error of plus or minus 3%." This is the result of Poisson statistics that comes with polling 1000 people. Why not poll 10,000 people and have an error of 1%? Because polling costs money, and the more people you poll, the more money it costs. Furthermore, both candidates and news organizations would like to do several polls over time in order to spot trends and try to discern the results of various strategies. For the same amount of money, they would rather do 10 polls of 1000 people each than one poll of 10,000 people.

That's all very well, but it is important to keep in mind the limitations of polls with a 3% error. If candidate A polls at 51% and candidate B at 49%, we're not really entitled to conclude that candidate A is ahead, since such polling numbers could easily arise with both candidates even or with candidate B ahead. If candidate A polls at 54% on Monday and then polls at 52% on Friday, we don't get to say, "OMG, candidate A is going down in the polls! Why is his campaign in such disarray?!" because it could well be that nothing whatsoever has changed about candidate A's standing with the voters, and the difference between the Monday poll and the Friday poll is normal statistical fluctuation.

Another type of measurement that is made by counting is tests for the efficacy of medicines. Such tests have their own inherent difficulties. Since both human and animal bodies are self-repairing mechanisms, the fact that someone took medicine and got better doesn't mean that they were cured by the medicine. To get around this difficulty, the test is done using two different groups: an experimental group that is given the medicine and a control group that is not.

To deal with the fact that researchers might unconsciously evaluate the control group differently than the experimental group, the researchers doing the evaluation are not permitted to know which test subjects are in the control group and which are in the experimental group. For tests involving human subjects, it's possible for their own biases to affect the outcome; so all the subjects are given something: the experimental group is given the medicine and the control group is given a fake medicine called a placebo, and no member of either group knows which they got. A test that uses all three of these protocols is called a controlled double-blind experiment.

But even in the controlled double-blind experiment, there is the issue of Poisson statistics. Even with a totally ineffective medicine, random chance can give rise to a case where the experimental group has a better outcome than the control group. How much better? Remember that $\sigma$ is only an average error: the Poisson probability distribution gives the possibility of errors of $2\sigma$ or $3\sigma$ or even more, though with smaller probability for larger errors. The (somewhat arbitrarily defined) notion of "statistical significance" is that an outcome is called statistically significant if it would occur by random chance only one time in 20.

OK, so suppose we have developed a medicine, and we subject it to a controlled double-blind experiment, and we get a statistically significant result that the medicine is effective. All well and good. We don't know for sure that the medicine really is effective, but odds of 19/20 are pretty good.

But suppose instead we had said, "our company is serious about developing a cure and have devoted vast resources toward finding one! We are going to test 100 medicines at once in 100 different controlled double-blind experiments!" Then even if all 100 medicines are totally ineffective, random chance is likely to give the result that around 5 of those medicines are deemed effective as judged by "statistical significance." Clearly, the (already arbitrary) notion of statistical significance is being misused here.

This result can be exceptionally dangerous, especially when there is a need for a medicine (as is happening in the current Covid-19 pandemic). As we write this, a large number of companies are testing a variety of treatments and vaccines. The desperation of the situation means that there is a desire for a quick solution. Therefore, anything positive or even potentially positive is given more weight than a properly mistrustful analysis would warrant.

One of the prime misuses of statistics is to offer a causal explanation for an outcome that could well be the result of random chance. Sporting events are full of stochastic processes: that basketball shot that almost went in, the football pass that was almost caught, the hockey shot that just barely missed the goal.

And yet, when was the last time you heard a sports commentator say something like the following? "Well Bob, the two teams were pretty evenly matched, so the fact that the home team won can basically be attributed to random chance. We could say a lot more, but none of it would be relevant, so let's just sign off now."

This tendency of commentators to make wise sounding pronouncements after the fact also occurs in politics; which was amusingly and cynically commented on by John F. Kennedy in 1960. Kennedy had just won the presidential election by a very narrow margin. One of his aides showed him a newspaper story that referred to Kennedy and his campaign team as "coruscatingly brilliant." "Yes," said Kennedy, "and if we had lost they would have called us coruscatingly stupid."

The way out of this difficulty is something called the **null hypothesis**, which is just a formal way of saying random chance. The idea is that given a set of experimental results, the very first thing one should do is to use the methods of probability to calculate the likelihood that such results would occur as a result of random chance. Only if that likelihood is small, should one look for some causal explanation.

We pointed out in our previous book, *Three Steps to the Universe*, that there is often a conflict between the need of science for clearheaded analysis and the need of the media for storytelling. Media reporters and commentators hate the null hypothesis, because it is not news. "New particle may have been discovered!" blares the headline. But then when you read the story, you find that this is a preliminary finding of a counting experiment where n is sufficiently small that the $1/\sqrt{n}$ error is large.

The results will get better when the experiment is run for a longer time resulting in a larger n. The null hypothesis is a likely explanation for the preliminary finding, but another possibility is that a new particle has been discovered. The story isn't wrong, but the headline is misleading. Sure enough, months later when n has become larger, we can confidently say that there is no new particle.

This problem is further exacerbated when the outcome is more directly important in people's lives as happens with medical research. In such situations, the tropish desire for explanation dovetails with the human longing for control and for problems to be solved. The care of science and the mistrust of mathematics look like obstacles, and the scientists and mathematicians can be alternately seen as saviors and enemies of the hoped for solutions.

The factual truth is that mathematical statistics and the scientific application thereof give us the ability to interact with and employ the stuff-happening of the universe, but they do not give us power over them. We hypothesize, test, observe, measure, and analyze the measurements. Then we go back and pick lines to pursue based on those results. No amount of wishing or hoping or demanding or yelling will change the fact that this is the most efficient system humanity has ever made for trying to plumb the depths of our own ignorance of the universe. And that the efficiency comes directly from the mistrust we use and our awareness of how likelihood resolves in reality.

We have come full circle from the start of the chapter. There we noted that probability is based on the fact that "stuff happens." We end by noting that proper use of statistics means first testing the null hypothesis. If it fits the data, then our best explanation in that case is "stuff happens."

# 10

## *Other Geometries: Not So Straight, These Sticks*

This chapter and the next form a pair of cautionary tales on the dangers of the lost inheritance and of the theoretical universe in general. The first tale concerns the blindness that comes from seeing too clearly. One of the subtle truths of human thought is that how we think about things affects how we perceive them. You can see this right at this moment while reading these words. You have been taught to see patterns of squiggles as words, taught so well that you probably don't even notice the letters themselves as shapes or wonder about the way they are created on page or screen. Your very facility with reading interferes with your ability to see these as anything but meaningful words.

This same form of blindness by recognition happens in other circumstances. You might instantly recognize a person you know and not notice any changes in him/her them (such as haircuts or new clothes). You might see a certain style of building and presume it's a house, not noticing a sign indicating that it's a shop or restaurant.

The more strongly we recognize, the more we do not see the things that do not fit our recognition. Strong, solid theories can create this very same unseeing sight. The more firm the theory, the more work is necessary to dislodge it.

The ancient Greek geometers created a graceful, beautiful, coherent, useful view of the universe that was applied everywhere, from the measuring of Earth to the building of cities to the mapping of the Heavens. It filled people's minds for millennia, and blinded them.

Euclid opened people's eyes and then stuck a stick in them. He made a world of lines and circles, of angles and edges, of perfect roundness, of cones (sold whole and in sections), of cubes and cylinders, a world of things that were obvious when he looked at them.

But he was wrong. What is true is not necessarily what's obvious, and what you see when you look depends on how you look at it.

*Euclid did all that on his own?*

**No, but in classic bad writing fashion we're putting the blame on the guy whose name is on the label. It's called Euclidean geometry even though he didn't invent it. He only wrote the books that codified it for centuries after.**

The most revelatory thing about Euclid's wrongness is how obvious it is if you open your eyes. Look at a tree and find me the lines and circles (Figure 10.1).

Look at a topographic map of Greece and find the straight lines and the circular arcs (Figure 10.2).

The world is not flat. Nor is it a sphere. It is bent and buckled, banged and burped out in many directions.

*So Euclid was wrong.*

**Except he was also right.**

He made a simple piece of the theoretical universe that doesn't model the shapes of everyday objects very well. But that piece is the starting point for our fundamental theories of the universe, and also enabled people to make ever more sophisticated theories that eventually went beyond Euclid.

It's not Euclid's fault, nor is it the fault of all the other Greek geometers we clump together under his name. If fault is to be found, it's among those who took Euclid at his word, the

**FIGURE 10.1**
Tree

ones who thought that what he was writing was what was obviously true. One person or a group of people laying down a good idea are not to be blamed when there come thousands after who follow mindlessly.

*Mindlessly is a rough judgment.*

**You're right. Mindlessly is an unfair word.**

Most of the people who worked in Euclidean geometry for the approximately 2,000 years in which it dominated geometric thinking were smart; some of them were brilliant. But they walked down the straight path Euclid created, looking to neither side, never questioning his assumptions about shape.

The unseeing awareness that gripped them came from being dazzled by a tool. Euclidean geometry has a grace and simplicity, an elegance that combines with its utility to make it not just good enough, but in its own way perfect. It is a neat, compact but expansive description of the observable universe. Indeed, it is sufficient for that universe, since any shape in the observable universe can be approximated as closely as the naked eye can observe by Euclidean objects and measurements. Yes, trees have funny shapes, but you can measure them by lines and circles, particularly if your purpose is to cut them up into columns and boards.

That's part of the cause of the error. Geometry as it was applied to the world was applied by architects who reshaped the things of nature to become things of humanity. They stripped the leaves from the trees and carved away the branches. They lined the shores with blocks of stone and piers of wood to make straight lines of the curious coasts.

**FIGURE 10.2**
Map of Greece

This ability to look at a tree and see lumber or look at a mountain and see blocks of stone meshed with the Euclidean worldview, physical tools and uses blending with the theoretical imagining. If you treat the real shapes of things as if they were merely nuisances to be removed, will you try to create a geometry for those shapes or will you get them out of the way to the "real" shape underlying?

It was not in looking at the things in front of their eyes that people first began to wonder about the rightness of Euclid. It was in looking at the elegance of the math itself. Only later did they discover that what they created applied to the real world and only later still did they see beyond even that to modern geometry and the view that axioms are not what you observe but what you choose them to be.

But first let's examine that nagging, misshapen lump amidst the beauty of Euclid's axioms.

## The Universe Is Bent

Non-Euclidean geometry was initially created because the parallel postulate annoyed people. It just wasn't as pretty as Euclid's other axioms, and it didn't have that air of the truly obvious the way *the shortest distance between two points is a straight line* does.

Let's look at the parallel postulate again with a slight rephrase:

> **Given a line and a point not on that line, there is exactly *1* line parallel to the first line passing through that point.**

If we draw a line and a point on a plane, we can quickly draw that one parallel line. Put a ruler down along the first line, then slide it carefully across the paper, maintaining its orientation until it touches the point you want to draw the parallel through. Now draw across the ruler. There you have it. That's not the way the Greek geometers would have done it. They would have carefully constructed the other line in such a way that it would have been provably parallel. But it shows the basic idea.

Geometric thinking and demonstration show that this postulate makes sense. It fits the everyday logic of road building and house making. But it bothered people because it sounded more like a theorem than an axiom. It wasn't clean enough for their thinking.

Many tried to prove the parallel postulate from the other axioms of Euclidean geometry. If, after all, it wasn't an axiom but a theorem, no one would complain about how ugly it was.

They tried to prove it. They failed. For century after century, more and more clever methods were created, but every proof had one or more holes in it, some gaping, some subtle. It didn't matter. The parallel postulate simply is not derivable from the other axioms.

Ultimately, two mathematicians, Johann Carl Friedrich Gauss, an 18th–19th-century German mathematician who contributed to so many parts of math that he can pop up in almost any field (mostly magnetic ones), and his contemporary, the Russian mathematician, Nikolai Ivanovich Lobachevsky, who is known, at least in name, to fans of comic songmaster and mathematician Tom Lehrer, each showed in their own way the independence of the parallel postulate. They did so by creating geometries that satisfied all the rest of Euclid's axioms but in which the parallel postulate was false.

Gauss did this:

> **Given a line and a point not on that line, there are exactly *0* lines parallel to the first line passing through that point.**

Figure 10.3 shows the equivalent drawing to the line and point diagram of the plane given above.

And Lobachevsky did this:

> **Given a line and a point not on that line, there are *more than 1* lines parallel to the first line passing through that point.**

And Figure 10.4 is his equivalent drawing.

*Hmm ... those "lines" don't look like lines. The first ones are circles and the others are curvy weird things going all over the place. And for that matter, a sphere isn't a plane, and neither is the whatever-it-is Lobachevsky was drawing on.*

**Well, maybe. Whether or not they're lines depends on what you mean by "line."**

*Well, according to Euclid, it's the shortest distance between two points.*

Let's look at that sphere. If you take two points and draw all the curves that connect those points, the shortest such curves are going to be what are called **great circle arcs**. A great circle on a sphere is a circle on the sphere with the same radius as the sphere. A great circle arc is a slice of a great circle (Figure 10.5).

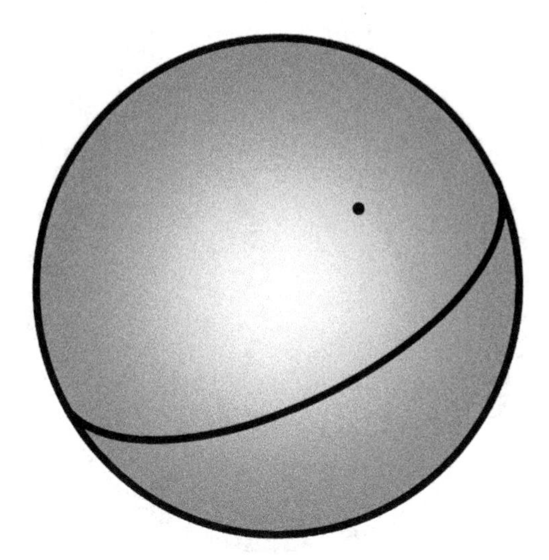

**FIGURE 10.3**
Sphere with Great Circle

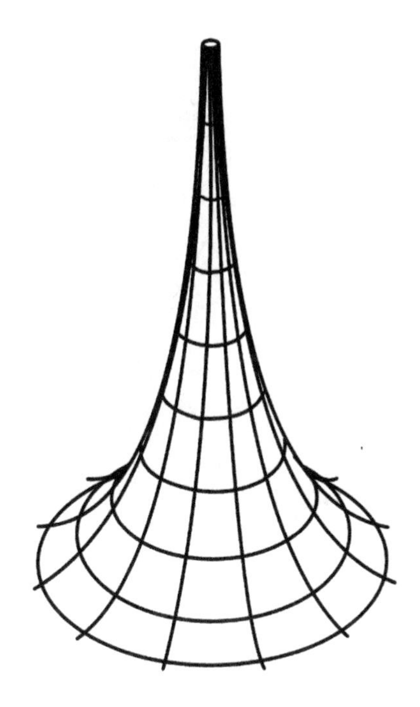

**FIGURE 10.4**
Pseudosphere

Take a look at two points on the sphere and various curves connecting those points. Notice that every such curve is longer than a great circle arc connecting them (Figure 10.6).

Of course that curve is not really a line, and we could get an actual line connecting the two points by going through the space inside the sphere. However, if we confine ourselves to only curves that stay on the surface of the sphere, then the great circle arc is the shortest way to connect two points and therefore satisfies the Euclidean axiom for a line.

**FIGURE 10.5**
Sphere with Great and Not So Great Circles

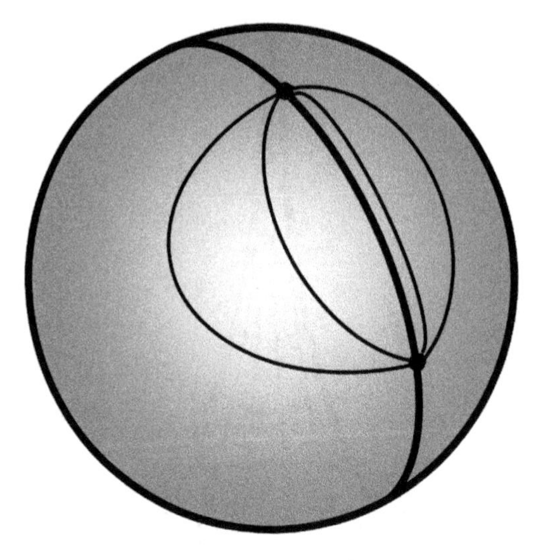

**FIGURE 10.6**
Sphere with Great Circle and Other Arcs

See, the fact of the matter is that we have two different ways of approaching the concepts for line and plane. One is to say that whatever objects of any kind that fit the axioms are perfectly valid objects to serve us as lines and planes; the other is a "I know it when I see it" notion. Sometimes these things fit together, and sometimes one interferes with the other. That's what happened for all those centuries. The shape "line" interfered with the axiomatic meaning "line."

But remember that the same problem existed with the idea of number. Fractions don't fit the number-of-stones idea that we had for counting numbers. Irrational numbers don't fit the ratio-of-numbers idea of rational numbers.

If we flip around the concept of an axiom, not as something that describes a thing that really exists but something that defines an abstraction, then changing the axioms changes what we are defining. We did this earlier in presenting analytic geometry and in examining operations to see if they were addition-like.

If we accept Euclid's axioms as they are, we have a constrained meaning of line and we can be content with Euclid's geometry. But remember that even in that case we found out that we could treat the real numbers as a line and the Cartesian plane as a plane. We discovered that objects from the realm of number fit Euclid's axioms so we could treat these sets of numbers and number pairs as if they were lines and planes.

We come up with even more possible fits if we start fiddling with the axioms, playing with them. Then we can find other geometries, other shapes to the world.

This might sound like a game, and it can be, but as with a lot of mathematical games, there is a real-world side to the play – in this case, the fact that the universe is not a flat space as Euclid described it. Thanks to Gauss, Lobachevsky, Riemann, and many others, many of whom did not know they were trying to reshape the universe, we have the mathematics to describe that new shape.

Let's take one more look at the parallel postulate and focus not on line but on the single number in it. The three geometries we've touched on are all defined in terms of number of parallels. Euclidean says 1, Gaussian says $<1$, and Lobachevskian says $>1$.

What we're dealing with here is one of the ways modern mathematicians look at the work they do. They examine the ways things are described and axiomatized and ask what happens if you remove or change one of the axioms, one of the conditions. What new kinds of objects do we find if we alter things a bit, or more than a bit? This kind of tinkering thinking allows for greater exploration of the theoretical universe.

*How can people who mistrust everything go wandering off like this?*

**By using mistrust as a guide.**

This kind of alteration is not done with unbridled flights of fancy. It's done by making individual changes and exploring the consequences. Sometimes that exploration leads to the conclusion that one is in an impossible place.

Consider the group axiom. There is a unique 0 such that $a + 0 = 0 + a = a$. Suppose we replace that with there are more than one 0s, each of which satisfies $a + 0 = 0 + a = a$. But we've already proven that that's impossible. So the existence of one identity means there's only one identity. We also showed that if + is associative, then each object can have only one inverse. So we can't change the group axioms to have multiple inverses as long as we have associativity.

Let's try something else. Let's take the axioms for an equivalence relation. We'll use R as a symbol for the relation.

Reflexive: aRa

Symmetric: if aRb, then bRa

Transitive: if aRb and bRc, then aRc.

Now let's replace the symmetric axiom with the following:

Either aRb or bRa, and if both aRb and bRa, then $a = b$.

Are there relations that satisfy these three axioms? Indeed there are. Consider $\leq$ (less than or equal to).

$a \leq a$

Either $a \leq b$ or $b \leq a$, and if both $a \leq b$ and $b \leq a$, then $a = b$.

And

if $a \leq b$ and $b \leq c$, then $a \leq c$.

By transforming one of the axioms, we've wandered into a different part of the theoretical universe of relations – that of ordering relations.

But what about the Euclidean geometry case? Did Gauss and Lobachevsky leave the correct geometry behind in creating their variants?

No, they opened up an area in which the modeling was better for the world around us. Every mental object that is abstracted from the real world is a model, an approximation of the nature and character of that object. It is perfectly possible that some related object may be a better approximation. Or, if one is dealing not with objects but with methods, then tweaking one or more characteristics of an object may produce something much more useful in different circumstances. Lengthen a knife and you get a sword. Enclose a fire and you get an oven.

The sphere and the pseudosphere above are only the beginnings of one branch of mathematical variation. They were the first of what were called "non-Euclidean" geometries, which is a pretty Euclid-centric way of looking at it.

To add to the peculiarity, let's be pedantic. The word "geometry" means "Earth measure." The Earth's surface is more or less a sphere (we'll do the less later in discussion of fractals). It's certainly more a sphere than it is a plane. So Gauss's geometry is closer to the reality of our planet than Euclid's.

Isn't it?

Well, yes and no. Yes, the Earth is a sphere, but on the everyday human scale we don't notice that. In a small enough piece – say, the space a human stands on and walks around in, the space of a house or a small farm – the curvature of the Earth doesn't come into play much. We're reasonably okay measuring distances across small stretches of the Earth using Euclid's straight lines and flat planes rather than Gauss's great circle lines on spheres.

The term for a shape that seems flat in small regions is **locally Euclidean**. In a locally Euclidean shape, we can approximate the geometry of a small region with Euclid's geometry.

This is actually less mysterious than it sounds. Let's think of Gauss's sphere in terms of its formula as a surface in space.

$$x^2 + y^2 + z^2 = R^2$$

where the constant R is the radius of the sphere. If we look only at the top half of the sphere, we can think of z as a function of x and y

$$z = \sqrt{R^2 - \left(x^2 + y^2\right)}$$

But now that we are thinking in terms of functions, we can easily generalize to much more general surfaces than the sphere. That is, we can consider a function $z = h(x, y)$ and look at the geometry of that surface. Here, the surface consists of points which at position $(x, y)$ are at a height $h(x, y)$ above the Cartesian plane.

For a function of a single variable, we have the tangent line at a given point, and the function is well approximated by the tangent line, as long as we are close to that point. Similarly, for a function of two variables, there is a tangent plane at any point, and the function is well approximated by the tangent plane as long as we are close to that point.

Thus, surfaces are "locally Euclidean" in exactly the same sense and for exactly the same reason that curves given by differentiable functions can be thought of as "locally a line."

The properties of these general surfaces were studied extensively by Gauss, and then abstracted and generalized by Riemann. One of the things Gauss wanted to know was how to go beyond the tangent plane approximation. For a function of one variable, the tangent line consists of the first two terms in the Taylor series, so to get a better approximation one simply adds the next term in the Taylor series. This yields the circle that best fits the curve, which is called the osculating circle.

But functions of two variables, which one needs to get a surface, are more complicated than functions of one variable. To get around this difficulty, Gauss imagined going along the surface in a particular direction to get a curve, and then defining the curvature $\kappa$ in that direction to be one divided by the radius of the osculating circle for that curve. This makes sense because high amount of curvature corresponds to a small circle, whereas a small curvature corresponds to a large circle, which means that the curve is almost a straight line. Here, the curvature is called positive when the curve goes inward and negative when the curve goes outward (Figure 10.7).

This different curvature for each direction sounds complicated, but Gauss noted that there must be some direction where the curvature is maximum and another direction where the curvature is minimum. He called these maximum and minimum curvatures the principal curvatures. Thus, each point of the surface can be characterized by the principal curvatures $\kappa_1$ and $\kappa_2$.

Gauss made a distinction between "intrinsic" properties of the surface, which can be deduced just from measurements of small distances within the surface, and "extrinsic" properties of the surface, which one needs the function h(x, y) to compute. Neither $\kappa_1$ nor $\kappa_2$ by itself is intrinsic, but their product $\kappa_1\kappa_2$ (now known as the Gaussian curvature) is intrinsic.

Let's consider a few examples of Gaussian curvature. For the sphere of radius R, all the osculating circles are circles of radius R. This makes both principal curvatures equal to 1/R, which in turn makes the Gaussian curvature equal to $1/R^2$. It turns out that the Lobachevsky plane has one principal curvature positive and the other negative, for an overall negative Gaussian curvature. The Euclidean plane has both principal curvatures zero, which yields a Gaussian curvature of zero. So we have a quick way of characterizing the three possible geometries that satisfy the truncated list of Euclid axioms where one leaves out the parallel postulate: they are geometries where the Gaussian curvature is positive (Gauss), zero (Euclid), and negative (Lobachevsky).

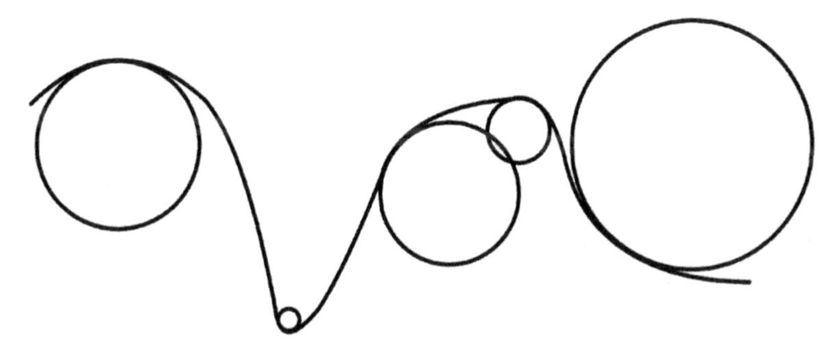

**FIGURE 10.7**
Osculating Circles

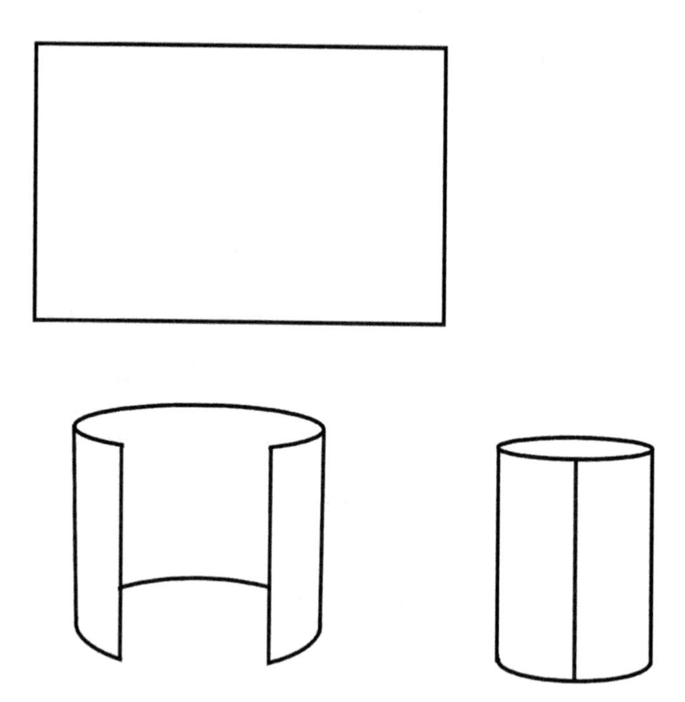

**FIGURE 10.8**
Plane to Cylinder

A surprising example is the cylinder. Let R be the radius of the cylinder. Then the osculating circle that gives $\kappa_1$ is the circle of radius R, which yields $\kappa_1 = 1/R$. However, the "osculating circle" that gives $\kappa_2$ is a line that goes along the cylinder parallel to the axis, thus giving $\kappa_2 = 0$. This means that the Gaussian curvature is zero, which means that Gauss is telling us that intrinsically the cylinder is the same as the plane! But Gauss is right. You can make a cylinder by taking a flat sheet of paper and curling it up. None of the local distances along the surface change when the paper is curled up, so as far as intrinsic local geometry is concerned, the cylinder and the plane are the same (Figure 10.8).

## Shortest Distance Between Two Points

In Euclidean geometry, the statement that the shortest distance between two points is a line is an axiom. But in the Cartesian plane, all facts of geometry become facts about numbers, so we should be able to calculate the lengths of curves and show that the shortest one is a line. Let's take one of our points to be the origin (0, 0) and the other point to be on the x-axis (D, 0). Then, the claim is that of all the curves that go from (0, 0) to (D, 0), the line segment on the x-axis is the shortest one. Let's think of another curve connecting those points given in parameterized form by (x(t), y(t)), where the curve starts at t1 and ends at t2. In order to start at (0, 0) and end at (D, 0), we have to have x(t1) = 0, y(t1) = 0, x(t2) = D, y(t2) = 0. We imagine chopping up the curve into little pieces, where t changes by a small amount dt and then adding up the results. The length of one of these little pieces is ds, where we have

$ds^2 = dx^2 + dy^2$, and dx is the amount x changes and dy is the amount y changes. However, since $x = x(t)$ and $y = y(t)$, we have $dx = x'(t)\,dt$ and $dy = y'(t)\,dt$. We then find

$$ds^2 = dt^2\left[(x'(t))^2 + (y'(t))^2\right]$$

and so the length of the curve is

$$L = \int_{t1}^{t2} dt\sqrt{[(x'(t))^2 + (y'(t))^2]}$$

Since $(y'(t))^2$ can't be negative, this means that the way to make the length of the curve as small as possible is to have $y'(t) = 0$ for all t (note that we can't do that for $x'(t)$ because x starts at zero and ends at D). But this just means that y(t) is a constant, and since $y(0) = 0$, that constant has to be zero. So the equation for our curve of shortest distance is $y = 0$, the line segment that connects (0, 0) to (0, D) (Figure 10.9).

Now let's see what happens if we try to do the same sort of thing on the surface $z = h(x, y)$. That is what curves are shortest distance curves along the surface. The surface is in space, so the distance ds between two nearby points satisfies $ds^2 = dx^2 + dy^2 + dz^2$. However, since $z = h(x, y)$, we have

$$dz = (\partial h/\partial x)\,dx + (\partial h/\partial y)\,dy$$

So to find $ds^2$ we should square the right-hand side of this equation and then add it to $dx^2 + dy^2$. This sounds messy, but we notice that the final expression for $ds^2$ can only consists of three kinds of terms: ones proportional to $dx^2$, ones proportional to $dy^2$ and ones proportional to dxdy. That is, we must have a formula of the form

$$ds^2 = g_{xx}dx^2 + g_{yy}dy^2 + 2\,g_{xy}dxdy$$

And carrying out the calculation, we find

$$g_{xx} = 1 + (\partial h/\partial x)^2$$

$$g_{yy} = 1 + (\partial h/\partial y)^2$$

$$g_{xy} = (\partial h/\partial x)(\partial h/\partial x)$$

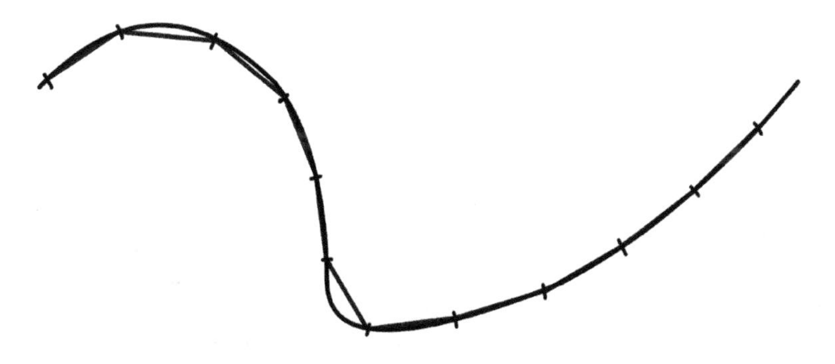

**FIGURE 10.9**
Segments Along Curve

Now, if we have a parameterized curve x(t), y(t) in the surface, we find that the length of the curve is

$$L = \int_{t1}^{t2} dt \sqrt{[g_{xx}(x'(t))^2 + g_{yy}(y'(t))^2 + 2g_{xy}x'(t)y'(t)]}$$

The quantities $g_{xx}$, $g_{yy}$, and $g_{xy}$ are called metric components.

We can use this expression of L to find the curve of the shortest length between any two points, but the result is not some simple rule like "the curve of shortest distance is a straight line." Instead, the curve of shortest distance, called a geodesic, is a solution of a differential equation, called the geodesic equation. The geodesic equation contains some very messy terms called Christoffel symbols (which one of David's professors liked to refer to as "Christ-awful symbols"), which are calculated using the metric components and their partial derivatives with respect to x and y.

The Christoffel symbols (along with their partial derivatives with respect to x and y) can be used to calculate the Gauss curvature. But the geodesic equation also gives us an intuitive way to understand the meaning of Gauss curvature. Remember that in the Euclidean plane, two nearby lines that start in the same direction remain parallel and never get any closer to or farther away from each other. In contrast, let's take two nearby points on the equator on the sphere, and for each point draw the line (great circle of constant longitude) going north. These lines get closer and closer together until they finally meet at the north pole.

Correspondingly, if we were to try the same thing with the Lobachevsky plane, we would find that the lines get ever farther away from each other. This sort of property is quite general: in any curved surface, two lines started in the same direction get closer together if the Gaussian curvature is positive, farther apart if the Gaussian curvature is negative, and stay the same distance apart if the Gaussian curvature is zero.

The intrinsic point of view of Gauss and Riemann notes that once we have used h(x, y) to calculate $g_{xx}$, $g_{yy}$, and $g_{xy}$, we can forget that these quantities came from an h(x, y) and just use them to calculate distances, geodesics, and Gaussian curvature. Gauss and Riemann then define a two-dimensional curved space to be one with a $g_{xx}$, $g_{yy}$, and $g_{xy}$. This definition was then generalized by Riemann to spaces of any dimension. In that way, Riemann could speak of an n-dimensional curved space without having to worry about what higher than n-dimensional space that space was curved in. This Riemannian study of general curved spaces is called differential geometry.

Riemann noted that since curved spaces are locally Euclidean, we might very well be living in a curved three-dimensional space, provided that the curvature was too small for us to notice its effects.

Riemann's remark was prophetic but not quite in the way he envisioned. In Einstein's theory of general relativity, three-dimensional space is combined with time to make a four-dimensional spacetime. It is spacetime, not space alone, that is curved. The geodesics of curved spacetime are the paths that objects take under the influence of gravity, so the "force" of gravity then just becomes a manifestation of the curvature of spacetime. And what about Newton's law that each mass makes a gravitational force that falls off as $1/r^2$? In Einstein's relativity, the corresponding statement is that mass density (and energy density too) causes the curvature of spacetime. One pithy summary of general relativity given by John Wheeler is that "spacetime tells matter how to move; matter tells spacetime how to curve."

## Metric Spaces

We asked the question, "What is the shortest distance between two points?" and answered it with metric components and integrals.

From the point of view of mathematical abstraction, distance is a function that takes two points and gives a number. Why do we have to use the old distance function we got from Pythagoras?

We already know we don't always use that distance function. The arc length integrals above, when used on geodesics, give us totally different distance functions for two points on a surface. Figure 10.10 shows two points in space that lie on three different surfaces, with three different geodesics and therefore three different distances.

So why don't we just substitute any function that takes pairs of points and gives us numbers and use that for distance?

You may object that distance has a real-world meaning. But so do a number of things that we have stretched into broader theoretical categories. In all cases of such pulling and twisting, we have isolated the important characteristics of things, turned those into axioms, and then figured out what other kinds of things fit those characteristics.

Distance is a real-world thing, but that doesn't mean we can't explore it and find more distance-like things.

First, a term: a function of pairs of points that gives numerical values and is meant to represent some kind of distance is called a **metric**. (Confusingly, **metric** is also the term for the thing whose components $g_{xx}$, $g_{yy}$, and $g_{xy}$ are used to compute lengths of curves. The "metric" of metric spaces that we are about to explore is not the same as the "metric" of differential geometry).

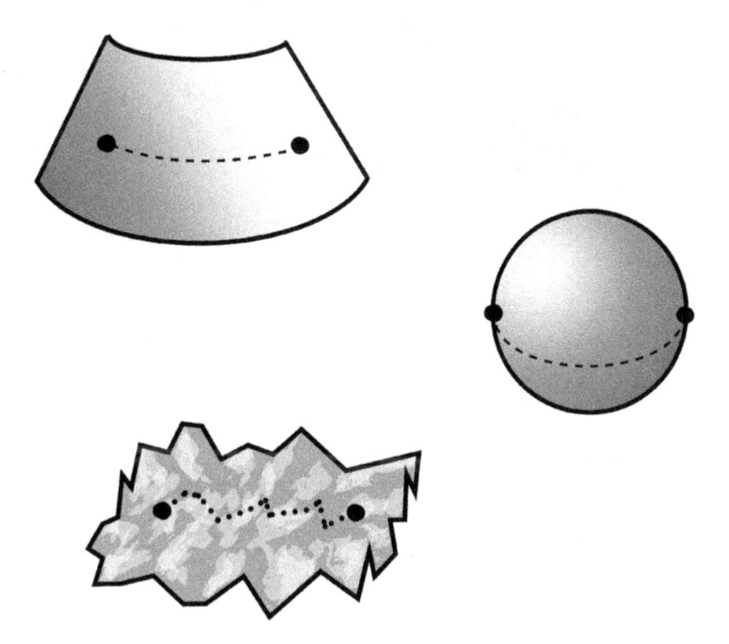

**FIGURE 10.10**
Geodesic Distance

For two points (x1, y1) and (x2, y2), the function d given by

$$d((x1,y1),(x2,y2)) = \sqrt{((x2-x1)^2 + (y2-y1)^2)}$$

is certainly a distance-like function, since it's the standard, everyday distance function.

But suppose we don't want to know the actual as-the-crow-flies distance between two points. Suppose we want to know the driving distance in a city.

For ease of calculations we'll use an abstracted American city in which everything is done in square blocks with two-way streets. There is no such real city, but cities like this exist in a mysterious other world called Math Problem World, which we'll visit again later (Figure 10.11).

What's the distance between these points as the car drives (assuming one drives straight, doesn't get lost, pulled over, confused, distracted by a beverage source, etc.)? It would be the distance across the map plus the distance up along it (Figure 10.12).

This distance is calculated by using the metric

$$d_t((x1, y1),(x2, y2)) = |x2-x1| + |y2-y1|$$

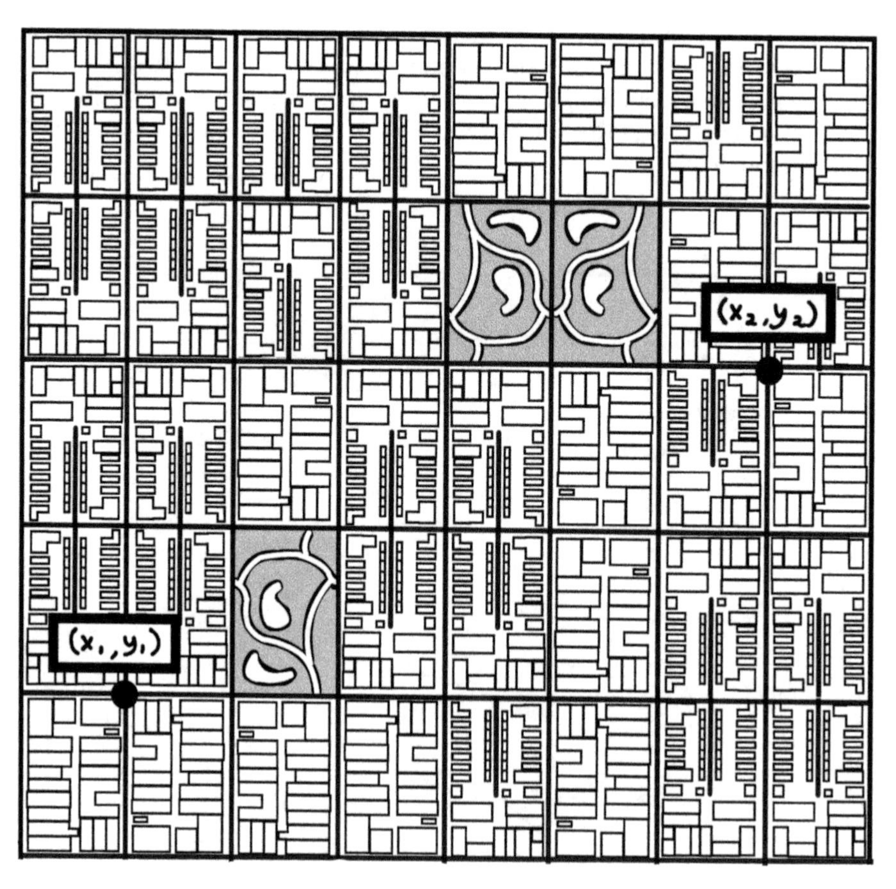


**FIGURE 10.11**
Gridded City Map

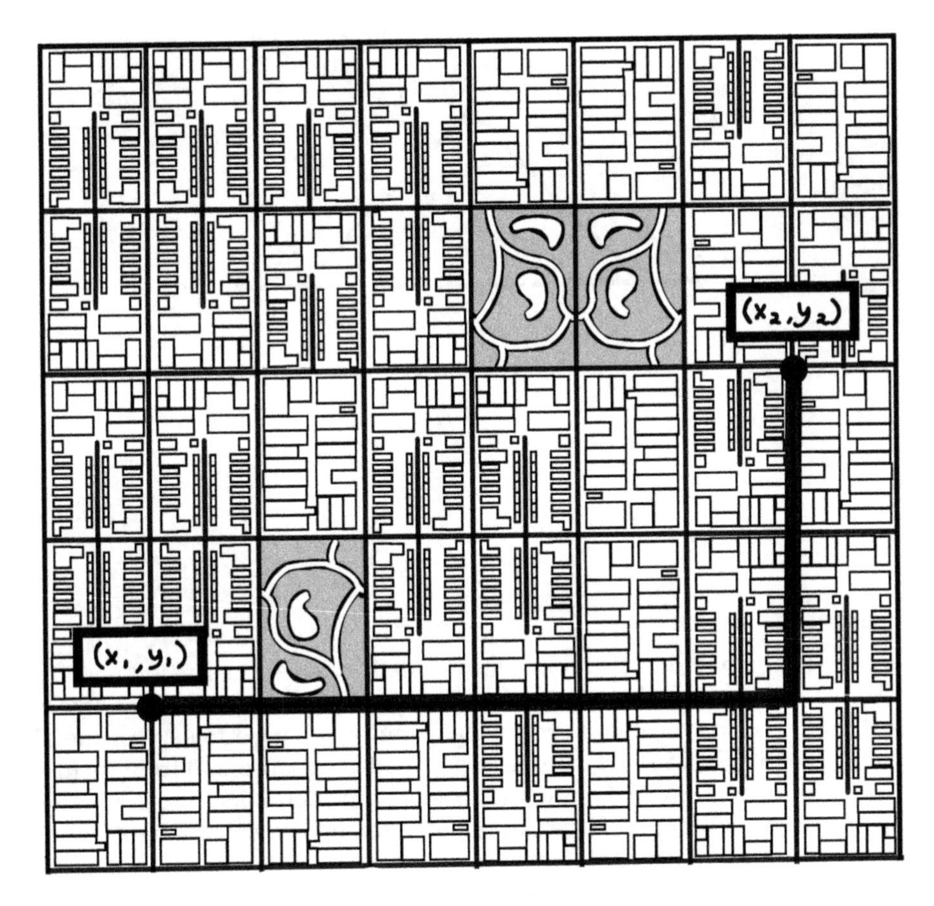


FIGURE 10.12
Gridded City Map with Route

This distance function is called the "**taxicab metric.**"

It works perfectly well as a distance function and is even of practical use. It also has one amusing oddity. Let's figure out what a circle is in the taxicab metric.

*A circle's a circle. What are you being weird about now?*

**A circle isn't always a circle.**

A circle is the set of all points a given distance, called the radius, from a given point called the center. This definition relies upon a means of determining distance. Change the distance function, change the shape.

If our distance is found using the taxicab metric, what shape do we end up with for the circle?

Let's say the center is the origin. What points (x, y) would be a given distance r from the origin?

Those would be the points for which $|x| + |y| = r$. A little testing and drawing gives us Figure 10.13 for a circle of radius 1.

You can also make more sophisticated kinds of metrics using, for example, the amount of energy it takes to get from one point to another. In this metric, a point 30 meters away straight up is farther away than a point 30 meters along the ground.

It's important to bear in mind that the space one is finding distance in need not represent a "real" space. After all, we created Cartesian planes and spaces with axes that did not represent position. We can create distance functions in these spaces easily enough.

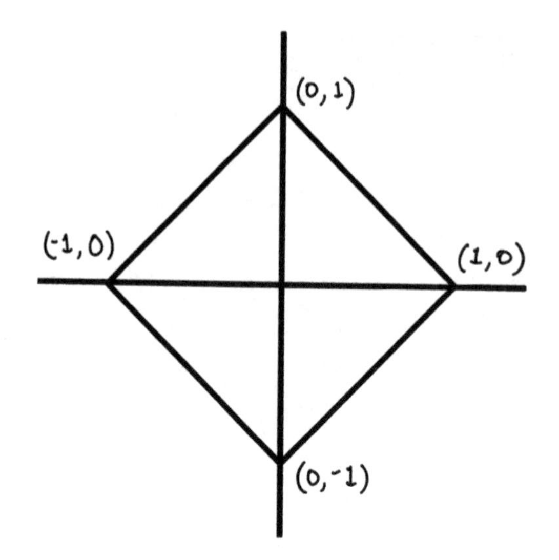


**FIGURE 10.13**
Square

But we can be even weirder than that. One can define a distance function on any set of objects. All we need is something distance-like that gives us a number for any two elements of the set. The six degrees of separation game can be used to express the distance between any two people (in this game, each person is a distance of 0 from themselves, a distance of 1 from any person they know, 2 from any person known to any person they know, etc.). With this kind of trick we can put a metric on any set, giving a rudimentary geometry to an unordered list of objects.

*That doesn't sound like geometry to me. Geometry is about shape.*

**And thereby hangs the complication. There's the shape you see, and the shape that is implicit in the measurement.**

If we shift our concept of geometry away from perception and on to measurement, we can create geometry without any ability to see any shapes.

Unfortunately, this kind of geometry sacrifices geometric thinking and hands everything over to analytic thinking. This shift is another reason that geometric thinking has fallen from mathematical fashion.

So far, we've been casual about what kinds of functions can be metrics. Can we really use any function that takes two objects and gives us a number as a metric? Aren't there characteristics that we want a metric to have and which we would use to say that if it doesn't fit these, it isn't a distance function? A list of such characteristics has been created.

A function from pairs of objects in a given set to numbers is called a metric if it satisfies the following:

For any two objects a and b from the set,

1. $d(a, b) \geq 0$. This is called **positive definiteness**. It means we never have negative distance.
2. $d(a, b) = 0$, if and only if $a = b$. In other words, each thing is at 0 distance from itself and nothing else is at 0 distance from it.
3. $d(a, b) = d(b, a)$. Distance going one way is the same as distance returning.

4. $d(a, c) \le d(a, b) + d(b, c)$. This is called the **triangle inequality**. It gives us a sense that distance really means shortest distance. We can't find a way to go from a to b and from b to c that is faster than going directly from a to c.

A set with a distance function of whatever kind is called a **metric space**. It is a very abstract kind of geometry. So what might we use this sort of abstract machinery for? How about to compare two functions? Suppose that we have two functions f(x) and g(x) and we want to have some idea of how far apart they are from each other. Consider the following distance function:

$$d(f, g)=\sqrt{\left[\int_{-\infty}^{\infty} dx(f(x)-g(x))^2\right]}$$

This is called the $L^2$ (pronounced "ell-two") distance. In order to make sure that this integral converges, our space of functions will have only those functions that go to zero in the limit of large (positive or negative) x.

We're not going to show that this distance satisfies the metric space axioms. Instead, we're going to show why this "distance" is enough like distances that we've seen before that we should be happy calling it a distance. Let's start by thinking about how we would evaluate the $L^2$ distance using a computer.

Computers can't add up an infinite number of numbers, so we would start by replacing the integral from minus infinity to infinity with the integral from a to b, where a is some large negative number and b is some large positive number. Then, we would divide up the interval from a to b into a large number N of equally spaced pieces of size $dx = (b-a)/N$. Our integral would then become a sum of terms of the form $(f(x_i)-g(x_i))^2$ multiplied by dx, and then in the end we would take the square root. But this "square root of the sum of the squares" is exactly the old Pythagorean distance.

Put another way, we can think of the space of functions as an "infinite dimensional vector space" and the $L^2$ distance as the "infinite dimension limit" of the usual vector space distance. We will return to this point of view, and its relation to quantum mechanics, in a later chapter.

Another sort of distance is the distance between a theory and an experiment that is supposed to test that theory. Suppose that you have a theory that relates two quantities y and x. You have calculated a function f(x) and your theory is that $y = f(x)$. Now someone decides to test your theory. For a bunch of different values of x: $x_1, x_2, \ldots, x_n$, they measure the values of y and get $y_1, y_2, \ldots, y_n$.

In the old elementary school notion of science as "hypothesis tested by experiment and disproved if it disagrees with experiment," it sounds as though we should go through the data and see if $y_1 = f(x_1)$ and $y_2 = f(x_2)$ and … and $y_n = f(x_n)$.

If we get even one disagreement, we should give up on the theory. But we know that can't be the right thing to do, because experimental measurements are not completely accurate and so the $y_1, y_2, \ldots, y_n$ that they measure are not the same as the $y_1, y_2, \ldots, y_n$ that are really there (or as Francis Crick put it, "any theory that agrees with all the experiments is wrong because some of the experiments are wrong").

So what should we do instead? We might think of defining a sort of Pythagorean distance d, where $d^2$ is given by the formula

$$d^2 = \left[y_1 - f(x_1)\right]^2 + \left[y_2 - f(x_2)\right]^2 + \cdots + \left[y_n - f(x_n)\right]^2$$

But this can't be quite right either, because each measurement can have a different expected error, so what we really want to know is how close theory is to experiment as compared to the experimental error. This is done by modifying the Pythagorean distance so that the squared distance $\chi^2$ is given by

$$\chi^2 = \left[y_1 - f(x_1)\right]^2 / (\sigma_1)^2 + \left[y_2 - f(x_2)\right]^2 / (\sigma_2)^2 + \cdots + \left[y_n - f(x_n)\right]^2 / (\sigma_n)^2$$

Here $\sigma_1$ is the expected error in measurement 1 and so on. The symbol $\chi$ is the Greek letter chi, which we were taught to pronounce like "kai" with the squared distance pronounced like "kai squared." However, our Greek colleagues tell us that $\chi$ should really be pronounced more like "hee."

One might then think that a theory should be judged solely by how small it can make $\chi^2$, with smaller $\chi^2$ meaning a better theory. But that can't quite be right as one can see by examining another old elementary school notion about science: that one "gathers data and then uses it to form a hypothesis."

So here's our theory: first our experimentalist friend makes his measurements and plots the points $(x_1, y_1),(x_2, y_2),\ldots,(x_n, y_n)$. Now, we play the old child's game of "connect the dots." We draw a line from $(x_1, y_1)$ to $(x_2, y_2)$, then a line from $(x_2, y_2)$ to $(x_3, y_3)$, and so on until we get to $(x_n, y_n)$. The set of line segments is our "theory" and it fits the data perfectly giving a $\chi^2$ of zero!

If "connect the dots" doesn't sound "science-y" enough, we can say that we are fitting the data with a "piecewise linear function," which is the technical mathematical term for "connect the dots." Alternatively, if we don't like a function with kinks, we can instead make our "theory" that f(x) is the $n - 1$ order polynomial that goes through the points $(x_1, y_1),(x_2, y_2), \ldots, (x_n, y_n)$. Once again, the "theory" agrees perfectly with the data and $\chi^2$ is zero.

Clearly, this sort of "theory" making is cheating. But since such cheating would be allowed if the rule were "make $\chi^2$ as small as possible," this means that we need a better rule. The better rule starts by defining the number of degrees of freedom to be the number of data points minus the number of free parameters in the theory that we used the data to fit. We have n data points, but both "connect the dots" and $n - 1$ order polynomial have n parameters. Thus, each of these methods of making a theory has zero degrees of freedom. The better rule says that a theory has to have more than zero degrees of freedom to be judged at all, and that for such theories we should compute $\chi^2$ per degree of freedom: that is, $\chi^2$ divided by the number of degrees of freedom. This better rule not only eliminates the sort of cheating described above, but it also rewards Occam's razor, since for a given $\chi^2$, a theory with fewer parameters gets a better score.

So how large do we expect $\chi^2$ per degree of freedom to be? In a correct theory, we would expect each measurement to differ from the theoretical answer by about the expected error of the measurement. Thus, each measurement should contribute about one to $\chi^2$, and therefore theories are judged to be in agreement with experiment if $\chi^2$ per degree of freedom is around 1. If two different theories are both in agreement with experiment, and if we have no other criteria to use to choose between them, then it is time to do more experiments.

The $\chi^2$ test draws upon the evolved concept of distance and lets us loop the mathematical principle of mistrust into an actual mathematics of mistrust. We can actually measure how much we should not rely on our own theories.

So what have we opened up with this foray outside the old geometry?

We've extended the ideas of distance, measured curvature, and found in effect how to have Euclid when we need him and give him up when he gets in the way. The world we can now see and measure is a curved one, with bends and swoops and graceful arcs upon which we can travel and measure the distance. It's also a world where distance need not connect with shape and indeed can supersede it, where geometry still measures, but it's not necessarily measuring Earth.

We'll come back to differential geometry later in the discussion of modern physics. For now we're going to leave distance and look at the other blind spot in Euclid's geometry – for if we see the world through his eyes, all we perceive are smooth, pretty shapes that are wholly unnatural.

## Nature Spiky in Coast and Leaf

**How long is this?** (Figure 10.14)

*Looks like 1 meter.*

**Look closer** (Figure 10.15).

*Okay, that's 10 pieces, each made up of a pair of 2 decimeter pieces, I guess it's 10 times 2 times 1 decimeter = 20 decimeters = 2 meters.*

**Look closer at one of those pieces** (Figure 10.16).

*So that's 10 times 2 times 10 times 2 centimeters = 400 centimeters = 4 meters.*

**Look closer.**

*Every time I look closer it's twice as long. Hey, when is this going to end?*

**It doesn't.**

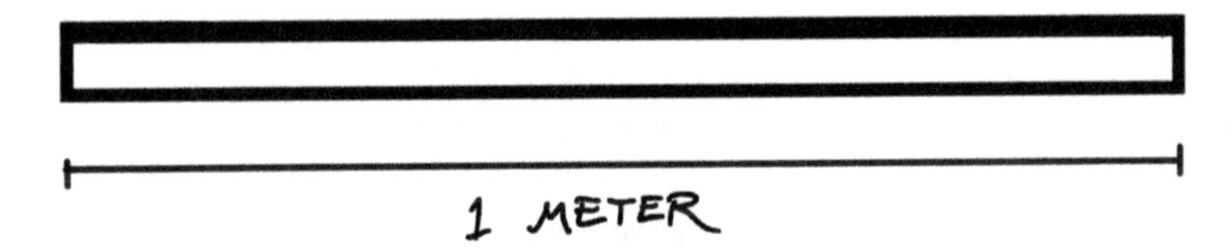


**FIGURE 10.14**
Sawtooth

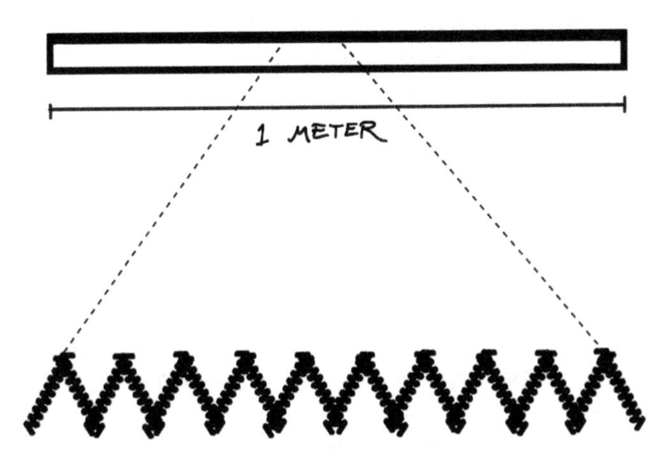


**FIGURE 10.15**
Sawtooth Zoomed in

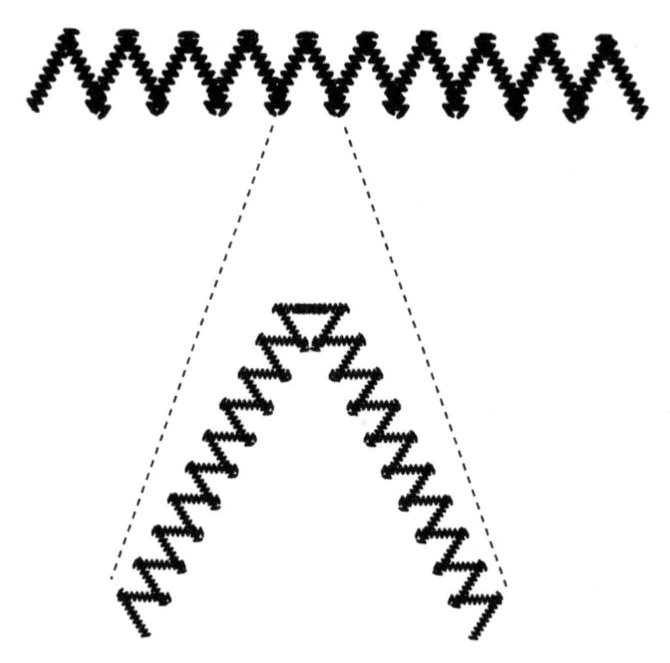

**FIGURE 10.16**
Sawtooth Zoomed in Even More

It keeps going. Each piece is made of 20 pieces, 1/10 the length of the piece's end-to-end length. So each time we zoom in, we measure it as twice as long.

*That means that stick has infinite length.*

**Yup.**

*What's gone wrong here? Our basic unit of measurement, the stick, has turned into some pathological monstrosity! Where can I get one?*

It hasn't, for three reasons: one dull followed by one mildly revelatory, which points toward the last one, which is quite deep. By the way, "pathology" and "monstrosity" have both been used in math as terms to describe things that are weird from the perspective of whatever theory is current.

*You mean mathematicians use them to describe things they think have cooties?*

Sometimes, but just as boys and girls stop using that term if and when they start to see beauty and humanity in each other, so mathematicians have found much to love in their pathological monsters.

Let's get back to the three reasons.

Dull: We defined the length of a stick as the length of the line segment that joins the endpoints. We didn't say the stick's length was its arc length. Just because the stick is infinitely long if you measure along, it doesn't mean it's an infinite distance from end to end.

Mildly revelatory: these different lengths represent the fact that the world is not the same at different levels of scale. The distance an ant has to walk across a bumpy field is longer than the distance a human has to walk across the same field, because the ant has to go up and down the bumps, whereas a human treats each bump as just a place to put a foot down (Figure 10.17).

All we've done with the saw-toothed line above is showed these different scales.

Deep: Shapes are not what we thought they were. When we started out with geometry, we made a batch of approximations so we could look at shapes, take measurements, and

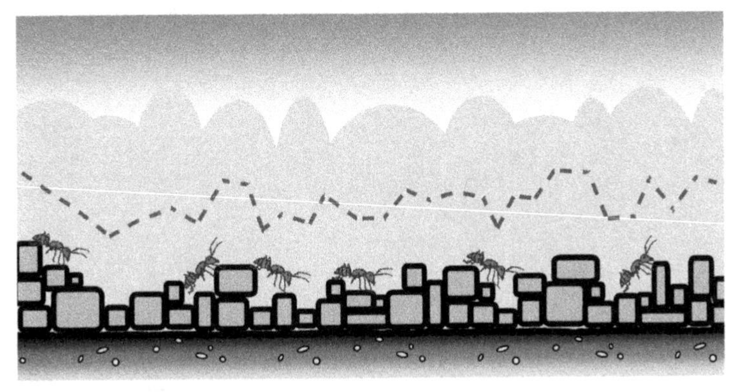

**FIGURE 10.17**
Human and Ant Distances

make drawings. We needed simplicity, and we cared less about nature than we did about things like architecture. We wanted angles for building and surveying. We made stones into blocks and wood into beams. We carved up the world into the shapes we needed for our purposes. But in the process we lopped off a large number of shapes that we never gave names to. Nor did we truly worry about how to describe these shapes mathematically. We had our tools, compasses, and straightedges, and we measured and drew according to the ways of our tools.

Later we added the curves of analytic geometry and we started making different approximations. We made our shapes rounder and odder. We added swoops and slides to our thinking. We made our mountains look like bell curves and our rivers sinuous (Figure 10.18).

But still we ignored a great deal. We did so, again, because of our tools. We had calculus, which wants smoothness. We treated breaks and corners as pathologies and relegated them to endpoints that we weren't going to worry about. Our spaces were smooth, differentiable, even infinitely differentiable, so we could approximate them with Taylor series. We loved the curviness of the world and we drew it with our functions. We even played it on our instruments, summing sines and cosines into Fourier series.

We did so with such a mental focus that the kind of thing at the beginning of this section with the wavy line was seen as a trick with nothing to do with reality.

Except that it stares us in the face when we look upon a beach, particularly if we look down (Figure 10.19).

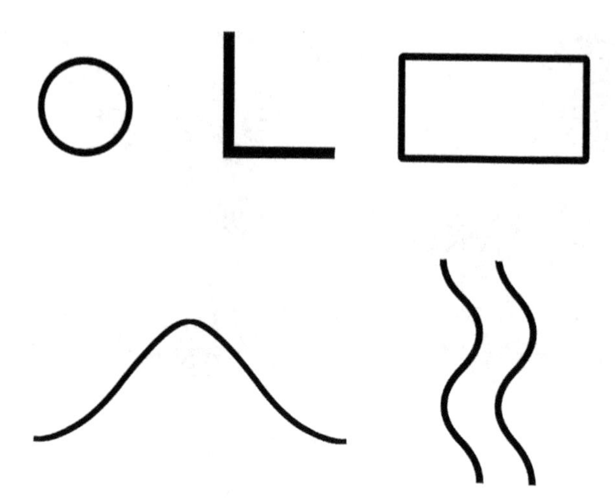

**FIGURE 10.18**
Simple Shapes

**FIGURE 10.19**
Map of Great Britain and Ireland

Benoit Mandelbrot, the 20th-century French mathematician and pioneer of fractals, famously asked a disingenuous question: "How long is the coast of Britain?"

The same trick that we did with a line, he showed happens in nature. A coastline has bays, and bays within bays, and juts and inlets within those bays and, so on down to the cracks in the stones and the lines in the sand. How long is the coast of Britain? Longer and longer the more accurately you measure it.

Mandelbrot goes on in his books to show that many shapes in nature share characteristics in common with this coastline. He shows that rivers and trees have a structure that grows in length and complexity as one examines it in smaller and smaller scales.

This is radically different from the Euclidean imagery and the smooth curves we've mapped before. A piece of a square is a line segment, that of a circle is an arc, and so on. Each of our locally Euclidean shapes that we used before gets simpler and straighter the smaller our scale. But these shapes, these coastlines and rivers and trees either stay as complex or grow in complexity as we get smaller.

Mandelbrot called shapes like this **fractals** (we'll get to the reason for the name later). Fractals are usually created by doing something, then doing it again at a smaller scale, and then again, and again, and so on. For example, take a line; poke out an angle from it; then poke out another angle from each of the line segments created; and then do the same on those segments; and so on (Figure 10.20).

In a sense, we've seen processes like this before. We've started with something, then done something on a smaller scale, and a smaller scale than that, and so on. All of our limit processes do this, so do the terms of a Fourier series. But there's something different about fractals: the stacking up of complexity, of putting one part on another, each as complicated as the one before.

Nature does many things this way. Some of them are as simple as erosion, where wind and water wear things down on scales from mountain meeting storm to water droplet soaking up dust. Some are more complex. The use of repetitive building instructions in DNA is an intriguing example of genetic efficiency. Consider the growth of trees. The trunk grows, branches spring forth, then more branches grow from the branches, and so on, all working off the same genes.

One must be cautious – as regrettably some people are not – when using such theoretical objects as models for nature. A mathematical fractal can be looked at as the limit of an infinite repetitive process, as we saw earlier with the stick made of sticks. But this does not hold up in nature. Eventually, in the process of getting smaller and smaller, the coast of Britain ceases to be modelable as a curve, since it eventually breaks up into molecules, which in turn consist of atoms and so forth. There comes a point in observing nature where fractal theory ceases to model correctly. Paradoxically, this model, which is built to work

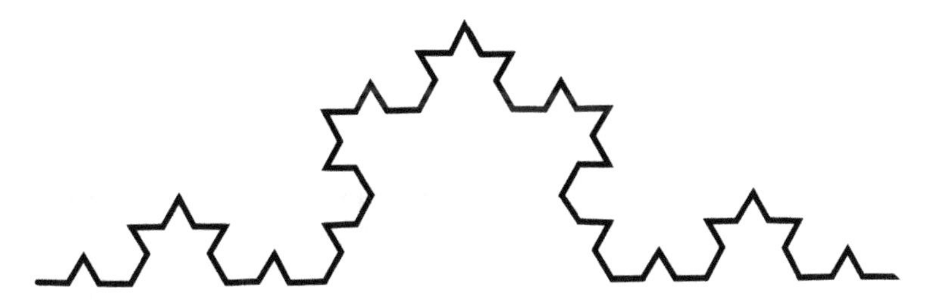

**FIGURE 10.20**
Piece of Snowflake Curve

on all scales, works only on a limited range of scale. But it does a good job on the scales it operates on, just as the simple stick works on its scale of distance between endpoints.

One of the earliest fractals to find its way into mathematics was made decades before Mandelbrot by Cantor. It is called a **dust**. To make **Cantor's dust**, take all the numbers from 0 to 1, [0, 1].

As a brief bit of notation, a way of representing the set of all numbers on the number line between two numbers a and b, including both, is [a, b]. This is called a *closed interval*. It's $\{c: a \leq c \leq b\}$. If we left off the endpoints, we would have an *open interval*, written (a, b), which is $\{c: a < c < b\}$ and unfortunately looks like an ordered pair. If we leave off a but include b, we have an *open-closed interval*, which is written (a, b], which is $\{c: a < c \leq b\}$. And if we leave off b but not a, we have a *closed-open interval*, [a, b) $\{c: a \leq c < b\}$.

So getting back to [0, 1] gives the piece of the number line shown in Figure 10.21.

To start the dust making, remove the middle third of this interval. That is, take away all the numbers from 1/3 to 2/3 [1/3, 2/3]. We now have two pieces: [0, 1/3), (2/3, 1] (Figure 10.22).

Now take each of those pieces and remove the middle third of that. In the case of the lower interval that would be [1/9, 2/9], and for the higher interval it would be [7/9, 8/9]. That leaves us with four intervals [0, 1/9), (2/9, 1/3), (2/3, 7/9), and (8/9, 1] (Figure 10.23).

Keep removing middle thirds of every remaining interval over and over again an infinite number of times. The remaining shape is very hard to depict on a number line because it's a totally disconnected set of numbers, but we can get a sense of it by graphing it in the Cartesian plane and putting a vertical line through each point on the x-axis that belongs to Cantor's dust (Figure 10.24).

This kind of process of doing something over and over again is called an **iterative process**. Many fractals can be made by iteration. Figure 10.25 shows a few standard examples.

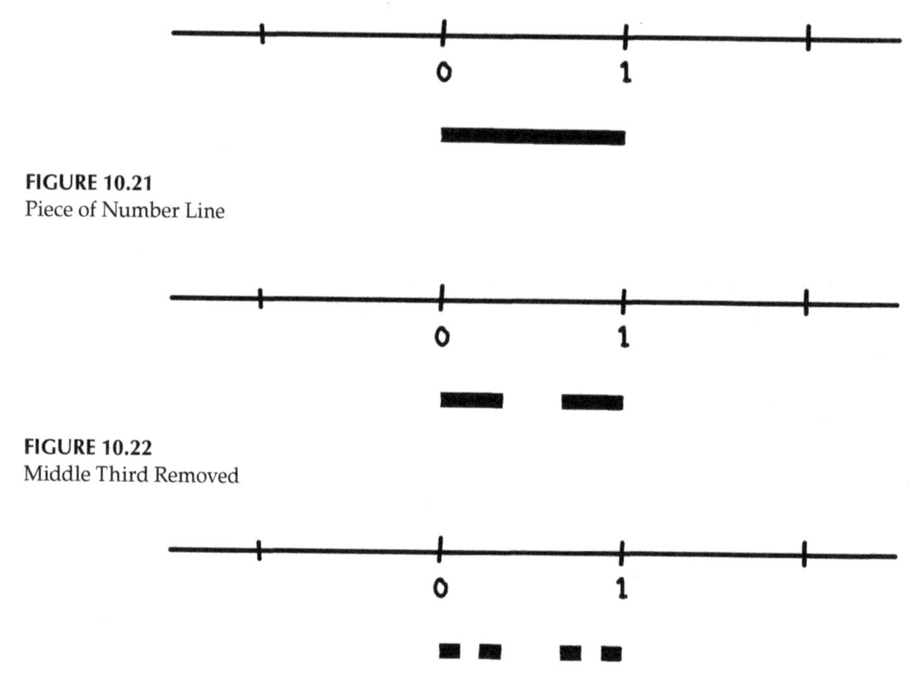


**FIGURE 10.21**
Piece of Number Line

**FIGURE 10.22**
Middle Third Removed

**FIGURE 10.23**
Repeated Removal

**FIGURE 10.24**
Cantor Dust with Vertical Lines

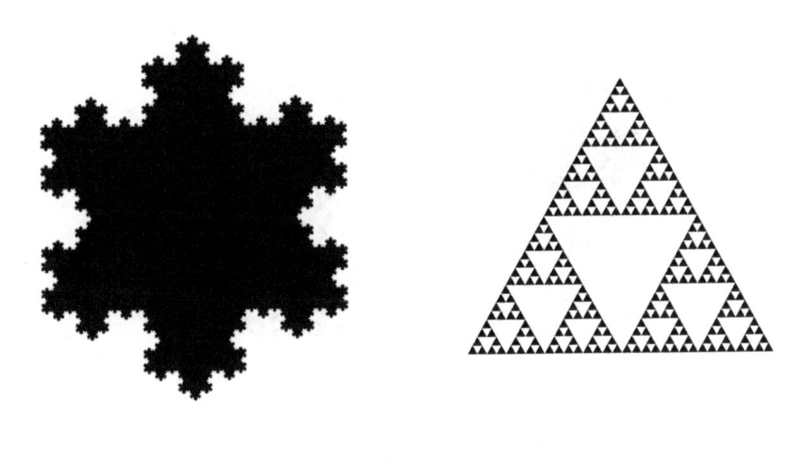

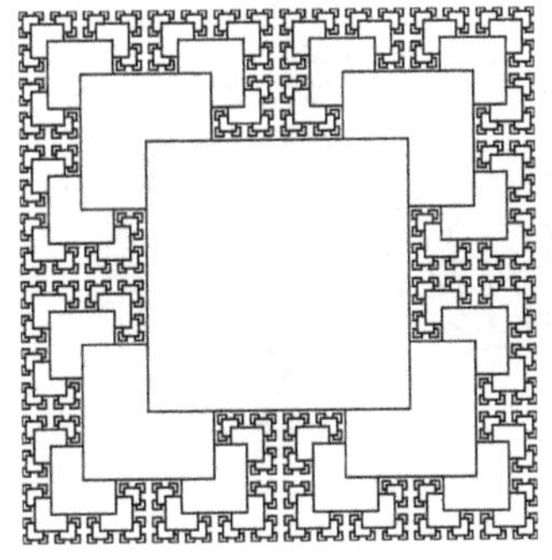

**FIGURE 10.25**
Iterated Fractals

Generating a fractal is usually a process of taking an initial state, describing the process of going from one state to the next, and then carrying out that process as many times as one wishes. The amount of calculation involved makes fractals incredibly difficult to generate without computers. With computers it is possible to use such iterative processes to create images that look startlingly realistic using fractal techniques (Figure 10.26).

*But why are these repetitive processes decent models for reality?*

**Because they are modeling the effects of real repetitive processes.**

As noted above, the growth of tree branches is actually done by growing a branch which then uses the same genetic instructions that grew the first branch to grow a new branch, branching off the old branch. This is efficient.

Furthermore, for a great many parts of living things a simple, repetitive process that creates a large surface area is phenomenally useful. The same trick that makes the coastline longer and longer at smaller scales gives a fractal surface a larger and larger surface area depending on how small a scale you examine it. Therefore, a natural process that creates fractals will create objects with very large surface areas for their volume.

For example, lungs are grown in such a repetitive process and they have a huge internal surface area. The process whereby oxygen is extracted from the air occurs at any place where an oxygen molecule touches the inner surface of the lungs. Therefore, the more the surface area, the more the oxygen transferred. This means that fractalized lungs get to grab more oxygen molecules and transfer them to the bloodstream than a simpler shape

**FIGURE 10.26**
Fractals Moving Toward Realism

of the same volume. Thus, fractalized growth in certain organ systems can have very high survival value.

That's all well and good, but it doesn't account for how such a complex shape evolved.

Herein lies a subtle point. Fractals like the lungs have complex appearance but simple methods of generation. They involve doing the same thing over and over again. Creating instructions to generate a fractal is not difficult if you already have the instructions to create each piece of the fractal. In computer programming, this kind of thing is done a lot using loops and recursion (see Chapter 14 for a fuller discussion).

The genetic code to make a piece of lung or a tree branch is complex, but once it exists, the genetic code to say do that again but smaller is relatively simple. A mutation does not need to change too much genetic code to produce an iterative growth process from a noniterative one.

But what about nonevolved natural fractals? The coastline we saw was made by erosion from the ocean. But if you look at the waves that cause the erosion, you see that the wave itself has a complex shape. The waves dig into the beaches at various different scales because the waves themselves come in at different scales. Examine the foam on the cresting waves to see another fractal.

Fractals work as models for repetitive natural process because they are repetitive geometric structures created by repetitive processes.

Fractal geometry has brought the artistic side of geometric thinking back, at least in part, to a branch of math that has long been overtaken by analytic and algebraic thought. The eye that can look at a shape and say "That's almost a tree, but not quite." (Figure 10.27)

This tree is not just making pretty pictures but is providing and refining mathematical models for objects in nature. Having those models, it becomes possible to begin to correlate the how and why of development of these things in nature with the how and why of their appearance. In short, fractal geometry may have made the geometric mindset important to science once again.

It may be that once again such people may stick their heads in the clouds (Figure 10.28) and discern how they came to be (Figure 10.29) without being accused of building … (Figure 10.30).

**FIGURE 10.27**
Fractal Tree

**FIGURE 10.28**
Real Clouds

**FIGURE 10.29**
Fractal Clouds

**FIGURE 10.30**
Castle in the Clouds

## Why Are They Called Fractals?

**What's the area of a line?**
*What?*
**What's the area of a line?**
*A line is one-dimensional. Its area is 0. Only two-dimensional objects have nonzero area. Why are you asking such a goofy question?*
**What's the volume of a disk?**
*A disk is two-dimensional. Its volume is 0.*
**All right, what's the length of a disk then?**
*A disk is two-dimensional. If you wanted to assign it a length, you'd have to say its length is infinite.*
**Yes. So, what's a finite measure for the amount of disk there is?**
*Area. Two-dimensional figures that don't extend forever in any direction have finite area. Are we done here?*
**We're just beginning.**

We defined dimension naively based on how many different parameters it took to describe something. One quantity makes for one dimension (hence a curve as a function of one parameter), two quantities makes for two dimensions, and so on. Dimension was one of those know-it-when-you-see-it sorts of things.

We also used the idea of what kinds of sticks were appropriate for measuring "quantity" as a marker for the number of dimensions a shape had. Actual sticks, one dimension; little squares, two dimensions; little cubes, three dimensions; etc.

But the stick as measurement fails for something like a coastline. The more carefully you measure it, the longer the result, until something as seemingly one-dimensional as a coastline has infinite length.

*So maybe we've been fooled. Maybe a coast is really two-dimensional.*

**Nope, sorry.**

If you try to cover it with little squares, the limit of the sum of the areas of the squares as you refine the area is 0. Infinite length, 0 area. More than one-dimensional, but less than 2.

*What kind of dimension do you call that?*

A **fractional dimension**. Hence the term "fractal" shape.

But what does fractional dimension mean? And how do we figure out what it is?

Here's one way, called the **Hausdorff dimension**, invented by German Jewish mathematician Felix Hausdorff who was born in 1868 and died in 1942 during the Holocaust. He was one of the founders of topology and contributed a great deal to set theory and to the 20th-century advances in analysis.

We're not going to use the exact definition Hausdorff used but the following should be enough to give a sense of the idea.

For ease of definition, let's assume the object whose dimension we want to measure does not extend infinitely in any direction. One way to look at the processes we've used before to measure the size of something is that we are covering the space the object occupies with our measuring shapes (sticks, squares, etc.). If the objects we are using are of smaller dimension than the shape, it will take an infinite number of such shapes. For example, it takes an infinite number of line segments to cover a square, but a finite number of squares can do it. There's going to be a lowest dimension for our measuring shapes that can cover the object with a finite number of such objects. We've seen this before. A disk is coverable with a finite number of squares (dimension 2), but an infinite number of line segments (dimension 1).

The naive definition of dimension that arises from this is that this lowest dimension is the dimension of the object itself. But if we examine our friend the coastline, we see that to fit in the nooks and crannies it needs to be covered by two-dimensional squares but has a 0 area.

There is a way to create a measurement that uses this process in an odd fashion. To do this, rather than covering with square-like shapes we use balls of the given dimension (remember that a 1-ball is a line segment without its endpoints, a 2-ball is the interior of a disk, a 3 ball is the interior of a sphere, etc.). We use balls of a given radius R and find the smallest number N of balls of that radius that will cover the object. The smaller the radius (R), the larger the number of balls (N). For any given shape, therefore, N is a particular function of R.

There is a quantity derivable from these two numbers R and N which for normal shapes gives the standard dimension and for fractals gives nonwhole numbers. Make a graph of ln(N) versus ln(R). For a given shape, this eventually smooths out to be almost linear as R decreases. The slope of this roughly linear shape is the Hausdorff dimension.

This is pretty messy and nonintuitive, but it does give us a number for the dimension of a fractal and it's the correct number for nonfractals.

To see that we get the correct answer for nonfractals, note that one property of ln is that $\ln(ab) = \ln(a) + \ln(b)$. So it follows that

$\ln(cR^n) = \ln(c) + n\ln(R)$. Now suppose that we cover a line segment with N tiny disks of radius R. Then, if we make R twice as small, we will have to make N twice as large. So, N is proportional to R, which means $N = cR$, which means that the slope of the graph of ln(N) versus ln(R) is 1. So a line segment has Hausdorff dimension of 1, just like its regular dimension. If instead we cover a square with tiny disks of radius R, then if we make R twice as small, we will need four times as many disks to cover the square. This means that N is proportional to $R^2$, which means that $N = cR^2$, which means that the slope of the graph of ln(N) versus ln(R) is 2. So a square has Hausdorff dimension 2, just like its regular dimension.

The fact that we get the right answer for the usual sorts of shapes is hardly a justification, since we could create an infinite number of calculations that give correct dimensions for standard geometric objects and other values for fractals.

But the question of meaning of dimension here is harder to get to. Suppose we have a shape whose dimension is between 1 and 2. There is a sense in which, if that shape's fractal dimension is closer to 1, it is more curve-like than one whose dimension is closer to 2, which is more surface-like. The Hausdorff dimension roughly fits this aesthetic, so it's an acceptable candidate for a more general definition of dimension.

Unfortunately, there are messy problems even with this. The same set of points might represent two different shapes of different dimensions depending on how one draws the shape.

Take a look at the curve in Figure 10.31.

We're going to put it through a fractal generation process where we replace each line with a smaller copy of itself. This is the most common way to generate simple fractals (Figure 10.32).

What happens when we carry out this process to infinity? We end up filling the entire square with a fractal whose dimension is between 1 and 2. If we try to measure this fractal's length, we end up with infinity. But its area is still 0 because that area is the limit of the sequence of areas of the successive generations that create this fractal. Each term of that sequence is 0, so the limit is 0.

In other words, if we look at the square as the limit of this fractal sequence, we end up with an object of less than dimension 2, but if we look at it as the square made normally by stacking up an infinite number of parallel line segments, we end up with the standard area. What does this mean?

It means that the set of points is not all the square is, that there is some other normally unnoticed structure to the set that tells us what we are looking at, that the shape is more than the set of the points that make it up, but is also the way the shape is made. And that even such a seemingly clear notion as dimension can be blurred by which process one uses.

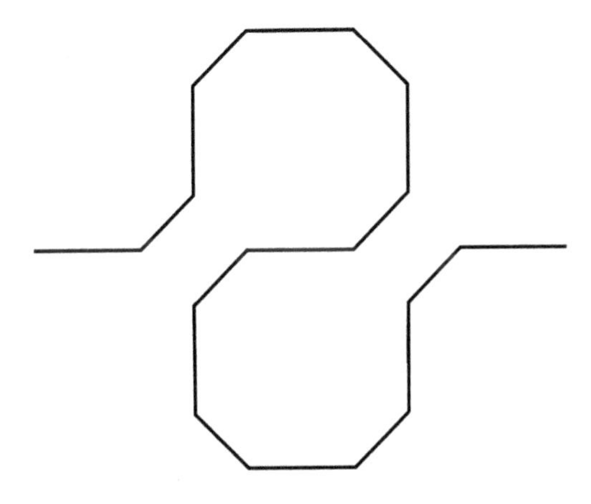

**FIGURE 10.31**
Peano Curve Generator

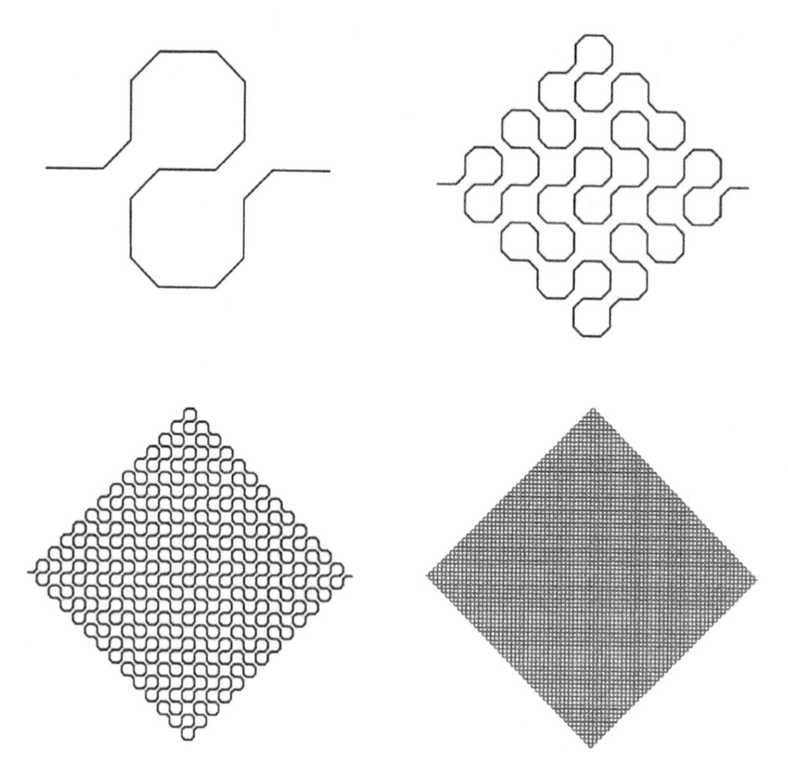

**FIGURE 10.32**
Iterations of the Peano/Square Filling Curve

Now we've reached a splitting headache. Our first definitions of shapes were like that of a circle, a set of points sharing a given characteristic. Now we're seeing that there's more to shape than the contents of the object.

There may be some reconciliation possible here between the nonshape shapes defined by metrics and the more-than-point-set shapes pushed out by fractals. But that's a matter of speculation, a place of possible research. The only thing that is clear is that there is tension in the idea of shape, and that tension is reflected in the presently unresolved status of geometry and geometric thinking.

Fractal geometry, like non-Euclidean geometry before it, is making it necessary to take a number of concepts one thought one could rely on and re-examine them, making them stricter and clearer and perhaps expanding them to other broader ideas. Where non-Euclidean geometry challenged our ideas of distance, fractal geometry does the same thing with size and its related concept, dimension.

Untouched on here is the tension between dimension and unit. When we did dimensional analysis, we constructed our units out of a whole number of base units. Yes, speed was distance/time, and energy was mass times distance squared/time squared, but that's still a whole number of component dimensions (2 for speed and 5 for energy).

Nothing may come of this question. It may simply be that for fractals, the question of units is meaningless in the same way that "how many" is meaningless for something that is answered with "how much."

But whether this is so or not, and indeed whether the dimensionality of fractals is anything more than a complex attempt to preserve an old idea in something that it does not quite fit, will depend on the work of a lot of mathematicians in the years to come.

## Iteration and Chaos

Fractal making and the earlier discussion of stochastic processes in the probability chapter are examples of iterative processes. So, of course, are addition and multiplication (to add 6 to 5, start at 5, add 1, add 1, add 1, add 1, add 1, add 1). Iteration is one of those things that math has incorporated from its beginnings. Usually it tries to find a way around the doing it over and over because it's too much work for humans.

*Of course, now that we have devices, we can dump the calculations on while we imbibe overcomplicated mixtures centered around coffee, tea, chocolate, or alcohol (if you are using all four in one drink, I don't want to hear about it); we have gotten a lot more casual about iteration.*

From that relaxed attitude an intriguing phenomenon has emerged: Chaos.

*Can't you hear the drama in the word, the glory of mad scientists and deranged teenagers shouting unto the heavens: Chaos!*

*It stirs the blood and wakens the heart. Who would not revel at the inherent disorder in that single word: Chaos!*

**Now, cut that out! This is a math book, not a cheap melodrama.**

Actually, chaos isn't all that chaotic. But it does reveal and model processes that otherwise defy mathematics.

Chaos arises in some circumstances where one takes a function f and an initial value x and creates a sequence.

$x, f(x), f(f(x)), f(f(f(x))), f(f(f(f(x)))), \ldots$

If for example, $f(x) = x^2$, we would get the following sequences for initial values of 0, 1, 2, 3, and 0.5, respectively:

0, 0, 0, 0, 0

1, 1, 1, 1, 1

2, 4, 16, 256, 65536

3, 9, 81, 6561, 43046721

0.5, 0.25, 0.0625, 0.00390625, 0.000015258789

Quick examples of this iteration can be generated by simply taking a pocket calculator, typing a number, and then hitting a single function key (such as cosine) over and over again until you get bored. By the way, for the cosine, make sure the calculator is set to radian mode, not degree mode.

All done?

Notice that some functions produce radically different sequences depending on which numbers you start with, and others don't really care. The cosine in the long run tends toward around 0.739. $x^2$ goes to infinity if the number is greater than 1 or less than –1; it goes to 0 if the number is between –1 and 1; and it goes to 1 if the number is 1 or –1.

A function is said to be chaotic if the long-term behavior of applying that function iteratively to two values that are close together produces results that after a number of generations are wildly disparate. There's a more exact definition, but that's what it boils down to.

The point is that with a chaotic process you could start with a value of, say, 0.5 and a value of 0.5003, and end up with long-term values of 0.0 and 1,000,000.0.

It's not hard to come up with chaotic functions. $f(x) = 10x - \text{int}(10x)$ (that is, multiply a number by 10 and chop off its integer so that you always have a number between 0 and 1) produces wildly disparate values for close starting values between 0 and 1. Consider the numbers 0.5 and 0.5003333888888…

Their sequences are

0.5, 0, 0, 0, …

and

0.5003333888888…, 0.003333888888…, 0.03333888888…, 0.3333888888…, …

The hallmark of a chaotic process is that the disparities grow explosively as one goes along the iterative process, so that two numbers close together as initial values grow vastly farther apart as one iterates from them.

Chaos may look like a cute trick to play with calculators, but it shows up in certain gravitational problems with a large number of interacting bodies (such as in the asteroid belt) and in the complicated interplay of air motion, solar power, and geothermal heat we call weather. One of the reasons we can't predict weather in the long term is because weather is a chaotic process. This is often exaggerated into the claim of the butterfly effect (that a butterfly flapping its wings causes a hurricane). Humans trying to pin the rap on lepidoptera, typical.

What is true is that small changes in wind speed, heat, solar radiation, and all the other contributing factors can produce radical differences in weather in only a few days.

Chaos looks like it should be a source of frustration for science because it shows processes that cannot be predicted in the long term. When we measure initial values, there is always some error in that measurement. For nonchaotic processes we accept the error and note that it will limit the accuracy of the results we calculate. But with chaotic processes, the error expands so quickly that our predictions are no more accurate than the weather report.

While this is frustrating for some people, it is fascinating to others. Thanks to chaos, we are seeing natural processes that cover their tracks, that are impossible to backtrack because we cannot determine which states led to our present state, and are impossible to predict too far forward for exactly the same reason.

Yet the chaotic nature of the process does give us vital predictive information, if we change our standard of prediction from saying what will happen to saying what can happen. Then, as in probability, we see that the chaotic model itself is a scientific prediction, one that can be tested by examining the outcome of the process with small variations of input.

Chaos, like probability, expands the concept of what the predictive character of science is, as well as expanding the ideas of what kinds of behavior nature creates.

Chaos and fractals are interrelated because if one has a function that takes points into points, and if that function is chaotic, graphing the long-term behavior of that function produces a fractal. This is not really surprising, since the primary characteristic of chaos is small change leading to large effect and the primary characteristic of fractals is small variation concealing large variation.

Chaos and fractals are relatively young branches of math. We have seen some, but clearly not yet all, of their applications. It is as if we were living in the time when the paintbrush had been invented. It would be hard for a person holding an early bristle-bound stick to predict what uses it would be put to.

So are we dealing here with lost inheritance? This is a young branch of mathematics, still growing. Who knows if it will bear fruit? It may be that fractal geometry will end up being largely a pursuit for computer artists. But that's still lost inheritance. Artists lost their mathematical legacy some generations back. Now they may be reclaiming at least some of it, and it may be that again mathematics will be part of artistic education and the geometric mindset regain some measure of equality with the other two.

What is true is that many shapes once ignored by mathematics are now visible through the lens of mathematical awareness. One can look at a tree and see the fractal growth pattern of its branches. One can examine feathers under a microscope and see how they relate to scales. One can create functions that mimic these bits of reality and therefore serve as models for the underlying genetic processes of their creation.

One thing is certain. For a while at least, a lot of new innovations and theories will attempt to explore and employ these young fields of mathematics.

It may be that fractals and chaos will produce valuable results in a number of fields, fulfilling the hopes of those who labor on them now, trying to find and refine the uses for them. We don't know yet. Not all elements of an inheritance are old established estates with known value and produce. Sometimes you inherit an unexplored mine and don't know what if anything will come of it until you start digging.

On the one hand, the attempts to prove the parallel postulate were an empty hole and there came a time when it was time to stop digging. On the other hand, at one point non-Euclidean geometry was an unexplored mine that when finally explored allowed us to dig down into the foundations of the universe and then dig deeper still.

Even if they don't know the outcome, mathematicians explore deep into the theoretical universe, abstracting ideas like distance, area, space, and dimension. Sometimes they find their ways back, and sometimes, as we'll see in the next chapter, they just keep going.

# 11

## *Algebra and the Rise of Abstraction*

If you'll pardon the redundancy, we'd like to restate the manner in which math approaches problems as the following process:

Step 1. Discern the relevant characteristics of the objects we need to work on.

Step 2. Abstract those characteristics into the most useful and most general theoretical objects we can.

Step 3. Manipulate the theoretical objects in mathematically proven ways that lead toward whatever goal we are seeking.

Step 4. Return the result achieved to the objects we started from.

What is not obvious looking at this process is that it can be applied repeatedly inside itself, like fractals building up by iteration. Look at steps 2 and 3, abstraction and manipulation, and consider the phrases "most useful and most general theoretical objects" and "manipulate the theoretical objects in mathematically proven ways."

"Most useful and most general" pushes mathematicians toward greater and greater levels of abstraction in order to lazily encompass larger and larger ideas. "Manipulate in mathematically proven ways" pushes toward the broadest kinds of proven manipulation. In the middle of an abstraction, one may realize that there is a more general abstraction hidden within the object one is working on, and in the middle of a manipulation one may see a more general manipulation with a more encompassing proof.

This push upward makes the estates of mathematical inheritance vaster than the universe and deeper than the deepest understanding of reality, because properly proven results apply to axiom systems with no apparent connection to the real world. Thus, there is a push away from the beginnings of the problems, away from those aspects of the real world that feed into mathematics.

It is this upward trend that has, despite all the utility of analysis and the possible reawakening of geometry, made algebraic thinking the dominant way in current mathematical thinking.

*This takeover was long in coming: centuries of plotting and scheming, of abstracting the heart from the other branches, of dangling useful tools and theoretical structures that would encompass the ideas and works of analysis and geometry. It was a masterwork of universal conquest.*

**Now, cut that out! This is a math book, not a historical melodrama.**

Algebra began as a means to an end, a way to make analysis and geometry easier by abstracting from the abstractions of number so that one could work with numbers in general, not particular numbers.

It is exactly for this reason that algebra has become dominant. Once one abstracts individual into general in one area of work, one can do it in all others.

Just as the sciences abstract into math, other parts of math abstract into algebra.

While this is extraordinarily effective, and has created incredibly general proofs and techniques, it also creates its own characteristic difficulties for learning and understanding mathematics.

Whenever we abstract, we lose some of the meaning of what we started with. If we abstract the mass from an object into a variable M, we lose everything about the object that is not its mass. We can no longer tell a 5-kilogram sack of flour from a 5-kilogram cat from a 5-kilogram block of plutonium. Similarly, if we abstract from an abstraction, we lose the meaning of what we abstracted into in the first place.

A linear function $f(x) = 3x + 2$ is a definite function producing definite values. Even though it's already an abstraction – we've turned whatever process gave us this into a function – it's a single process. We can use it to solve specific problems, such as for what value of x does $f(x) = 8$.

We can solve that as follows (pardon the belaboring):

$$8 = 3x + 2$$
$$6 = 3x$$
$$2 = x$$

But suppose we abstract the linear function further into $f(x) = mx + b$.

We can carry out a variation of the above process by asking the question, for what value of x does $f(x) = c$. The process we did above abstracts to

$$c = mx + b$$
$$c - b = mx$$
$$(c - b)/m = x \text{ (assuming m is not 0).}$$

Now we've done something more mathematically impressive than the first example, since we now have a general solution to all such equations. But if we needed to know that $x = 2$ because we needed to know how much of something we had to use, then we've lost that information.

We can reclaim it by substitution of 3 for m, 2 for b, and 8 for c. We might even need to reclaim what units we were talking about to understand the meaning of the answer we get. If $f(x) = 3x + 2$ is a statement about the relation between two masses, we need to remember that when we make our substitutions. In order to properly use the general result, we need to recover the lost meaning.

But each time we climb the ladder of abstraction, we lose more meaning. We need to remember more and more things as we go back down. Otherwise, we end up simply manipulating mathematical objects without any sense of what they're about.

It might seem that we have reached the limit of abstraction from our initial equation with the general solution above. And we have reached that limit, if our concern is the ability to solve linear equations. But we can generalize further, abstracting from a specific function to a function in general, and let $f(x)$ be any function of one variable. Then we can ask the question, for what value of x does $f(x) = c$?

Obviously, we can't answer this. Since we know nothing about $f(x)$, we've got nothing to calculate from. But mathematics doesn't have to stop with the inability to solve. We can start proving theorems about the characteristics of solutions to the above, even if we don't know what the solutions are.

Suppose we narrow the characteristics of f(x) from a general function to, say, a polynomial. We can prove a lot about the ability to solve f(x) = c. That general solution would give us machinery that could be used in any such problem. But to reach these heights, we left our initial problem far behind and our function has lost all its meaning.

Accompanying this loss of meaning is the danger of mission creep. Let's look a little more closely at the problem f(x) = c if f(x) is a polynomial.

First, a little terminology. The highest power of the variable (in this case x) that shows up in f(x) is called the **degree** of the polynomial. If f(x) = 3x + 2, the degree of f(x) is 1, since the highest power term is 3x which is $3x^1$. If f(x) = $4x^2 + 7x + 6$, the degree is 2, since $4x^2$ is the highest power term. If f(x) = $x^{1469} + 8x^{56} + 2$, the degree is 1469.

We know that we can solve f(x) = c if f(x) has degree 1 since that's just mx + b = c, which has x = (c – b)/m as a solution. We can find solutions, if solutions exist, if f(x) has degree 2, since that would make it a quadratic and we've already solved those. There are also formulas that give us the solutions if f(x) has degree 3 or 4. We're not putting those in because they're messy and unenlightening.

Above degree 4 there is no general formula for finding solutions of f(x) = c. This does not mean that one cannot solve particular problems, only that the abstraction into general equations affords no solutions.

**Does this mean that the mathematicians gave up?**

*Not them, they've got too much chalk for that.*

No, what they did was change the question. Instead of creating the means of solution, they started proving theorems that determined if solutions existed. That is, they proved under what conditions there were solutions without actually producing means of finding the solutions.

*What good is that?*

**Quite a lot, actually.** If you can prove that the particular problem you are working on has no solutions, you can give up on it. This has been important in the history of science as it has made it easy to abandon hypotheses that required solving unsolvable problems.

Nevertheless, this change of focus, this mission creep came about because in trying to lazily solve the larger problem of general polynomial solutions, it was discovered that the larger problem could not be solved. Rather than abandon the abstraction and the laziness, there was a change in the question being asked. Enlightening results were found from changing the question, but people don't necessarily want their questions changed.

The power and the danger of algebraic thinking in this regard is heightened by the fact that abstracting from abstractions can be done recursively, with each level of abstraction leading upward into a new level that is more general, more powerful, less meaningful, easier to lose track of in the inheritance, and more subject to creeping around the questions.

Ironically, the process of thought that led to this spiral out of the world probably began with an invisible process of making sure that everything that was measured or was calculated from measurements was a number. In other words, abstraction arose from trying to keep mathematics concrete.

This problem, making sure that numbers are numbers, may not sound like much, but it was very tricky at various times. The problem arises from two demands we make on numbers:

1. What we count and measure are numbers.
2. What we calculate from numbers are numbers.

The first is obvious.

The second is an implicit demand we make whenever we mess with an equation, trying to find a solution. We want to be able to change both sides of the equation in the same way and thus preserve truth. But in order to do that we have to be sure that when we transform the numbers on both sides, we end up with numbers on both sides. In order to do this, we end up increasing the things we think of as numbers.

Let's start from the beginning.

When dealing with stones we go: 1 rock, 2 rocks, 3 rocks, 4 rocks, and so on. We get the positive counting numbers and are content.

Well, what about no rocks? How do we count the absence of rocks?

Okay, okay, we'll put in a number for nothing, so we have 0 rocks. After all, we can use a number that means the coyotes ate all the sheep and none are left.

But then the accountants got into the act. And they got in early. Many of the oldest surviving writings are inventories of who owned what. They cared a lot about stones, those ancient listers of who owned what and who owed what to whom. It's the owing they wanted to deal with. If I had 6 rocks and owed you 5, then they could say I had 1 net rock. That was okay: $6 - 5 = 1$.

But suppose I had 5 rocks and owed you 6. How many rocks do I have in my net? What's $5 - 6$?

To answer this, the concept of number needed to be expanded to include having less than nothing. Negative numbers were added to the numerical repertoire.

After 0 this may seem innocuous, but it represents a major jump in abstraction. A pile with some rocks is obviously a real-world object. A pile with no rocks is pretty close to real; one can certainly imagine the blank spot where the rocks would go. But a pile with −8 rocks is an accountant's invention representing what is owed. Debt is a social, not a physical object.

**Still, what's a little abandonment of reality if it makes problems easier to solve?** (*Whoops, watch that slope. It's slippery.*)

Seriously though, negative numbers are not hard to deal with if you remember that subtraction is really addition and vice versa, since

$$a - b = a + -b$$

Meanwhile, the geometers were making their own demands on what was and was not a number. Some of them were completely reasonable, others caused minor problems, and still others caused problems that would not be solved for centuries.

To start with, they wanted to cut up sticks into smaller sticks that were the same size. Well, no problem there: 12 inches = 1 foot; 100 centimeters = 1 meter. We're okay with that.

But they also wanted a number to represent how many feet an inch was. If 12 inches = 1 foot, then 1 foot is made up of 12 parts called inches. So one inch is one twelfth part of a foot, or 1/12 of a foot. That 1/12 did not represent a number so much as a process, a function for making new smaller sticks out of old big ones. Chop into 12 equal parts and take 1 of them.

Turning processes into objects is one of the things the analytic mind does. The function "divide by 12" was not too much of a stretch for them, particularly if they had knives and axes.

The strange transformation wasn't the creation of the function, it was saying that the function "make 12 pieces out of 1" was a kind of number.

Remember that 12 itself is really a process: take a rock, put on another rock, and so on to 12. Maybe this strangeness is actually one we've been doing all the time and didn't notice.

*I see you palming that card. We're going to use this process into number trick again, aren't we?*

**We'll be doing nothing else for a while.**

With our new numbers, we have a lot of questions to answer that we didn't worry about before, particularly questions of equality. When we had whole numbers, we were only dealing with the processes of addition and subtraction, and those just amounted to putting on or taking away rocks.

We could say without trouble that

$3 + 4 = 7 = 10 - 3$, because those are all variations on the same simple piling up.

But with our new pieces-of-things numbers, we're not all always doing the same kinds of slicing up. For example, if we divide a 1-foot stick into six parts, we can notice that each part is the same length as 2 of those 1/12 pieces. We are tempted to say that $1/6 = 2/12$, and we succumb to the temptation just as later analysts will succumb to the idea that the function $f(x) = 2x/2$ and the function $g(x) = x$ are the same function.

This process may look like it makes sense in the world, but in fact we are leaving reality farther and farther behind us. If we switch from sticks to the favorite commodity of the Indo-European peoples, cows, we discover just how unworldly our fractions are: $1/2 + 1/2 = 1$, but 1/2 cow + 1/2 cow does not give you a whole cow. It just gives you as much meat, bone, and leather as a whole cow.

*Its milk and methane-producing qualities are severely curtailed, not to mention that it produces fewer moos and has a radical reduction in grazing needs.*

If we press on in our mathematical way, disregarding these dead cows as mere earthly inconveniences, we quickly discover numerical problems as well. How can we tell if one of these new fractions of numbers, or fractions as they are called in elementary school (rational numbers is the standard mathematical term because each fraction is the ratio of two numbers; the ratio of a to b is $a/b$), is equal to another or whether one is greater than another?

We want to know this because we want to preserve as many useful characteristics of our previous mathematical system as we can. And since we created these rational numbers for their measuring purposes, we want to preserve the "bigger" and "smaller" characteristics we cared about when we started measuring in the first place.

As said before, math almost always solves problems by reducing them to problems one already knows how to solve. We already can tell if two whole numbers are equal or if one is greater than the other, so all we need to do is turn our fractions into whole numbers and then compare the whole numbers in order to determine how the fractions stack up against each other.

We start with the following idea: We know that $2/12 > 1/12$ because we really mean that 2 sticks of length 1/12 are longer than 1 stick of length 1/12. More broadly, we can always tell which, if either, of two fractions with the same denominator is larger. $a/b > c/b$ if $a > c$ (as long as $b > 0$). Similarly, $a/b = c/b$ if $a = c$. If we have two fractions, $a/b$ and $c/d$, we need to transform them into fractions with the same denominator. We can find such a new denominator by using the trick of multiplying by 1 in an interesting fashion:

$a/b = (a/b)1 = (a/b)(d/d) = ad/bd$. Similarly, $c/d = cb/bd$.

So $a/b > c/d$, if $ad/bd > cb/bd$. This is true if $ad > cb$.

For an example, 2/3 is greater than 1/5 because $2 \times 5 > 3 \times 1$.

Notice that this is the numerator of the first fraction (a) times the denominator of the second (d) compared to the numerator of the second (c) times the denominator of the first (b). This process for comparison is called **cross-multiplication** because you multiply across, not because lots of people get grumpy in learning it.

Okay, we've made fractions part of our numbers. Let's move on.

*Not so fast, pal. How do we add and multiply these things?*

**Oh, that. Yes, that is important.**

After all, we don't just want to preserve the ordering we had, we want to be able to do the basic calculations we need to do.

Multiplication of fractions is pretty easy. If we look at multiplying by a fraction as a two-step process (to multiply by a/b, first multiply by a, then divide by b), we can see that (a/b)(c/d) = (a/b) times c divided by d, which is ac/b divided by d, which is ac/bd. In other words, to multiply two fractions, we just multiply the numerators to get the new numerator and multiply the denominators to get the new denominator.

This process makes sense if we look at multiplication geometrically. Let's take two sticks and make a mark ¼ of the way up one and another ½ the way up. Then make a rectangle of these two sticks (Figure 11.1).

The product of ¼ and ½ will be the area of the smaller rectangle marked out by those marks (Figure 11.2).

If you look at the area we've cut out and you break the whole rectangle up accordingly, you'll see that our new rectangle has an area 1/8 of that of the big rectangle (Figure 11.3).

$$\frac{1}{2} \times \frac{1}{4} = 1 \times \frac{1}{2} \times 4 = \frac{1}{8}$$

Addition of fractions is trickier. We can only add things that are alike, and in the case of fractions, likeness means having the same denominator (i.e., they are measured in the same units like adding inches to inches, rather than inches to meters). We can use the same trick as comparison of fractions, although we will do better.

Before we do so, it's worth looking at a very common error that comes upon people learning fractions. It is an error that comes from the same human capacity that fuels mathematics, the ability to see and create patterns.

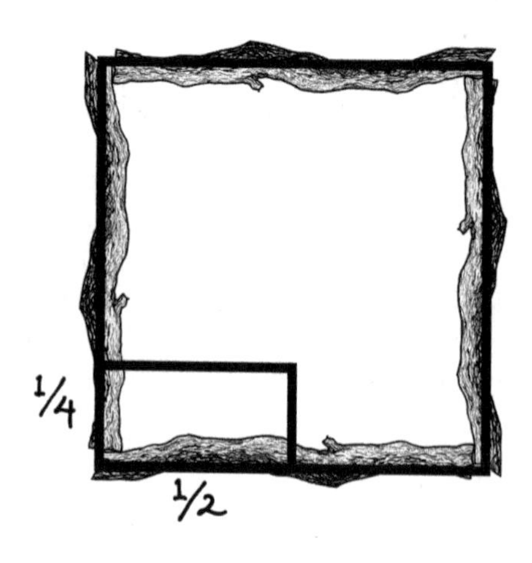


**FIGURE 11.1**
Rectangle from Sticks

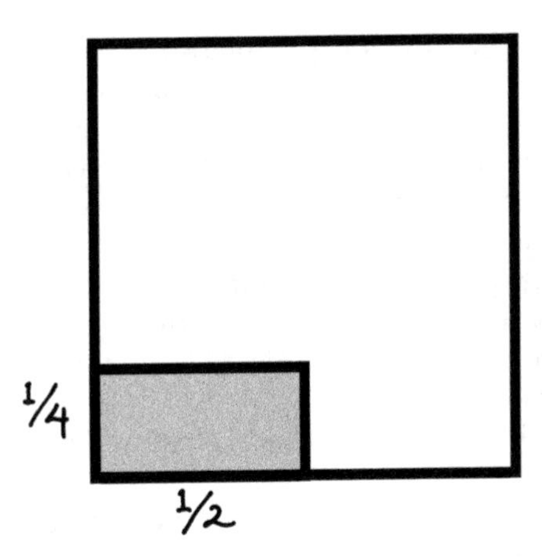


**FIGURE 11.2**
Area of Rectangle

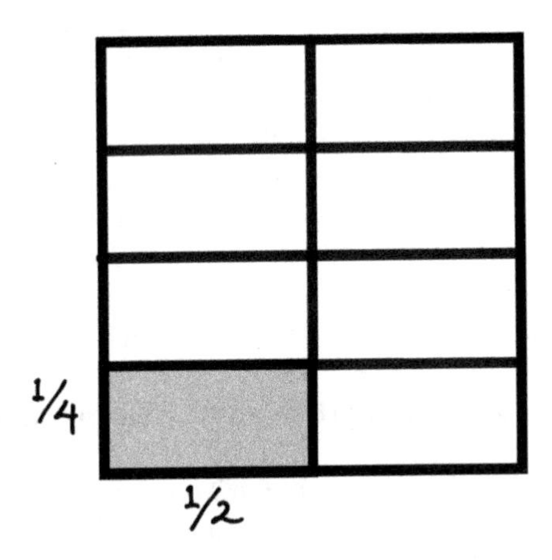


**FIGURE 11.3**
Small Area = 1/8 of Large Area

A lot of people see

a/b + c/d and conclude it must be (a + c)/ (b + d).

They think that to add fractions you add the numerators to get the new numerator and the denominators to get the new denominator. This fits into a simple aesthetic pattern. It's pretty, but it's wrong. Its prettiness is not connected to how fractions actually work. Mathematical aesthetics requires connecting the beauty of the patterns with the underlying purpose and meaning of the mathematical activity.

And therein lies the problem that will dog us all through this chapter. As we climb the ladder of abstraction, more and more meaning is lost and therefore more and more intuition of how things have to work will be lost. As a result, more and more people fall away, unable to connect their sense of what makes sense to what they are being asked to do.

This can be overcome, but in order to do so, it is necessary at each rung of the ladder to take a look around, get a sense of how things work at this rung, and become firmly grounded in that manner of work before climbing up to the next. The way to do that is exactly what we're doing right now, going through the process of turning what we knew how to do before into its equivalents at the new stage.

So to add a/b and c/d, we need to turn them into equivalent fractions that have the same denominator and add their numerators: a/b + c/d = (ad + cb)/db.

For example, we can't directly add 1/3 and 1/2, but we can turn them into 2/6 and 3/6, respectively, and add them to get 5/6.

We would like to do better than this, if for nothing else than the excess work in problems like 2/4 + 3/2. If we follow the prescription above, this would be (4 + 12)/8 = 16/8 = 2. But what we want to do is note that 3/2 = 6/4 and add 2/4 + 6/4 = 8/4 = 2, which has less multiplication and easier division.

We certainly want to avoid adding 1/100 and 1/1000 and having to turn that into (1000 + 100)/100,000 = 1100/100,000 = 11/1000.

This leads us to the concept of **least common denominator**, which is the lowest whole number that is a multiple of both denominators. If we can find this number, we can transform both fractions so that they use the least common denominator as their denominator, then add the numerators and clean up if need be.

*Hmm ... maybe we'd better say that a little more slowly and clearly.*

**OK. This is a matter of laziness.**

We don't want to do unnecessary multiplication, so we want to find the smallest numbers we need to multiply. We need to translate each of the fractions we want to add into equivalent fractions with the smallest possible denominators.

In a lot of cases least common denominators can be found by looking. The least common denominator of 100 and 1000 is 1000, that of 2 and 3 is 6, that of 5 and 4 is 20, and so on. But a general method needs to be determined.

The process of finding these least common denominators leads deep into the field of **number theory.**

**That's arithmetic in a fancy imported suit, wearing a monocle and smirking superciliously.**

*Yes, but it's also the mathematician's version of arithmetic: namely, you don't worry too much about how to calculate things and instead concentrate on proving theorems about numbers.*

Another way to think about the least common denominator is that it is the **least common multiple** of the two denominators.

In strict definition, the least common multiple of two whole numbers a and b is the smallest whole number c, such that there are whole numbers d and e such that c = ad and c = be. In other words, it's the smallest number that a and b both divide without leaving any fractions behind.

To figure out the least common multiple (**lcm**) of two numbers, we end up figuring out another number, the **greatest common divisor (gcd)** of those numbers. The gcd is the largest whole number that divides both of the numbers. The gcd of 6 and 27 is 3, the gcd of 40 and 55 is 5, and so on.

Then we rely on a theorem which says that ab = lcm(a, b)gcd(a, b).

*That was a pretty fast move.*

**You're right. Let's back up a bit.**

Each whole number has numbers it divides into – that is, whole numbers which can be obtained by multiplying the first number by another whole number. These numbers are called **multiples** of the original number. For example, the multiples of 3 are:

0, 3, 6, 9, 12, etc., an infinite list.

Each whole number has numbers that divide into it to give whole numbers. These are called the **divisors** of that number. For example, the positive divisors of 30 are

1, 2, 3, 5, 6, 10, 15, and 30, a finite list.

The greatest common divisor of two numbers is the largest number common to the list of divisors of those numbers. Since every number is divisible by 1, the gcd is always at least 1. For example, suppose we want to find the gcd of 30 and 24. We can list the divisors of 30, which we already did, and the divisors of 24, which are

1, 2, 3, 4, 6, 8, 12, and 24.

Common to this list are

1, 2, 3, 6. Since 6 is the highest number on the list, it's the gcd.

Finding lcms can be done in a similar fashion. Write down the infinite lists of multiples and spot the lowest number common to both lists. There will always be such a number, since if the numbers are a and b, then ab belongs to both lists, as does 2ab, 3ab, etc. Every pair of numbers has an infinite number of common multiples.

The proof that $ab = \text{lcm}(a, b)\text{gcd}(a, b)$ is pretty but requires more algebraic apparatus than is useful at this time. Let's go back to finding gcds.

*You mean back to making lists and checking them twice?*

**No way. We need a better process than that.**

The process for finding the gcd uses the process of **remainder division**. This is division of whole numbers into whole numbers that gives whole number quotients and remainders. It's a process of division used when you don't want to break things up into smaller-than-whole pieces. For example, if you are dividing 100 cows into lots of 30 each, you end up with 3 lots of 30 and 10 cows left over. The 3 is called the **quotient** and the 10 the **remainder**.

If we are dividing a into b, this will give us the largest whole number q (the quotient) of a's that b can be broken up into, and a whole number r (the remainder), representing the number less than a remaining after we've made as many a's out of b as we can. What this means is that $b = aq + r$, where all of a, b, q, and r are nonnegative whole numbers and $r < a$.

The process of finding the gcd can be done as follows:

Suppose we have two numbers, a and b, that we want to find the gcd of.

Suppose further that $b > a$.

Then doing integer division, we get

$$b = aq + r$$

If $r = 0$, then $a|b$ (read as "a divides b") and a is the gcd of a and b.

If not, we would seem to be stuck – except for one crucial point. Notice that $r = b - aq$. So if we have a number c such that $c|a$ and $c|b$, then $c|r$. In other words, any divisor of a and b is also a divisor of the remainder left when we divide b by a. Furthermore, notice that r is smaller than a and smaller than b.

Let's examine gcd(r, a), the greatest common divisor of r and a. What can we tell about it? First, it must divide both r and a. But since b = aq + r, this new gcd must also divide b. We want to prove that gcd(r, a) = gcd(a, b).

We know that gcd(r, a)|a and gcd(r, a)|b, so we know it's a divisor of a and b, which means that gcd(r, a) ≤ gcd(a, b). (Since it's a divisor of a and b, it cannot be more than the greatest common divisor of a and b. That's what greatest common divisor means.)

But we know that gcd(a, b) divides r since it divides b and a and r = b − aq. This means that gcd(a, b) is a common divisor of a and r, which means that

$$\gcd(a,b) \le \gcd(a,r)$$

So gcd(a, b) is both greater than or equal to and less than or equal to gcd(a, r), which means that

$$\gcd(a,b) = \gcd(a,r)$$

*OK, now they're all either nodding off or wondering what the point of all that is.*

**Well, here's where it gets really good. We're going to introduce them to one of the ultimate achievements of mathematical laziness: recursion.**

The point is that since what we are after is finding the gcd(a, b), we've just proven that finding that is the same as the problem of finding gcd(a, r). That doesn't sound like we've made any progress, but we have because instead of working with the numbers a and b, we're working with the smaller numbers r and a. Now, we keep going, getting easier and easier problems until we come to a case where the remainder is zero, and we have our solution: the smaller of our two numbers is the gcd. This process of doing the same sort of thing over and over again until it stops is called recursion.

As with many abstract things, it becomes clearer when we do an example.

Let's try this process on 35 and 84:

$$84 = 35 \times 2 + 14(r = 14)$$

So now the numbers we work with are 14 and 35:

$$35 = 14 \times 2 + 7(r = 7)$$

So now the numbers we work with are 7 and 14:

$$14 = 7 \times 2 + 0(r = 0)$$

so the gcd = 7

To find the lcm, we need to multiply 35 by 84 and divide by 7, or we could divide 35 by 7 (getting 5) and then multiply by 84, producing 420.

Recursion, the method of turning a complex process into an iterative process, is a vital tool in computer programming, and we'll discuss it in more depth in the appropriate chapter.

*Okay, that's just ridiculous, all that work to find what number to use as a common denominator. You call that lazy?*

**No. I call it something that leads to insights into the multiplicative structure of whole numbers and that makes the process of finding least common denominators easy.**

We can think about it like this. If you have whole numbers a and b, there is always a whole number c such that $a + c = b$. But there is rarely a whole number d such that $ad = b$. Being a **summand** (something which can add to) is commonplace. Being a **divisor** is rare. (As long as you are talking about whole numbers. For rational numbers, being a divisor loses its strange character. Rationality democratizes divisibility. That's social progress for you.)

Some mathematicians were fascinated by this rarity and created number theory around it. Ancient Greek mathematicians praised certain numbers, calling them **prime.** (A number is prime if it has only two divisors, 1 and itself. 2, 3, 5, 7, 11, and 13 are the first six prime numbers.) They even called some numbers **perfect**. (A number is perfect if it is the sum of its divisors apart from itself. For example, 6 is divisible by 1, 2, 3, and 6, and $1 + 2 + 3 = 6$.)

Perfect numbers are a mild amusement. Prime numbers are the roots of the multiplicative structure of numbers.

Every whole number is the product of a unique set of prime divisors. For example, 36 is $2^2 \times 3^2$ and 190 is $2 \times 5 \times 19$. No two numbers have the same list of factors with the same powers. This concept has the kind of elegant beauty that can enrapture mathematicians, leaving them to poke at and mess with prime numbers, creating theorems and unsolved problems that have delighted, frustrated, and sometimes been useful.

In this case, the prime factorizations of two numbers point toward finding common divisors.

For example, we can look at the factorizations of 36 and 190 above and see that they share only 2 as a common factor, so the $\gcd(36,190) = 2$.

This piece of number theory expands much further into deeper results in other fields. One of the invisible parts of human thinking is that once a concept is created, the mind tends to try to apply it elsewhere. To a large extent, that's what we're doing right now as we expand the idea of number, the concept of prime expanded to the idea of something you can't divide and end up whole, factorization carried over into the idea of expressing something as a product of primes, whatever the primes are.

It's helpful to think of an analogy with chemistry. There are many complicated molecules, but they are all made of atoms. Atoms are much simpler than molecules, so one step in understanding a particular molecule is to ask for the list of atoms that it is made of. Similarly, all positive integers can be thought of as being built up out of prime numbers through multiplication: prime numbers are the atoms of number theory. For any area of mathematics then, one road to understanding and insight is to ask and answer the question: "for the objects that we are studying, are there simpler objects that they are all made out of? And if so, what are these simpler objects?"

Let's now apply that attitude to a different kind of object we might want to factor: polynomials.

Since polynomials are functions, their factors would presumably also be functions. Remember that a lot of the time in the sciences, we end up with problems like find all the numbers x for which $f(x) = 0$. We saw problems like this when dealing with falling objects. But suppose $f(x) = g(x)h(x)$ and suppose that g and h are simpler functions than f.

Now if

> $f(x) = 0$, then either $g(x) = 0$ or $h(x) = 0$ because the only way a product can be 0 is if one of the things it's a product of is 0. Furthermore, suppose $g(x) = 0$. Then, $f(x)$ will be 0, and it will also be 0 if $h(x)$ is 0.

In other words, every solution to $f(x) = 0$ is a solution of $g(x) = 0$ or of $h(x) = 0$ and vice versa.

In set theoretic terms,

$$\{x : f(x) = 0\} = \{x : g(x) = 0\} \cup \{x : h(x) = 0\}$$

In practical terms, this means that if we can turn a complicated polynomial into a product of simpler polynomials and we solve each of those simple polynomials, we'll have solved the complicated one.

It turns out that every polynomial has a factorization made up of linear terms (that is, terms that look like (mx + b)) and quadratic terms (terms that look like $ax^2 + bx + c$). It's an odd piece of cross-fertilization that will reappear in a different form later.

Be warned, just because every number and every polynomial have prime factorizations doesn't mean that those factorizations are easy to calculate. Math likes the universal results because they're lazy, but it doesn't always worry about the work needed to implement them.

Back to the expansion of number. We started with stones, added 0, then negatives, then rational numbers. We can add, subtract, multiply, and divide all that we count and measure. That should take care of everything.

*Ahem. Mr. Pythagoras has a sacrificial bone to pick with you.*

**Oh, that irrational transcendentalist.**

## Construction and Its Limits

If it weren't for geometry, numbers might have stayed rational.

The geometers wanted every stick they could construct to have a length. That's reasonable. After all, what sense does it make if only some parts of our buildings have numerical dimensions?

But it's important to recognize the sources of each of our additions to what constitutes a number. Our earlier work could be seen as algebraic and analytic. What follows is very insistently geometric.

Let's make a square that's 1 × 1 and draw its diagonal. How long is that diagonal? (Figure 11.4)

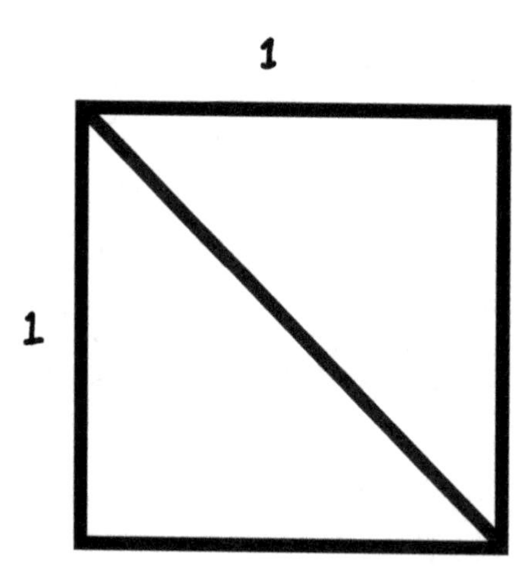


**FIGURE 11.4**
Square with Diagonal

$a^2 + b^2 = c^2$, so if a and b are 1 each, then $c^2 = 1 + 1 = 2$, so $c = \sqrt{2}$.

But what number is the square root of 2?

One mathematical legend attributes a proof of what follows to Pythagoras, though we'll do it a bit differently than he would have.

At the time of Pythagoras, the only numbers known were whole and rational numbers. Since every whole number $a = a/1$, every whole number is also a rational number. So the question was, what rational number is $\sqrt{2}$?

Every rational number is the ratio of two integers, so assume that $\sqrt{2} = a/b$. We can further assume that a/b is in **lowest form**, which means that a and b have no common divisors except 1. (Two numbers that have this property are called **relatively prime.**)

We're going to give a brief digression on a bit of math that people learn early and then don't bother with because it seems to have no importance.

The integers are divided into two sets:

Even numbers: 0, 2, 4, 6, 8, etc.

and

Odd numbers: 1, 3, 5, 7, 9, etc.

The definition of evenness is that an even number a is one for which there is another whole number c such that $a = 2c$. Every whole number that is not even is odd. An odd number will be $2c + 1$ for some c.

Suppose a is even, so that $a = 2c$ for some c. Then $a^2 = 4c^2 = 2$ times $2c^2$. If a is even, then $a^2$ is even. Square an even number, you get an even number.

What happens if you square an odd number? If $b = 2c + 1$, then $b^2 = 4c^2 + 4c + 1$, which is $2(2c^2 + 2c) + 1$, which is 2 times an integer plus 1. In other words, if b is odd, $b^2$ is odd.

Now back to Pythagoras.

If $\sqrt{2} = a/b$, then

$2 = a^2/b^2$

$a^2 = 2b^2$

This means that $a^2$ is even. Which means that a is even (since only an even number squares to an even number). So there is some c for which $a = 2c$, and $a^2 = 4c^2$, so

$4c^2 = 2b^2$

$2c^2 = b^2$

$b^2$ is even, which means b is even. But if both a and b are even, they share a common factor (2), which means that a/b is not in lowest form, which contradicts our hypothesis.

**Thus, we have a proof by contradiction.**

*You do know that won't be satisfying to a lot of people.*

**True. Proof by contradiction is something that most people don't have a feel for.**

To a mathematician, schooled in mistrust, the demonstration of contradiction is enough. But in everyday life, there are lots of contradictions. We can hate tomatoes but love tomato sauce. That is not a mathematical contradiction, but it feels like a contradiction. Things like that, contradictions of taste, are so commonplace as to be chalked up as simple eccentricities.

In math, they are far more than that. Remember that a mathematical proof relies on starting from a hypothesis and performing actions that preserve truth. In a direct proof, we are trying to show that because we started with truth and preserved truth, we must have ended up with truth. A proof by contradiction works in the opposite fashion. We preserved truth and ended up at falsehood. Therefore, we must have started at falsehood. It's sort of like knowing what tree you planted the seed of by examining the fruit that grows from it. If it's apples, you planted an apple tree.

Back to the proof we just finished.

We've shown that we cannot find two whole numbers a and b for which $\sqrt{2} = a/b$. In other words, $\sqrt{2}$ is not a rational number.

Please feel free to insert as many jokes as you wish connecting the concept of a nonrational number to that of an irrational person. Digress as far as you want into anyone whose political views you disagree with, whose style of dress you abhor, any of your friends and relatives, and, of course, those annoying people who cut you off in driving because they don't realize that you and you alone own the highway.

All done? Good.

When confronted with this fact, Pythagoras, so the story goes, made a great sacrifice to the gods. Whether he did so because they had made a world in which such a weird thing could be, or because they had let him discern this truth, or because it was the done thing in those days is not clear.

There's another story that it wasn't Pythagoras who proved this, but that whoever did was killed by the Pythagoreans for messing up their view of reality. Take what lessons you want from both the myths.

Pythagoras, or whoever it was, then made a choice that he probably did not even think of as a choice. He could either treat $\sqrt{2}$ as not a number at all, or he could expand what numbers were. He would not likely have seen this choice because the number was sitting in front of him. It was the length of the hypotenuse of the simplest right triangle there is, the diagonal of a $1 \times 1$ square. The stick was in front of him. Its length had to be a number.

But what did this add to the set of numbers? Just one new weird, irrational number? No, because you could as easily get $\sqrt{5}$ from a triangle that had one side of length 1 and another of length 2, or $\sqrt{13}$ from a triangle that was 2 by 3. Furthermore, the sides of a triangle need not have integer lengths. You could end up with $\sqrt{(a/b)}$ for any positive whole a and b.

Was this the limit of stuff being added? Do we only need to have rational numbers and their square roots to have a complete set of numbers?

That depends on the point of view you wish to take. There is a set of numbers that is mostly what we've just described. They are called the **constructible numbers**, and they have a curious quality of looking totally artificial while seeming reasonable.

The constructible numbers work like this: The Greeks drew their geometry using two tools: the compass, which allowed them to make circles, and the straightedge, a ruler without markings. Using a batch of procedures from these tools and starting with a line segment of given length (1, for example), they could produce line segments of various lengths. They could, in particular, make segments of any rational length, and they could create a segment that is the square root of any segment they already had.

The arbitrary quality of this process is that they acted as if the two tools they used, the compass and straightedge, somehow comprised all legitimate ways of making sticks. The reasonable quality is that they only demanded the status of number be given to those sticks they could actually construct. They kept a real-world connection between the sticks and the numbers that represented their lengths.

The trouble with the constructible numbers is that there are numbers the Greeks wanted to get at but couldn't with their small toolkit. Here are three famous problems they could not solve because of this:

1. To square the circle, which means to create a square with the same area as that of a given circle.
2. To duplicate the cube, which means to create a cube whose volume is twice that of a given cube.
3. To trisect an angle, which means to construct an angle that is 1/3 of a given angle.

We're not going to bother with trisecting the angle, and we'll come back to squaring the circle, but let's look at duplicating the cube.

If we have a cube the side of which has length a, then the volume of that cube is $a^3$ ("a cubed," *no surprise there*). For another cube to have twice its volume, the side of that new cube (call it b) would have to satisfy $2a^3 = b^3$ or $b = a\sqrt[3]{2}$ (or "b equals a times the cube root of 2"). But construction only gets us square roots; no cubes need apply. Fundamentally, the absence of this construction process is an absence of numbers as well.

What came from this did not happen in ancient Greece. It took a number of centuries. Some of what we're going to talk about later in this chapter happened earlier in history. But the way of thinking about numbers that follows is worth looking at as if it were a separate creation because it points a long way toward even later developments.

Let's pause for a second and consider the causes we have followed to extend our ideas of number.

We created whole numbers by piling up rocks.

We created negative numbers to deal with the idea of debt.

We created rational numbers because of wanting to make smaller sticks out of bigger ones.

We created the constructible numbers because we could literally construct sticks of their lengths.

This next set of numbers arises in part from a common kind of equation and in part from the idea that certain things should be numbers. Consider the graph of $y = x^3 - 2$ in Figure 11.5.

This graph touches the x-axis at the point where $0 = x^3 - 2$ (since $y = 0$ is the x-axis). The graph clearly shows a point where the graph intersects the x-axis. We want that point to correspond to a number. We want the x value there to be a number. That number has to satisfy

$$0 = x^3 - 2$$

or

$$x^3 = 2$$

In other words, we want the cube root of 2 to be a number. More generally, if we graph a function f(x), where f is a function the values of which we can directly calculate by addition, subtraction, multiplication, and division, we want the points where $f(x) = 0$ (that is, the points where the graph intersects the x-axis) to correspond to numbers. What this means

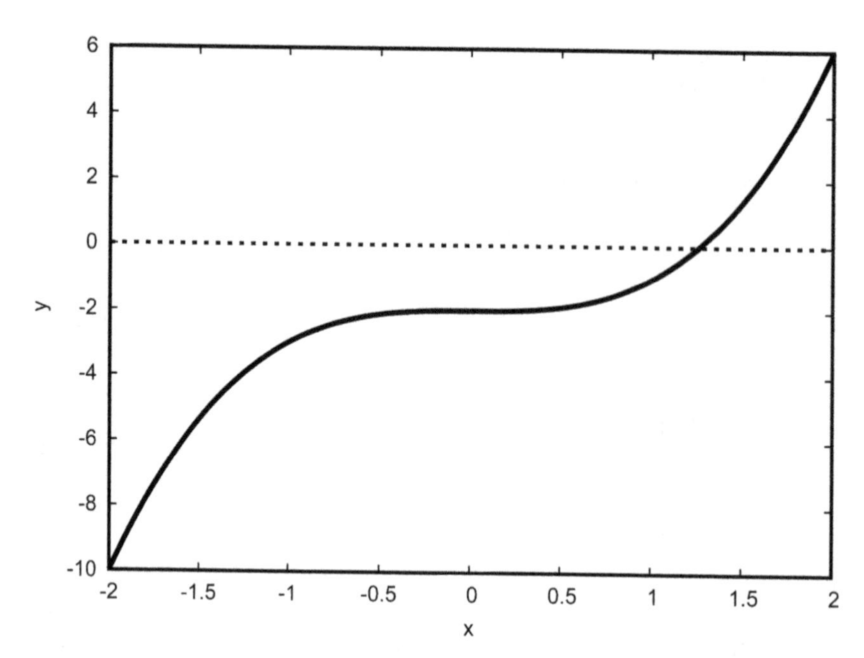


**FIGURE 11.5**
Graph of $y = x^3 - 2$

analytically is that we want the solutions of equations made from numbers and variables to be numbers.

What kinds of functions are we talking about? What can we directly calculate?

We can add subtract, multiply, and divide rational numbers. So if f(x) is a polynomial with rational coefficients, we can, at least in theory, calculate it. In practice, if f has more than half a dozen terms or is higher than 5th degree, we'd rather chew our own feet off than actually stick numbers in it and compute.

*Hmm, if only there was a way to automate calculations. Hmm.*

**Stop with the foreshadowing! We'll get to computers later.**

This, by the way, is part of laziness. Mathematicians care that the calculations can be done, but they're not interested in doing them. That, as the late great Douglas Adams said, is somebody else's problem.

What this new set of numbers boils down to is taking the rational numbers and putting in all the nth roots of positive rational numbers (square roots, cube roots, fourth roots, etc.), and then allowing addition, subtraction, multiplication, and division of the results. This way we can have $\sqrt[27]{1{,}000{,}005} + 3 + \sqrt[1{,}00{,}005]{27}$ and other wacky fun things.

This set of numbers is called the **algebraic numbers,** which turned out to be a naive name since algebraicists weren't content with them.

The constructible and the algebraic numbers can be looked at as attempts to add some sense of rationality to the irrational numbers, to try to hold the lid down on how weird numbers could get and still be numbers. Unfortunately (or fortunately, or indeed without any regard for luck or fate), it was too late for that.

This battle was lost the moment someone picked up a compass, drew a circle, and wanted to know how long that sweeping arc was. In order to figure out this number, the circumference of the circle they had drawn, they created a limit procedure, the one with the inscribed polygons we showed earlier, that led them to a decent approximation for $\pi$.

And there it sat, happy and contented, ticking away, waiting innocently for the rest of math to catch up until someone could prove that $\pi$ was not algebraic.

Not only can't you square the circle (since you can't find an algebraic number a such that $a^2 = 2\pi$), you can't find any polynomial with rational coefficients for which the solution is $\pi$.

Now there's a problem. We've got this number we absolutely cannot do without, that really ought to be a number. But how can we add this one thing in, this indispensable $\pi$ – and worse yet, is it only one thing?

*You mean there might be more of those non-algebraic numbers out there, lurking in the number line?*

**Yes. In fact, not only isn't it one thing, there are more of these things than there are algebraics.**

There are as many of these **transcendental numbers** as there are real numbers. There are only as many algebraics as there are integers (we're not going to take the time to prove that one).

But how do we get a grip on these numbers? So far we have only $\pi$ and whatever wackiness we can do with it (multiples of $\pi$, roots of $\pi$, etc.).

Oddly enough, we are rescued by notation. If we write our numbers in decimal form, rather than as ratios of numbers, we have a way to get to all the real numbers.

If we look at, for example, 0.11137… (arbitrary number of digits)
this means

1/10 + 1/100 + 1/1000 + 3/10000 + 7/100000 + …

which is an infinite series. It should be a convergent series because the sequence of sums

1/10
1/10 + 1/100
1/10 + 1/100 + 1/1000
1/10 + 1/100 + 1/1000 + 3/10000
1/10 + 1/100 + 1/1000 + 3/10000 + 7/100000
…

get arbitrarily close together as we go farther along. A sequence that "should" converge because its members get arbitrarily close together is called a **Cauchy sequence** after 19th-century French mathematician Augustin-Louis Cauchy. Cauchy made contributions to several areas of mathematics, in particular the study of functions of complex numbers. There's an exact definition of Cauchy sequence, which is another one of those pick a number epsilon things that's exact but unedifying. What matters here is that not every Cauchy sequence of rational numbers converges to a rational number and not every sequence of algebraic numbers converges to an algebraic. For example,

3
3.1
3.14
3.141
3.1415
etc.

This sequence would converge to π, but π is not an algebraic number.

What can we do?

We can add in a batch of new numbers which we define the Cauchy sequences as converging to. In effect, the numbers are the Cauchy sequences. Actually, to be messier, they're really equivalence classes of Cauchy sequences. We would say that two Cauchy sequences converge to the same number if the differences between the corresponding elements of the sequence converge to 0. In other words, if we have two Cauchy sequences.

a1, a2, ...

and

b1, b2, ...

If

|a1 – b1|, |a2 – b2|, ... converge to 0, then the sequences are equivalent. Proof that this is an equivalence relation is not too hard. Reflexive and symmetric properties are easy to see. What about transitive?

Suppose the differences between

a1, a2, ...

and

b1, b2, ...

converge to 0

and those between

b1, b2, ...

and

c1, c2, ...

converge to 0. Let's consider the differences between

a1, a2, ...

c1, c2, ...

So the sequence

|a1 – c1|, |a2 – c2|, ...

Now we're going to yank something in from left field, actually from metric spaces. In the number line, distance between two numbers is equal to the absolute value of one number minus the other. In other words,

|a – c| is the metric on the line. The metric obeys the triangle inequality, which says that, oh, for example

$$|a1 - c1| \leq |a1 - b1| + |b1 - c1|$$

$$|a2 - c2| \leq |a2 - b2| + |b2 - c2|$$

and so on for every term in the sequence.

The terms on the right side of this inequality converge to 0 and therefore the terms on the left side converge to 0. (Since the terms are all absolute values, they cannot be lower than 0. So they are squeezed between 0 and 0 and must have a limit of 0.)

Let's stop once more. We've proven more than what we said. The above proof could be rewritten with any metric space that obeys the standard axioms of a metric because the only properties we called upon were those axioms. We didn't need the fact that the a, b, and c values were numbers. We've shown that we can define a Cauchy sequence in any metric space and can prove that we can make equivalences of those sequences based on whether or not they would converge to the same value if there were such a value.

So what do we do? We say that those equivalence classes of Cauchy sequences are numbers. The space so created is said to be **complete** since every Cauchy sequence in it converges.

The next bit of work we need to do is to show that the numbers we created above can be added, subtracted, multiplied, and divided and that they end up with the results we expect. With a little fussiness, we can show that adding the corresponding elements of a Cauchy sequence creates a sequence that converges to the sum of the numbers the two sequences converge to, and so on for the other operations.

*Just an infinite process dragging out minute there. Are you telling me that the real numbers are "really" equivalence classes of infinite sequences of rational numbers?*

**No, I'm telling you that we've just sketched out a process for creating something which is a model for our intuitive idea of the real numbers.**

What we mean by the real numbers is every possible length for a stick and its negative. What we mean is that any way we can get a number that makes sense on a number line, we want to be a number. All the above apparatus does is to tell us that yes, there is such a set and yes it makes sense at the deepest level of mistrust. But no, we don't have to think about the real numbers as really being these complicated mathematical objects.

What we need to know is that every stick now has its length and we can rest content.

*Hmm. Hmm! Happy with the numbers, nothing bothering me, nothing more to do.*

## Imaginary and Complex Numbers

One of the most radical additions to the numbers came about in a very indirect way, through the general solution of the cubic equation. That is, for any numbers $a_0$, $a_1$, $a_2$, and $a_3$, we want to find all the numbers x for which $a_3x^3 + a_2x^2 + a_1x + a_0 = 0$. This is a more complicated problem of the sort that we have already looked at for linear equations and quadratic equations. This problem was solved in the 16th century by two Italian mathematicians named del Fero and Tartaglia. However, the solution is called Cardano's formula after a third 16th-century Italian mathematician who was the first to publish the solution (or to quote Tom Lehrer, "My name in Dnepropetrovsk is cursed when he finds out I publish first!").

We're not going to show the method used to obtain the solution, because the algebra is messy and complicated. However, the method uses the quadratic equation and its solutions to find the solutions of the cubic equation. Recall that the quadratic equation takes the form $ax^2 + bx + c = 0$ and its solutions are

$$x = \left[-b \pm \sqrt{b^2 - 4ac}\right]/(2a)$$

Since the square of any number is positive, we are (so far) not allowed to take the square root of a negative number. So when $b^2 - 4ac < 0$, the quadratic equation doesn't have solutions. And indeed that's what analytic geometry says too, since the parabola given by $y = ax^2 + bx + c$ never crosses the x-axis when $b^2 - 4ac < 0$. However, when the cubic equation has three solutions, the quadratic formula part of Cardano's formula requires us to take the square root of a negative number.

To do this, Cardano introduces a quantity i whose square is minus 1. That is $i = \sqrt{-1}$. In the final answer that the formula gives, all the occurrences of i cancel each other out and the answer is correct. However, the intermediate state of the calculation seems to require introducing a new type of number that doesn't appear to make any sense.

The modern terminology reflects this 16th century ambivalence. What before were just called numbers are now called "real numbers." In contrast, both i and any real number multiplied by i are called "imaginary numbers." A number made by adding a real number and an imaginary number ($a + bi$, where a and b are real numbers) is called a "complex number." Here "complex" doesn't mean "complicated" but rather "made from more than one piece." In the complex number, $a + bi$, the number a is called the real part and the number b is called the imaginary part. (Note that zero times i is still zero, so a complex number with zero imaginary part is real.)

The manipulations in Cardano's formula require that these complex numbers obey the same associative, commutative, and distributive rules as real numbers. In particular, this means that for any two complex numbers $a + bi$ and $c + di$, we have

$$(a+bi)+(c+di)=(a+c)+(b+d)i$$

$$(a+bi)(c+di)=ac+(bi)c+a(di)+(bi)(di)$$

Let's look at that last term: $(bi)(di) = bdii$. $ii = -1$, so this equals $-bd$. Therefore,

$$(a+bi)(c+di)=(ac-bd)+(bc+ad)i$$

Note that the rule for adding complex numbers means that the real parts add to form the real part of the total, and the imaginary parts add to form the imaginary part of the total. But this is just the way a two-dimensional vector (a, b) behaves. So we can think of the complex number $a + bi$ as being equivalent to the vector (a, b), which is itself equivalent to the point (a, b) in the Cartesian plane. The real numbers are just complex numbers of the form $a + 0i$, which are the points (a, 0) which is the x-axis of the Cartesian plane. In other words, in the Cartesian plane of the complex numbers (called the complex plane), the x-axis is just our old friend the number line, the imaginary numbers are given by the points on the y-axis, and all other points correspond to complex numbers.

Division of imaginary numbers is more tricky than multiplication. If we look at $(a + bi)/(c + di)$, there seems to be nothing we can do to simplify this into an $e + fi$ form – unless we remember that messy divisions are nearly always solved by finding some way to multiply the ratio by 1 in a tricky form (that is, z/z for some specially fancy z). If we can transform this so that the denominator is a real number, we're fine. We don't have any problems dividing a complex number by a real one, since it's pretty clear that $(a + bi)/c = a/c + (b/c)i$.

Our goal is to find some complex number that we can multiply by $c + di$ to get a real number. Then we only have to multiply $(a + bi)/(c + di)$ by that number divided by itself and we've done it.

Rather than belabor this, we're going to use the number $c - di$. This number is called the **complex conjugate** (or just conjugate) of $c + di$. Same real part, imaginary part multiplied by $-1$.

Now $(c + di)(c - di) = c^2 + d^2 + cdi - cdi = c^2 + d^2$. In other words, a complex number multiplied by its conjugate is always a real number.

So we can turn $(a + bi)/(c + di)$ into
$(a + bi)/(c + di)$ times $(c - di)/(c - di)$, which equals
$(ac + bd + (bc - ad)i)/(c^2 + d^2)$, which equals
$(ac + bd)/(c^2 + d^2) + i(bc - ad)/(c^2 + d^2)$

Having introduced a new number i to take the square root of −1, you might worry that we might have to introduce yet another kind of new number to take the square root of i. Remarkably, that does not happen. Consider the number $z = (1 + i)/\sqrt{2}$. That is, $z = a + bi$ with $a = b = 1/\sqrt{2}$. The rules for multiplying complex numbers then give

$$z^2 = \left(a^2 - b^2\right) + 2abi = i$$

But it turns out that complex numbers are much more versatile than that. They can be used to factor any polynomial as a bunch of linear factors of the form (x-r) where r is a (possibly complex) number. Every polynomial equation of order n has n complex roots.

We're not going to show how to do this in general, we're just going to show how it works for quadratic polynomials.

Let's take a quadratic polynomial $ax^2 + bx + c$ and suppose that it is possible to factor it into linear factors. That is, we suppose that there are numbers $x_+$ and $x_-$ for which

$$ax^2 + bx + c = a\left(x - x_+\right)\left(x - x_-\right)$$

Then the polynomial is zero whenever $x = x_+$ or $x = x_-$. So the only thing that $x_+$ or $x_-$ can be are the solutions of the quadratic equation. That is,

$$x_+ = \left[-b + \sqrt{b^2 - 4ac}\right]/(2a)$$

$$x_- = \left[-b - \sqrt{b^2 - 4ac}\right]/(2a)$$

That's fine as long as $b^2 - 4ac$ is a positive number, but what if it is a negative number. It turns out that we can still use the formula above to factor the polynomial, only now the numbers $x_+$ and $x_-$ are complex.

Let's see how this factorization works for the quadratic polynomial
$x^2 + 2x + 7$. In this case, $a = 1$, $b = 2$, and $c = 7$. So we get

$$b^2 - 4ac = 2^2 - 4(1)7 = 4 - 28 = -24$$

$$\sqrt{-24} = i\sqrt{24} = \left(2\sqrt{6}\right)i$$

We then find

$$x_+ = \left[-2 + \left(2\sqrt{6}\right)i\right]/(2) = -1 + i\sqrt{6}$$

$$x_+ = \left[-2 - \left(2\sqrt{6}\right)i\right]/(2) = -1 - i\sqrt{6}$$

Note that $x_+$ and $x_-$ are complex conjugates of each other, and a little computation shows $x_+ + x_- = -2$ and $x_+ x_- = 7$. So sure enough, we find

$$(x - x_+)(x - x_-) = x^2 - (x_+ + x_-)x + x_+x_- = x^2 + 2x + 7$$

which is our original quadratic polynomial factored.

Now let's look at a consequence of the general result (which we have asserted but not proved) that a general polynomial can be written as a product of linear factors, possibly with complex roots. (Here "root" means the constant c in the factor (x – c) or equivalently a number for which the polynomial is zero.) If a polynomial has a complex root a + bi, then it must also have its complex conjugate a – bi as a root; otherwise, the coefficients of the polynomial will not be real numbers. But now if we look at the product of both linear factors, we find

$$[x - (a + bi)][x - (a - bi)] = x^2 - 2a\,x + a^2 + b^2$$

This is the hidden reason behind our earlier assertion that any polynomial can be factored into linear and quadratic factors.

Now let's think about complex functions. In particular, if we have an ordinary function f(x) and we consider the complex variable z = x + iy, can we figure out what the complex function f(z) should be? In principle, we would like to take the formula for f(x) and just change every "x" in the formula for f into a z. However, since so far all we know how to do is add, subtract, multiply, and divide, the only functions that we do this "x turns into z" trick for are polynomials and ratios of polynomials.

Or maybe not. A Taylor series is a sort of infinite polynomial. So maybe we can make a complex function out of a real function for which we have a Taylor series. Let's try this for $e^x$.

First a quick digression on powers of i:

$i^0 = 1$

$i^1 = i$

$i^2 = -1$

$i^3 = -i$

$i^4 = 1$

So the powers of i cycle through these four values. Furthermore, notice that every even power of i is real and every odd one is imaginary.

It turns out (though we won't give a proof) that the property $e^{a+b} = e^a e^b$ continues to hold for complex numbers, so $e^z = e^{x+iy} = e^x e^{iy}$. Since we already know how to find $e^x$, all we need to figure out is how to find $e^{iy}$. Using the Taylor series for $e^x$, we find

$$e^{iy} = 1 + iy + (iy)^2/2! + (iy)^3/3! + (iy)^4/4! + (iy)^5/5! + \cdots$$

$$= 1 + yi - y^2/2! - y^3 i/3! + y^4/4! + y^5 i/5! + \cdots$$

Separating into real and imaginary parts, we get

$$1 - y^2/2! + y^4/4! + \cdots + i\left(y - y^3/3! + y^5/5! + \cdots\right)$$

The real part of this is the Taylor series for the cosine of y and the imaginary part is the series for the sine of y.

$$e^{iy} = \cos(y) + i\sin(y)$$

This is the kind of interrelationship between apparently unconnected things that mathematicians are fascinated by. Nearly every person of any mathematical mindset who has seen that equation has stared at it, intrigued by the implications of such an interconnection.

The next equation is a monumentally simple result that comes from this and yet has an even greater elegance. Let $y = \pi$.

$$e^{i\pi} = \cos(\pi) + i\sin(\pi) = -1 + 0i$$

or

$$e^{i\pi} = -1$$

Results like this, an interconnection between the two most famous transcendental numbers (e and $\pi$) with the alien taste of i, starts mathematical minds going, setting them poking and exploring. This is one of those cases where mathematics is the motivation for mathematics and whether anything useful to anyone else comes out of it doesn't matter.

A great deal more, an entire field called **complex analysis**, having to do with the calculus of complex functions, arises from the useful and the beautiful that grew out of simply adding one special number, one scintilla of imagination to the real numbers. Despite its imaginary origins, complex analysis is a useful field, with physical applications that justify in many respects the aesthetic leap that led to its creation.

The ways of making new numbers we've talked about above are by no means the only expansions of numbers, but beyond this the systems get stranger and the work of explanation is a bit more involved than we need to be. We started from stones and have ended up with complex numbers, adding things along the way for various reasons and changing the meaning of the word "number."

Were this the end of the process, then it would be analysis that reigns over all. But it isn't. This continuous stretching of number has always been accompanied by the desire to make sure that whatever we called numbers can be added, subtracted, multiplied, and divided. The great leap of algebra lay in the realization that it is the operations that delimit the numberness of numbers, not the numbers themselves.

So what have we inherited at this point? We have a number of different kinds of numbers, each a justifiable extension of our initial stone and stick numbers, each added for a purpose, and each permitting the abstraction and manipulation of more and more real-world (and not so real-world) characteristics.

Along the way the work, the proofs needed each time we added to the storehouse of numbers, has been done to preserve the character of the way numbers interact, to allow for the operations we rely on to be done with these new numbers. Along the way, algebraic thinking grew, as the minds of the mathematicians worked to make the four simple operations – addition, subtraction, multiplication, and division – still be available for every kind of number.

This expansion and focus on the operations gave them an abstract vitality that grew alongside their worldly practicality. Addition of complex numbers is not piling up stones, but the additionness of both meanings of the + sign is clear in both cases.

Taking up the heritage of new kinds of numbers is not all that has been elaborated in the growth of number kindred. There is also a vast wealth in the heritage of +x–/.

## Algebraic Structures: Abstraction Spreads Wide Its Arms

We looked before at one abstraction of operation, the group. Now we'll look both more closely and more broadly.

In all the expansions of number we have done, there is always an implicit assumption. Whatever things we say are numbers, we must be able to add and multiply them and end up with new numbers.

The reason, for example, that we can solve all polynomial equations in the complex numbers isn't just that we stuck i into the set of real numbers. We insisted that we be able to multiply i by any real number and then add any real number to that. The complex numbers are not so much the result of sticking an i in as the demand that i be treated as a number, and that means addition and multiplication with i must be possible.

The algebraic mindset looks at these and sees that "numberness" does not lie in being a member of a particular set, it lies in being subject to two operations, + and x.

The first great leap in this abstraction, the one that leads to the endless ladder of abstractions to come, is to see that what matters in a set which is kind of like what we call numbers is that it has two operations, which we will call + and × that obey certain rules.

The algebraic mindset further sees that these rules vary from set of numbers to set of numbers, that each of the sets we've been dealing with, Z (the integers), Q (the rational numbers), R (the real numbers), C (the complex numbers), and others, has its own distinguishing features in the way its operations behave.

To create a Z-like algebraic structure, it is necessary to have a set A with two operations + and × such that the following rules hold true:

If a, b are members of A, a + b and ab (we'll use the same shorthand for this multiplication as we do for regular multiplication) are members of A.

| | |
|---|---|
| $a+ (b + c) = (a + b) + c$ | Addition is associative |
| $a + b = b + a$ | Addition is commutative |

There is an element 0 of A such that

| | |
|---|---|
| $a + 0 = 0 + a = a$ | Additive identity |

For each a there is a –a such that

| | |
|---|---|
| $a + -a = -a + a = 0$ | Additive inverse |

These make the structure an **abelian group** under addition.

Now on to multiplication.

| | |
|---|---|
| $a(bc) = (ab)c$ | Multiplication is associative |

$a(b + c) = ab + ac$

| $(b + c)a = ba + ca$ | Multiplication is distributive over addition |
|---|---|

A structure that has these properties is called a **ring**.

Notice two things: we ask a lot of addition, but little of multiplication; and we have a single axiom that connects the way the two operations interact, the distributive law. Without the distributive law we would have no way of doing anything with formulas involving both addition and multiplication. It is this single axiom that creates an interconnection.

There are already things we can prove from this. Here are two quick ones.

$a0 = 0$

Proof:

$a0 = a(0 + 0)$ (since $0 = 0 + 0$)

$a0 = a0 + a0$ (distributive law)

$a0 + -a0 = a0 + a0 + -a0$ ($-a0$ is the additive inverse of $a0$)

$0 = a0 + 0$ (since $a0 + -a0 = 0$)

$0 = a0$. QED.

*QED. Is that some strange sneaker brand?*

**It's short for "quod erat demonstrandum."**

*Which means what, that mathematicians also know Latin?*

**No, it means that which was to be proven. It's sort of like taking a bow at the end of a proof.**

*Is there also a Latin acronym for a standing ovation?*

Moving on to the second theorem:

$a(-b) = -ab$ ($-ab$ is the additive inverse of $ab$)

Proof:

$a(-b) + ab = a(-b + b)$ (distributive law)

$a(-b + b) = a0$ (because $-b + b = 0$)

$a0 = 0$ (previous theorem)

Therefore, $a(-b)$ is the additive inverse of $ab$, so it is $-ab$. QED.

*Will you quit it with the curtain calls?*

**Sorry.**

What have we just done? We've proven two theorems about the relationships between two operations that are enough like addition and multiplication to satisfy the ring axioms. So the theorems above are true for any such pair of operations. No matter where and in what circumstances we pull them out, we can use these theorems.

We can make our rings more and more number-like by adding some other axioms.

A ring that satisfies

$ab = ba$ (multiplication is commutative)

is called a **commutative ring.**

A ring that has an element 1 such that

$a1 = 1a = a$ for every a in the ring (multiplicative identity)

is called **a ring with identity.**

A commutative ring with identity in which for every a except 0, there is an $a^{-1}$ such that

$aa^{-1} = a^{-1}a = 1$ (possession of multiplicative inverses)

is called a **field.**

Of our previously examined sets of numbers, Z is a commutative ring with identity, and Q, R, and C are all fields (so are the algebraic and constructible numbers).

We could go over theorems proven about numbers in the thousands of years of mathematical history and check which of the axioms were used in their proofs. Any that used nothing more than the axioms above would also hold true for any ring or field depending on which axioms were employed.

Not only can we create new general theorems using this abstract algebra, we can transform old theorems that we thought applied only to number and extend them to all number-like algebraic structures. It's like discovering that some moderately useful food crop actually has wide medicinal uses.

Let's dig a little deeper and see what else can be gleaned. Let's look at R and C. The thing that makes C most useful is that in C every polynomial can be factored into linear terms. In other words, every polynomial can be solved.

But what's a polynomial?

A polynomial is shorthand for a process of successive multiplication and addition. After all, $x^5$ just means xxxxx *(If you feel jokes about a succession of x's are necessary, please tell them in the privacy of your own home.)*

So let's take a ring with identity, an element a of the ring and a variable x that varies over the elements of the ring. Let's also take a positive whole number n. *Note*: n is a whole number, an actual bona fide, pile of rocks whole number.

The monomial

$ax^n$

is a meaningful object in relation to this ring. It's a function from the ring into itself that takes any element of the ring (x), multiplies it by itself n times (using the multiplication in the ring), and then multiplies a by that result. If we further define $x^0 = 1$ (where 1 is the 1 in the ring, not the number 1), we can write out as a meaningful object a polynomial in the ring. So, for example,

$ax^2 + bx + c$ is a perfectly reasonable function to generate using elements of the ring and variables ranging over it.

Polynomials can be created in any ring with identity. Furthermore, the process of multiplication of polynomials depends only on the commutative, associative, and distributive laws for addition and multiplication, so there's no reason that that can't be done in any commutative ring with identity.

We can therefore ask the question of whether or not a polynomial can be factored in a given ring or field. By factored we mean that it can be written as a product of linear terms ($ax + b$, where a and b are elements of the ring).

Let's look closely at $ax + b$. Suppose we want to know what value of x satisfies $ax + b = 0$ (where 0 is the 0 in the ring).

$$ax + b = 0$$

In the ring we can still perform the same operations to both sides of an equation and preserve truth. We can still do algebra in our algebra.

$$ax + b = 0$$
$$ax + b + -b = 0 + -b$$
$$ax + 0 = -b$$
$$ax = -b$$

At this point we are stuck unless our ring is really a field, in which case there is an $a^{-1}$ for each nonzero a. We can multiply by $a^{-1}$ and get

$$a^{-1}ax = a^{-1}(-b)$$

but $a^{-1}a = 1$

So $x = a^{-1}(-b)$

Notice that this is nothing more than the proof of the solution of a linear equation. Translated into field terms, it extends the concept of solving a linear equation and applies it to any field. If we look at our polynomials derived above, we can see that just as we could sometimes factor polynomials into linear terms in the real numbers and can always do that in the complex numbers, we might be able to do that in some arbitrary field. And just as the solutions to a polynomial equation are the solutions of its factors in the real and complex numbers, so they are the solution in any other field.

In other words, we can transplant the process we used to figure out the solutions of polynomials from numbers to any field where we can factor polynomials.

A field in which all polynomials can be factored into linear terms is called a **splitting field**. R is not a splitting field, but C is.

The above allows one to transport a large number of results and methods of polynomial solutions into any field, and especially into any splitting field. We can take every theorem proved about polynomials that relies only on the field axioms and know that they also apply to any other field. The abstraction of properties of addition and multiplication into the field axioms allows us to vastly expand our regions of laziness. In one swoop we've gobbled up books worth of theorems and techniques and spread them around.

We've expanded our inheritance just by looking at the characteristics of numberlikeness that can be found in addition and multiplication. But we've also left other characteristics behind and we have to be careful not to try to use them.

There is no sense of ordering in our rings and fields. We do not have an inherent <, so no theorems that rely on that can be used. Any sense of numberness that relies on ordering cannot be used in the structures above. One can have sets with ordering as their algebraic structure and many of the theorems that come from ordering can be translated over to those structures, but there is no inherent relation between addition and ordering as there is in the real numbers.

We have to separate our intuitions of + and < in order to not be trapped by flawed thinking in our algebraic inheritance. For example, there is a kind of ring that is derived from the integers called **integers modulo n**. There is one such ring for every whole number greater than 1. This set is written $Z_n$

The members of this set are all the nonnegative numbers up to but not including n. The way addition on this set works is that if you have two members a and b, then a + b = the remainder of dividing a + b by n. Similarly, ab = the remainder of dividing ab by n. For example, in $Z_5$.

$$2 + 1 = 3$$

But 2 + 4 = 6, which has a remainder of 1 after division by 5. Similarly, 3 times 4 is 2.

It's pretty easy to prove that for any n, $Z_n$ is a commutative ring with identity (its mathematical operations are derived from standard operations). It also has a real-world analogy. **Modulo arithmetic** (as this is called) is like adding time as one goes around a clock. If we numbered the hours as 0–11 instead of 1–12, then the hour hand of a clock would be $Z_{12}$ and if we numbered the minutes as 0–59, the minute hand would be $Z_{60}$. (Note: for those who have only seen digital clocks, ignore the word hand and substitute numbers instead.)

$Z_n$ inherits a lot from Z, but it loses ordering since

a + 1 need not be greater than a, especially if a = n − 1.

Indeed, the negative numbers here are positive numbers. For example, in $Z_5$, 3 = −2 since 2 + 3 = 0.

All ordering is gone from these rings, but they still retain a great deal of the arithmetic we are used to, if you don't mind bending a line into a circle.

One more algebraic structure before we jump to the next level of abstraction. A **vector space** V over a field F is an abelian group with an operation called + which also has a special operation called **scalar multiplication**. This operation is a little tricky. It takes an element **v** of the vector space and an element a of the field and creates a new element a**v** of the vector space.

Now please watch carefully, because the axioms below are rather strange. Scalar multiplication has to satisfy

$$(a + b)v = av + bv$$

$$a(v + w) = av + aw$$

$$a(bv) = (ab)v$$

These look pretty normal as axioms go, except that we're actually using two different additions and two different multiplications. Let's look at the first axiom.

$$(a + b)v = av + bv$$

On the left side we are adding a and b. This addition takes place in the underlying field F because a and b are members of F. But on the right-hand side we are adding a**v** and b**v**. These are elements of V, so the addition there is vector addition.

The second axiom is not so bad. It's vector addition on both sides.

The third axiom, however, is messy again.

$$a(bv) = (ab)v$$

**bv** is a member of V, so a(**bv**) is scalar multiplication. But on the right-hand side ab is multiplication in F. This axiom says that successive scalar multiplications are the same as doing field multiplication and then doing scalar multiplication.

Vector spaces, unsurprisingly, are meant to abstract the algebraic characteristics of vectors in the Cartesian plane, Cartesian space, and so forth. If this sounds like algebra is creeping in on geometry, it is. We'll come back to this after we've explored the next level of abstraction above these already abstracted operational thingies that we took from abstracting our way up from sticks and stones to complex numbers.

## Morphisms: Preservatives Added and Multiplied

Suppose you have two groups, A and B, and a function h from A to B. Such a function takes elements of one set and gives elements of the other. (Such functions are called **maps.**) This function does not care at all about the group structure of the sets. In order to make a map that is connected to the algebraic structure, we need to demand something of the map that relates to that structure. The simplest such demand is to ask that it preserve the structure for every $a_1$ and $a_2$ in A,

$$h(a_1a_2) = h(a_1)h(a_2)$$

Caution is advised in reading this. The multiplication on the left is multiplication in A; that on the right is multiplication in B. A map that fits this condition is called a **group homomorphism.**

Here's an example. R is a group under addition. $R_+$ (the positive real numbers) is a group under multiplication. $h(x) = e^x$ is a homomorphism from R to $R_+$ since $e^{x+y} = e^xe^y$. On the left side of this, we are adding, which is the group operation in R; on the right side, we are multiplying, which is the group operation in $R_+$.

Here's another example. There is a very small group containing just two elements {1, −1}. The operation here is multiplication. Since 1 × 1 = 1, 1 × −1 = −1, and −1 × −1 = 1, this is a group. Now let's make a map function h(n), which maps (Z, +) (the integers as a group under addition) into this group. The map sends every even number into 1 and every odd number into −1. This map is a homomorphism. Here's why:

Let's try to figure out h(n + m) for any n and m.

The result of this mapping depends only on whether n + m is even or odd. If n + m is even, h(n + m) = 1; if not, h(n + m) = −1.

But n + m is even if n and m both are even or both odd and is odd if one is even and the other is odd.

Suppose both are even.

$$h(n+m) = 1 = 1 \times 1 = h(n)h(m)$$

If both are odd

$$h(n+m)=1=-1\times-1=h(n)h(m)$$

If n is even and m is odd

$$h(n+m)=-1=1\times-1=h(n)h(m)$$

And if n is odd and m is even

$$h(n+m)=-1=-1\times1=h(n)h(m)$$

So the map is a group homomorphism.

There are also **ring homomorphisms**, maps between rings which preserve structure. They must satisfy:

$$h(a_1+a_2)=h(a_1)+h(a_2)$$

$$h(a_1a_2)=h(a_1)h(a_2)$$

Again, left-hand side operations are those of the ring mapped from, right-hand side operations are those of the ring mapped to.

Broadly, a homomorphism is a mapping that preserves algebraic structure (homomorph means "same-shape"). Algebraists discovered that they could learn a lot about algebraic structures by examining the homomorphisms between them. (There is a special term for a homomorphism of vector spaces. It's called a **linear transformation**.)

Suppose we have two of the same kind of algebraic structure (two groups, rings, fields, …), A and B. And suppose there was a homomorphism i from A to B which was also a one-to-one correspondence between the underlying sets. (Remember that if there is a one-to-one correspondence between two sets, they have the same number of elements, so this is only possible if A and B have the same number of elements.). Such a mapping is called an **isomorphism** (isomorph means "equal-shape"). We ran into this idea in an earlier chapter, and here it is in its native habitat, abstract algebra.

For any isomorphism i, there will be another isomorphism $i^{-1}$ (called the *inverse map*) from B to A such that

$$\text{if}\quad i(a)=b,\ \ i^{-1}(b)=a$$

As far as algebraic structure is concerned, if there is an isomorphism between A and B, they are effectively the same. They may have no elements in common, their operations may look completely different, but they have the same number of elements and they act the same.

The mapping above from R to $R_+$ we saw before is such an isomorphism. Its inverse map is ln(x). Algebraically, the real numbers as a group under addition are the same as the positive reals as a group under multiplication. These maps are, however, not ring or field homomorphisms. For that matter, $R_+$ isn't even a ring, since it has no additive inverses and no 0.

Also note that the map before from Z to {1, –1} is not an isomorphism, since the two groups do not contain the same number of elements, nor is there an inverse operation.

Isomorphism is a vital concept because everything that can be proven in one algebraic structure that depends only on that algebraic structure holds true for every algebraic structure isomorphic to it. This means that sometimes in order to prove something in algebraic structure A, one finds an isomorphic structure that is easier to do the proof in. For example, it's easier to add than multiply, so our isomorphism between R and $R_+$ means that we can prove algebraic truths about the multiplication in $R_+$ using addition in R.

This concept of equivalence by finding a mapping that preserves structure has become wildly popular in algebra, allowing algebraic structures to be broken up into equivalence classes and to have their theorems spread all over the place.

*Wait a second. That definition of sameness is darn strange. Why is the existence of a function that carries over what you want a way things are the same?*

**Actually, it's a very old way of seeing things as the same.**

It's as old as the pile-up-the-rocks-unpile-the-rocks method of seeing if numbers are the same. Make a process that goes from one thing to another that preserves what you care about. You can also see it in the relation between a scale model and a building. You preserve the ratios of measurements in scaling up or scaling down, and that's what you care about.

The isomorphism is a vehicle the way a scale factor is, a means of translation. If you can create a true translation from one thing to another, then in whatever way the translation preserves truth, the two are truly the same – but only in the ways that the translation preserves truth.

While this may sound weird, it's simply another mistrustful version of that fundamental method of human thought, the analogy. We analogize things and insofar as they work the same way, we can treat them as the same. Things work the same if their processes can be analogized. Isomorphism is a tool for making sure the processes are analogous on the level that they need to be analogous on for the analogy to work.

## Algebraic View of Geometry: Slouching Toward Topology

Algebra and geometry began paired together, since sticks were numbers and numbers sticks. They drifted apart while geometry and analysis came together. Algebra still had a role to play, but it was in the calculations that came from the analysis rather than in the shapes themselves.

Algebra asserted dominance by examining what objects mattered in the analysis of geometry and abstracting those. We've seen one example of this already, the metric space, which is a set of points and a metric that obeys certain properties. If we look at the metric as a distance function, as we have done, then it seems fundamentally geometric and analytic. But if we look at it simply as a function which takes pairs of elements in the set and gives a number, and we note the axioms given for it, we see little to distinguish a metric from the operations in groups, rings, fields, and vector spaces. In particular, if we see the underlying set as not structured until the metric gives it structure, we are firmly in the algebraic mindset.

Let's run through that slowly:

Geometry is concerned with shape. Therefore, it looks at the space and sees it as having shape.

Analysis is concerned with process and characteristics. Therefore, it looks at the space and the metric and sees the latter revealing the characteristics of the former.

Algebra is concerned with operation. Therefore, it regards the space as largely unimportant. It's the metric that gives structure to the set.

The progression of abstraction in this case went:

Space has distance.

Distance is determined by a function.

That function has certain characteristics.

Any function with those characteristics is distance-like.

Any set with such a function can be seen as like unto a geometric space because it has a distance-like function, a metric.

This transforms a fundamentally geometric view into an algebraic one by a series of abstractions.

But this was not the end of the abstraction process. Even metric spaces were deemed too concrete. An attempt was made to create a concept of space that abstracted most of the desired characteristics of geometry without actually needing shape or distance or anything that most people would think of as necessary to the geometric mindset.

In fact, the types of spaces created, **topological spaces**, are not susceptible to geometric thinking, nor especially to analytic. They are an algebraic substitute for geometry. Topology did this by focusing not on the distance between things but on the ability to separate them in space.

Hold on to your hats ye of geometric mindset:

A topological space is a set T with a set O of subsets of T, called the **open sets**. O has the following characteristics: the empty set and T are members of O. The union of any list of elements of O is a member of O, and the intersection of any finite list of elements of O is an element of O. Any member of O is said to be open. Any subset C of T for which there is an open set A such that T − A = C (T − A is called the complement of A, it is the set of all elements of T that are not in A) is called **closed**. In other words, C is closed if there is an open set A such that C consists of all the points in T that are not in A.

These terms open and closed sound like they have meaning. And they do, in the spaces where the ideas originated. In $R^n$, there are open and closed sets. The simplest open sets are the **open balls**. They consist of the interiors of spheres, but not the surfaces of the spheres. There are also **closed balls**, which are both the interiors and the surfaces. The open sets in $R^n$ are created by taking arbitrary unions and finite intersections of the open balls. The closed sets can be found by taking complements of the open sets (that is, finding the sets that consist of every point not in a particular open set).

The utility of open sets lies in being able to say that two points are separated if there is an open set that contains one but does not contain the other. This abstracts an aspect of geometry we think of as connected with distance (being away from) without requiring a distance function.

In some respects this is weird. We've traded a function for a messy set of sets that interact in a certain way. It looks like we've lost the simplifying benefits of abstraction. And we have, but we've gained a lot in the benefit of generalization.

This highly abstract definition of topological space is largely a starting point from which various other desirable characteristics can be added until one gets to the kind of spaces one really wants to study.

Most of the concepts in topology are abstractions of characteristics of shapes in space. But their definitions in topology are very confusing, because they are phrased only in terms of open and closed sets. Here's one that bothered us:

A subset C of T is said to be **compact** if every set of open sets whose union contains C (such a set of open sets is called an **open cover** for C) has a finite subset that is also an open cover of C (called a **finite subcover**).

If you work through this and apply it to real space, you discover that in this case compact is equivalent to closed and bounded. That is, the set C is closed and does not in any direction extend to infinity. You can't use that definition in topology because "extend to infinity" is meaningless in most topological spaces.

Compactness is an important concept because of the idea of convergence of sequences. Any sequence of points that lies entirely in a compact set will have at least one subsequence that converges to a point in that compact set.

Notice what's happening here. In order to create the most general idea of space, one has to define the space in nonspatial ways. This is a more extreme abstraction than that used in groups, rings, and fields, where we tried to preserve the operational character of numbers in order to understand the numberness of number.

In creating topological spaces, we are abstracting a vital characteristic of spaces (their open set structure), but it's not one that people connect to easily. Even a metric space, abstract though it is, contains an intuitive concept: distance. Openness is itself already an abstract geometric idea.

Topology has been and continues to be very useful in studying aspects of the spatial structure of spaces, but it is very hard to grasp since there are no shapes in these spaces.

The sacrifice of geometric thinking in the service of geometry seems acceptable to the topologists, since they are concerned with the utility and breadth of the proofs they make rather than their comprehensibility to those who look at space. Algebra has sacrificed understanding that connects to other ways of thinking for the sake of breadth of proof. The algebraists can point to the value of that breadth of proof and justly claim to have made results that cross a vast expanse of possible structures and systems.

*What's the harm in that?*

**The problem is that the algebraists do not necessarily have a good sense of what problems need to be addressed, particularly when they abstract from things they may not have a good sense of.**

This would be acceptable if the teaching of these branches of math did not freeze out the very people who might best be able to find the problems that the algebraists should be solving. The algebraists reached abstract algebra by leaping out of the structures analysts created to solve their problems, and topology from the spaces of geometers. But they may have severed the connection to the sources of the work that needs doing.

*Well, we theoretical physicists have our own list of mathematical problems that need to be worked on, and our own way of looking at those problems.*

**But, don't you also want to talk to the mathematicians who work on the abstracted and generalized versions of those problems? And doesn't the level of abstraction make that difficult?**

*Yes, that's certainly true. Talking with mathematicians is often like learning a foreign language.*

So although they are exploring and extending the realm of mathematical inheritance, they may not be doing so in ways that will be helpful to people who use mathematics.

We'd like to finish this chapter with perhaps the ultimate level of abstraction in mathematics, a sort of abstraction of mathematical abstraction.

## Category Theory

At some point, somebody noticed the tendency from algebraic thinking to have objects (like groups) paired with structure-preserving morphisms. In algebra, these are homomorphisms. In topology, a function from one topological space to another that sends open sets into open sets is called a **homeomorphism**. Differential geometry has **diffeomorphisms**, and so on.

So someone decided to abstract this entire concept into the idea of a **category**. A category is a set of algebraic structures coupled with the set of structure-preserving morphism for those objects. The category of groups consists of two sets: G, the set of all groups, that's all groups, every single group there is, and H, the set of all group homomorphisms, that's all group homomorphisms, every structure-preserving one of them. There is similarly the category of vector spaces and linear transformations, the category of topological spaces and homeomorphisms, and so on.

Category theory has as objects these vast categories. It also has maps between them which are called **functors** and yes, basically functors are morphisms between categories.

The theorems of category theory are abstract analyses of the connections between whole collections of algebraic structures and how they interact.

The most impressive thing about category theory is that it actually proves valuable results that help in the study of all objects that can be classified as categories. The most frustrating thing about it is how far it is up the ladder of abstraction.

There's no necessity to stop at category theory. If enough results are proved about the interrelationship of categories and the properties of functors, it might be useful to create a further abstraction in which category theory itself is but one object.

This shows how algebra has become the top mathematical way of thinking. It can always put itself above any currently used description of a problem. It acts like a general, and geometry and analysis become soldiers following orders and knowing that their own chance of promotion comes from taking up the way the general thinks.

The triumph of algebra is a difficult one to deal with from the perspective of the lost inheritance. Algebra has greatly expanded the breadth of mathematical understanding and stretched the theoretical universe vast distances. Yet its results have become difficult for many people to understand and it has extended its way of thinking into fields that might be easier to learn from the analytic or geometric perspective.

It's like owning a company and having a brilliant manager who runs it. The manager may be increasing your wealth beyond measure, but you don't necessarily understand how it's done or even how you can lay your hands on the money and what you could buy with it if you did.

But if you dig in a little, if you insist on always trying to connect what is being spoken of to some part of reality or some process derived from some part of reality, things become easier to understand. Despite the vast generalities algebra has created, each and every one of them come from somewhere real. Each is an abstraction and generalization of an aspect or an operation as strongly rooted in fact as any other mathematical object. To understand what a particular aspect of algebra is or what it's trying to do, what is needful is to find the things it is rooted in and the problems it has been created to solve.

Algebra has always been best at handling the middle of the mathematical process, of taking hold of the abstracted results and messing with them in ways that preserve truth. The ladder of abstraction has allowed algebra to create more and more general, sophisticated tools, but in doing so it has overshadowed the beginning and end of the process, the first abstractions, the ones that connect by geometry and analysis to the real world.

Algebra is not wrong in doing this. The utility of these abstractions upon abstractions is the very reason algebra has done so well. What is important is for the other fields to not assume that the most abstract means of approach will be the best. In other words, one needs to know how high to climb the ladder.

Nearly every problem has an optimum distance from which to examine it. This is the distance where the important characteristics are the clearest for the problem being worked on. Climbing the ladder to that height is the vital skill needed for those who want to employ algebraic thinking for their own applications. It may be that there will be tools on a higher level that would help. The trick is to reach up and bring them down rather than to climb up to where the tools are, leaving the most useful view behind.

We realize that this is an abstract analogy of the process of abstract analogizing. That's the way the ladder rises.

# Part III

# Toolkit of the Theoretical Universe

The lost inheritance of mathematics is a strange inheritance. It isn't really an estate or a sum of money or a portfolio of investments. It's a batch of tools, a set of means that can be applied to various different ends. The mathematicians are the makers of those tools and everyone else, particularly scientists, are the users of them.

Makers and users may share their tools, but they see them very differently. The relation between these two views of things, mathematical and the manners in which those tools are made, used, and taught, forms the final part of our exploration.

# 12

## *The Smith and the Knight*

Consider a sword. In cultures that were centered on a weapon-using nobility, swords were vital symbols of power. Because of that, we're going to exploit the bright shiny special effects of swords to talk about math and science. We could as easily talk about can openers, stoves, sewing machines, flutes, or any other tool, but swords have more drama (Figure 12.1).

A sword has two basic parts: the hilt and the blade. The hilt is the part that you grip in order to use the sword. The blade is the part that does the actual cutting and/or stabbing.

A sword also has two people fundamentally involved in it: its maker, the swordsmith, and its user, the knight (or social equivalent). The smith and the knight have radically different views of the sword, but they share two common views, or rather they seem to share them. First, they want the sword to do its job as well as possible; the smith because it is the smith's creation, the knight because the knight's life depends on it. Second, each regards the sword as theirs and theirs alone.

If one looks at mathematics as a collection of tools, of swords to attack and defeat the problems one encounters in life and work, then the mathematician is the smith, the maker of the tool; the scientist (or architect or engineer or accountant …) is the knight, the user of the tool.

These two viewpoints see and interact with the tools in decidedly different ways, which we will explore in a moment. They are not the only views, and it's worth looking outside of the entire smith-sword-knight interplay to examine the external perspective on the whole matter.

In sword-using societies, there were a number of cultural and legal barriers that kept swords out of the hands of people deemed unworthy or dangerous (dangerous to those the society did not want to endanger with its swords). Swordsmanship was a difficult skill to learn, and there were many barriers in the training as well as the social blocks. Usually such barriers only encouraged the desire for swords on the part of those who were denied them because of the glamour and power the sword conferred.

Think how different it would have been if smiths and knights had conspired to get everyone else to think that having swords was nerdy and only "geeks" knew how to fight with them. In such a case we might be writing about the lost inheritance of fencing and trying to convince people that swordsmanship was the common heritage of all.

We note that our extended metaphor of the smith and the knight is definitely **not** one that mathematicians would use. They would be more inclined to view themselves as artists than artisans, makers of works of great beauty and insight, rather than great utility. If their results can be used by people in other fields, that's good, but it's not the reason they do mathematics.

But this is a book about mathematics written by users of mathematics rather than mathematicians. To emphasize this point of view, we go a little overboard, portraying the users of mathematics as valiant knights, while the mathematicians are relegated to the status of doughty smiths whose purpose is to give us tools.

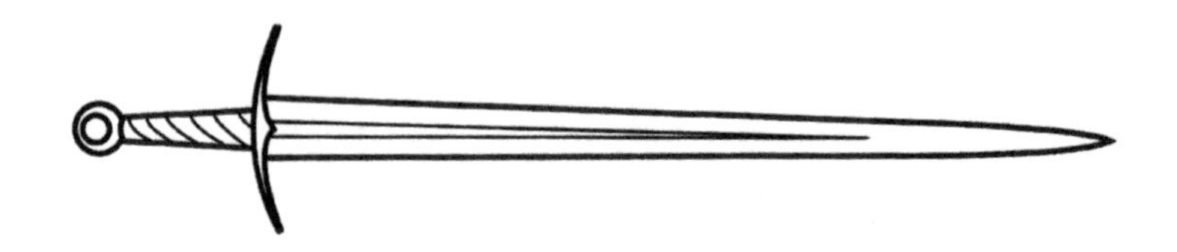

**FIGURE 12.1**
Sword

## How the User Sees the Tool

The blade must do its work. The hilt must be comfortable to use. Weight must be properly distributed through the weapon so as to be conducive to a good swing. Those are the user's concerns.

*It helps if the tool has a cool name ... and looks good ... and sounds good, and if I can use it for a lot of things, but it should be easy to use, unless it's so easy anyone could use it, because you know, then, why would I bother – but it has to be convenient, unless it requires specialized ability which I can show off with, but most of all it has to help me do my job.*

*Or make me look cool.*

Efficacy, versatility, ease of use, and an aesthetic of what makes a tool cool are the concerns of the user. Each user varies in what matters most among these three. Some will take up a hard-to-use, inefficient tool just because it has chrome on it. The aesthetic side of all this we'll leave to the end of the chapter. Here we focus on efficacy and ease of use.

Efficacy is the answer to the question, how effective is the tool?

To which one can ask, effective at what?

The knight has a purpose to which the sword is put, the sewer a purpose for the sewing machine, the cook for the oven, and so on. Efficacy is always relative to the purpose to which the tool is being put. A tool may be effective for more than one purpose (an oven can be used to dry clothes, for example, or a sword to chop vegetables), so efficacy must be judged in relation to the user's intent each time it is used.

To a knight, purpose can be embodied in the sentence, "That which I want should happen when I act to bring it about." An artist of a knightly mindset thinks that what he or she has conceived as an artwork should be made to exist. A knight of knightly mindset thinks that the person he or she is fighting should be defeated. This desire to cause a thing to be or cause an event to happen shapes the knight's view of tools. The efficacy of a tool is measured in how much it facilitates the coming to be of the desired outcome.

In sum, the knightly view is concerned with ends, and tools are seen as means to that end.

If a knightly mind seeks to do only one thing, then its sole interest as far as tools are concerned is to find and learn to use the tools that make doing that one thing possible. But most people, regardless of walk of life, have more than one goal in mind. Even if one is focused on a single profession, following that profession will require the ability to achieve multiple ends. Because of this, a knightly mindset is concerned as much with versatility as with efficacy. Versatility is a measure of how broadly applicable a tool is.

There's a saying that when all you have is a hammer, the whole world looks like a nail. This is obviously a matter of misapplied tools, which we'll get to later.

If we look at a hammer, the plain blunt end of a hammer's head, we find a tool that has a myriad of uses. A tool that drives nails can be used to make various pieces of furniture, build houses, hang up paintings, create patterns for stringing lace, and so forth. Without nails hammers can still be used for such purposes as tenderizing meat and breaking up rocks and dirt.

There are two ways to generate versatility: making a single tool more versatile or creating a toolkit – that is, a collection of tools dedicated to a single purpose. A sensible knight has a toolkit full of versatile tools. Old, established craft professions tend to have such kits, as do soldiers, and, as we will see, scientists.

The last knightly consideration is ease of use. This is the desired quality in the hilt that the hand uses to employ the blade of efficacy.

The hilt is another one of those apparently simple concepts that conceals a lot of nuance within its simplicity. A tool may be easy to use only after a long time practicing its use (such as most musical instruments) or it may be easy to use straight out of the box (such as a drinking glass).

Tools that are difficult to learn become the province of professions rather than of commonplace use. If anyone could pick up a guitar and play it as well as a professional, there would be few professional guitarists. One of the greatest changes in modern society came about when computers stopped being tools with long learning times and became usable out of the box.

Of course, computers didn't really become easier. It's that new programs were written that had short learning times. Programming still takes a lot of time to learn, but the tools programmers made became easier to use and more versatile.

Some tools have a learning threshold. They take work to learn but become easy once learned. Some tools you can keep learning forever, getting better and better at them the more you practice.

Tools, once learned, can still vary in ease of use. A submarine is an immensely complicated object to control, no matter how skilled and practiced the crew.

Other tools, such as a bicycle, become second nature and are used without conscious effort once one has mastered the process of using them.

The differences in ease of learning and ease of use, and what prerequisite physical or mental abilities are necessary or helpful in learning and use, create differences in how tools are perceived. Tools that are easy to acquire and easy to use become invisible parts of people's lives, the way locks with keys have become in the modern world.

Rare tools and difficult-to-use tools can become symbols of status or symbols of strangeness, depending on how the user is seen. Consider again a society where only dweebs use swords.

If the needful prerequisites are valued by the society (the way strength and determination are valued by military societies), then the tools are often associated with professions of respect. If they are devalued or considered weird or strange, the associated professions may be looked down upon.

But none of these considerations focus on the utility of the tools. The question of whether or not a particular tool is worthwhile to learn tends to fall by the wayside in the face of all the social pressures for or against its learning. In these difficulties of getting hold of the tools we see how it came about that the inheritance of mathematics was lost. Once the learning became hard and the learned became "nerds," who but nerds willing to work at it would become users of the swords of math?

## The Maker's View of the Tool

So much for the knight, the user. What does the smith, the maker, think about the tools made? What concerns go into the creation of the tool and how do they differ from the view of the user?

A user's concern is the ability to use the tool to meet the ends the user seeks to achieve. A maker's concern is the way the tool works, not so much what it does, but how it does what it does. The difference between the "what" focus of the user and the "how" focus of the smith is the difference between driving a car and looking under the hood, of using a computer and computer programming, of cooking and food chemistry. For every tool we use, there is a smith's view of it and there are smiths looking at it that way.

So how is "how" looked at?

There is no single answer to this question for the same reason that there is no single way to get from one place to another place. There are many different alternate routes and many different methods. Indeed, there is a direct correlation between the breadth of what one wants to do and the multitude of ways one might get there.

Let's just consider modes of transportation. Fly, take the train, be driven by someone else, sail, drive oneself, use a motorized boat, an oar-powered boat, bicycle, walk. Depending on where you want to go to and where you are leaving from some of these modes may not be possible (walking across oceans, for example, is impractical). Others may be silly (such as airplane travel within a single city). Each practical method has its advantages and disadvantages. It may be that one needs to use a way that requires one or another of these (it may be necessary to get from A to B at airplane speeds, for example).

Given their choice, people will tend to pick modes of travel that fit their needs, their personal preferences, and their fears. There are people who will not fly because of airsickness or fear of crashes or fear of hijacking or any of a number of other reasons. There are people who are willing to go by car only if they can drive because they don't trust other drivers, but some of those same people will happily travel by train or plane even though they know nothing about the engineer or pilot.

This choice among ways applies as well to the constructions of ways. If there is only one option in the making of things, people will try to make that option work. If they have options, they will tend to opt for ways that fit their personal ways of thinking.

People who make tools make them to mirror their own mental processes. This is not so much a deliberate choice as a series of biases, often expressed unconsciously. The ways a smith thinks push toward ways of seeing problems and therefore toward ways of solving those problems.

Ways of thought are partially idiosyncratic, a matter of personality, and partially a matter of the ways one has been taught to think. (Yes, it's both nature and nurture. Get over it.) While elements of this apply to users as well as makers (since one tends to use tools one is comfortable with), it stands out more strongly in the process of making. The more complex the tool one is creating, the more its workings will conform to the ways of thinking of the maker.

But there is a limit to the eccentricity that a maker can bring to a tool. If the tool is only to be used by the person making it, then it can be outfitted to that person's thinking completely. One can see this sort of thing in a person's private workspace. Such a place is a tool made to make the work one does easier. Cooks arrange their kitchens to their convenience, people with the freedom to do so arrange their offices to their needs, artists arrange their studios. These are circumstances where the person regardless of his or her job is acting as a smith, making the tool of his or her workspace.

Sir Isaac Newton was once noted for making his own tools, and supposedly said, "If I had staid for other people to make my tools and things for me, I had never made anything." Note that he is also one of very few physicists to design his own mathematical tools.

If, on the other hand and the other and the other, the tool being made is meant for many hands, then the maker has to determine how to make it practical for others to use. To do that the maker has to think to some extent like a user, to consider the characteristics of efficacy and ease of use. This can be strangely difficult. A tool made according to the way one thinks will seem natural in its uses and clearly easy to employ. This issue arises constantly in the notion of "user friendly" computer software.

In making and then handing a tool over to another, a question may arise that stumps the smith.

"How does it work?" the knight asks.

The knight, seeing the thing for the first time, has much that needs to be learned. The smith may know the answers, but only intuitively. The tool may fit the smith's mind so perfectly that the question of how it works sounds as nonsensical as asking how one breathes.

The question of what is necessary to learn its use, what practices show the way, what difficulty there is in fitting hilt to hand, and revealing how blade does its work, these can all be blind spots to the smith.

A smith can also have difficulty seeing ways that might improve the tool, if those improvements require ways of thinking the maker is not comfortable with. A person who flies everywhere might try to find an air route from point A to point B that might involve five different planes, and might not notice a route that flies from A to C, then takes a short train from C to D, and then a plane from D to B. If the person's ways of thinking were confined to air travel, the alternatives might never be considered.

The biases a smith brings to their toolmaking come both from individual personality and how they came to be a smith.

There are three basic sources of smiths: smiths trained by smiths, dissatisfied knights, and smiths coming in from the outside.

Smiths who are trained by smiths embody one of the important consequences of the way-of-thought principle of toolmaking. For such smiths the teaching of particular ways of thought is the most vital element in the growth of particular fields of toolmaking and tool use.

If makers are selected from among those whose ways of thinking most fit the current tool sets, then they are most likely to expand those toolkits and refine the currently available tools by looking at the paths currently being followed and seeing what shortcuts might be available, or thinking up new ways to use the old tools, or tweaking a blade here, a handle there.

Dissatisfied knights who become smiths do so because they are frustrated enough with the tools they are given that they try to learn the making of them in order to bring about innovations of better use. Dissatisfied knights are not really outsiders to the field, they are simply switching perspectives. They bring the prejudices of the users over to the smiths.

Sometimes they discover that what they want in their tools is not really possible. "You mean I really can't make a sword that will fight for me?" And sometimes they bring vital considerations. "Your hilt design means we have to break our knuckles to parry."

The outsiders who become smiths are usually either crackpots or people who bring an outside perspective that sees a concern with the tools unmet by those who grew up within the traditions.

There will, of course, be those among the makers of all sources who tend to strike out in their own directions, creating tools as yet unseen. But if they are selected and educated in the particular ways currently in fashion, they will still be striking out from previously laid-down paths. They will, mostly, innovate from current usage or against current usage (that is, they will create new tools derived from seeing the flaws or opportunities in the ways the present tools work). Rarely will they create entirely new ways that have no obvious or direct connection to the toolkits presently known and employed. Rarely – not never, just rarely.

## Beauty in the Use of Tools

If people chose tools "rationally," then users would pick tools based on efficacy for the job at hand and makers would make them to produce maximum efficacy. But in reality that is not how things are done. People tend to choose tools that are comfortable for them one way or another. They will choose tools with the least learning required (often because they are variants of tools they already are comfortable using), or from sources they trust or which they like the look of. As noted, makers will usually make tools in conformity with their ways of thinking. In both the case of the maker and the user, there is a guide to their decisions that has little or nothing to do with the practicality of the tool. That guide is the beauty of the tool.

Users tend to see the beauty of a tool's appearance or some characteristic they have learned to think of as beautiful.

*You mean the way powerful engines appeal to be–*

**Careful. We don't want to be accused of gender bias or cynicism.**

*Appeal to a certain segment of the population.*

This kind of appeal can be a good thing. If one's sense of beauty for tools is tied to their utility, then that sense can be a good guide. A chef who likes the feel of a knife or a knight who likes the balance of a sword is using such a properly aligned sense of beauty.

But it takes training to align ones sense of beauty to actual utility. People who have not learned to see this tend to fixate on the trivial or the external.

*Buy this. It's got cool decals. Buy this. It's covered in chrome. Buy this. It's got a designer label.*

But even for those properly aligned in seeing, there is another sense of beauty which can get in the way of picking or making a good tool. The very ways of thinking that have been taught, the sense of rightness that guides good choice, can fail if the idea of what is right does not extend to the tools that would actually do the job.

This last is particularly tricky, since people taught to do a certain job or to employ or make certain kinds of tools develop traditions of right and wrong and pass those traditions on. Their students accept the traditions, with some modifications, and pass them on. This works well as long as the tools and methods are appropriate for the work being done. If they cease to be so, the traditions become an impediment to doing the work.

More problematically, the traditionalists select as students, assistants, and heirs those who are best capable of taking in and following their traditions. Again, as long as the tradition works, this is just fine. But eventually the tradition will reach its limits.

Then a range of possibilities emerge: either to change the tradition, abandon the tradition, or say that what lies beyond those limits is not part of the job. In various circumstances any of these might be the right decision. But figuring out which is the best course is difficult, and if those considering the matter regard the tradition as the only way they are unlikely to make the right choice unless there is some external force making them do so.

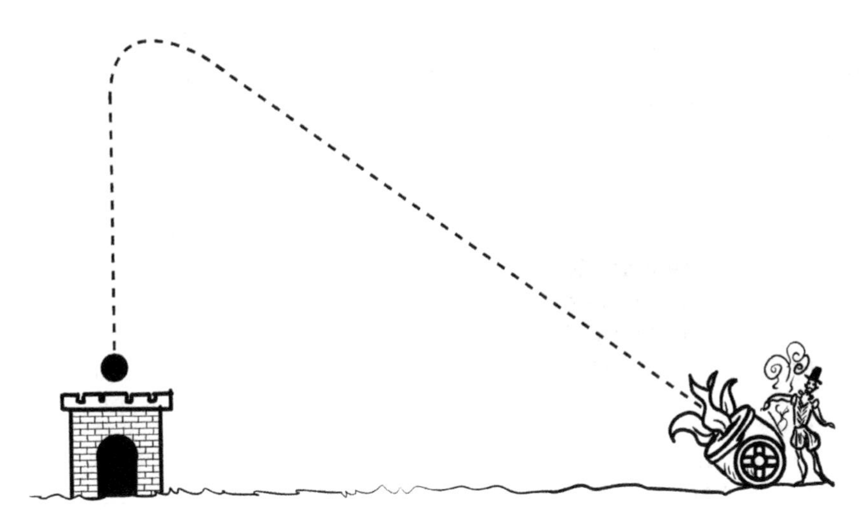

**FIGURE 12.2**
Aristotelean Cannon Shot

*Sort of a social version of Newton's First Law of Motion.*

**Ironically so. It was just such a tradition and change that led to Newton, and the force was the force of guns.**

For a long time the accepted physics in Europe was the physics Aristotle expounded without ever having done an experiment. In this physics, earthly objects moved in straight lines and tended to slow down to a stop. A cannon shot in a world according to Aristotle would go as shown in Figure 12.2.

By the time cannon fire became accurate enough to be used in siege warfare, the cannoneers had determined that this was not the true path of deadly metal flight. The cannoneers worked for people with guns and swords who demanded (sometimes with drawn sword) a knightly practicality in the outcome of the shots. As a result, there was a fair bit of force put in place to create a new physics of motion, which culminated in the cannon functions we have already detailed. One can argue as to whether the cannoneers were dissatisfied knights who became mathematical smiths or whether they were complete outsiders, but regardless, they pushed for the innovations that eventually led to the Newtonian universe.

Had it not been for this practical consideration, it is very hard to say how long the false Aristotelian beauty would have clouded reality.

## Beauty in the Making of Tools: Artists and Artisans

Makers tend to see the beauty in the way a thing works and will tend toward making tools that fit that sense of beauty.

*That sounds good. They extend the ways, so the tools will do a better job.*

**Not necessarily.**

A swordsmith might be fascinated by the springiness and flexibility of blades to the point of a creating a sword that is so springy and flexible as to be next to useless in real fighting. For example, a fencing foil.

*But people use fencing foils.*

**For sport, where they're not trying to kill people, which, if you remember, was the original purpose of swords.**

The point is that tools created this way may not be useless, but they can certainly have drifted far from their initial purpose. New uses can be found for the new tools, but that does nothing to ameliorate the fact that they can't serve their original function.

*You're not claiming that foils were created this way.*

**No, I just swiped them as visual metaphor. Remember, we're mostly talking about abstract mental tools, not physical ones.**

There is a tension in the mind of a smith, two overlapping ways of thinking, both of which are needed in a good smith. They are the artisan and the artist. The artisan's concern is the practical use of the tool made. The artist's concern is the way of the use and the beauty of the way.

Both exist in the smith's mind because both spring directly from the contemplation of the way of the tool. Some smiths see themselves purely as artisans and do not see that they still judge the tools they make in aesthetic terms. Others see themselves purely as artists and reject, sometimes resent, the idea that their creations serve purposes.

Many of the great master painters and sculptors of long ago were evenly balanced between the two. Their artworks were made for the purposes of their patrons. The Sistine Chapel ceiling, for example, was done not as art for art's sake but to make vivid the stories the Catholic church wanted to teach.

Modern and postmodern artists, on the other hand, have tended to see art as an end in itself and rarely concern themselves with what purpose a mind that perceives the art might use it for.

## Mathematics as Toolkit

Our metaphor of smiths and knights is an abstraction of common ways and difficulties humans have in dealing with tools. It is, like an algebraic structure, a generality meant to illuminate the shared aspects of a particular situation. We reached that abstraction by leaving behind the basic identifications:

Mathematician = Smith

Scientist = Knight

Having explored the abstract realm, it's time to go back to the particulars of this problem and examine them in light of what we found through that abstract exploration.

Math is a collection of tools used by various professions. Scientists place the highest demand on the tools and are the ones in need of the most sophisticated tools. Mathematicians are the toolmakers, the ones who create and refine the tools.

Therein lies the fundamental gap between science and math, the gap between the user and the maker. Please note, the same person can be knight and smith; it's not as if having one way of thinking excludes the other. Newton, as noted above, was one such. Possession of both ways often makes one better at both. Of course, it also makes one susceptible to the distractions of both. In addition, it is rare to have the skill to do both well.

In the history of mathematics and science, particularly physics, we can see the ways in which the two ways of thinking intertwine. There are two sources for the development of new tools: the needs of knights and the exploration of smiths. We will look at this process in more depth in the next chapter, but a little summary won't hurt.

The real world presents numerous things that people have sought to understand and to act upon. These things create a pressure for tools and theories by which they can be understood and acted upon. This pressure creates regions in the theoretical universe.

By way of example, let us consider the region of motion. Theories of movement are some of the oldest theories we would call scientific, and the study of movement has been of both theoretical and practical concern.

This region brought forth a number of tools, and each tool created its own pressures for more tools. The measurement of movement led to the geometry of movement, which led to the analytic geometry of movement. Ideas of straight-line motion led to ideas of curved motion, which led to calculus, which led to differential equations.

This pressure to develop tools was not a matter of one user coming to one smith and asking for a specific tool. Few tools in any field are developed like that. It's much more common for a problem to be floating around, known to everyone interested in the field, with various smiths interested in it trying to come up with tools or applications of tools to solve it.

Meanwhile, while the users are creating pressure for tools to fit problems, the smiths are fiddling around with the applications and consequences of the tools that already exist. Because of the pursuit of beauty, some of these paths wander far away from anything that might seem useful. Number theory, topology, group theory, and various other esoteric pursuits have no clear connection to the world (or at least had no such clear connection to the world at the time they were first studied). But sometimes these esoteric explorations come back in surprising ways. For example, group theory has become very important in quantum mechanics.

There is, it must be admitted, a certain artistic pride in creating something that seems beautiful, fascinating, and useless, and a subtle letdown in discovering that one's abstract, unworldly work actually connects to reality.

*That's just weird.*

**Yes, but when you are spinning in the dizzy heights of beauty, to be told that there's a practical side can seem a little shameful.**

It's like a character in a romance novel besotted with someone for their grace and appearance who then discovers that the person is wealthy or has important connections that might help the lover's career. It gives a certain, um …

*Sordid quality?*

**That's it.**

*I'm sure the mathematicians got over it as easily as the lovers who discovered they had wealth to go with the beauty.*

**I can't tell if that's practical or cynical.**

Moving on. It may seem like a strange coincidence that a connection between a tool created purely for beauty and a practical use should appear out of the rarified air of the theoretical universe. And it would be, if the math being done did not ultimately derive from the real world in the first place. No matter how abstract or deeply theoretical, all of our mathematics is built on a foundation of earthly sticks and stones. We should not be startled if in following circuitous paths we find ourselves back in the real world once in a while.

## Using the Toolkit

The surest signs of skill in any field is the ability to pick out and apply the right tool or tools for the job at hand. This is also the ability that creates the impression that a job is easy. One watches skilled users pull things out of their boxes or their minds and swiftly apply them, and one is amazed that the task they are doing is even considered work, until one stops and thinks and realizes that one does not have any idea oneself which tools are the right ones or how to use them. Then the opposite impression arises that the field is impossible for anyone who is not somehow naturally gifted in the field.

The first impression, that anyone can do it, is common among fans of music who play air guitar and dream of themselves on stage. The second, that only the talented need apply, is common among watchers of science and mathematics who think that it's solely the province of those weird people in tweed jackets, white coats, and t-shirts with obscure science fiction jokes on them. But of course neither of these views is correct.

There are two skills in tool use: the knowledge of what tool applies to what situation and the understanding of how to apply the tool. In certain aspects of science, these jobs can be done by separate people (for example, one scientist who understands a theory might come up with a concept for an experiment to test a hypothesis, and another scientist who understands the equipment involved would construct the equipment and carry out the experiment). In applying mathematics to science, it's generally regarded as necessary to be able to do both of these. One must be able to figure out what mathematical model applies to a situation and then do the actual application.

These two skills are commonly taught simultaneously in the following manner: first a mathematical procedure is taught (such as solving quadratic equations) with enough problems to develop facility with the process, then a batch of "word problems" are given, which artificially create situations in which a pseudo-real-world situation is meant to be turned into this kind of process (for example, a distance, rate, and time problem that can be turned by abstraction into a quadratic that needs to be solved for time) and then solved.

This process of learning tends to go on all through elementary and high school math. The next step in learning these skills (or filtering out those who cannot learn them) usually occurs in calculus. Here one is confronted with having to find derivatives and integrals of various functions and one is taught a variety of methods of doing so. Now the student has to figure out which tools apply to the particular problem and how to apply those tools in the more abstract situation of working not upon abstractions that model the world but abstractions that modify the abstractions that model the world.

This is a more subtle shake-out point than the earlier barriers for those who can do the math needed for science. Here the struggle comes not in understanding concepts but in picking out tools. Survival in calculus classes is the first part of the acid test for this necessary tool in the branches of science that rely strongly on math.

*You mean physics.*

**Survival in differential equations is the second.**

*Definitely physics.*

Those who do survive generally find that practice gives them greater ease in figuring out when to apply a tool. They may have trouble learning the tools that follow, but they have developed the mindset that looks at a problem and tries to find the tool that fits it, then uses the tool. This is a vital mental process for connecting the mathematical aspects of the theoretical universe to the observable and detected universes. It's also the tool that permits scientists to refine the tools they are given, or to bug mathematicians (either their inner mathematician or the person who works down the hall) to refine those tools or to build new ones to solve problems that there are as yet no tools for.

## Expanding the Toolkit

While the student of science (certainly physics but also chemistry and biology) learns what tools to apply and how to apply them in math classes, a student of math is learning something different in the selfsame classes. The two students will be studying the same books, but they will focus on different aspects. Where the science student will mostly see the means to solve the problems, the math student will focus on the manner in which things are proved.

Recall from the calculus chapter, we gave a general description of what a proof for figuring out a derivative using limits is like.

*Yeah, yeah, you do all the calculations first, then take the limit.*

All the proofs follow that rubric, and a budding mathematician will see that. The proofs and the ways of proof will be that person's main area of interest. The same problems will still be done and the same methods learned that the budding physicist learns, but they will be seen as of less importance than calculus as an idea.

Mathematics professors sometimes say that math begins after calculus, and there is a certain truth to that. In every postcalculus math class either of us has had, with the exception of differential equations, the emphasis had shifted over from problem-solving justified by proofs to problems that were themselves proofs. The change was an insidious one. No more did we care about what we got as a result. What mattered was how we got there and how carefully we got there. From that point onward math classes were not about the uses of math but about the math itself.

Beyond calculus the task of the student is to learn enough of the ways math is done, as well as to gain an understanding of what has been done in the past in order that one might eventually be able to expand the understanding of the ways of math. The apprentice smith learns to create new tools or explore old tools in order to make them more versatile or apply them elsewhere or simply to create and refine new and related beautiful mathematical forms and proofs. The classes taught cover the currently extant mathematical theoretical universe, showing doors into different areas one might want to specialize in should one, as is expected, go on to graduate school in mathematics and eventually do research.

Mathematics is not presented to mathematicians as a toolkit. It is seen as a thing unto itself, a purer world in which things make logical sense and can be made to make sense if one has the wit and skill to learn and to explore. Beyond calculus there is little attention to the aspects of math that connect to the sciences. And yet, despite this disconnect, the work being done does connect in enough places that rarified areas of mathematics still provide tools to scientists.

The divergence of views that happens during and after calculus is largely responsible for the separation between mathematicians and scientists. It is true that their views of tools are fundamentally distinct, but it's not true that one mind can only have one of these views. Every human is a tool user (even if the tool is nothing more than the TV remote control) and every human is a tool maker (sometimes the tool is just a sandwich that fits one's tastes in sandwiches, but nevertheless it is a made tool). Training someone to be a skilled tool user in one area or a skilled tool maker in one area need not be a means of separating one person from another.

Furthermore, the classes that seek to perform this separation create a third group as well: those who feel they can be neither and simply disappear from math and science altogether.

That might sound acceptable if the goal is solely the creation of mathematicians and scientists who regard each other with amused bemusement, but in that third group are those who will regard both groups with distrust and alienation.

There's no reason why the teaching cannot produce smiths, knights, and people who have some understanding of the ways of smiths and knights without their devotion to the sword.

Indeed, this understanding is the living heart of the lost inheritance. Mathematics is a mental endeavor, specifically a human mental endeavor. While its works are presented as outside of human thinking, as existent on their own, it is only in human minds and human works that math as math exists. We can argue philosophically about whether or not math exists in the laws of the universe, but even the most ardently devoted adherent of that view would not say that the math as we see it and understand it is what exists in the world. It's not as if the equations are graven into the universe. They may underlie how it works, but they weren't written down somewhere in the underpinnings of reality.

Mathematics as understood is a human thing. And therefore the ways in which humans develop it and explore it, as well as the perceptions and blind spots they have in regard to it, are a vital part of understanding what it is and how it grows over time, passing from mind to mind and becoming a shared but unique to that mind perspective in the person who has just learned it.

It is as unwise to see mathematics as if it were divorced from human thinking as it is to see a sword in a museum and not know that it was made by a human and used by another human in human ways for purposes both human and inhumane.

It's not just that mathematicians aren't aliens. It's that you can only really understand the math humans do by seeing their humanity and the human purposes of their tools.

# 13

## *Building the Theoretical Universe*

The invention of non-Euclidean geometry marks a parting of the ways between math and physics. Before the work of Gauss and Lobachevsky, mathematicians saw themselves as searching for the "right" axioms and working out their consequences, much as physicists saw themselves as searching for and working out the consequences of the laws of nature.

But Gauss and Lobachevsky made mathematicians realize that they had complete freedom in the choice of axiom systems to investigate. Math could be whatever the mathematicians wanted it to be as long as the axioms were stated clearly and their consequences worked out logically: mathematics need not have anything whatsoever to do with nature – it was a formal game to be played by mathematicians for their own amusement and edification.

This newfound freedom did not result in complete chaos: the systems investigated by mathematicians were abstractions and generalizations of the previous mathematics, which was in turn motivated by the study of nature. However, the connection of mathematics to nature was becoming ever more tenuous.

Or so it seemed. But then in the 20th century something remarkable happened. The new physics theories of relativity and quantum mechanics needed to make use of much of the exotic mathematics that was developed by the mathematicians for their own purposes.

This break in physics was so radical that pre-20th-century physics is now called "classical physics" with relativity and quantum mechanics termed "modern physics."

We have already treated much of classical physics in Chapters 6–8. In this chapter, we will complete our coverage of classical physics with two topics: fluids, and electricity and magnetism. In Chapter 15, we will cover modern physics.

**Well, looks like this is where we part ways. I'll be over here wandering through all possible worlds, while you're slogging around in this four-dimensional backwater.**

*Not so fast. We're going to make use of the infinite dimensional function spaces and the different axiom systems, and you're going to have to keep coming back to this spacetime since you all live in it.*

**Oh sure, but I can fictionalize the others and play around out there as much as I want.**

*But the people you're writing the fictions for live here too. So get back here and get your feet wet.*

### Fluids

Liquids and gasses behave in very similar ways: so much so that they are collectively termed "fluids" and their study in physics is called fluid dynamics.

That may sound weird because they don't feel the same to us and we have trouble swimming in air or breathing water. But that's the observable universe. In the detected universe they behave similarly and so in the theoretical universe we model them the same way.

At each point of space a fluid is characterized by three properties we have encountered earlier: density $\rho$, velocity $\mathbf{v}$, and pressure P. Density (mass per unit volume) and velocity (how fast the fluid is moving and in what direction).

Pressure has to do with the forces that fluids exert on their surroundings. When a fluid is confined, it exerts a force on the walls of its container. The pressure is the force per unit area exerted by the fluid on the walls of the container. Each little piece of fluid is surrounded by other fluid, so the surrounding fluid exerts forces on it. But these forces are in different directions and completely cancel out if the pressure is the same on all sides. It is only if the pressure is a little stronger in one direction that the fluid feels a net force. This is similar to the case of the stretched guitar string: each little piece of string feels a small net force due to the fact that the string pulling on either side of it pulls in nearly, but not exactly, opposite directions.

We have several senses that register pressure. Our sense of touch is largely concerned with pressure and our sinuses (especially if stuffed) can feel air pressure. Our sense of hearing also works in part by the effect of waves of air pressure on our eardrums.

**Oh goody, Bessel functions.**

*Not right now. We have other equations to solve.*

So just like in the case of the guitar string, we should apply F = ma to each piece of fluid and get differential equations that describe how ρ, **v**, and P change with time. However, all the partial differential equations we have dealt with so far (wave equation, heat/diffusion equation) have been for the behavior of scalar quantities (guitar string height, temperature, concentration). The velocity **v** is a vector, so we're going to need a little more mathematical machinery to describe how to use vectors in partial differential equations.

We will begin with a way of "multiplying" vectors called the "dot product." Recall that a vector **A** is written in boldface, with its magnitude A denoted by the same letter without the boldface. We can also denote the magnitude by $|\mathbf{A}|$. Now, for any two numbers a and b, we have

$$(a+b)^2 = a^2 + b^2 + 2ab$$

which we can solve for ab as

$$ab = (1/2)\left((a+b)^2 - a^2 - b^2\right)$$

The idea of the dot product is to do the same thing with vectors. That is, for any two vectors **A** and **B**, we define the number **A·B** by

$$\mathbf{A}\cdot\mathbf{B} = (1/2)\left(|\mathbf{A}+\mathbf{B}|^2 - A^2 - B^2\right)$$

This is a very weird form of multiplication. Normally, we think that if we multiply two like things, we will end up with another thing of the same kind. Multiply integers to get integers, real numbers to get real numbers, etc. But here we multiply two vectors to get a scalar.

We can argue that we already have a weird vector multiplication wherein we multiply a scalar by a vector to get a vector, but this is a different kettle of fish.

The reality is that the dot product gives vital information about the geometric relationship between the two vectors. Start by drawing the vectors **A** and **B** starting from the same point, and note that together they make an angle θ (Figure 13.1).

Then, the geometric formula for the dot product is

$$\mathbf{A}\cdot\mathbf{B} = A\,B\cos\theta$$

$$\text{Or } \cos\theta = \mathbf{A}\cdot\mathbf{B}\,/AB$$

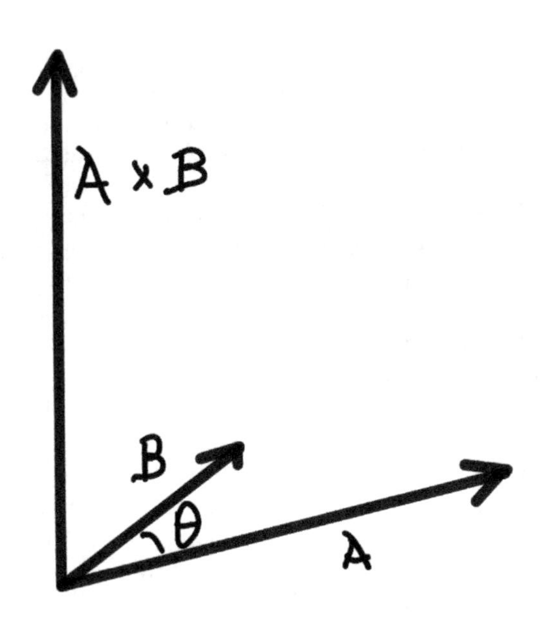


**FIGURE 13.1**
Vectors with Angle Between

In other words, we have a way to find the angle between two vectors using only fairly simple calculations (they'll get even simpler in a moment). That's great, and if we had written it as a function of two vectors instead of as a form of multiplication, there would be nothing to complain about. But this is the symbolism we're stuck with, so gripe away if you feel like it.

There is also a formula for the dot product in terms of components of the vector:

$$\mathbf{A} \cdot \mathbf{B} = A_x B_x + A_y B_y + A_z B_z$$

Now that's easy to calculate, but it contains something curious. The lengths of the vectors and the angle between them don't depend on the basis we're using, but the components do. So, this very simple component-based calculation is actually not dependent on which basis we're using. It's invariant under change of basis, which means that we can use whatever basis is convenient to us.

We're now going to introduce some ways of taking derivatives of functions and vectors called divergence and gradient, also known as div and grad. The basic idea is to treat the partial derivatives $\partial/\partial x$, $\partial/\partial y$, and $\partial/\partial z$ as though they were components of a vector. This is one of those tricks of formalism that can cause headaches in people learning the math and physics. We've written out vectors as ordered lists of numbers. Now we're making an ordered list of operators $(\partial/\partial x, \partial/\partial y, \partial/\partial z)$ that looks like a vector but isn't so (although there are ways in which it is).

From a function f, we produce the vector $\nabla f$ f (gradient of f or grad f) whose components are $(\partial f/\partial x, \partial f/\partial y, \partial f/\partial z)$.

From the vector **A,** we produce the function $\nabla \cdot \mathbf{A}$ (divergence of **A** or div **A**) given by

$$\nabla \cdot \mathbf{A} = \partial A_x / \partial x + \partial A_y / \partial y + \partial A_z / \partial z$$

Note that the operator $\nabla^2$, which we used in the wave equation and the heat equation, is just the divergence of the gradient: that is, $\nabla^2 f = \nabla \cdot \nabla f$.

We've called these derivatives, but we know the meaning of derivative as the rate of change of one quantity with respect to another.

What do gradient and divergence mean? Their names are evocative. Gradient is a term related to slope (i.e., derivative), and divergence implies that there is something to diverge from. So what do these terms mean?

To get a feel for gradient, let's go back to an example we used in the chapter on non-Euclidean geometry: a curved surface given by $z = h(x, y)$, where x and y are the usual coordinates in the plane and h is the height above the plane. Let's consider the two-dimensional vector $\nabla h = (\partial h/\partial x, \partial h/\partial y)$. This vector points in the direction in which the slope upward is steepest, and the magnitude of $\nabla h$ is the steepness of that slope. So the gradient really is a type of slope. Another way to think of $\nabla h$ uses the dot product. Let's pick a two-dimensional vector **s** of unit length. Then, the quantity $\mathbf{s} \cdot \nabla h$ is the slope of the curve that we would get by going along the surface in the **s** direction. In other words, $\nabla h$ contains the information that we need to find the slope in any direction.

To understand divergence, we return to the fundamental theorem of calculus. Or rather we get explicit about the claim that we made in Chapter 6 that there are other more complicated versions of the fundamental theorem. One of these more complicated versions is called Gauss' theorem and is written as follows:

$$\int_V \nabla \cdot \mathbf{A} = \int_S \mathbf{A} \cdot n$$

This theorem says that if we have a closed surface S, and if V is the volume enclosed by S, then for any vector **A**, the integral over the volume of the divergence of **A** equals the integral over the surface of the dot product of **A** with the unit vector **n** that is perpendicular to the surface. Or put another way, the integral over the volume of $\nabla \cdot \mathbf{A}$ is the integral over the surface of the outward-pointing component of **A**. Gauss' theorem is like the fundamental theorem of calculus because it relates the integral over a region (in this case the volume V) of a derivative (in this case $\nabla$ **A**) to the behavior on the boundary of the region (in this case the surface S) of the quantity whose derivative we're taking (in this case **A**).

Now let's apply Gauss' theorem to the case of a fluid. If we pick the vector **A** to be $\rho\mathbf{v}$, then $\int_S \mathbf{A}{\cdot}n$ is the rate (in kilograms per second) at which fluid is flowing out of the volume V. Since this is also $\int_V \nabla{\cdot}\mathbf{A}$, we find that $\nabla{\cdot}\mathbf{A}$ is the rate per unit volume of outward flow. Of course, if **A** is not $\rho\mathbf{v}$, then we cannot attribute that particular physical meaning to $\nabla{\cdot}\mathbf{A}$. Nonetheless, an intuitive way of thinking about what $\nabla{\cdot}\mathbf{A}$ means is that it is the rate per unit volume of outward flow for whatever quantity **A** is the flow vector of.

This is extremely important from an experimental and engineering perspective. The rate of outward flow (i.e., the rate at which a fluid is leaving something like a spigot, for example) can be measured. Using Gauss' theorem, we can relate that outward flow to what's going on inside the pipes and cisterns of hidden plumbing.

Now, with all this machinery of vector calculus, we can finally write down the equations of fluid dynamics, which are the following:

$$\partial\rho/\partial t + \nabla\cdot(\rho\mathbf{v}) = 0$$

$$\rho\partial\mathbf{v}/\partial t + \rho(\mathbf{v}\cdot\nabla)\mathbf{v} = -\nabla P$$

These equations, called Euler's equations, are actually a simplified model of fluid dynamics, so before we look at solving them, we will discuss what simplifying assumptions are being made.

Some fluids, like water, flow quite freely, while others, like honey, tend to flow in a more sluggish manner. This quality of flowing sluggishly is called viscosity. Euler's equation leaves out the viscosity and is thus assuming that viscosity is small enough to be negligible. Euler's equations do not have any expression for $\partial P/\partial t$, because they assume that P is a known function of $\rho$, so that if we find how $\rho$ depends on time, we will also have found how P depends on time.

To see why pressure and density depend on each other, imagine taking a fluid and putting it under increased pressure. This will tend to compress the fluid, so that a given amount of fluid occupies a smaller space and therefore has a higher density. Nonetheless, the assumption that P is a function of only $\rho$ is a simplification, because in a more general situation pressure is likely to depend on temperature as well.

Euler's equations look complicated, but conceptually they are fairly simple. The first equation simply says that fluid is not being created or destroyed but only moved around. This is easier to see if we rewrite the equation as

$$\partial\rho/\partial t = -\nabla\cdot(\rho\mathbf{v})$$

Since $\nabla\cdot(\rho\,\mathbf{v})$ is the rate per unit volume at which fluid is flowing outward, this means that $-\nabla\cdot(\rho\,\mathbf{v})$ is the rate per unit volume at which fluid is flowing inward. Thus, within a given small volume, $\rho$ increases only because fluid is flowing into the volume and decreases only because fluid is flowing out of the volume.

The second Euler equation is really just $\mathbf{F} = m\mathbf{a}$ in disguise. The momentum $\mathbf{p}$ of an object is $\mathbf{p} = m\mathbf{v}$, so another way of writing $\mathbf{F} = m\mathbf{a}$ is $F = d\mathbf{p}/dt$.

The momentum per unit volume is $\rho\,\mathbf{v}$, and the second Euler equation says that there are two ways for the amount of momentum in a little volume to change: momentum can flow into (or out of) the volume, or the momentum can change due to the action of a net force that exists because the pressure has a gradient.

There are a number of equations in physics that form elaborations of more basic principles. These elaborations produce more complex equations that are more precise models for particular situations.

So how do we solve Euler's equations?

We're looking at pressure changes within a fluid or gas. We know that we experience such things as sound-like, so maybe the mathematics of sound will work for us.

This tempts us to use Fourier's trick, but we immediately run into a problem: Fourier's trick involves finding the behavior of each individual note and then adding up the notes. But this assumes that adding two solutions results in a solution. Differential equations are called linear if we can add solutions to get a new solution, and are called nonlinear equations if we can't. The wave equation and the heat equation are linear. Euler's equations are nonlinear.

But we can get a linear equation from Euler's equation by using the Taylor series trick: we start with some known solution and then write a Taylor series for the equation keeping only the first two terms of the Taylor series. This is the analog of approximating a function by its tangent line, and it gives us a linear differential equation.

We will choose as the known solution $\rho = \rho_0$, where $\rho_0$ is a constant and $\mathbf{v} = 0$. Then, our approximation is $\rho = \rho_0 + \delta\rho$ and $\mathbf{v} = \delta\mathbf{v}$, where $\delta\rho$ and $\delta\mathbf{v}$ are small quantities. The Taylor series linear approximation for Euler's equation then gives

$$\partial\delta\rho/\partial t + \rho_0\nabla\cdot\delta\mathbf{v} = 0$$

$$\rho_0\partial\delta\mathbf{v}/\partial t = -(dp/d\rho)\nabla\delta\rho$$

The $dP/d\rho$ comes from the fact that P is a function of $\rho$ and we are expanding everything in a Taylor series. We know that compressing fluids makes them more dense, so $dP/d\rho$ is a positive number. We will write this number as $c_s^2$ for reasons that will become clear in a little while. We start by working out the dimensions of this quantity. The dimensions of $dP/d\rho$ are pressure/density = (force/area) divided by (mass/volume) = (force/mass)(v olume/area) = (acceleration)(length) = (length/time)$^2$ = (velocity)$^2$. Therefore, $c_s$ has dimensions of velocity.

We will now apply Fourier's trick to these linearized Euler equations. But having learned something about complex numbers, we are now in a position to apply an even more user friendly version of Fourier's trick: our individual notes will be of the form $\exp[i(\mathbf{k}\cdot\mathbf{r} - \omega t)]$. Here, $\mathbf{k} = (k_x, k_y, k_z)$ and $\mathbf{r} = (x, y, z)$ so that $k\cdot r = k_x x + k_y y + k_z z$. Applied to such notes, $\nabla$ becomes multiplication by $i\mathbf{k}$ and $\partial/\partial t$ becomes multiplication by $-i\omega$. Thus, we look for solutions of the form $\delta\rho = \rho_1 \exp[i(\mathbf{k}\cdot\mathbf{r} - \omega t)]$ and $\delta\mathbf{v} = \mathbf{v}_1 \exp[i(\mathbf{k}\cdot\mathbf{r} - \omega t)]$ for some (complex) constant $\rho_1$ and some (complex) constant vector $\mathbf{v}_1$.

On the face of it, this doesn't seem like a good strategy for finding solutions of Euler's equations. We'll get solutions, sure; but they will be complex solutions. Fluids have density, pressure, and velocity that are real numbers, not complex ones. So it seems like we will find mathematical solutions of the equations that have no chance of describing nature.

But appearances are deceiving in this case. Once we have the complex solution, we can take its real part, and that will still be a solution of Euler's equations. The exponentials of imaginary numbers will then give rise to the sines and cosines that we are used to from Fourier's trick.

Putting our Fourier note into Euler's equations, we obtain

$$-i\omega\rho_1 + \rho_0 i\mathbf{k}\cdot\mathbf{v}_1 = 0$$

$$\rho_0\left(-i\omega\mathbf{v}_1\right) = -c_s^2\rho_1\left(i\mathbf{k}\right)$$

Solving the second equation for $\mathbf{v}_1$ we find

$$\mathbf{v}_1 = c_s^2\left(\rho_1/\rho_0\right)\left(\mathbf{k}/\omega\right)$$

This means that we can pick $\rho_1$ to be anything we want and the equation gives us $\mathbf{v}_1$. Furthermore, since $\mathbf{k}$ is the direction that the wave is moving in, the solution tells us that the fluid velocity is in the direction of the wave motion. Such waves are called longitudinal, as opposed to waves like the guitar string (which are called transverse waves), where the motion of the string (up and down) is perpendicular to the motion of the wave (along the string).

Putting the solution for $\mathbf{v}_1$ in the other Euler equation, a little algebra gives us

$$\omega = kc_s$$

So once we pick k, Euler's equation tells us what $\omega$ is. But for wave motion, $\omega/k$ is the speed of motion of the wave. So these waves of disturbances of the fluid must move with speed $\sqrt{dP/d\rho}$.

These waves of fluid disturbance are the sounds that we hear in the media. When we speak, we disturb the pressure of the air, and these disturbances propagate outward in all directions, in accordance with the mathematics of Euler's equations. If we know how the pressure depends on density, we can figure out how fast the sound will go.

Fluid dynamics is a very impressive use of classical physics. Remember that we started with Newton's laws of motion, which model the motion of individual objects. Fluids are the epitome of nonindividual stuff. They are one of the root causes of "How much?" rather than "How many?" questions.

But the basic trick of calculus (break big up into small, operate on small, and then go back to big) works very well here. We can treat a little bit of fluid as if it were an object, apply Newton's laws, and then scale up to get equations covering the whole how muchness of the liquid.

Fluid dynamics is a convenient methodology to model many different real-world phenomena. It's mathematically lazy and works by abstracting diverse things into one model.

We're now going to turn to something that developed in the opposite direction. Physicists used math to prove that many diverse phenomena that seemed to need different models because they behaved unlike each other were actually the same underlying physical process.

## Electricity and Magnetism

From ancient times on there were a number of strange phenomena that people had no scientific explanation for (lightning, lodestones, some metal objects being attracted to others, amber and cats giving off sparks, etc.). We could spend a very long chapter describing the processes and experiments that lead to the basic math that turned into the involved math to follow.

The most important early result was the hypothesizing of the concept of electric charge. Charge was kind of like mass in that an object would have an amount of charge and two objects would affect each other's movements based on how much charge each had and how close to each other they were. But unlike mass, charge could be either positive or negative, and two objects with like charges would repel each other, whereas those with unlike charges would attract each other.

We now turn to the properties of electricity and magnetism. Electric charges exert forces on each other given by Coulomb's law, named after 18th-century French physicist and military engineer Charles-Augustin de Coulomb.

$$F = k_e |q_1 q_2| / r^2$$

Here $k_e$ is a constant, $q_1$ and $q_2$ are the charges, and r is the distance between them. Since charge can be either positive or negative, the force is a repulsion if the charges have the same sign and an attraction if they have opposite sign.

Magnets also have a force of attraction and repulsion. But what they work on was kind of strange. Each magnet has two poles (called north and south). The north pole of one magnet attracts the south pole of another and vice versa. Magnets also attract some kinds of metals and are attracted by the Earth or at least the Earth's magnetic poles.

That these two forces were related was by no means obvious. Indeed, the phenomenon required that people first develop ways to generate electricity and convey it along wires.

Once those advances were made, a shocking (insert the rest of the obvious jokes here) discovery was made.

A magnet exerts a force on an electric charge, but only when the charge is moving. Electric current (charges moving in a wire) exerts a force on a magnet. A magnet sitting near a coil of wire does nothing; but if the magnet is suddenly moved, an electric current will start to flow in the coil. Clearly, there is a connection between electricity and magnetism, but what is it and how do we make sense of it?

Michael Faraday, a 19th-century English physicist and chemist, conceived of electric charges and magnets as being surrounded by "lines of force" that pervaded all space. Electric lines of force would emanate outward from positive charges and be gathered inward by negative charges, on which they would end. Magnetic lines of force would emanate outward from the north pole of a magnet and be gathered inward by its south pole.

Nowadays we refer to Faraday's lines of force as electric fields and magnetic fields. For Faraday, Coulomb's law should be written in two parts. Part 1: a charge Q makes an electric field, which at a distance r from the charge has a magnitude

$$E = k_e |Q| / r^2$$

The electric field points directly outward from the charge if Q is positive and directly inward if Q is negative. Part 2: a charge q in an electric field **E** feels a force given by

$$\mathbf{F} = q\mathbf{E}$$

Before we can discuss the force exerted on a charge by a magnetic field, we must introduce one more piece of vector machinery: the cross product.

The cross product is a way of combining two vectors **A** and **B** to get a new vector **A** × **B**. The geometric description of the cross product starts with the figure below and specifies the vector **A** × **B** by giving its magnitude and direction. The magnitude is given by (Figure 13.2)

$$|\mathbf{A} \times \mathbf{B}| = AB \sin \theta$$

The direction is perpendicular to the plane containing **A** and **B**. This doesn't quite specify the vector since there are two directions perpendicular to any plane. So which of the two directions is it? The answer is given by something called the "right hand rule": put a screw in the plane and turn the screwdriver from the direction of **A** to the direction of **B**. Then the screw is driven in the direction of **A** × **B**. Note that we would get the opposite direction if we started at **B** and then turned toward **A**. This means that this sort of multiplication does not commute: **B** × **A** = –**A** × **B**. There is also a formula for the cross product using the components of **A** and **B**, but the formula is messy, so we're not going to write it down.

The force law for a particle acted on by both an electric field **E** and a magnetic field **B** is

$$\mathbf{F} = q(\mathbf{E} + \mathbf{v} \times \mathbf{B})$$

Here q is the charge of the particle and **v** is its velocity.

This equation is the first unification of these two otherwise disparate physical effects. It's very impressive, but it gets overshadowed by what comes next.

Note that the force due to the magnetic field is perpendicular to both the magnetic field and the direction of motion of the particle. Since a wire carrying an electric current has

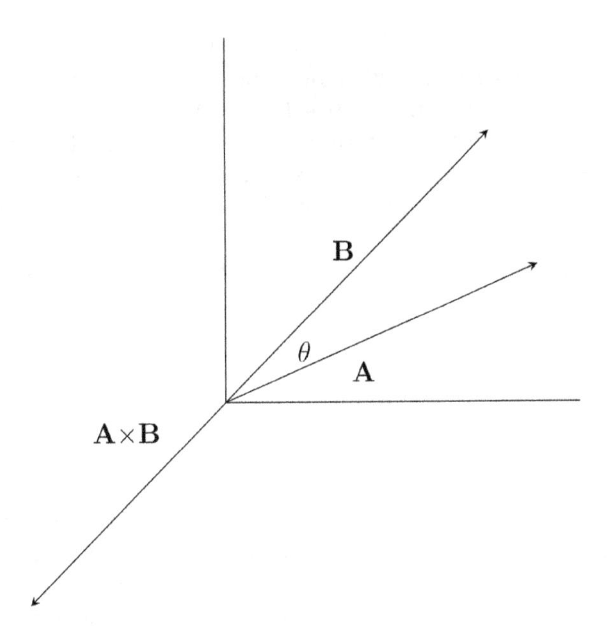


**FIGURE 13.2**
Cross Product

moving electric charges, this force law means that magnets can be used to turn coils of current carrying wire, which is the principle behind the electric motor.

But the force law is only part of the story: we also need laws that tell us how charges produce electric fields, how moving charges produce magnetic fields, and how changing magnetic fields produce electric fields. These laws were produced by James Clerk Maxwell, a 19th-century Scottish mathematical physicist who set about to codify in mathematical language the electric and magnetic phenomena explored by Faraday. This effort succeeded so well that it is regarded as the paradigm for the development of all subsequent physics. Einstein kept pictures of Faraday and Maxwell (along with a picture of Newton) in his study for inspiration. When asked if he stood on the shoulders of Newton, Einstein replied "no, on the shoulders of Maxwell."

In order to make his equations consistent, Maxwell found that he had to add one extra term not anticipated by Faraday: a changing electric field gives rise to a magnetic field.

To display Maxwell's equations, we will need one more piece of vector calculus called the curl. Just as div comes about by treating the operator $\nabla$ like a vector in the formula for the dot product, so curl comes about by doing the same for the cross product. From the vector **B**, we produce the vector $\nabla \times \mathbf{B}$ (curl of **B** or curl **B**) given by taking the messy component formula for $\mathbf{A} \times \mathbf{B}$ and substituting $\partial/\partial x$ for $A_x$, $\partial/\partial y$ for $A_y$, and $\partial/\partial z$ for $A_z$ (resulting in another messy formula which we're not going to write down).

To get a feel for what curl means, we are going to start with another of the more complicated versions of the fundamental theorem of calculus called Stokes' theorem. As a formula, Stokes' theorem says

$$\int_S (\nabla \times \mathbf{A}) \cdot \mathbf{n} = \int_C \mathbf{A} \cdot \mathbf{dl}$$

In words, the theorem says that we start with a surface S whose normal vector is **n** (this isn't one vector, **n** represents the vector perpendicular to the tangent plane of the surface at each point), and whose boundary is the closed curve C. Then, for any vector **A**, if we integrate $(\nabla \times \mathbf{A}) \cdot \mathbf{n}$ over the surface, that is the same as integrating around the closed curve C, the component of **A** that points in the direction of the curve.

As with Gauss' theorem, Stokes' theorem is a version of the fundamental theorem of calculus because it relates a derivative (in this case $\nabla \times \mathbf{A}$) integrated over a region (in this case S) to the quantity we are taking the derivative of (in this case **A**) on the boundary (in this case C).

To clever this up a bit, there is something called the boundary operator, which takes a shape and returns its boundary. The symbol for the boundary operator is $\partial$. So in this case, $\partial S = C$. So if you look at Stokes' theorem and Gauss' theorem with this notation

$$\int_{S} (\nabla \times \mathbf{A}) \cdot \mathbf{n} = \int_{\partial S} \mathbf{A} \cdot \mathbf{dl}$$

This shows an interrelationship between the geometric boundary and the analytic derivative. You can trade a geometric boundary for an analytic derivative or vice versa, which again is another way of looking at the fundamental theorem of calculus.

Now let's apply Stokes' theorem to fluids by choosing our vector **A** to be $\rho\mathbf{v}$. Then, $\int_C \mathbf{A} \cdot \mathbf{dl}$ will be positive if the fluid is flowing in a counterclockwise circle and negative if it is flowing in a clockwise circle (and zero if there is no net circular flow). In other words, $\int_C \mathbf{A} \cdot \mathbf{dl}$ measures the extent to which the fluid has whirlpools. Through Stokes' theorem, this makes $\nabla \times \mathbf{A}$ a sort of "whirlpool-ness per unit area" that is usually called vorticity. As with divergence, if **A** is not $\rho\mathbf{v}$, then this particular physical interpretation in terms of fluid whirlpools doesn't hold. Nonetheless, to get an intuitive feel for curl, it is helpful to think of $\nabla \times \mathbf{A}$ as the vorticity of whatever it is **A** is the flow of.

Now we have all the mathematical tools for Maxwell's equations:

$$\nabla \cdot \mathbf{E} = \rho/\varepsilon_0 \qquad \text{(M1)}$$

$$\nabla \cdot \mathbf{B} = 0 \qquad \text{(M2)}$$

$$\nabla \times \mathbf{E} = -\partial \mathbf{B}/\partial t \qquad \text{(M3)}$$

$$\nabla \times \mathbf{B} = \mu_0 \mathbf{J} + \mu_0 \varepsilon_0 \, \partial \mathbf{E}/\partial t \qquad \text{(M4)}$$

We will now discuss each of these equations in turn. Equation (M1) says that charges make electric fields. In this equation, $\rho$ means charge density (that is electric charge per unit volume) not mass density. The constant $\varepsilon_0$ is related to the constant $k_e$ in Coulomb's law by $k_e = 1/(4\pi\varepsilon_0)$.

Equation (M2) says that magnetic lines of force do not end. Here, "line of force" does not have the direct physical meaning that Faraday attributed to it, but simply means the curve you would get by starting at some point in space and then moving by always going in whatever direction the local magnetic field is pointing. In a bar magnet, the magnetic lines of force come out from the north pole, go out and around, come back in at the south pole, and then go through the magnet to the north pole, making a closed curve.

Equation (M3) says that changing magnetic fields make electric fields. This is the principle behind electric generators. Spin a coil of wire in a magnetic field and the wires of the coil encounter an ever-changing magnetic field, which gives rise to an electric field, which makes an electric current move in the wire.

For centuries humans have used the power of wind and water to spin wheels. The 19th century added coal-powered steam engines, which also spun wheels. Now those spinning wheels could be used to generate electricity, which could be used to power light bulbs, and also to turn electric motors, which could be used to spin more wheels. It might seem silly to spin a wheel, to generate electricity to power a motor just to spin another wheel. But the electric motor made it easier to spin those wheels at just the speed and with just the power desired. The clunky steampunk apparatus of the 19th century gave way to its sleeker 20th-century successors.

Equation (M4) says that electric current makes a magnetic field. Here **J** is the electric current per unit area, sometimes called a current density. One way to get a feel for this quantity is to note that a charge density $\rho$ moving at velocity **v** has a current density $\mathbf{J} = \rho\mathbf{v}$. Equation (M4) says that we don't have to rely on bar magnets: we can make our own electromagnets just by running an electric current through a coil of wire. These electromagnets can be used in both generators and motors to enhance or replace the ordinary magnets that these machines needed in order to do their jobs.

Just like the spinning wheel to spin a generator, to run a motor, to spin a wheel, equation (M4) along with equation (M3) gives rise to its own Rube Goldberg apparatus that nonetheless is very useful. Take two coils of wire that are in close proximity, say by being wound around the same object. Now run an alternating current (AC) through coil 1. AC has a sine wave behavior: first running in one direction through the coil and then turning around and running in the other direction. Thus, equation (M4) tells us that coil 1 will make an ever-changing magnetic field, and equation (M3) tells us that that changing magnetic field will give rise to an electric field in coil 2, which will create an electric current in coil 2. In fact, calculation using equations (M3) and (M4) shows that a sine wave of current in coil 1 will result in a sine wave of current of the same frequency in coil 2. AC in coil 1 gives rise to AC in coil 2. This Rube Goldberg apparatus with the two coils is called a transformer.

But what is the point of a transformer? We already had AC. What is the point of running that AC through some apparatus that just gives us AC back? The point is that the voltage of AC in coil 2 is not the same as the voltage of AC in coil 1. By choosing the ratio of number of windings (circles of wire) in coil 2 to those in coil 1, we can turn whatever voltage the AC had in coil 1 to whatever voltage we desire it to have in coil 2. Thus, generated power can be transmitted at high voltage (which leads to less loss of power in the wires) along power lines, and then changed by a transformer to a less dangerous and more usable lower voltage before being sent to the outlets in our homes. This power from the outlets can in turn be changed by a transformer to the even lower voltage needed to run our computers or charge our phones.

**The history here is fascinating because the math, the physics, and the technological engineering happened simultaneously, each pushing the other two. There's a joke that has two characters: a physicist and a British prime minister. The physicist has just shown off a piece of electronic development. The PM says, "And what use is this?" the physicist replies, "Prime minister, in 20 years you will be taxing it."**

*Anyway.*

So far we have looked only at the first term on the right-hand side of equation (M4), which says that magnetic fields are made by currents. But there is a second term, the one that Maxwell added, which says that magnetic fields are also made by changing electric fields. This term, along with equation (M3), indicates that electricity and magnetism together should give rise to waves.

To see why, think back to our treatment of sound waves in fluids. There gradients in pressure led to fluid being moved around, which led to fluid bunching up, which led to gradients in pressure, and so on. Now if we consider equations (M3) and (M4) together, we find that changing magnetic fields can give rise to changing electric fields, which then give rise to changing magnetic fields, and so on. So the same sort of conditions that gave rise to sound waves in fluids seem to be present in the equations of electricity and magnetism.

But before we investigate wave motion, we are going to look into a mystery involving a missing equation. Recall that the first Euler equation says that fluid is not created or destroyed, but only moved around. But electric charge is also not created or destroyed, but only moved around. So shouldn't we have something like the first Euler equation for electric charge? Did Maxwell forget to write down an equation?

It turns out that the missing equation isn't really missing: it's hidden inside Maxwell's equations. To find it, we are going to start with some properties of dot product, cross product, div, and curl. Recall that $\mathbf{A}\cdot\mathbf{B} = A\ B \cos\theta$. But the cosine of 90 degrees is zero. So any vector that is perpendicular to **A** has a dot product with **A** of zero. Now recall that $\mathbf{A} \times \mathbf{B}$ is perpendicular to the plane containing **A** and **B** and is therefore perpendicular to both **A** and **B**. It then follows that for any two vectors **A** and **B**, we have $\mathbf{A}\cdot(\mathbf{A} \times \mathbf{B}) = 0$. But now recall that div is made using $\nabla$ in the dot product and curl is made by using $\nabla$ in the cross product. So we also have that for any vector **B**,

$$\nabla \cdot (\nabla \times \mathbf{B}) = 0$$

Or in words, the divergence of a curl is zero.

Now we apply our newfound knowledge of divergence and curl to Maxwell's equations. We take the divergence of equation (M4) and find

$$\nabla \cdot (\nabla \times \mathbf{B}) = \mu_0 \nabla \cdot \mathbf{J} + \mu_0 \varepsilon_0 \nabla \cdot \left(\partial \mathbf{E}/\partial t\right)$$

But the left-hand side is zero, and dividing by $\mu_0$ to simplify we find

$$\nabla \cdot \mathbf{J} + \varepsilon_0 \nabla \cdot \left(\partial \mathbf{E}/\partial t\right) = 0$$

But $\nabla\cdot(\partial\mathbf{E}/\partial t)$ is the same as $(\partial/\partial t)(\nabla\cdot\mathbf{E})$ because it doesn't matter what order we take derivatives in. So we have

$$\nabla \cdot \mathbf{J} + \varepsilon_0 \left(\partial/\partial t\right)\left(\nabla \cdot \mathbf{E}\right) = 0$$

But equation (M1) tells us that $\nabla\cdot\mathbf{E} = \rho/\varepsilon_0$, so we have

$$\nabla \cdot \mathbf{J} + \partial \rho/\partial t = 0$$

This is precisely the missing equation. It tells us that charge is not created or destroyed, but only moved around: the only way for $\rho$ to increase in a given small volume is for charge to flow in, and the only way for $\rho$ to decrease is for charge to flow out. The missing equation was there all along, hidden inside Maxwell's equations.

**Watch closely folks. He's about to pull a very tricksy rabbit out of the hat of electricity and magnetism. Keep your eyes peeled for one more bit of unification.**

*Is that what you call foreshadowing?*

We now turn to electromagnetic waves. Though electric and magnetic fields are made by charges and currents, they can propagate away from the regions where they are made; so we start by considering waves in regions of space where there are no charges and no currents present. The electric and magnetic fields are described by Maxwell's equations, but with $\rho$ and **J** both equal to zero. That is, we have the following equations.

$$\nabla \cdot \mathbf{E} = 0$$

$$\nabla \cdot \mathbf{B} = 0$$

$$\nabla \times \mathbf{E} = -\partial \mathbf{B}/\partial t$$

$$\nabla \times \mathbf{B} = \mu_0 \varepsilon_0 \, \partial \mathbf{E}/\partial t$$

Now we apply Fourier's trick to these equations. That is, we look for a single note of the form $\mathbf{E} = \mathbf{E}_0 \exp[i(\mathbf{k} \cdot \mathbf{r} - \omega t)]$ and $\mathbf{B} = \mathbf{B}_0 \exp[i(\mathbf{k} \cdot \mathbf{r} - \omega t)]$, where $\mathbf{E}_0$ and $\mathbf{B}_0$ are complex constant vectors. Now we can replace every $\nabla$ with an $i\mathbf{k}$ and every $\partial/\partial t$ with a $-i\omega$. This gives us the following equations:

$$i\mathbf{k} \cdot \mathbf{E}_0 = 0$$

$$i\mathbf{k} \cdot \mathbf{B}_0 = 0$$

$$i\mathbf{k} \times \mathbf{E}_0 = i\omega \mathbf{B}_0$$

$$i\mathbf{k} \times \mathbf{B}_0 = -i\omega \mu_0 \varepsilon_0 \mathbf{E}_0$$

The first two equations tell us that $\mathbf{E}_0$ and $\mathbf{B}_0$ are perpendicular to **k**. That is, electromagnetic waves are transverse waves. Both the electric field and the magnetic field point in directions at right angles to the direction in which the wave is moving. The third equation tells us that once we have picked $\mathbf{E}_0$, then $\mathbf{B}_0$ is given by

$$\mathbf{B}_0 = (1/\omega)\, \mathbf{k} \times \mathbf{E}_0$$

So, electric and magnetic fields are at right angles not only to the direction of the wave, but also to each other.

Plugging the result for $\mathbf{B}_0$ into the last of our four equations we obtain

$$\mathbf{k} \times (\mathbf{k} \times \mathbf{E}_0) = -\omega^2 \mu_0 \varepsilon_0 \mathbf{E}_0$$

To deal with this equation we will invoke (without proof) the following property of the cross product: for any vectors **A**, **B**, and **C**, we have the following:

$$\mathbf{A} \times (\mathbf{B} \times \mathbf{C}) = (\mathbf{A} \cdot \mathbf{C})\mathbf{B} - (\mathbf{A} \cdot \mathbf{B})\mathbf{C}$$

Applying this general rule we have

$$\mathbf{k} \times (\mathbf{k} \times \mathbf{E}_0) = (\mathbf{k} \cdot \mathbf{E}_0)\mathbf{k} - (\mathbf{k} \cdot \mathbf{k})\mathbf{E}_0 = -\mathbf{k}^2 \mathbf{E}_0$$

where in the last step we have used the fact that $\mathbf{k} \cdot \mathbf{E}_0 = 0$. We then find

$$-k^2 \mathbf{E}_0 = -\omega^2 \mu_0 \varepsilon_0 \mathbf{E}_0$$

So we can choose $\mathbf{E}_0$ to be anything we like, but $\omega$ and k must satisfy

$$\omega/k = 1/\sqrt{\mu_0\varepsilon_0}$$

Remember that $\omega/k$ is the speed at which the wave travels. So electromagnetic waves must travel at speed $1/\sqrt{\mu_0\varepsilon_0}$. How fast is this? Maxwell knew the values of $\varepsilon_0$ (since this is related to the amount of electric field that a charge makes) and $\mu_0$ (since this is related to the amount of magnetic field that a current makes); he used these values to calculate $1/\sqrt{\mu_0\varepsilon_0}$ and found that it was equal to the speed of light. He concluded (correctly) that light is an electromagnetic wave.

**That is extremely weird. Electricity and magnetism seem to be restricted in what they interact with. But light is everywhere and seems to interact with everything.**

In just four equations, Maxwell had found a complete description of electricity and magnetism that also turned out to be a complete description of light. But electromagnetic waves can come in any wavelength, and visible light is just a narrow range of wavelengths. In particular, the sort of apparatus that people used in the 19th century to study the flow of electricity could be modified to produce and detect electromagnetic waves.

This leads to one of the most surprising things to emerge from the entire history of science: that not only is there light we do not see by, but also that most light is invisible to us. Mathematically, this led to a subtle revolution in the detected universe. Mistrust was extended even further. Not only did we need to measure carefully, there were whole regions of measurement of sight that we could not directly measure because most of the universe was invisible to us. From this point onward, science would need to create means of detection that translated the invisible universe into the visible.

The sort of experiment that proved this was first done by Heinrich Hertz: the first demonstration of radio waves. It wasn't long before radio was used to make a wireless version of the telegraph, and the rest, as they say, is history.

Whether for sending messages at high speeds, providing light, or exerting fine control over machinery, electricity proved itself superior to the steampunk valves, gears, and kerosene of the 19th century, leading eventually to that most modern of machines, the electronic computer.

# 14

## *Computers*

**What's that weird glowing thing on my desk and in my lap and in my pocket and on my wrist and in my oven and my car and, and, and...?**

There are many tools so common that no one is surprised that a particular person owns one: hammers, screwdrivers, cars, ovens, refrigerators. People more or less understand how the simpler tools work. For the more complicated ones, they have at least some vague idea how they work (e.g., refrigerators essentially pump heat out to keep things cool).

But there is one tool that is now so common that most people don't even realize how many of them they own. It is a tool that many people take pride in not knowing how it works, or think they know it works because they think it thinks. We're talking, of course, about the computer.

The pride people feel in computer ignorance is an extension of the "those people are weird" idea that we talked about at the beginning of the book. In the case of computer geeks, this view has become exaggerated beyond even the weirdness of mathematicians, and has gone so far as to become a point of pride among those of us in the arcane siblinghood of computer programmers.

*Arcane siblinghood?*

**Sounds better than "software nerds."**

The idea that computers think is the fault of science fiction. Long before computers actually existed the idea of a thinking machine was around. There had been automata in stories for millennia, machines that carried out tasks seemingly without supervision because they were built to carry out those tasks. Although fiction with clockwork thinking machines was not unknown (the character of Tik-Tok in L. Frank Baum's Oz books, for example), computers and robots as ideas appeared to most people to be different from these clockwork mechanisms because they could carry out multiple tasks, seemingly all tasks.

There was a stretch of time in science fiction where it was common to portray computers and robots as superior in thought to humans but lacking in sense or emotion or something else that allowed lantern-jawed heroes to overcome them.

Developing in parallel with these "humans are really better" stories were the "machines are really better" stories. In these tales, devices were made that were smarter and more capable of thought and action than humans. These machines would examine our frailties (including things like starting wars and starving each other) and would eventually take over.

In all of these stories, the computer or robot is portrayed as having a mechanical or electronic brain. That is, an artificial version of what we do our thinking in. The truth of computers is both simpler and odder than this.

First of all, despite the stories of fearsome effort needful to create these devices, the truth is that it is not too hard to create things that are confident that they are more capable and smarter than we are and will eventually take over. Every generation of humans does that in bringing about the next one.

The reality of computers is, however, that of high-speed artificial will-less stupidity controlled by instructions created by brave, dashing, brilliant, egotistical, self-righteous, overconfident ...okay, it's people like us.

The machines we have cannot choose to take over because they cannot choose, and they cannot think better than we do because they don't really think.

But that isn't the reality underlying the reality. While it's tempting to describe the evolution of devices that eventually led to modern computers (Jacquard looms, Babbage engines, hardwired computers, microchips, etc.), the important process, for our purposes at least, is not where computers came from, but the mathematics that underlies them. We will examine how the implementation of computers was built up structure by structure to be the seemingly user-friendly hardware and software we rely on.

*And get really mad at when it doesn't work right.*

**Well, yes. That goes without saying.**

Missing in all the mythology of computers is the basic truth that they are an impressive manifestation of the cycle of abstraction, theoretical work, and return to the world. The computer is a physical embodiment of a mathematical process.

It's by no means the first. The pile of rocks has it beaten by who knows how many millennia. The ruler and the compass also have it on grounds of age.

But the computer does something no earlier embodiment does. It serves not just as a tool of embodied mathematics, but as a tool that allows for the embodiment of mathematics. It is in some ways an artificial theoretical universe. Instead of abstracting into one's own mind and doing math work therein, one can learn to abstract into the computer's non-mind and let it do the math work.

How is it possible for a nonthinking thing to carry out what looks like the work of real thought? Let's once again look at the process of mathematics.

1. Abstraction from the real world into mathematical objects.
2. Transformation of those objects by truth-preserving actions.
3. Return to the real world of those transformed objects.

Manifest as computers, these three steps work as follows:

1. Computers, as we will see later, deal only with certain abstracted concepts, those that can be represented as numbers. There are a variety of means of getting those numbers into the computer (see the discussion of input below).
2. Computers can be given and will carry out specific precise instructions that take the numbers they have stored and produce new numbers.
3. Those new numbers can be presented to the world in a variety of forms (see output below).

Every computer program works more or less to this pattern. It is a set of instructions to the computer that carries out these steps. The computer seems to be performing an action of thought, but it is not. The real thought, as we will see later, comes in creating the procedures by which the computer carries out the work. In other words, the program is a lazy solution. Create the means by mistrust and laziness and let the computer do the carrying out of the problems.

The scale at which computers embody mistrust and laziness is phenomenal. Modern computers can be carried in a pocket and computer programs can be copied from system to system in a flash (*or a flash drive*). As a result, the principle of solve the problem once (write the program) and implement it an arbitrary numbers of times (run the program) is no longer confined to creating a method and then teaching the method to other people. The method can be embodied into software (*if you call that embodiment*) and passed around without even teaching the method.

Looked at from the mathematician's perspective, a computer is only a substitute for one of those people who takes the way of solving the problem and then solves it.

*So computers are substitutes for scientists and engineers and all those other people who just use the math?*

**It's a tempting thought, but I doubt most mathematicians would think of it that way.**

The computer is the substitute for the steps of math application where calculation is done, not the part where someone determines through their own knowledge and understanding what means of calculation to use. Computer programmers have their own brand of laziness in recycled calculation methods called code reuse.

**If I write a program and hand the code (that is, the actual written instructions) to another programmer, that programmer can alter the code if they think of a better way of carrying out part of the process and can make such a change easily, assuming I haven't programmed in an annoying and idiosyncratic fashion.**

*Hmm ... "an annoying and idiosyncratic fashion" does sound like a good description of your computer programming style.*

**Me?! What about you?! Didn't one of your colleagues say, "if you showed David's programs to your friends in the Computer Science Department, they would kill themselves so that they could roll over in their graves"?**

*Well, yes ... but he was joking ... sort of.*

**I rest my case.**

Regardless of these considerations of how computers are not really thinking, it is a fact that somehow this piece of silicon can embody, literally embody, processes of mathematics. How did that come about?

## Logic Embodied

In an earlier chapter, we talked about logic as a branch of mathematics and philosophy that focuses on truth and meaning. There is another approach to logic which is less concerned with those and focuses instead on algebraic abstraction of the processes of logic: **Boolean algebra**, named after its creator the early 19th-century English mathematician George Boole.

Unlike most branches of algebra, which concern themselves with multiple different sets and operations that obey the same axioms, Boolean algebra concerns itself with one set and all the possible operations that that set can have. The set is {T, F}, a two-member set. Any operation (let's call it +) on this set has to specify only four results to completely define which operation it is. We need to know

$$T + T, T + F, F + T, \text{ and } F + F$$

If we know the results of these four, we know everything about the particular operation. This means we can write out a table with all possible operations for this set. Each column of this table (after the A and B columns, which specify the values in A + B) will specify these four results:

| A | B | 1 | 2 | 3 | 4 | 5 | 6 | 7 | 8 | 9 | 10 | 11 | 12 | 13 | 14 | 15 | 16 |
|---|---|---|---|---|---|---|---|---|---|---|---|---|---|---|---|---|---|
| T + | T | T | T | T | T | T | T | T | T | F | F | F | F | F | F | F | F |
| T + | F | T | T | T | T | F | F | F | F | T | T | T | T | F | F | F | F |
| F + | T | T | T | F | F | T | T | F | F | T | T | F | F | T | T | F | F |
| F + | F | T | F | T | F | T | F | T | F | T | F | T | F | T | F | T | F |

From a purely algebraic standpoint, this is interesting on its own. We've taken the smallest possible set you can really have operations on (since a 1 element set only has one possible operation) and written out a complete list of those operations. Combinatorically, we can figure out how many such operations there have to be. We need to specify four different actions of the operation, and we have two choices for each of those four different actions. So the number of possible operations is $2^4 = 16$, which is what we have above.

Logic takes Boolean algebra and gives meaning, or tries to give meaning, to the elements T and F, saying that they mean true and false, and to the operations as the equivalent of symbolic logic operations. The kind of logic that does this is called **truth-functional logic**. It says that all that matters about a statement is whether it is true or false.

What we're about to write is not a standard notation but is a way of elaborating the idea of truth-functional logic with a little bit of precision, but not too much.

Let us imagine the set of all logical propositions (which has problems as a set, but let's not worry about that now). Call this set P. Then truth value is a function from P into the set {T, F}, which is supposed to assign the value T to all true statements and F to all false statements.

Let's write this function as V(p). Truth-functional logic wants to find operations on {T, F} which correspond to the logical connectives. So, for example, it wants to find one of the 16 above operations, give it the name "and" (&) such that

$$V(p \& q) = V(p) \& V(q)$$

It also wants operations "or" (given by the symbol|) and "not" (given by the symbol ~) such that

$$V(p \mid q) = V(p) \mid V(q)$$

and

$$V(\sim p) = \sim V(p)$$

In effect, it wants to find operations that make V a homomorphism, a function that takes logical connectives (ways to conjoin statements) into the operations listed above.

For example, suppose we want to find the equivalent of & (and). We want A & B to be true only if A is true and B is true, so we look at the table and find the column which is true only at the first row. That's column number 8. For | ("inclusive or"), we want A | B to be true if A is true or B is true. That's column 2. For Exclusive Or we want truth if A is true or B is true but not both. That's column 10. For ~ ("not"), we want ~A to be true if A is false and we don't care about B. That's column 13.

So far this sounds pretty reasonable. If we accept symbolic logic's idea that we can preserve truth while breaking statements up using logical connectives, we seem to be okay in transforming that into Boolean algebra.

But we've really gone as far as we can. The other logical connectives are not really truth-functional.

There was an attempt for a long time to declare that "If A then B" (A → B) was column 5, since we would want it to be possible to have "If True then True," "If False then True," and "If False then False," but not "If True then False" (falsehood should never follow from truth). Column 5 actually is B | ~A and the debate amounted to the question of whether or not this was equivalent to If A then B. This caused a few generations of trouble in symbolic logic which is still being straightened out.

But there are a few statements that are pure nonsense, but which would be true if we accept Column 5 as meaning →. Here's one Bertrand Russel pointed out.

$$(p \rightarrow q) \mid (q \rightarrow r)$$

If you try this with the above table, you'll find out that this is true regardless of the truth values of p, q, and r. But what this means in real language is that the following piece of errant nonsense is true.

"(If Fred is drunk then Wilma is sober) or (if Wilma is sober then the moon is made of gorgonzola)."

Boolean algebra was wedded to logic early on, and indeed it works reasonably well for "and," "or," and "not." The meaning of these symbolic logic terms remains more or less the same in Boolean algebra, and those meanings end up being the ones that matter in computers, which are very bad at cause and consequence.

While logic based on Boolean algebra was being devised, other mathematicians were creating a theory of automata. An automaton is an abstract machine that can be controlled by instructions. Two of the early 20th-century pioneers of this field were Hungarian-American mathematician John von Neumann whose work graces many branches of math. He's also responsible for the idea of self-replicating machines which grace many an SF and horror story. And English mathematician/WWII code breaker extraordinaire Alan Turing. Turing's prosecution by the UK government for being gay led to his suicide and more than half century of covering up how much he had done for the UK during WWII. He was eventually posthumously pardoned in 2013 and a law passed in his name retroactively pardoning every other man in the UK convicted for being gay.

Von Neumann and Turing's initial work on automata was purely theoretical. It was created mostly to see what such machines could and could not do from a purely theoretical perspective. We don't want to go too far in the discussion of this theory, for while its creators were the founders of computer science, what ended up being the theory and practice of computing came about less from their foundational work than from the development of particular pieces of physical circuitry.

This is the world-of-fact component that allowed computers to jump from the theoretical into the observable universe. And it is an illustration of the curious power of simplicity.

If you look at the table of Boolean operations above, you can see that each operation requires only three things: A, B, and A + B, each of which can be in either of two states.

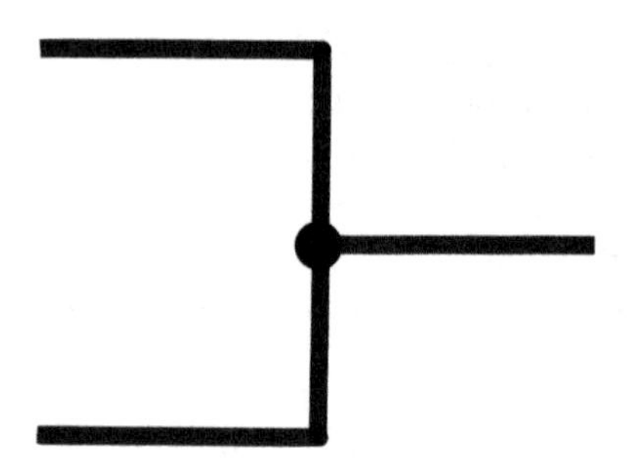

**FIGURE 14.1**
Junction of Two Wires

We can physically make such a thing by taking three cups and three balls. The cups are labeled A, B, and A + B. If we have a ball in a cup, its value is T. If there is no ball in the cup, its value is F. Each operation then is a rule that is used to say whether or not there is a ball in the A + B cup, based on whether or not a ball is in A and a ball is in B.

The physical object just described won't do the job of a computer because a person still has to check A and B and follow the rule to put a ball into A + B. But suppose there were some piece of hardware that has a different way of expressing whether or not A is T or F and B is T or F; and then suppose some natural process would automatically create the correct T or F value for A + B.

That may sound ridiculous, but it's not if instead of balls and cups we use wires and currents.

Consider a junction between two wires with one wire coming out of it (Figure 14.1).

In the circuitry, we are imagining every wire has either a strong or a weak current flowing through it. (Relatively strong. We're not talking electric chair or taser or even lick-a-battery strength current here.) Let's call low current 0 and the high current 1 (we could call them L and H, but we've got too many letters as is).

Imagine that at this junction a piece of hardware is placed, which using the currents coming in from the two entry wires determines the current coming out of the lone exit wire.

A piece of hardware that does this job is called a **gate**. The physical process of a gate is an electrical engineering matter, but gates aren't difficult to make. They basically work as special switches which determine on or off for a final state based on on or off of two initial states. A gate which will send a high current through only if both entry currents are high is called an **AND gate**. A gate that sends a high current through if either current is high is called an **OR gate**. A gate that sends a high current through if one current, but not both currents, is high is called an **exclusive OR** or **XOR gate**. A gate that sends a high current through only if the first current is low is called a **NOT gate**.

*Gee, I wonder where those terms came from.*

Here's how an AND gate disposes of incoming currents:

| enter1 | enter2 | exit |
|---|---|---|
| 1 | 1 | 1 |
| 1 | 0 | 0 |
| 0 | 1 | 0 |
| 0 | 0 | 0 |

Here's what an OR gate does:

| enter1 | enter2 | exit |
|---|---|---|
| 1 | 1 | 1 |
| 1 | 0 | 1 |
| 0 | 1 | 1 |
| 0 | 0 | 0 |

Here's what an Exclusive OR (XOR) gate does:

| enter1 | enter2 | exit |
|---|---|---|
| 1 | 1 | 0 |
| 1 | 0 | 1 |
| 0 | 1 | 1 |
| 0 | 0 | 0 |

And a NOT gate:

| enter1 | enter2 | exit |
|---|---|---|
| 1 | 1 | 0 |
| 1 | 0 | 0 |
| 0 | 1 | 1 |
| 0 | 0 | 1 |

These resemble the charts above, but there is one crucial difference. These gate structures are not abstract. They are physical pieces of electronic hardware. When first created, they were the size of small boxes. Nowadays billions of them can be etched into silicon chips.

The real-world existence of such circuitry components is what makes computers possible, because at the root we can attach the meanings and operations of Boolean algebra to the passage of current through these electronic gates and we can chain the gates together to create passages that might represent ~(A & B | (~C | A)) just by building multiple components or finding ways to channel the results of an exit into an entry. We're not going to bother with circuit diagrams. The thing to remember is that at this level of computing, at the level of gates and currents, we are dealing with reality, not abstraction.

Yet the abstraction of T and F is close to the reality, as close as numbers to sticks and stones. In some respects, this process is not unique. There have been many tools created that mirrored mathematical ideas in embodied reality, written numbers, for a start. The creation of such tools has been done often when a mathematical process was too difficult to do in mind without outside aid. In effect, real-world aids were created that helped with the middle of the mathematical process (the transformation by truth-preserving processes).

Indeed, the major purpose for all the mathematical notation we've seen is to serve as this kind of tool. Rather than starting in the world and building up before returning, this process starts in theory and builds down a single step from a wholly abstract concept into a real-world object that mirrors that concept. In this case, Boolean algebra is connected to a solidly concrete (*Silicon, you mean*) object: gate circuitry.

The thing that is different here is not the creation of the tool, but that the tool is not something kept in the human mind. Computers are not simple assistants to human mental processes. They take care of some of those processes on their own.

There is another perspective in which this weird arising is a natural consequence of the level of abstraction mathematics had obtained by the time these objects were first created. The easy way in which mathematicians had learned to abstract ideas so that even truth and falsehood could be translated into simple symbols coupled with the principles of substitution and equivalence by isomorphism led to this simple equivalence.

Gate circuitry is isomorphic to Boolean algebra because the objects are in one-to-one correspondence:

high current <--> T

low current <--> F

And the operations of Boolean algebra correspond directly to the manner in which the gate circuits work. In other words, from an algebraic perspective, the two are for all relevant purposes the same. Because the algebraic mindset can flow easily from one object to anything isomorphic to it, all the Boolean algebra and logic thinking could be imposed on these hapless pieces of electronic reality. They did not know they had been given meaning. They still don't. They don't think, but they can be treated as if they were logic embodied.

## Binary Arithmetic Embodied

We can employ another embodiment of meaning using the same hardware because we can pull down into the world a different isomorphic algebraic structure, one that can be elaborated into a much more complex structure than the two elements we begin with in Boolean algebra. We can look at 0 and 1 not as F and T, but as the numbers 0 and 1. The operations produced are all the functions that take two numbers and give a third number, provided all three numbers are members of the set {0, 1}.

Two numbers alone are not terribly useful. But now that we're thinking in terms of numbers, not truth and falsehood, we can extend our operations beyond the narrow confines of this two-element set. Remember that in base 10, we can express any number with just 10 digits if we string them together. In base 2 (also called **binary**), we can do the same with just two digits, 0 and 1.

Let's take a look at addition in binary. First, let's see what the place values are. In base 10, they are the powers of 10: 1, 10, 100, 1000, etc. In base 2, they are the powers of 2: 1, 2, 4, 8, 16, 32, etc. In base 10, the numbers that take up those places are called **digits**, which means fingers. In base 2, they are called **bits**, which is short for binary digits.

Let's slowly and carefully sum two four-bit numbers. We'll add 1001 (9 in base 10) and 0011 (3 in base 10).

1001 + 0011

First we add the numbers in the 1s place. That's 1 + 1 = 2. 2 in base 2 is 0 carry a 1. So we look at the 2s place. We are adding 0 + 1 + 1 (carried) = 2, which is again 0 carry a 1. So we look at the 4s place, which is 0 + 0 + 1 (carried) = 1. Finally, we look at the 8s place, which is 1 + 0 = 1.

Thus, 1001 + 0011 = 1100 or, in base 10, 9 + 3 = 12.

One of the most important things about digit-by-digit addition is that at any given point in the process, you never need to do more than add two single-digit numbers together and make note of any carrying that might need to be done. (By the way, if you spotted that base 10 digit-by-digit addition is addition in $Z_{10}$, you've definitely got a handle on the algebraic mindset.) The same applies to binary addition (which is digit-by-digit addition in $Z_2$), but in this case we only need to know the sum of each possible set of two bit values. In other words, we only need to know 0 + 0, 0 + 1, 1+ 0, and 1 + 1.

Hmm. That sounds like we're close to building binary addition out of one of our TF operations.

Unfortunately, that's not quite true. We actually generate two pieces of information out of adding two bits: the sum and what, if anything, is carried. We need to embody the following:

| First Bit | Second Bit | Digit | Carry |
|---|---|---|---|
| 1 | 1 | 0 | 1 |
| 0 | 1 | 1 | 0 |
| 1 | 0 | 1 | 0 |
| 0 | 0 | 0 | 0 |

Look at those two columns. The Digit column is First Bit XOR Second Bit, and the Carry column is First Bit AND Second Bit. In other words, we can physically embody the underlying process of binary addition using two gates that we can physically build, an XOR and an AND gate.

Well, fine, but does that actually get us to addition?

Not quite. We need to do one more thing and then we can actually get our computer to compute something. We need to be able to represent more than a single bit so that we can set up a real addition problem and then dump it through our gate setup one bit of addition at a time. We need, further, to have some place to put the results so that the outcome is somehow stored, so that when we go on from the 1s place to the 2s place, we have stored the value of adding the 1s places of the two numbers we are adding. In short, we need some way for our computer to not just transmit 1 and 0 through gates; we need to be able to hold 1 and 0 in something, something that can interact with and/or produce electrical current.

Let's pause for a second. We went too fast from the ability to represent addition of a single bit to suddenly wanting to create a mechanism that allows for real binary arithmetic.

Why make the jump? Because let's face it, a machine that can do your addition for you is darned convenient.

How did they make the jump? This is where the theory of automata came in. The theory of automata led to the consideration of calculated values and how to store the results of those values somewhere for use in other calculations. But more importantly, those considerations also came from the everyday practice of mathematics, from the simple idea of doing a calculation and writing it down to use later on. What was needed here was some way to write things down in a way that would be easy for our electronic calculation device to read.

Fortunately, the universe has given us such somethings: electricity and magnetism. As we saw in the last chapter, electricity and magnetism interact. Electricity can easily be used to create binary states (object charged or not) and so can magnetism (piece of material magnetized or not). All we would need to make a magnetic device for storing bit values is to create a set of capacitors or magnets that we can flip between an on and an off state, using an electrical current. On would then represent 1 and off would represent 0.

We can physically build a place to store the 1s and 0s in a binary number using a row of these capacitors or magnets. Then we can use those stored values to place currents into the paths going into our gates, and using the results of those gates we can build an electromagnetic device that we can interpret as doing addition.

We just slopped over some fairly tricky engineering problems that the early computer hardware makers had to solve. They knew it was doable because they'd been moving current into and out of electromagnets for quite some time. They had needed to do this since electric motors and other odds and ends in power generation and usage used exactly this process.

What happened in the early engineering of computers was not so much a breakthrough of engineering but a change of meaning and emphasis in processes they already used. They shifted mind and physical structure from a purpose of generating power to a purpose of storing information, from real moving around of electricity to the appearance of doing addition.

It is vital to remember that the device is not doing addition. We are getting it to do something which our human minds see as addition. The machine is not adding. It is flipping magnetic fields and running currents through wires and paths. We are doing addition by creating the machine to carry through certain processes and interpreting the results as addition. There will be more about this distinction between what the computer is doing and what we see it doing later in the chapter.

Here's an image of a piece of this process of addition. We have four sets of four storage areas. Each one is an electromagnet. Each set represents a place value:

| | | **8s place** | **4s place** | **2s place** | **1s place** |
|---|---|---|---|---|---|
| A | bits | 1 | 0 | 0 | 1 |
| B | bits | 0 | 0 | 1 | 1 |
| Carry | bits | 0 | 0 | 1 | 1 |
| Result | | 1 | 1 | 0 | 0 |

So in the 1s place we have 1 + 1 = 0, but we carry 1 to the 2s place. In the 2s place, we first add 0 + 1 = 1, but then we add the 1 we carried to get 1 + 1 = 0 carry a 1. In the 4s place, we then add 0 + 0 = 0 but then add the 1 we carried to get 0 + 1 = 1 carrying 0. Finally, in the 8s, place we add 1 + 0 = 1.

Using similar processes of breaking down operations into combinations of gates, we can get a computer to do binary subtraction, multiplication, and division.

That's pretty impressive. We've tamed the lightning to do our arithmetic. Not bad.

But we are not limited in the meaning we give to our high and low currents. Remember the electromagnets don't care about meaning or about anything for that matter. We can shift out of binary and go back to truth and falsehood. We can use gates as switches, so that one operation, such as addition, is activated if the resultant current from a gate is T, but another operation, such as multiplication, is activated instead if the result is F.

Here's the important thing: because we are talking about paths and currents, we can put the same gates in front of the instruction choices that we use to carry out the instructions. In other words, we can treat the instruction controls the same way we treat the numbers they work on. This will become vital later.

Notice how fast we are slipping around between the meanings of our currents. Seen one way it's a number, seen another it's a truth value.

From the algebraic perspective, this doesn't matter; they're isomorphic. From the hardware perspective, this doesn't matter because the hardware doesn't have a perspective. It's just magnets and wires that change the electromagnetic field.

But we who are interacting with it? We care. We want to be able to push one button and have two numbers added and push another button and have them multiplied. We want the meaning we are putting into this system to correspond to the meaning we get out. In other words, we want this to work the way mathematical abstraction is supposed to work. We do that largely by labeling the buttons and making sure they are hooked up to the correct gates.

But remember, the machine does not know the labels and does not understand the meanings. If you miswire the + button to the – operation, the machine will not complain, it will subtract where you want it to add and be none the wiser.

We have made a basic calculator. But a computer is so much more. After all, early calculators couldn't do word processing, make spreadsheets, create and play presentations through projectors, communicate over long distances, show people's personal web pages, post to social networks, send and receive e-mail, filter spam, etc., etc. There is some qualitative difference between calculators and computers, isn't there?

No, there isn't. What there are are multiple layers of human ingenuity laid over this basic process. Underlying all are those same binary gates and binary storage systems.

What happens in the rising up from these mathematical and physical depths to the everyday world of computing is an exaggerated mirror of the same process that leads to the loss of mathematical inheritance. A great deal of work has to be done to let someone who has no deep understanding of the workings of computers use a desktop or laptop computer or a cell phone or tablet.

And yet, because of that lack of understanding, the mythology of computers as thinking machines that want to replace us, far from being debunked, has only grown. Ignorance has risen up in parallel with the vast presence of computers so that people rely more strongly than ever on their mathematical inheritance but are still spared the burden of understanding it.

## Computer Programming

At heart, a computer consists of a device that can carry out the kinds of basic instructions (gate processing and binary arithmetic) discussed above, and devices to store binary values (that is, objects that can be in one of two states). The device that carries out instruction is called a **processor, microprocessor, or CPU** (central processing unit). The storage device(s) is/are called **memory** (a misleading name, for reasons to be dealt with later). There also usually need to be devices that can give data and/or instructions to the computer. These are called **input devices**. Devices to display in some form the results of instructions are called **output devices**.

In order to rise up from the binary depths to the bug-ridden, spam-filled, always at risk from hackers heights of computer use, it might seem that incredible sophistication must be used. But in truth there are only six tricks used over and over again in layers built upon layers to reach the dizzying (*really just low in oxygen but high in caffeine*) regions of current computing.

The six tricks are as follows:

*Assignment*: Placing a value in a specified memory location.
*Calculation*: Performing a mathematical calculation from supplied data.
*Branching*: Using a value to determine which of several instructions to carry out.
*Looping*: Performing the same instruction over and over again.
*Data Structuring*: Creating complex data objects by assembling simple data objects.
*Input/Output*: Taking in data from outside sources and displaying data on outside display systems.

Before we go into this in depth, look at this list again. Notice that the first four are methods of the theoretical universe part of the mathematical process (stating that some variable has some value is assignment, calculation is calculation, branching is deciding which of various tools to use, looping is doing the same thing over and over again). The last two are part of the process of abstraction and return. Input abstracts into data structures, output takes data structures and produces real-world effects from them.

A computer program consists of a list of instructions, each of which will be a use of one or more of these tricks. The program is put into the computer's memory and the computer carries it out. It will follow each instruction in turn until the program is done. The computer acts without choice, without will, without awareness, and without caring. Remember the lack of these later when we look at computer "thought." For now, let's look at how these tricks can build the computerized world up from a few two-bit calculations.

To start with, each computer has one or more processors that can do their basic calculations provided we provide them with appropriate batches of bits. But the bits are at the moment a disorganized mess. We have this vague mass of "memory" consisting of who knows how many bits. They need organization. Thus, we perform our first act of data structuring in order to turn the computer's memory into something useful.

To create this, we need to stop looking at individual bits and start grouping them into bigger structures. Let's face it, a bit can only have two possible values, 0 or 1. We want to deal with larger ranges of numbers than that. Unfortunately, we can't jump from bit to number, though we pretended we could earlier. The truth of the matter is no computer can handle all numbers. Even a computer with a quadrillion ($10^{15}$) bits of memory cannot handle a number larger than $2^{1,000,000,000,000,000}$ (which is about $10^{3,000,000,000,000}$) without cheating. We'll get to the cheating later. So we can't use our general concept of number, which has an infinite range of possible values, as a data structure; we need a more limited arrangement of bits from which we can build other arrangements.

Early on, for whatever reason, it was decided to group bits into lists of eight at a time. Eight bits is called a byte. *Well, that tells us everything we need to know about sense of humor in computer work.*

A single byte can represent any number from 0 to 255, since binary 00000000 = 0 and binary 11111111 = 255. That's pretty limited.

But two bytes can be used to represent any number from 0 to 65535, and 4 bytes can be used for any number up to 4294967295. This use of multiple bytes to represent larger and larger numbers is the basic trick of data structuring: Put together smaller objects in a single arrangement and from it create a new object.

Note that it's not enough to just stick the bytes together. We have to assign each byte a role for what part of the number it represents. If we look at a four-byte number,

byte 1 represents the first 8 binary places (the 1s through 128s place),
byte 2 represents the second 8 binary places (the 256s through 32768s place),
byte 3 represents the third 8 binary places (the 65536s through 8388608s place),
byte 4 represents the fourth 8 binary places (the 16777216s through 2147483648s place).

So a 32 bit binary number such as 11100111010101010101011100000100 would be broken up as follows:

byte 4 byte 3 byte 2 byte 1
11100111 01010101 01010111 00000100

Yes, we know that's eye-blurring. Early computer programmers actually had to deal with this by hand. And no, that's not why we're all mad enough to host tea parties for Alice. Most modern programmers never have to look at things like this and we're still nuttier than an almond grove.

One early attempt to make this easier to read was to translate from binary into another base that would allow the preservation of the byte structure. Unfortunately, base 10 did not work. The base had to be a power of two to evenly break up our new representations into bits. Two different ideas were and still are used: base 8 (octal) and base 16 (hexadecimal). The latter is more popular because in base 16, any byte is a two-digit number. Hexadecimal does have the slight inconvenience of having more than 10 digits, so its "digits" are 0, 1, 2, 3, 4, 5, 6, 7, 8, 9, A, B, C, D, E, F, where A = 10, … F = 15. Thus, one sometimes sees hexadecimals numbers like 3A4E, which is $3 \times 16^3 + 10 \times 16^2 + 4 \times 16 + 14 = 14{,}926$ in base 10.

Don't worry, that's the last bit of hexadecimal we plan to dump in here. We just wanted to show what is sometimes done to make this stuff nominally readable.

In clustering bits into bytes we've done some organization, but we've still got an inchoate mess of bytes in our computer's memory. We still can't find or get at anything.

The solution to this is called **addressing**. Every byte in a computer's memory is assigned a number that represents where it is in memory. That way we can assign values to memory by address and pull out the values by address. We can tell the central processor to pull in the value at address 3A4E F782 (okay, just a little more hexadecimal), and it will do it. We can also tell it to store the results of a calculation in that address.

Now we have reached one of those weird points in computers that will happen over and over again, where a source of confusion is also a sign of power. Notice that in order for the CPU to be able to obey instructions about addresses, the addresses have to be in binary (which we shortened to hex in order to save all of our eyes). But that address represents a byte or set of bytes, which means that the value at the address is also a number in binary. The address and what it represents are the same kinds of things. This is one of those points that trips up programmers all the time, particularly when they are learning.

Think of it like this: suppose that the inventors of algebra had decided that the ideal way to write down variables was to use numbers instead of letters. So instead of saying $x = 4$, we would say $3 = 4$. But the 3 would mean a variable while the 4 was the value of the variable. That's really confusing. Or rather, it's confusing to intelligent beings. The computer is too stupid to be confused.

We'll solve this headache in a little while. We have one more piece of pain to drop into the mix.

We've spoken of sending instructions to the CPU, but how do we do that? The CPU is a batch of gates. The only way it can determine what instructions to carry out is to have current flow through particular gates, which send currents out to other gates, which pull down from addresses and other gates, and so forth. So the only way commands can be sent is in the form of current arrangements; in other words, batches of bits; in other words, sets of bytes; in other words, the same kinds of numbers that are being stored in memory.

To use these numbers as commands, there has to be a correspondence between numbers and commands, in effect a secret code that says that 14AF means multiply or whatever.

Here's the mess we're in. Binary numbers are used as data, instructions, and addresses. It's as if x = 3 + 4 was written as 5 37 3 48 4, where 37 represents "=" and 48 means "+" and while 3 and 4 are meant to be 3 and 4, 5 is an address representing the variable x. And by the way, we were being kind by writing all that in base 10.

This living nightmare, where everything is written out in binary numbers, is called **machine language**. There used to be (and still are for all we know) some programmers who thrived on it, but the rest of us want to get away from it as fast as possible. We don't want to have to juggle this many different essentially meaningless numbers.

*Hey, doesn't this make us as math phobic as anyone else?*

**Not really.**

The thing about machine language is that it isn't lazy. It takes too much mental work to do anything in it. Programmers flee machine language because of an adherence to the principles of math.

*You mean we take pride in our laziness?*

**Yes we do. What's your point?**

The next step up from machine language is called **assembly language.** It's something of an improvement, but not by a huge amount. In assembly language, short strings of letters are used to represent the machine language commands. So a typical command might be:

```
5A7F
STO 44AA
```

That means take the value 5A7F and store it in memory location 44AA. We're not going to linger over assembly language much, because what matters here isn't the language itself but the existence of a special kind of computer program called an **assembler**. An assembler takes assembly language and translates it into machine language by substituting the machine language binary for the appropriate sequences of letters. Assemblers make it possible to write computer programs and then have the computers themselves turn them into real computer commands. Assemblers begin the process of separating the programmer from the binary gate guts of the machines.

*Isn't that what we said was the danger in the way people treat computers?*

**No, the danger isn't in the separation.**

The danger is in the ignorance of what's going on at the base. A programmer may never look at machine code, but they know that's what's really going on.

*Hold on one more second. How does the computer recognize the letters in assembly language commands?*

**ASCII me no questions and I'll tell you no lies.**

*Is that another one of those programming witticisms?*

**About half of one.**

**Remember, computers have no sense of meaning. Numbers are just patterns of currents. They don't distinguish commands from data from addresses.**

*So?*

**So, we can rely further on their meaningless thinking and treat some numbers as letters.**

*How?*

**In the same way cryptographers and schoolkids have for centuries. We create a lookup table, a correspondence between letters and numbers.**

*Oh, one of those A = 1, B = 2, ..., Z = 26* things.

**Yes.**

One of the most common, although now outmoded, codes is called the ASCII code (ASCII stands for American Standard Code for Information Interchange). It used a single byte to represent the various characters on a keyboard. Since a byte can represent any of 256 different values, there was room in the table for all the upper- and lowercase letters, the digits, and the punctuation marks used in standard American keyboards with room left over for some special characters. Nowadays ASCII is considered too limited and there is a new standard called Unicode which uses 4 bytes and can encode every symbol from every language on Earth with a lot of room to maneuver.

This kind of coding made it possible for computers to take strings of letters as input and produce strings as output. Note: strings, not words. To a computer "Mom" is 3 bytes (in ASCII) or 12 bytes (in Unicode). It has no meaning. It does not conjure up an image of anyone's mother. It's just three pieces of numerical data found on a lookup table. M O M.

*Wait. It already looked up M the first time. Why look it up again?*

**Because it's too stupid to know it already looked it up.**

Thanks to this ability to type in and print out numbers, assemblers could be made so that programmers could write their code in assembly language and then strive to get out of writing in assembly language by creating what are called **high-level languages**. A high-level language is a means of creating computer commands closer to the level at which humans act and think. Writing in high-level languages is a less arduous journey from intention to manifestation. Think of it as something like the difference between 3 + 4 = 7 and moving actual piles of rocks around to do addition.

In order to make this refreshing jump, other programs had to be written called **compilers** that turned high-level programs into assembly language programs, which in turn were assembled into machine language which the computer could actually run.

Before we go on to talk about high-level languages, there are a couple of issues to address. The first is the use of the word "language."

Computer languages are not languages. They are useless for communication. You can't write poetry in them. They can't describe things or embody meaning. They do not grow naturally. They cannot handle concepts, let alone new concepts. They are methods of writing out orders to be carried out by a machine that does not think. The only reason they are called languages is because we have a sense of using them as a written means to bring something across from one being, the programmer, to another, the computer. But they are no more languages than driving a car is a language. It's not as if "push down on the brake and turn the wheel" means "make a right turn." Those actions cause the right turn, but they don't mean it.

In the same way,

```
5A7F
STO 44AA
```

is not a communication of meaning that a certain number is at a certain address. It is an instruction to store that number at that address.

The second issue is the other reason, beyond ease of writing, that high-level languages are so popular: laziness – we mean, **machine independence**. Machine language and assembly language depend on what is called the **architecture** of the CPU. These languages need to be crafted specifically so that the commands correspond exactly to the underlying path architecture of instructions that the CPU can carry out. Does the CPU do multiplication by a built-in path or by a multiple use of addition? Can it do division or must that be coded? How many bytes does it use to code addresses?

Machine and assembly languages need to have these things built into them and programmers in those languages have to know them. But high-level languages don't need to know this. They are "machine independent" in that the same code (more or less) will work on different machines with different underlying architectures.

The compiler that translates high-level to assembly must be written in such a way as to translate properly from the high-level instruction to the specifics of each machine. Note that by each machine we don't mean each individual computer, but rather each make and model of computer chip. The same machine code can run on every computer with the same chipset. Therefore, a compiler run on one of them can translate one set of high-level code into a piece of compiled code that can be sent out to run on each of those machines. There's a few more factors based on operating systems that can mess up what we've just said, but it's basically correct.

High-level languages are like higher order abstractions. They let you solve more general problems than the low-level languages. Sometimes the solutions need low-level tweaking, but in general one is solving the bigger problem. The unusual thing here is that unlike most algebraic abstraction, the higher level the language, the easier it is for people to understand, because the abstraction is moving up to the human level from a level far below it rather than beyond the human level as most algebraic abstraction does.

We now have enough foundation to take a look at high-level languages and how they work.

The first jump upward is the ability to not worry about where a particular piece of data is stored. This is done by creating a stand-in symbol for some block of memory. Rather than specifying that a number is stored at a particular address, we can use a letter or string of letters to stand for the address. For example, we might want to have a symbol "n" for a particular address in which we can store a number. In most high-level languages, assigning a value of 5 to that symbol would be done with a statement that looks like

```
n = 5
```

That looks like an equation, but it isn't. It's an instruction. It's the same kind of instruction as

```
5A7F
STO 44AA
```

But it's easier to write and it looks more like a piece of math. Computer science came out of mathematics. The underlying thinking is mathematical (algebra with a dash of analysis). So the comfort of programmers is in mathematical structures, albeit turned and twisted in strange ways.

Let's look at the following lines of code:

*Umm, before we do that, maybe we should tell them that "code" is computer programmers' jargon for "computer program."*

**Right. It's certainly quicker to say "code" than "computer program," so this piece of jargon saves time. But it's an unfortunate piece of terminology since in ordinary language "code" is what you write in to make your message look like gibberish to anyone who doesn't have the secret decoder ring.**

*Right, although to be fair a lot of computer code does look like gibberish.*

**Anyway, let's look at the following lines of code:**

```
n = 5
n = 8
```

If these were two equations, then we would be specifying two equations that together had no solution. But that's not what's happening here. Before the first line we do not know what value n has stored at its address. After the first line, but before the second line, n has 5 stored at its address. After the second line n has 8 stored.

Let's look at these lines in combination with other things.

```
n = 5
m = n + 7
n = 8
m = n + 7
```

The line m = n + 7 isn't an equation either. It's actually three underlying instructions:

Retrieve from memory the value stored at location "n"
Add 7 to that value
Store the result at location "m"

When the program is compiled, assembly language instructions that do these three things will be created by the compiler.

Let's go back to the code.

```
n = 5
m = n + 7 (after this line is executed, m has the value 12)
n = 8
m = n + 7 (after this line is executed, m has the value 15)
```

Even more extreme is the following line that looks like abject nonsense:

```
n = n + 1
```

As an equation it's raw silliness, but as an assignment it adds one to whatever value n represents and stores it back in the location represented by n.

In a way it's strange that people whose training was in mathematics would create computer languages that look like messed-up algebra. An equation is usually a question, a thing to be solved, or a statement, an assertion of what is so. But a line of code is a command.

It's like the difference between "Fred runs." and "Run, Fred!." In "Fred runs" Fred is running. "Run, Fred!" is an order for Fred to run, not a statement about whether Fred is running.

Or try this one:

```
x = 45.72
```

*Hey, wait a minute! Where did the real numbers come from? So far we've only been able to assign integers and ASCII characters to locations.*

**First of all, we're going to stop calling them locations and addresses and start calling them variables.**

*But that will just add to the confusion.*

**True, but it's what they're called in programming. We're just going to have to live with it.**

*OK. But that still doesn't explain the 45.72*

**Look at it. What do you see?**

*The real number 45.72.*

**Really? I see two integers, 45 and 72, with a marker between them. I can build a data structure that can hold a real number by using two integers.**

*Is that how it's done?*

**Not exactly. But the real way is messier and hardly worth going into.**

Besides, what matters is the ability to take simple data structures and build complicated ones out of them. We can even make a data structure that holds very large and very small numbers by using scientific notation. For example, let's look at $3.672 \times 10^{12}$. There are only two pieces of data needed for this number: 3.672 for the multiplier and 12 for the power. This is often written in computing 3.672E12 (E being short for exponent). We can also write really small numbers like $1 \times 10^{-18}$ as 1.0E-18.

*That's all well and good, but your processor can only add integers made up of a small number of bytes.*

That's what programming is for, to compel the computer to do what it can already do in such a way as to get it to do what we wanted it to do. Let's suppose we want to add the real numbers 45.72 and 66.54. We write a piece of code (that is, a short program) that adds the numbers after the decimal point (72 and 54) to get 126. We carry the 1, leaving 26. Then we add 45 and 66 with a 1 carried to get 112, thus giving us 112.26 as the answer. Written properly, this procedure will allow us to add any two real numbers. More work is needed to add numbers in scientific notation, but it can be done without too much effort.

*Ah, the old "reduce it to the previous problem" trick of mathematicians only now used for computers.*

The main thing to understand here is that it is possible to interpret a batch of bytes in a lot of different ways. We can look at them as integers, real numbers, scientific notation, ASCII or Unicode characters in a string, and so on. Each interpretation makes the bytes subject to different pieces of code written to deal with that particular interpretation. In most high-level languages, for example, the symbol "+" means a lot of different things. It can mean add two integers, add two real numbers, add two exponentials, or stick one string of characters after another.

*Huh?*

**"Fred " + " and " + " Ginger" gives "Fred and Ginger"**

*Oh.*

This can be particularly confusing if one has a string of characters containing numbers. For example,

5 + 7 gives 12

but

"5" + "7" gives "57"

*So how do you know what you're dealing with? How does the computer know how to interpret the batch of bytes at some location?*

There is a concept called **typing**, (and no it's not what fingers do on the keyboard). A **type** is a declaration of what kind of value a variable will hold and therefore what the operations on that variable will do.

In a way we've done this before. We've used + as a symbol for integer addition, rational addition, real addition, complex addition, vector addition, not to mention each special addition in groups, rings, fields, and vector spaces. We've known what the + sign meant because we knew which set of objects we were working in. Typing is kind of like that.

At this point, it would probably be good to pick a particular family of computer languages and use the notation of those languages. There's a lot of diversity out there and the different languages handle things each in their own way. The currently most successful family of languages arose from a language called C.

*C? What a boring name.*

**True. But it's easy to remember, a bit easier than BASIC, Fortran, COBOL, SNOBOL, ALGOL, Pascal, Modula-2, PL/I, PL/C, Forth, LISP, APL, and many more.**

C has a lot of variations which are used in many areas of programming, C++, Objective-C, Swift, Java, and JavaScript, to name five.

In C every variable that is used must be declared with its type.

For example:

```
int i,c;
```

This is an instruction to the computer that means set aside two pieces of memory with labels "i" and "c" and treat them as integers. The semicolon just means "end of line of code." Every statement in C must end with a ;

Here's another type declaration.

```
float x,y,z;
```

This means that x, y, and z are declared to be real numbers.

*"float"? Why float? In Fortran the declaration for a real number is "real."*

**It's short for floating point number. Don't ask. And please don't mention Fortran again.**

*But it's used by a lot of physicists.*

**And only physicists. No one else likes Fortran. It's too primitive.**

```
char a;
```

This means that a is a single character, such as "n."

Knowing the type of the variable makes it easy for C to know how to interpret commands and functions for that variable. C is what is known as a **strongly typed language,** which means that it insists on knowing exactly what the type is for each variable and will only accept assignments of the correct type of value. There are also **weakly typed languages** where a variable takes on the type of the last value assigned to it. JavaScript is weakly typed. So

a = 5

in JavaScript makes "a" a variable of type integer. But if a later statement said

a = 65.4

then a would become a variable of type float.

*Enough with the assignments, calculations, and data structures already. What about the other tricks?*

Two of the others, branching and looping, control the order in which the computer carries out instructions. Unless otherwise told to, a computer does one command, then it does the next in the program, then the next, and so on until it reaches the end of the program. Branching and looping are ways to jump out of this one-foot-in-front-of-another process.

In a lot of cases, both looping and branching depend on what are called **conditionals.** A conditional is a calculation that determines if some condition holds. For example, 5 < 3 is a conditional that is false, while 7 > 6 is a conditional that is true.

*Wait, that's a completely different trick. It's not on our list.*

**Yes it is. These are calculations. To determine if a > b, subtract b from a and see if it's positive.**

*See if it's positive? That doesn't sound like a calculation.*

**Nonetheless it is.**

When computers store numbers, they have what's called a *sign bit,* which is either 0 for positive numbers and 1 for negative numbers or vice versa. It doesn't matter which. The point is that when subtracting a from b, the result of the comparison is determined by the sign bit.

*Wait a minute. Are you saying that comparisons give bit values as answers?*

**Yup.**

In particular, 1 is true and 0 is false. Look at the following weird statement:

a = (7 < 5);

7 < 5 evaluates to 0 (since it's false), which means that after this statement a will be equal to 0.

*Yuck. I thought high-level languages were supposed to make things more math-like, easier to comprehend for programmers.*

**Yes, but once it's understood it can be twisted and abused.**

Besides, remember the isomorphisms we relied on in building our computers in the first place. 1 = T, 0 = F. We started out slipping around between these meanings. We didn't abandon our slipperiness when we went to high-level languages; instead, we built up from it.

Unfortunately, sometimes we add to the confusion. In particular, assignment and comparison can be confused because they both use = signs. In C-type languages, the expression == means "test for equality." This does little to alleviate the weirdness.

a = 5

assigns 5 to a, but

a == 5

tests to see if a is equal to 5 and gives 0 or 1 as its answer accordingly.

Consider the following statement:

a = (a == 5).

If, before this statement, a had the value 5 stored in it, then the a == 5 would evaluate to 1 and this would then be assigned to a (which would have a value of 1 from then on until it was next changed). If, on the other hand, a had the value 17 before this statement, a == 5 would evaluate to 0, and a would be 0 thereafter.

One of the most common errors in C programming is putting in one too many or one too few =, accidentally programming a conditional when you meant to program an assignment or vice versa.

*Can we break out of this and into loops?*

**OK.**

The simplest form of looping is a command that is carried out over and over again as long as some condition is true. This kind of loop is called a **while loop**. Here's an example:

```
a = 2;
i = 0;
while (i < 4)
{ a = a * 2; (the * here means multiplication)
i = i + 1;}
a = a * 5;
```

The instructions within the { } ("braces") are all performed if the while condition is true. The { } in effect make the computer treat all the commands inside the braces as if they were a single command for the purposes of the loop.

Let's look at what happens when this loop is processed.

First, before the loop is started, a is set equal to 2 and i is set equal to 0.

The first time the computer goes through the loop, it checks to see if i < 4. Since i starts out at 0, this evaluates to true and the instructions inside the loop are carried out. First a = a * 2 is carried out. So a, which was 2, becomes 4. Then i = i + 1 is carried out, so i becomes 1.

We've reached the end of the body of the loop (the instructions in the { }), so we go back to the top and check the conditional again. i is now 1, which is still less than 4, so we execute the body again. (*That sounds kind of disturbing, doesn't it? In programming "execute" means carry out and "body" means a block of code and ...yeah, it's kind of disturbing.*)

So a is doubled again to 8 and i is incremented to become 2. Since 2 < 4, we go through again. a becomes 16 and i becomes 3. We go again since 3 < 4. a becomes 32 and i becomes 4. Now when we reach the top of the loop, i's value, 4, is not less than 4, so we don't execute again (*4 times on one body is enough*), and we go on to the first line after the loop, which is a = a * 5. Now a left the loop with a value of 32, which we then multiply by 5, giving 160.

Suppose in setting this code we had mistyped, and instead of writing "while (i < 4)," we had typed "while (1 < 4)." 1 is always less than 4, which means that the loop would go on and on over and over again, doubling a and adding 1 to i.

This kind of loop is called an **infinite loop** and it's a bad thing. Eventually, a would grow so large that no integer variable could hold it and what's called an **overflow error** would emerge.

This kind of difficulty is not as dire as you might think; well-written operating systems can spot infinite loops and terminate the programs before something annoying happens.

For those of you who are fans of old Science Fiction TV shows, it was quite common to find far future computers (often Artificial Intelligences) that had no checking for infinite loops. This is one of the big differences between speculative fiction and reality. In reality, people need to solve problems when they show up.

While there are a batch of loop variations (**for** loops, **do until** loops, and some others), most loops are essentially **while loops**, carrying out a specified set of instructions over and over so long as a particular conditional is true. The above should give enough of an idea of how they work.

*I'm already feeling loopy. I guess branching is next. You're going to go out on a limb aren't you?*

**Yup, thanks for showing the right sense of humor for programming.**

Branching has two basic forms: **if-then-else** and **invocation**. If-then-else is another use of conditionals. First, something is tested. *If* it's true *then* one command is carried out, otherwise (*else*) a second command is performed.

So consider the statement:

```
if (a < 5)
a = a * 2;
else
a = a - 3;
```

If a is less than 5, it will be doubled; if not, it will have 3 subtracted from it. This process is called branching because it involves using different paths of code (branches from the main trunk of the program). If-then-else is a very useful testing process and is one of the primary ways that programs can appear to be intelligent by seeming to make choices based on circumstances.

In fact, no choice is being made. Automatic tests are being performed and one or another set of commands is followed based on the test results. There will be more about the illusion of intelligence later.

On to invocation, which needs a section all its own.

## Modular Programming, Procedures, and Functions

A complicated program can run to hundreds of thousands of lines of code. It would be an unutterable mess if it was necessary to write all of that in a single file and have to recompile the whole thing every time a programmer changed a single line. Fortunately, it is possible to name a block of code and invoke that name within another block so that in the middle of the outer block the inner block is run and then control passes back to the outer block. This is a form of branching, since one leaves one list of code behind and calls another list.

The style of programming that breaks up code into these kinds of blocks is called **modular programming,** and it's one of the things that makes the insanity of coding a little less insane (*that and good quality chocolate*). There are two kinds of modules: **functions** and

**procedures**. The distinction between them is that a function produces a value that can be assigned to a variable or used in a test or any other way one might use a value, and a procedure does not return a value. Otherwise, there's no difference.

To be flexible enough, functions and procedures need to be able to take in parameter values, preferably sent from the body of code that is invoking the function. Thus, like mathematical functions, computer functions have argument or parameter lists called **argument** or **parameter lists**.

*Wow, computer programming just does not stray far from math.*

**Thanks for the sarcasm.**

No it doesn't and that's mostly a good thing. Remember that math exists to abstract things from the real world, mess with them, and then return them to the real world. Computers exist to take in input, mess with it mathematically, and then output it, an exact image of the mathematical process, so the better it fits math, the better the programs will work.

Here's an example of a function definition which will do some of what one would want it to, in this case calculate $a^b$ if one is given values for a and b (a can be any real number, b has to be a whole number). The process of turning it into a definition that does all we might want will take a little time.

```
float pow(float a, int b) {
int i = 1;
  float answer = a;
while (i < b)
{
answer = answer * a;
  i = i + 1;
}
  return answer;
  }
```

The top of the function is the line float pow(float a, int b). The first "float" means this function will return a value of type float. "pow" is the name of the function which needs to be used each time the function is used. The () enclose the parameter list just as they would for a mathematical function, except that each parameter is specified by type. The body of the function is within the outermost { }. C-type languages always enclose the bodies of loops and functions and if-then-elses in { }s, assuming they have more than one line. One of the most common errors is not supplying an } to go with each {.

*Hey, what's with the variable called answer? Why are we using words instead of letters?*

Variables in programming can have lots of different kinds of names, not just the single letters that are used in algebra. It makes it easier to keep track if you give them mnemonic names. The computer doesn't care, but it's easier on the programmer.

This function is meant to calculate $a^b$. Note that it only works when b is an integer. The return statement at the end is used to send back a value so that one could invoke this function in a statement like

```
x = pow(7.5, 3);
```

This statement looks like a calculation and an assignment, but it's really a branching statement and an assignment. The use of "pow(7.5, 3)" invokes the code above. It assigns 7.5 to the variable a and 3 to the variable b inside the function then carries out the code and at the end returns the value of the variable "answer." The statement that called the function (called means invoked) assigns the returned value to the variable x.

So does this function work?

Yes and no. It will calculate pow(7.5, 3) just fine. But suppose we try to do pow(7.5, –3). Then it will never go through the loop because i starts out with a value of 1, the test used in the while loop is (i < b), and 1 is not less than –3, so the function will return 7.5, which is the wrong answer.

We have two solutions to this problem. We can either restrict this function to positive integers, or we can put in a little refinement as below.

```
float pow(float a, int b) {
int i = 1;
int c = abs(b);
float answer = a;
while (i < c)
{
answer = answer * a;
  i = i + 1;
}
if (b < 0)
answer = 1/answer;
  return answer;
}
```

Two lines have been added and one changed. We've added a new variable, c, which is equal to the absolute value of b. We did this by invoking a built-in function, abs, which returns the absolute value of its argument. We then use c in the loop in place of b, so we're calculating $a^{|b|}$. We then check later to see if $b < 0$. If it was, we take 1 divided by the answer and return that (because $a^{-c} = 1/a^c$).

Does this do what we want?

Not quite. What if b is 0? Then we don't go through the loop and return the wrong answer a. But since $a^0 = 1$, we can solve this with one line early on. We should also deal with the possibility that a is 0, in which case we should return 0 regardless of what b is.

```
float pow(float a, int b) {
if (a == 0.0)
return 0;
if (b == 0)
return 1;
int i = 1;
int c = abs(b);
float answer = a;
while (i < c)
{
answer = answer * a;
i = i + 1;
}
if (b < 0)
answer = 1/answer;
return answer;
  }
```

Most C-type languages have built in implementations of pow, abs, and a bunch of other functions like sine, cosine, etc.

*How do they do that? Computers can only add, subtract, multiply, and divide. How can they do trigonometry?*

**It's all thanks to Taylor and Maclaurin.**

Taylor series let us do good approximations of functions like sine and cosine using just +, *, –, and /. The approximations are bulky, annoying polynomials that no human would want to calculate, but the computers don't care.

Because a function or procedure can be invoked by any piece of code, it can also invoke itself. This is called **recursive programming** and is a form of looping.

Consider this:

```
int factorial(int n) {
if (n <= 1)
  return 1;
return n * factorial(n - 1);
  }
```

This function calculates n! based on the fact that if n > 1, n! = n(n – 1)!

So to calculate, say, 3! we do 3 times 2! which becomes 3 times 2 times 1! which becomes 3 times 2 times 1 = 6.

**We ran into a recursive procedure before when we were calculating gcds of pairs of numbers. Translating such procedures into this kind of programming came quite naturally.**

*To people who think that way, you mean.*

Recursive programming is incredibly useful, particularly for drawing fractals and calculating chaotic effects. One does have to take care that one does not create an infinite loop by invoking an invocation which invokes itself again and so forth, never-ending until too many invocations are present in the system and something crashes.

Once we have functions and procedures, all the analytic thinking developed before can be brought over and a large number of mathematical tasks handed to computer systems by properly coded functions which, if you remember, are most like unto machines following commands. It's no coincidence that the machine = function metaphor works so well, nor is it strange that analytic thinking is vital to computing.

*Wait a minute. We've shown functions. What about procedures? What's the point of calling something that doesn't return the call?*

**Because it can have other effects. For example, suppose you have a word processing document that you want to print. You would need a procedure to send it to the printer and have it printed.**

*How did the printer get into this?*

**I'm glad you asked.**

**Input and Output: computers talk to the world and to each other.**

Last trick: Input/output, or I/O for short.

**Input** is the trick of taking something from the outside world and from it generating data. This is happening as this chapter is being typed. Each of the characters in the phrase "Each of the characters" is generating a single piece of stored information (probably its Unicode value, depending on the word processor).

**Output** is the trick of taking data and from it generating an appearance of something. This is also happening as this chapter is being typed since each character as it is typed is used to select and draw little squiggly black things on the white image in the window on the computer screen, so as to create the appearance of type on white paper. The particular squiggles are selected from a set of images called a **font** which has a single picture for each letter of each acceptable size. These images are then placed in appropriate places on the screen. Exactly where on the screen is calculated based on characteristics of the document and the window. As each new line is typed, the image scrolls upward, drawing the visible part of the document again and again as each line of text appears.

The image generated on the screen is mostly created from two sets of data: the string of Unicode characters generated from the keys tapped on the keyboard and the font data stored for the particular font selected. The same data can show up in a variety of different ways with a different choice of fonts.

Font. Font. Font. Font. Font … . This is of course one of the charming distractions of computer use.

It is this very ability to separate the information particular to a piece of data from the information necessary to display it that allows computers to communicate efficiently. If one can minimize the amount of data necessary to transfer, one can ease the process of handing files from system to system. On a computer a picture is worth a lot more than a thousand words because a thousand words can be reduced to a few thousand Unicode characters, but a picture is usually millions of pieces of data about which pixel has what color.

*Go a little slower. You just dropped "pixel" in there.*

A **pixel** is a special storage location. It represents an abstracted location in an image. The location is talked about as if it were a point (and often specified by x, y coordinates), but it's really a small area, meant to be drawn on screen or on a piece of paper with a particular color.

Color is one of those tricky things that are partly a property of objects and partly a property of human perception. Visible light comes in different wavelengths, which are perceived as different colors: the colors of the rainbow, with red being the longest wavelength light and violet being the shortest. So in principle the "color" of an object should be how much light it gives off at each wavelength: a whole function worth of information. However, we perceive color using specialized cells called cones in the retina of our eyes. There are three different types of cone, each sensitive to a different part of the spectrum of light, so the amount of information needed to specify what color we see in a particular object is just three numbers. The sum of these numbers is basically brightness, leaving only two numbers to represent color.

Thus, a computer representation of a color only needs to get these two numbers right. A representation of color and brightness needs to get three numbers right. For display on a computer screen, this means thinking of each color as a mixture of red, green, and blue, the colors used in the display screen. Each color's value can be specified on a scale from 0 to 1 (0 is none of the color, 1 is pure color). These are specified as RGB values.

For example:

A dark blue could be R:0.0, G:0.0,B:0.5

Yellow would be R:1.0, G:1.0,B:0.0

and a rich magenta would be:R:0.75,G:0.0,B:0.5.

For printing, however, one should think of each color as a combination of the colors of the printer ink: cyan, magenta, yellow, and black, listed as CMYK (K is black, because they ran out of color names that begin with B and black ends with K).

Those same colors would have the following values:

Dark Blue C:1.0, M:1.0, Y:0.0, K:0.5
Yellow C:0.0,M:0.0,Y:1.0,K:0.0
Rich Magenta C:0.0, M:1.0,Y:0.33,K:0.25

In principle, the amount of information needed to output a picture is enormous, because we need three numbers for each pixel, and we need an enormous number of pixels because each pixel must be so small that we don't notice that the picture is just a bunch of dots. In practice, we can get by with less information because real pictures don't have details on scales as small as a pixel, so nearby pixels have nearly the same color. The algorithms used to go from enormous data files without losing essential information are called **file compression.**

Outputting an image file to a screen or printer requires a lot of memory and a lot of processing. That's why most computers these days have video cards, which are themselves little computers dedicated just to displaying images on the screen.

Forms of output include video, printing, computer-to-computer data transfer, and sound generation.

**Wait a few beats. How do computers do sound generation?**

Let's go back to how math models sound. We use Fourier Series which are sums of sines and cosines. But sines and cosines can be represented by Taylor series which are polynomials. And polynomials are just multiplication and addition which computers can fake using gates.

This is an absurdly inefficient process from a human perspective. It's ridiculous to make one set of series and then make another series from that in order to generate each tone in turn.

But modern computers can do incredibly large numbers of boring calculations faster than a human can even notice, so all that number crunching to play an $A^{\#}$ is just fine. Digital music provided by math.

While music and these other kinds of data may be expensive in time and processing power to generate, they are conceptually not too hard to handle. In each case, there is a kind of stored data (all in the form of bits and bytes) and a process of generating output from that stored data. The process may involve sophisticated hardware engineering and some specialized pieces of code to carry out the actual translation, but once those are made programmers can largely rely on them to take care of the translation to output.

Input is trickier.

How to give instructions to computers, how to hand them the data they need, is a field that is constantly evolving, particularly as computers get smaller, since the same kinds of data need to be input from smaller and smaller devices. Keyboards and mice are the primary methods of input used in desktop and laptop computers today. Smaller keypads are common in handheld systems, with specialized software and thumb-based typing to make input tolerable but not great. Touch screens work well for small systems with a lot of choice involved, and there are some voice-controlled systems, which are getting better over time. A few video game systems are controlled by motion sensors, and there are specialized virtual reality rigs that involve "reaching and grabbing." There are also drawing pads and handwriting recognition software, which work only so well.

There is not yet a satisfactory input system for anything except text, where keyboards do decent work (and they were adapted from typewriters). Mice and menus systems work reasonably well, but are unsatisfying for complex work. This is an ongoing field of development, and we hesitate to make any predictions about it.

There is one more element of programming that needs to be touched on, but despite its importance we need not go into it in detail.

Every computer needs programs that regulate the running of other programs. The program/programs that do this are called the computer's **operating system**, or OS. Different OS's regulate programs in different fashions and to a large extent dictate what programs can run on a given system. Right now the major differences in user experience and software availability are found in the choice of OS. Each OS tends to have its own adherents (sometimes with fanatical devotion to that OS and even more fanatical antipathy to various alternative OS).

We don't want to dip too far into this owing to the theological character of the dispute and the likelihood that any present-day discussion of the particulars of operating systems will become irrelevant before too long. There's too much change to make predictions and too much going on to waste one's time touting one side or another. We're not neutral on the subject, just not evangelizing.

So what good is knowing all this if one is not going to program? Because things go wrong. Things go wrong with computers every day. Sometimes it's an actual error in programming and sometimes it's a mistake caused by assuming the computer knew what you meant even though it can't actually tell that you couldn't possibly mean that you sold five hot dogs and made $100,520.46 on the sales. It took in the numbers, worked on them, and spat out the result.

On a slightly more sophisticated level, the vital thing to remember is that the computer is just carrying out its programming and whatever the labeled intent of that program may be, it's the real instructions that are being followed. Those instructions are not matters of words and meaning, but an arrangement of tricks to create the illusion of obedience to meaning. Remember that no matter how good the interface, it's all a trick.

This awareness should be had by everyone using computers. Beyond this minimal understanding, we do think it's worthwhile for people to learn how to program. We think this for two reasons: First, it's better to understand your tools than use them in ignorance; and second, there are advantages in carefulness that come to people who have to instruct incredibly stupid machines every day.

Because computers are dumb as rocks, they can't figure out what you mean if you make a mistake. You need to use the exact, correct words in the exactly correct order to create the effect desired. This necessity for precision is a very useful tool for dealing with other areas of exactness like math and science.

People often have trouble with mathematical and scientific ideas because the ideas are narrowly circumscribed and given exact meanings that do not slip over into other meanings.

In math the word "dimension" tends to mean the number of parameters; it does not mean the place your evil twin with the goatee comes from. "Group" does not mean the same thing as "set." Chaos means a certain condition of some processes; it does not mean the result of hosting a birthday for 5-year-olds. The way words and concepts are used in the sciences can frustrate and confuse people outside the fields (and no we don't mean the algebraic structure "field"). Learning computer programming can inculcate one to this exactness of meaning because the computer is too stupid to understand what a person "means," it only does what it's programmed to.

## Speed, Memory, Bandwidth, Price, Size, and Efficiency

The complexity of programs that can run on a given system depends mostly on the underlying hardware. It is the rapid change of power and sophistication of that hardware that has spurred the radical changes in where and how computers are used. When we were young, computers took up entire rooms. In our teens, the first pocket calculators came out to replace slide rules as the computing tool of choice.

Richard's high school had a teletype that connected over a phone line to a university mainframe. In college he bought a personal computer that had 64k of RAM (64,000 bytes of memory) and used a cassette tape player for long-term storage. A few years later, he bought a system that used 5¼" floppy disks; later still a system that used 3½" disks and had a hard drive. Time wore on and system replaced system.

Richard's current machine has effectively eight processors, making it what would once have been called a supercomputer. It runs at 3 gigahertz, which means it can perform 3 billion instructions per second. It has 6 gigabytes of RAM and hard drives adding up to around 4 terabytes (4 trillion bytes) of storage, not to mention the ability to burn CDs and DVDs for long-term storage. It has a printer that can print out this entire book in less than an hour and an internet connection fast enough to make watching streaming video practical.

What has been happening is a startling, ongoing growth in a single technology which over a period of little more than a generation has radically shifted what computers are in society, what they are used for, and how programming is done.

Let's start with the basics, speed and memory. The faster a computer chip, the more instructions it can execute in a second. The more instructions it can execute, the more complex a program it can run. Consider that a program that takes 10 hours to run on one system takes an hour on a system 10 times as fast, 6 minutes on a system 100 times as fast, 36 seconds on one 1000 times as fast, and 3.6 seconds on one 10,000 times as fast, 0.36 seconds on one 100,000 times as fast, and 0.036 seconds on one a million times as fast.

Programs that once took all day now can run without anyone even noticing. This also means that a procedure that once took all day, and therefore had to be the be-all and end-all of a program, can become part of another procedure, which in turn can be nested in still yet another procedure and so on for a while before we even notice that time is being taken up in calculation. This makes more and more complex work possible. In short, yesterday's day's work has become today's moment's work, and yesterday's life's work has become today's all-in-a-day's work.

Accompanying the speed demonization of computers has been the memory glut. There is so much memory available that things that would once have been considered ludicrous (like storing and watching full-length movies on a handheld device.) are now commonplace. Memory is dirt cheap. You literally cannot buy a computer with as little memory as a room-sized mainframe had when we were young.

Furthermore, size has decreased. Computers can now be sewn into clothing. Flash drives are used as keychains. Telephones, ovens, music players, so many things we carry and use in the home are really computers dedicated to one or a few tasks.

Think about it. Computer use is now so casual that they are made to be disposable, to be treated with utmost indifference.

Never mind familiarity. Speed, strength, and miniaturization have bred contempt.

The only thing growing fast, but not as fast as the rest of these things, is bandwidth, the speed at which computers can communicate with each other. This still takes time. But it is possible now to buy full-featured software through downloads, to keep your software updated through internet connections, to casually reach out and talk to systems across the world without even noticing it.

And for the last 20 years, prices of everyday computers have been relatively constant. A desktop system now costs what it did then, as does a laptop. You get a lot more power for that same money, but what you can get costs about the same. The smaller computers, such as one finds in phones, are cheaper, and the prices are dropping while features are becoming more diverse.

There have been two effects of this process. One has been the growth in underlying complexity of the software used every day. The second has been a growth in sloppiness of programming.

There's too much reliance on computer speed and memory. Programs have become processor- and memory-hogs. When computers were slower, people tweaked their code to gain every advantage. Nowadays there is a lot of reliance on code-writing software to generate programs automatically, and typically nobody is checking to see if the system does its job as efficiently as possible.

The paradox, of course, is that having reached the point where no person actually knows what's going on at the core level in any system, we have also come to have user interfaces that have grown friendly enough that people use their computers for everyday things. The toolmakers are growing careless and the tool-users carefree.

Despite the impressive presence of embodied mathematical inheritance, everyone involved is losing their awareness of the math at their fingertips, with perhaps the ultimate irony that many retail establishments are using computers with bar code scanners or preprogrammed buttons on cash registers so that the users of the systems don't even know what prices they are charging, nor do they need to know how to add them up or subtract them from the money they are given (if they are given money rather than a credit card, it's all taken care of with one computer talking to another).

Rather than computers creating a greater mathematical presence, they've enhanced its absence. It's just a few steps from the system to the understanding, but thanks to the system, fewer people try to get the understanding. The mathematical laziness that created computers has brought forth more of the everyday laziness which keeps people from mathematics.

And yet, despite this, mathematical awareness is sneaking out in odd ways. It used to be difficult to teach the concepts of functions and parameters, because it was hard to connect the ideas to people's lives.

But anyone who has ever played a computer-based role-playing game knows that their ability to hit and hurt the pixelated monsters they are cutting into depends on parameters like their character's strength, level, and equipment. The same people who do not learn how to do addition in their heads have been learning about functions from the moment they picked up game controllers.

Despite all of this growth and presence of computers, the strangest thing of all is that the myth of the thinking machine is still there in the minds of people who deal with computers, even though they are confronted with computer stupidity every moment of their waking lives.

## Human Thought and Computer "Thought"

The myth persists, and myth it is for many reasons. Let's start by looking at memory.

Human memory is associative. Ideas and events are connected to each other by chains of association, the way the smell of a favorite food can bring up a feeling of comfort, or a particular song can bring up an entire relationship from beginning to end.

This may sound like the way computer memory uses addresses for memory locations, but it isn't. The things connected in a human mental association are not content and address of content, they are pieces of content, each of which in a sense acts as an address to pull in all the others.

Imagine an event in your life that mattered greatly. Think about the place where it happened, the sounds, the smells, whatever you can recall. Each thing there can serve as a memoriam for the entire circumstance. Each piece of the memory accesses the whole.

Now here's a trick that has been used to create false memories. If you put one more object that was never there in that scene – say, add an image of a person or a particular thing that mattered to you to your memory of the scene – if you concentrate and reconstruct the scene in your mind over and over again with that new thing added, you will likely come to recall the scene with that new addition, and you might even end up recalling the event as if those things had been there. We don't recommend this for important memories for obvious reasons.

*So we shouldn't suggest they associate this book with their favorite activities, not to mention their favorite people?*

**Stop that.**

If the object or person you added has other associations in your mind, then those will become connected to the memory of the event and the memory of the event can be used in your mind to reach those memories.

This trick can be used beneficially to create artificial memory structures. Before literacy became widespread, such processes were used by learned people and in simpler form by nearly everyone. It gave them the ability recall a great deal. The whole practice of this was called the **Art of memory.**

The underlying point is that if you put two things together in your memory, they can stick together and form associations. In computer memory, this is not the case.

If you have two adjacent memory locations in a computer and one contains your phone number and the next your mother's phone number, the two will not gain any connection to each other. Furthermore, neither of them will act as connections to memories of you or your mother in any way.

Computer memories don't even stick in the places they are stored.

```
a = 5;
a = 7;
```

After the first line of code above, the number 5 completely occupies the memory location accessed by a. After the second line, 7 completely occupies that location. There is no sign that a 5 had been there before. Computer memory does not leave tracks once overwritten. Human memory is nothing but tracks.

Human thought also has this sort of associative and wide-ranging character. We can bounce ideas around all over our minds, playing with them in trivial ways or placing them deep within our understanding to craft art and science from them. Our brains are full of processors and processing power.

In contrast, all computer "thinking" goes on in the processors, and only what is explicitly called for to be worked on is actually worked on. Furthermore, it is only worked on in the manner in which the program currently running works on it.

A case can be made that in humans there is no distinction between memory and thought, that our associative memory is always thinking and that our thoughts are actions of connecting memories to create new things.

Nonetheless, a great deal of time, effort, and thought has gone into the field of Artificial Intelligence (or AI for short), the effort to conceive and eventually construct some type of thinking machine.

Part of the motivation for AI comes from the observation that both the transistors in computers and the neurons in the human brain are physical systems. The architecture of the connections between neurons is very different from the wiring between transistors in a computer chip. But it is still tempting to think of neurons as the "hardware" and some aspects of their interconnection as the "software" and thus (in this analogy) to think of the brain as something like a computer.

Any human mental task that we understand completely, down to the last detail, we can program a computer to do. Thus, AI can be thought of as a two-pronged approach combining research in computer science with research in psychology. On the computer side, the idea was to create ever more sophisticated sensors, actuators, pattern recognition algorithms, goal-seeking algorithms, and other computer hardware and software to enable computers to do the sort of things humans do. On the psychology side, the idea was to gain an ever more sophisticated detailed and complete understanding of how human beings perform their mental tasks.

Unfortunately, the promise of AI has not been fulfilled (or at least has been very slow in coming). Even recent advances tend to reinforce the huge gulf between human thought and computer "thought." One of the most successful approaches is something called machine learning, which has been used to make computer programs to play chess and go better than human players.

In machine learning, the computer plays an enormous number of games of chess against itself, and it notes which configurations of the chess pieces on the board tend to lead to victory. Then when playing against a (human or machine) opponent, at each move it chooses whichever of the available moves leads to what its experience tells it is the most favorable configuration.

What is most striking is how this computer play looks to human onlookers. It looks as though the computer has somehow recreated in a flash strategies discovered painstakingly by humans over centuries; and has in addition created new strategies, so offbeat and outrageous that no human would have ever thought of trying them. But these apparent characteristics of computer play are pure anthropomorphisms, for the machine is doing none of that. Instead, it is drawing on vast self-generated experience to choose favorable positions. No canny strategy, no daring originality. Those are purely human characteristics that we are mistakenly tempted to attribute to the computer.

Belatedly some people are coming back to the notion that the difference in wiring between transistors in a computer chip and neurons in the brain is important. To the extent that neurons resemble anything in computer world, it is not transistors in a chip but users in a social network: constantly being yelled at by hundreds of other users, and then in their turn shouting out to be heard by hundreds of others.

There is therefore an ongoing effort, usually subsumed under the name **neural nets,** to build a network of elements that is wired together more like neurons than transistors. But the ability to make such a network with its wide-ranging and semichaotic wiring on the scale used in computer chips would require great advances in engineering and nanotechnology. Furthermore, the wiring of neurons is not truly chaotic, but presumably obeys some fractal-type laws that we have not yet discerned. Thus, the neural net approach to AI would also require great advances in theory. Even if this approach does eventually succeed, the machines that it produces will be nothing like our current computers.

## Computer Use in Math and Science

So far we've talked about computers as tools of embodied mathematics, but what use do they serve in math and science?

Remember that the fundamental mathematical embodiment of computers is not logic but laziness. Computers are tools that can do our calculations for us if we program them correctly. Modern-day computers are capable of doing so many calculations so fast that problems that could be solved in theory but not in practice by human beings (because they would take the entire lifetimes of thousands of skilled people) can now be done in the background, leaving the mathematician or scientist free to plan an attack on the next theorem or calculation.

A lot of procedures that are human impractical are computer practical, because you can have a computer do the same thing over and over again and it won't want to rise up and hit you over the head with a keyboard for the unending boredom of it all. Nor will it grow sloppier and lazier with the brain-dribbling dullness.

Both of these are also evidence against the computer = human idea. The computer doesn't remember or care that it's done the same thing again and again. It does what it's programmed to. If you want it to do a Riemann sum with 4,000,000 calculations, it will do it. If you want it to draw a fractal through a hundred iterations it will do it.

...Provided you program it correctly. This brings up a worrisome area. If humans are doing the calculations to solve problems, then humans can look them over and make sure they were done correctly. Lewis Carroll pointed out that there was an uncomfortable paradox in this process, since my checking over the accuracy of your results depends on the accuracy of my checking procedure. Computers make this paradox even worse by doing more calculations than a person can handle in a lifetime. How can one make sure that the work was carried out properly? How can mistrust be made manifest in this vast expanse of laziness?

By checking the program. If the procedure is mathematically correct, then the result should also be.

This has made some mathematicians uncomfortable. It does not seem like real mathematical mistrust. It's more like scientific caution, the need to examine apparatus and run experiments more than once.

There was particular worry in 1976 when the **four color theorem**, one of the famous unsolved problems in mathematics, was proven using computers. The four color theorem states that any map on a plane can be colored with only four distinct colors so that no two adjacent areas have the same color. Part of the proof was done by 20th- and early 21st-century American mathematician Kenneth Appel and German mathematician Wolfgang Haken, who demonstrated that the problem could be reduced to a finite number of cases which could then be proved by checking each case.

This is a perfectly ordinary proof method, proof by exhaustion. It's not uncommon to end up with half a dozen or so individual instances that need checking in a proof. Except that in this case they had more than a thousand cases to check. They used a computer

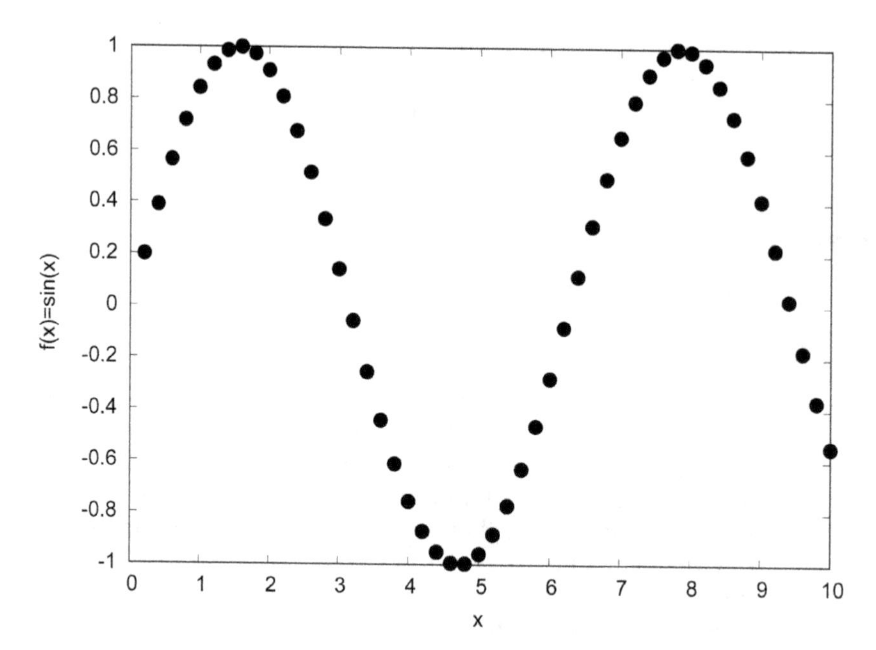


**FIGURE 14.2**
Sample of Function Values

program to do the checking. This meant that by some standards of proof, no human being really saw the four color theorem being proved.

What happened to deal with this difficulty was not a mathematical solution but a generational shift. The current crop of mathematicians are in general more comfortable with computer-based proofs than the older mathematicians. It seems likely that this discomfort will simply fade over time, but the caution should be remembered since these days few people keep track of what is actually going on in the guts of their computers.

The scientific uses of computing are widespread. For starters, there is the issue of so-called **big data**, collected data so voluminous that it is beyond human capacity to process it all: the sequences of bases in the human genome, the stars observed by the Hubble Space Telescope, the temperature readings of all the world's weather stations.

But what is beyond human capacity is trivial for today's computers with their vast memory and enormous processing speed. Big data can be collected, sorted, averaged, subjected to statistical tests, compared to various theoretical models, all in the comparative blink of an eye.

As we mentioned in previous chapters, scientific laws often come in the form of differential equations, and there are computer methods for solving such equations. We will end this chapter by expanding on this theme.

Earlier we presented the Newton quotient as a computer approximation to the derivative. We will now show how Taylor series can be used to make a more accurate approximation. A computer can't store a whole function f(x), but it can sample that function by giving its values at the points $x_i = i\,\Delta$ for some small $\Delta$ and for i = 1,2,3,…,n. That is, we start with the numbers $f_i = f(x_i)$ (Figure 14.2).

Now consider a Taylor series for f(x) expanded around $x_i$.

$$f(x_i + a) = f(x_i) + af'(x_i) + (a^2/2)f''(x_i) + (a^3/6)f^{(3)}(x_i) + (a^4/24)f^{(4)}(x_i) + \cdots$$

Here the ... refers to all the terms with higher powers of a. Now apply the formula to a = Δ recalling that $f(x_i) = f_i$ and $f(x_i+\Delta) = f_{i+1}$.

$$f_{i+1} = f_i + \Delta f'(x_i) + (\Delta^2/2)f''(x_i) + (\Delta^3/6)f^{(3)}(x_i) + (\Delta^4/24)f^{(4)}(x_i) + \cdots$$

Then in this notation, the Newton quotient is $(f_{i+1} - f_i)/\Delta$ and our formula gives

$$(f_{i+1} - f_i)/\Delta = f'(x_i) + (\Delta/2)\, f''(x_i) + \cdots$$

where, as before, the ... stands for all the terms with higher powers of Δ that we haven't written down. So sure enough the Newton quotient is a good approximation to the derivative: the error we make is of order Δ, so we can make the error small by choosing a small Δ.

But we can do even better. Let's go back to the Taylor series formula, and this time put in minus Δ for a. Then, we get

$$f_{i-1} = f_i - \Delta f'(x_i) + (\Delta^2/2)f''(x_i) - (\Delta^3/6)f^{(3)}(x_i) + (\Delta^4/24)f^{(4)}(x_i) + \cdots$$

Subtracting the minus Δ formula from the Δ formula, we get

$$(f_{i+1} - f_{i-1})/(2\Delta) = f'(x_i) + (\Delta^2/6)\, f^{(3)}(x_i) + \cdots$$

So now our error is of order $\Delta^2$. But if Δ is a small number, then $\Delta^2$ is a very small number, so now we are making a much smaller error than before, so this method is more accurate than the Newton method. The Taylor series also gives an accurate expression for the second derivative. Adding the Δ equation to the minus Δ equation and subtracting $2f_i$, we find

$$(f_{i+1} + f_{i-1} - 2f_i)/\Delta^2 = f''(x_i) + (\Delta^2/12)f^{(4)}(x_i) + \cdots$$

These approximations to derivatives are called finite differences, and their use in a computer simulation of differential equations is called the finite difference method.

Fourier's trick also gives rise to a computer method. Recall that Fourier's trick involves expressing the initial configuration of the system as a sum of "notes," finding the development over time of each note, and then adding up all the notes to find the time development of the system. But a function is typically made up of an infinite number of notes, so applying Fourier's trick involves summing an infinite series. In a few special cases, we can explicitly sum the infinite series and find the answer.

What should we do in all the other cases? Recall that since the infinite sum converges to the answer, this means that a finite sum gets pretty close to the answer, and the more terms we include in the sum, the better the approximation. So the approximate version of Fourier's trick is to only use some of the notes. At this point, there is nothing that says that we have to use a computer to apply this method. But since we want the approximation to be highly accurate, we had better include a lot of notes. So we have a method that involves a lot of computation: let's use a computer. Computer methods that use Fourier's trick are called spectral methods.

Computer simulations have become so ubiquitous in science that mathematicians have taken to calling them "experiments" and even some scientists have come to think of people doing scientific computing as having a sort of third specialty somewhere in between experimentalist and theorist.

However, both these points of view are mistaken. An experiment always involves asking a question of Nature. Any computation, whether done by a human or a computer is only asking a question of theory. Thus, people doing scientific computing are theorists. They are just theorists using methods that require so much computation, that it needs a computer.

**Or, as the saying goes, "To err is human. To really screw things up requires a computer."**

*Yes, that's true. Though to be fair, we scientists can screw things up pretty well even without computers.*

# 15

# *The Theoretical Universe of Modern Physics: Toolkit Included*

Physics and mathematics had gone their separate ways, with physicists still investigating nature but mathematicians looking into formal systems of axioms of ever more rarefied abstraction, moving further and further from the mathematics of nature. Or so it seemed. But then in the 20th century all that changed. The revolutions of relativity and quantum mechanics needed exotic mathematics that had never been used in physics before. And yet much of that mathematics had been developed over the years by mathematicians pursuing their own abstract ends.

This seems strange, but abstraction still needs to abstract from something and the human mind is still firmly rooted in human life living in reality. Our flights of fancy and our deep delving do not go as far away from the real world as we might think.

This is especially true when the real world turns out to be stranger than we think it is.

## Relativity

By the end of the 19th century, physics was firmly in the grip of a steampunk mindset. It was a coal-powered time with great control over mechanical processes and practical applications of F = ma in everything from ships to guns to factories.

It was also polluted and colonialist and statistically disease ridden and bad for people in other ways, but let's stick to the physics for the moment, because things were not going along as smoothly as the technology implied.

It was all very well for the mathematics to work out, but physicists didn't feel that they had a real understanding without a mechanical picture. Sound waves were fine since they had to do with pressure and density and velocity, real measurable properties of real fluids.

But Maxwell's newfangled electromagnetic (EM) waves? Wave behavior just because the equations said so? That didn't sound like real physics. Dash it all, waves were made by some medium well, waving sort of thing. Waves in water, waves in air, waves in puddings, those all made sense. But what were these EM waves waving in?

Faraday's lines of force at least sounded more real: like stresses and strains of some underlying medium. But what medium? They didn't know, and so they just invented one and called it the ether: the underlying medium in which light and other electromagnetic waves propagate, just as sound propagates in air. The model they were using insisted that there must be a medium, so they posited one.

They didn't know how much of their models would be destroyed with that casual assumption.

Whatever this ether was, it had some pretty strange properties. For starters, it was so insubstantial that no one had noticed it before. But recall that in fluids waves propagate at a speed of $\sqrt{dP/d\rho}$, where $dP/d\rho$, called the stiffness, is a measure of how hard the fluid is to compress. But sound takes 5 seconds to go a mile, whereas light can travel around the world seven times in a second. Since light is much, much faster than sound, the ether must be enormously stiff compared to air, or water, or steel, or any substance people had encountered. How could something be so insubstantial that we don't even notice that we are moving through it, and at the same time many millions of times stiffer than steel?

No one knew the answer to those questions, but at least they could try to measure our motion through the ether. Since the Earth moves around the Sun, it is always in motion and in different directions at different times of the year, so the Earth had to be in motion relative to the ether at most or all times of the year. Michelson and Morley tried to measure this motion using a clever piece of equipment called an interferometer. This is an L-shaped apparatus that at the corner of the L takes a light beam and splits it in two. Each half of the light beam travels down its arm of the L, where it is reflected by a mirror back the way it came. The two light beams meet back at the corner of the L, where they combine to form a beam that is either especially light or especially dark depending on differences between the motion of the light in the two arms (Figure 15.1).

Why would there be differences in the motion of the light? Well, suppose the apparatus is moving through the ether in the direction of one of the arms. Then the light in that arm has an upstream and then downstream journey, while the light in the other arm has a cross-stream journey each way. This difference can be detected by the interference of the two light beams. In particular, as the orientation of the apparatus is changed, so the two arms switch which one is the upstream-downstream one and which one is the cross-stream one, the combined light beam should change its overall brightness.

This, by the way, is an excellent example of how good experimental apparatus should be designed. They were trying to measure an invisible object (the ether) that had consequences into visible phenomena. The visible consequences were refined and controlled until they would reveal exactly not only what was desired from it, but also the unknown, unlooked for, unthought of possibility that they were just plain wrong.

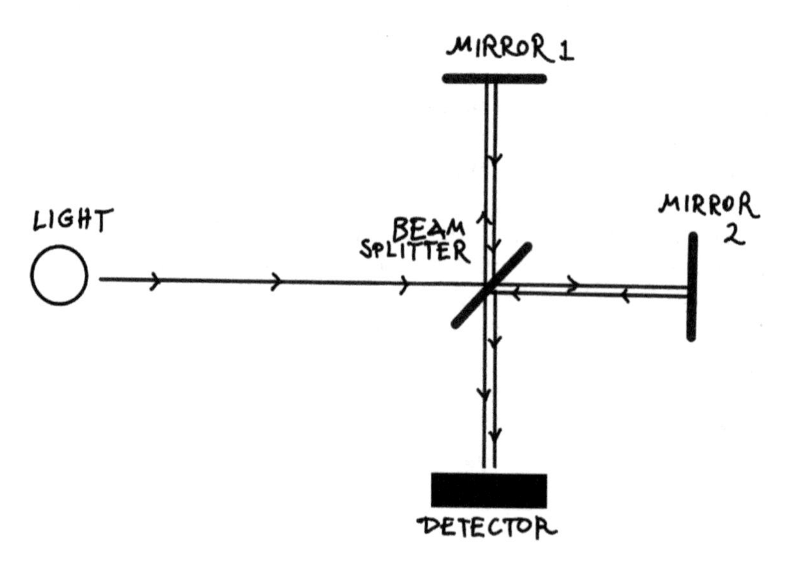


**FIGURE 15.1**
Interferometer

Michelson and Morley did their experiment and found no effect. No dependence on the orientation of the apparatus. No dependence on the time of year. No motion through the ether. No ether at all.

The response to this amounted to, "what the?"

That wasn't the only strange thing about light as EM waves. Other experiments showed that the speed of light didn't depend on the speed of the object emitting the light, as it would if light acted like a projectile. If I'm standing in a train moving at 70 miles/hour and I throw a baseball in the direction the train is moving at 30 miles/hour, then the ball moves at 100 miles/hour, at least as measured by someone standing by the train tracks. But light just moves at the same speed c (approximately $3.00 \times 10^8$ m/sec) no matter who emits it or who measures it.

How to account for this strange behavior of light? Albert Einstein found the answer through a remarkable generalization of the Pythagorean theorem. To understand this generalization, we begin by recalling that when we graph a function $y=f(x)$, the graph is a curve in the x-y plane. But we also often graph motion by plotting x as a function of t, and are thus producing a curve in the t-x plane. If we were to imagine making a corresponding graph of motion in space, we would have a curve in a four-dimensional (t, x, y, z) spacetime.

Each point (called an event) of this spacetime is a single point of space at a single instant of time. The Pythagorean theorem tells us that the distance ds between two nearby points in space satisfies $ds^2 = dx^2 + dy^2 + dz^2$, where dx is the difference between their x-coordinates, and correspondingly for dy and dz.

That's the flat Euclidean metric. But remember that there's more than one possible metric.

Einstein tells us that in spacetime we are supposed to use the following expression:

$$ds^2 = -c^2dt^2 + dx^2 + dy^2 + dz^2$$

The presence of the speed of light c in this formula is not surprising. We can add only things that have the same units, so to turn dt into something with the same units as dx, we must multiply by something with the units of speed. However, the minus sign is very surprising. This spacetime metric is not a metric at all in the sense of the axioms for metric spaces given in Chapter 10. $ds^2$ can be zero or even negative. What would that mean? An imaginary distance?

This is a situation in the development of the idea of metrics that looks much like the development of kinds of numbers. We were comfortable with our positive definite metrics just as we were comfortable with positive integers, integers, rational numbers, and real numbers, but problems and physical needs made it necessary to expand the idea of number.

So what about expanding the idea of metric?

This uncomfortable metric is useful in characterizing the behavior of light. To see why, consider a light ray that starts at one event and goes to a nearby one with elapsed time dt and change in spatial coordinates (dx, dy, dz). Then the distance covered by the light ray is $\sqrt{dx^2 + dy^2 + dz^2}$, but since the light ray is traveling at the speed of light, the distance covered must also be c multiplied by the elapsed time. In other words, $cdt = \sqrt{dx^2 + dy^2 + dz^2}$, which means that $ds^2 = 0$. So, another way of saying that light always travels at the same speed is to say that for the funny looking metric given above, light rays have $ds^2 = 0$.

To explore the properties of this strange four-dimensional metric, we will start with the metric of the two-dimensional Cartesian plane: $ds^2 = dx^2 + dy^2$. The choice of direction for the x-axis is completely arbitrary, and so in particular we could have chosen an $(x', y')$

coordinate system with the x′- and y′-axes rotated at an angle α from the old x- and y-axes. In this case, the new (x′, y′) coordinates of a point would be related to their old (x, y) coordinates by

$$x' = x\cos\alpha + y\sin\alpha$$

$$y' = y\cos\alpha - x\sin\alpha$$

But distance is an actual physical quantity, so it shouldn't depend on our arbitrary choice of where to point the x-axis. Or, to put it another way, we should get the same distance using (x′, y′) coordinates that we did with (x, y). Let's see if we do. From the formulas above, we get

$$dx' = dx\cos\alpha + dy\sin\alpha$$

$$dy' = dy\cos\alpha - dx\sin\alpha$$

and then taking the squares of these two formulas, we find

$$(dx')^2 = dx^2\cos^2\alpha + dy^2\sin^2\alpha + 2dx\,dy\cos\alpha\sin\alpha$$

$$(dy')^2 = dx^2\sin^2\alpha + dy^2\cos^2\alpha - 2dx\,dy\cos\alpha\sin\alpha$$

Now, adding these two quantities, we find

$$(dx')^2 + (dy')^2 = (\cos^2\alpha + \sin^2\alpha)(dx^2 + dy^2)$$

This doesn't look the same as $dx^2 + dy^2$, but it is because no matter what angle α is, we have

$$\cos^2\alpha + \sin^2\alpha = 1$$

To see this, consider the old staple of trigonometry textbooks, a right triangle with sides a and b and hypotenuse c, and let α be the angle for which a is the adjacent side (Figure 15.2).

Then, we have cos α = a/c and sin α = b/c, and therefore

$$\cos^2\alpha + \sin^2\alpha = (a^2 + b^2)/c^2 = 1$$

Einstein applied the same sort of reasoning to the spacetime metric with the strange minus sign. Could there be some sort of "rotation" between t and x that would leave $ds^2$ unchanged? This sort of "rotation" couldn't use cosines and sines because of the minus sign in $ds^2$. Instead, we would need some other type of functions where the square of one function minus the square of the other was always equal to 1. To see what sort of functions we should use, recall the connection between sines, cosines, and exponentials of imaginary numbers:

$$\exp(ix) = \cos x + i\sin x$$

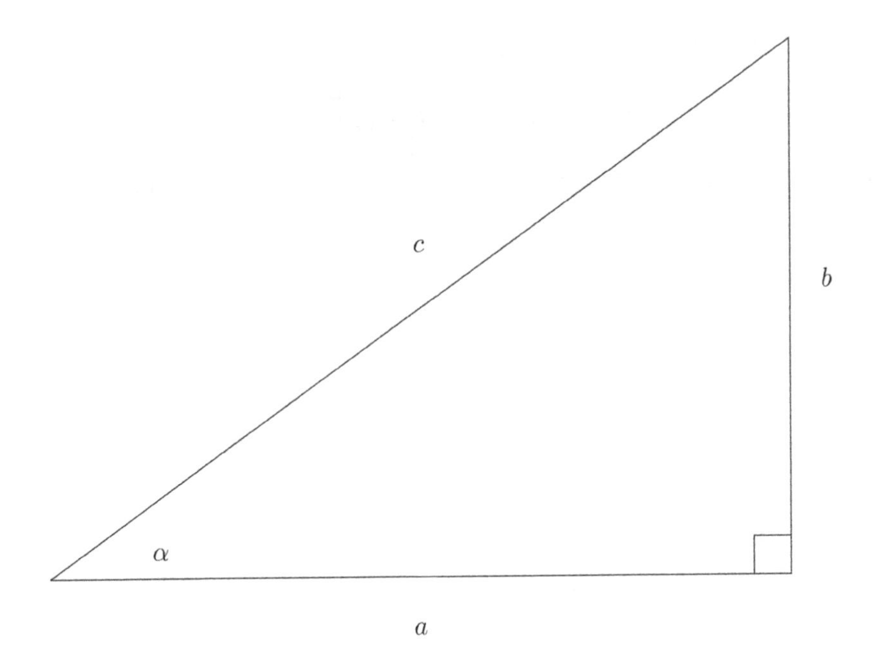


**FIGURE 15.2**
Right Triangle

which also means that

$$\exp(-ix) = \cos x - i \sin x$$

Adding these two formulas and dividing by 2 we find

$$\cos x = (1/2)\left[\exp(ix) + \exp(-ix)\right]$$

Similarly, subtracting the two formulas and dividing by 2i, we find

$$\sin x = (1/2i)\left[\exp(ix) - \exp(-ix)\right]$$

The new functions that we need to make Einstein's "rotations" are called cosh (short for hyperbolic cosine) and sinh (short for hyperbolic sine), and their formulas look like the expressions above for cosine and sine but with real exponentials rather than imaginary exponentials. That is,

$$\cosh x = (1/2)\left(e^{x} + e^{-x}\right)$$

$$\sinh x = (1/2)\left(e^{x} - e^{-x}\right)$$

We now check that these functions have the property that we need. Note that $e^{x}e^{-x} = e^{x-x} = e^{0} = 1$. So we find

$$\cosh^2 x = (1/4)\left(e^{2x} + e^{-2x} + 2\right)$$

$$\sinh^2 x = (1/4)\left(e^{2x} + e^{-2x} - 2\right)$$

Subtracting the second equation from the first, we have

$$\cosh^2 x - \sinh^2 x = (1/4)(2-(-2)) = 1$$

which is just what we need. Einstein's "rotation" between t and x (usually called either a Lorentz transformation or a boost) is

$$ct' = ct \cosh\psi - x \sinh\psi$$

$$x' = x \cosh\psi - ct \sinh\psi$$

Here the parameter $\psi$ plays the same role for boosts that the angle $\alpha$ plays for rotations. In a little while we will find the physical meaning of this parameter. But first we will check that the transformation does what it needs to: namely, make $ds^2$ in the new coordinates the same as $ds^2$ in the old coordinates.

From the formulas above, we have

$$cdt' = cdt \cosh\psi - dx \sinh\psi$$

$$dx' = dx \cosh\psi - cdt \sinh\psi$$

Taking the squares of these formulas, we find

$$c^2(dt')^2 = c^2dt^2 \cosh^2\psi + dx^2 \sinh^2\psi - 2\,c\,dt\,dx \cosh\psi \sinh\psi$$

$$(dx')^2 = c^2dt^2 \sinh^2\psi + dx^2 \cosh^2\psi - 2\,c\,dt\,dx \cosh\psi \sinh\psi$$

Subtracting the first formula from the second one, we find

$$-c^2(dt')^2 + (dx')^2 = (\cosh^2\psi - \sinh^2\psi)(-c^2dt^2 + dx^2) = -c^2dt^2 + dx^2$$

So, the spacetime metric $ds^2$ is the same in the new coordinates as in the old ones.

So, we abstracted, we calculated, and we found something that fits together, but what happens when we bring this back to reality. For that we need to consider what a curve in spacetime represents. Our graphs of something versus time show how that quantity changes over time. The graph is a visual representation of the history of that changing quantity.

When we're graphing position in spacetime, we're showing the history of where an object was as it moved around.

We can imagine ourselves as an observer sitting still at the origin of the old spatial coordinates, so that $x = 0$ for all time. Similarly, we can imagine some other observer sitting still (or so it seems to him) at the origin of the new spatial coordinates, so that for this observer $x' = 0$. But when expressed in our coordinates using the boost, $x' = 0$ becomes

$$x \cosh\psi - ct \sinh\psi = 0$$

which in turn becomes

$$x = c(\sinh\psi / \cosh\psi)\, t$$

To make this formula less messy, we will use the abbreviation
$\tanh\psi = \sinh\psi/\cosh\psi$ (in analogy to the formula $\tan\theta = \sin\theta/\cos\theta$) so that we have

$$x = (c \tanh \psi)t$$

But this is just the formula for an object moving in the x direction at speed v, where $v = c \tanh\psi$. So the new observer at rest in his own coordinate system is to us an observer moving at a constant velocity. This turns the statement that $ds^2$ is unchanged by a boost into the statement that the physics of light is unchanged as measured by an observer moving with constant velocity.

The statement that physics is unchanged by uniform motion of the observer is called the principle of relativity. This principle is usually associated with Einstein; however, Einstein was quick to point out that the principle of relativity goes back to Galileo and was an essential part of the theories of Galileo and Newton.

Galileo devoted much effort to defending the theory of Copernicus that the Earth revolves around the Sun, and that the apparent motion of Sun and stars around the Earth is simply due to the Earth's rotation. But the rotation of the Earth makes someone at the equator move at a speed of over 1000 miles/hour: an enormous speed compared to the usual modes of transportation of Galileo's time. If Copernicus is right, people wondered, how could we be moving at these enormous speeds and not even notice it?

Galileo's answer is that we don't notice constant velocity at all, only changes in velocity, i.e., acceleration. In Newton's laws of motion, this idea is made explicit: an observer moving with constant velocity **v** sees everyone's velocity changed by adding **–v** to it. But the derivative of a constant is zero, so this constant change to velocity results in no change to acceleration. Same **a** means same **F**=m**a**. Newton's laws of motion are the same for all uniformly moving observers. So far, Einstein's ideas while weirdly expressed seem to be firmly grounded in good old pre-steampunk mechanics.

Now let's see if we can express the boost in terms of the observer's velocity. We have $v = c \tan h\psi$, so we want to use this to express $\cosh \psi$ and $\sinh \psi$ as functions of v. We have

$$1-\left(v^2/c^2\right) = 1-\tanh^2\psi = (\cosh^2\psi - \sinh^2\psi)/(\cosh^2\psi) = 1/\cosh^2\psi$$

which means that

$$\cosh\psi = 1/\sqrt{1-v^2/c^2}$$

It's clumsy to have to keep writing $1/\sqrt{1-v^2/c^2}$, so we introduce the abbreviation $\gamma = 1/\sqrt{1-v^2/c^2}$. With this notation, we have $\cosh \psi = \gamma$. It then follows that

$$\sinh\psi = \cosh\psi \tanh\psi = \gamma v/c$$

So in the expression for the boost, we can replace each $\cosh \psi$ with $\gamma$ and each $\sinh \psi$ with $\gamma$ v/c, which gives

$$t' = \gamma\left(t-\left(v/c^2\right)x\right)$$

$$x' = \gamma(x - vt)$$

Note that c is such an enormous number that most velocities we encounter are much less than c. This means that v/c is very close to zero and $\gamma$ is very close to 1. In this case, the boost formula is $t' = t$ and $x' = x - vt$. That is, time doesn't change at all and positions change by having vt subtracted from them. This is the relativity formula of Galileo.

This is a basic feature of all new physics theories that supplant old physics theories. The old physics was tested by experiment and therefore must be approximately correct in the situations that the experiments probed. The new physics must therefore give approximately the same answers as the old physics, at least for those cases tested by the old experiments.

This point is sometimes missed by people steeped in the Great Books approach to education. They say things like "Aristotle's physics was shown to be wrong by Galileo and Newton, but then Newton was shown to be wrong by Einstein. So what's the difference?" The difference is that Newton's theories were tested by experiment, and Aristotle's weren't.

**Aristotle's physics look completely different if you actually try to make them work. For example, cannon shots go up along a line, stop in midair, and then fall.**

At this point, all our physics theories are approximations that come with warning labels: "these are the conditions under which this theory works well, and these are the conditions where it doesn't." In this good approximation sense, Newton will always be a part of physics, and Aristotle never was.

Now, let's look more closely at the metric. We know that $ds^2=0$ for light, but what about when $ds^2$ is not zero? What does that describe? For objects moving more slowly than light, $ds^2 < 0$, and in fact Einstein tells us that all material objects must move slower than light since otherwise the formula for $\gamma$ would not make sense. For moving objects, Einstein showed that the quantity $d\tau$ given by

$$d\tau = (1/c)\sqrt{-ds^2}$$

is the time elapsed on a watch attached to that object. This elapsed time is called proper time. Finding the total elapsed time is just like finding the total length of a curve: we call the curve through spacetime describing the history of the object the world line of the object. Now we divide the world line into small pieces, calculate $d\tau$ for each piece, and then add up the results.

This is the next result that goes, wait what? Proper time is not the same for each observer. Different observers moving at different relative velocities will measure different durations for the same event.

Proper time gives rise to a famous result of relativity, known as the twin paradox (Figure 15.3).

One twin stays on Earth, while the other twin travels at high speed to a nearby star and then returns at high speed. Upon his return, the traveling twin is younger than the twin who stayed on Earth. This "paradox" is not really a paradox at all. It only becomes paradoxical if one tries to set up the problem using something like formal logic: "by definition twins are always the same age. But Einstein tells us that these two twins are not the same age. OMG, how paradoxical!"

Instead, we suggest approaching the twin "paradox" with the slogan "clocks are like odometers." If two people were each to drive from Detroit to Chicago but they took different routes, their car odometers would read a different number of elapsed miles. This is because car odometers don't measure "the distance from Detroit to Chicago" but only the distance along the route traveled by the car. Similarly, Einstein tells us that clocks don't

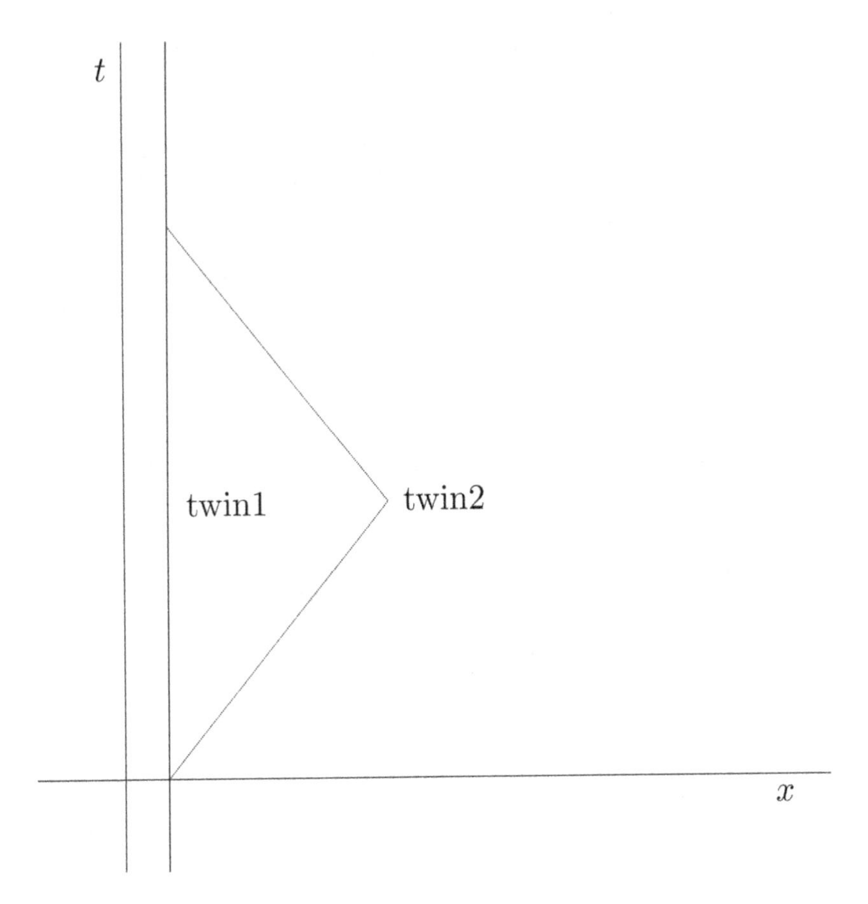


**FIGURE 15.3**
Twin Paradox

measure "the time between two events" but rather the elapsed proper time along world lines. Since the twins have different world lines, they can have different ages, as measured by the clocks that each of them carries.

Note that it is the traveling twin who is younger: in the spacetime diagram, the curve has smaller proper time than the straight line. This is a consequence of the minus sign in the metric. We can see this by using t as a parameter for the world line in the formula for the proper time. We then find that the total elapsed proper time $\Delta\tau$ is given by

$$\Delta\tau = \int_{t1}^{t2} dt\sqrt{1 - v^2/c^2}$$

Here t1 is the time at which we start, t2 is the time at which we end, and $\mathbf{v} = d\mathbf{x}/dt$ is the instantaneous velocity of the object whose proper time we are calculating.

There is no shortest proper time between two events. However, there is a longest proper time, and the twin who stays on Earth has it. In relativity, we replace "the shortest distance between two points is a straight line" with "the longest proper time between two events is a straight line in spacetime." Note that a straight line in spacetime means not only that the path in space is straight, but also that the velocity is constant. Thus, Newton's first law of motion can be replaced with the statement that an object subject to no force travels a straight line in spacetime.

Relativity ushered in a radical new understanding of the nature of space and time. But there is a price to be paid for this new understanding: it forces us to go right back to the beginning and rewrite all the equations of mechanics. For the most part, this rewriting leads to small corrections to Newtonian mechanics, but there is one case where the corrections are not small leading to the most famous equation in all of physics: $E = mc^2$.

Note that fame is not the same as utility. $F = ma$ is still the most used equation. But $E = mc^2$ has mathematical elegance and can be put on a bumper sticker or incorporated into a meme.

We will now explore some of this rewriting, including the famous equation. Recall that we started mechanics by thinking of position x as a function of time t and then taking derivatives to get velocity and acceleration. But this approach doesn't sound sensible from the point of view of spacetime, where t is just another coordinate, much like x, and one that depends on choice of observer in much the same way that x depends on choice of direction of the x-axis.

To facilitate this point of view, we will temporarily adopt a system of units in which $c = 1$. One way to do this is to measure length in light seconds (the distance that light travels in 1 second) instead of meters. This change will make our formulas simpler since we won't have to write all the powers of c. For example, the formula for $\gamma$ becomes $\gamma = \sqrt{1 - v^2}$.

This is computationally convenient but does rather mean that we're not working problems useful for any travel on Earth, but that's fine. On Earth, Newton takes care of most of our needs.

At the end of our calculations, we will put the powers of c back using dimensional analysis. There is also a simplification in spacetime diagrams, where now the paths of light rays are lines at a 45 degree angle to the vertical (Figure 15.4).

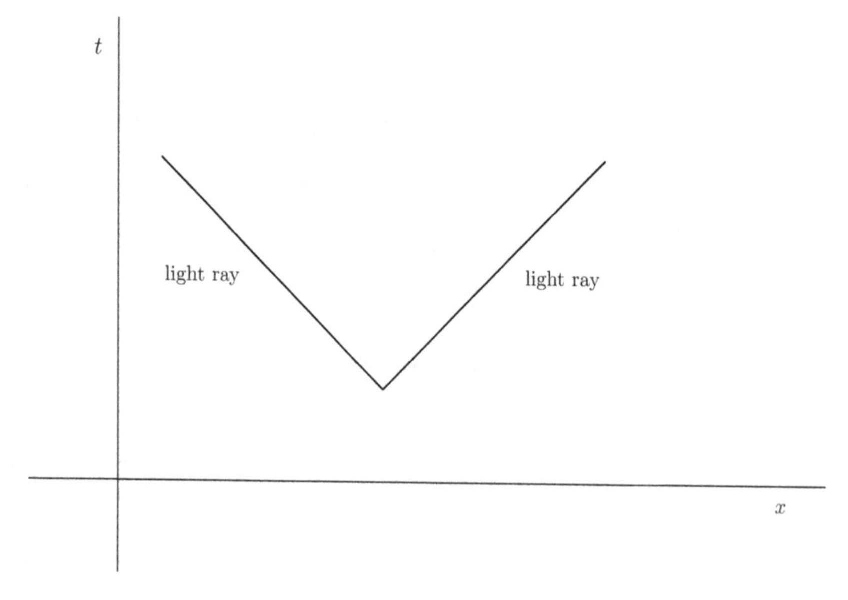


**FIGURE 15.4**
Simplified Spacetime Diagram

Since time and space are on equal footing, instead of a spatial position vector with three components – x, y, and z – we should have a spacetime position vector with four components: t, x, y, and z. The notation for this four-vector is $x^\mu$. Here $\mu$ is an index that goes from 0 to 3, so $x^\mu$ is a collection of four numbers: $x^0 = t$, $x^1 = x$, $x^2 = y$, and $x^3 = z$. Or put another way, $x^\mu$ is a four-tuple of numbers $x^\mu = (t, x, y, z)$. Put yet another way, since we are already used to the notation where a boldface **r** stands for the triple (x, y, z), we can think of a four-vector as a pair consisting of a number (the time component) and an ordinary vector (the spatial components). That is, we can use the notation $x^\mu = (t, \mathbf{r})$.

**Wait a minute, you just stuck that μ in where we place exponents. That doesn't look like vector symbolism. It looks like x raised to the μth power.**

*I'm afraid there's only so many places to stick symbols and we're going to need the subscripts for other stuff. We'll be using the superscripts for this and try not to conflate them with powers.*

**Fine.**

We now want to take the derivative of the four-position to get a four-velocity. But what should we take the derivative with respect to? Certainly not t, which is just another one of the coordinates. Instead, we take derivative with respect to $\tau$, a real physical quantity representing the time elapsed on a clock carried by the moving object. That is, we define the four-velocity $u^\mu$ by $u^\mu = dx^\mu/d\tau$. From the formula for $d\tau$ we find that $u^\mu = (\gamma, \gamma\mathbf{v})$. Similarly, we can take another derivative to get the four-acceleration $a^\mu = du^\mu/d\tau$. Newton's second law then becomes $F^\mu = ma^\mu$.

In Newtonian mechanics, the momentum is $\mathbf{p}=m\mathbf{v}$, so the analogous quantity in relativity is the four-momentum $p^\mu = mu^\mu$. The spatial part of the four-momentum is $\gamma m\mathbf{v}$. This looks reasonable, since for small velocity, $\gamma$ is approximately 1, so we find that at small velocity, the spatial part of the four-momentum is approximately equal to the Newtonian momentum. This is what we would expect from the notion that at small velocity relativity gives us the results of Newtonian mechanics.

But what is the time component of the four-momentum? It turns out that it is the energy. This gives $E=\gamma m$, or putting back the correct factors of c using dimensional analysis, we have $E = \gamma\, mc^2$. Therefore, at zero velocity, we have

$$E = mc^2$$

This is the famous formula. Since c is such a large number, this formula tells us that locked inside each innocuous bit of matter is a truly staggeringly huge amount of energy. Chemical reactions release only a tiny fraction of that intrinsic energy. Nuclear reactions are millions of times more powerful than chemical reactions and still release less than 1% of this stored energy. Matter and antimatter can annihilate releasing all the energy. This sounds pretty dangerous, but luckily there is very little antimatter in the universe.

Matter falling into a black hole can release a fair percentage of its stored energy. But again, we're in no danger: all black holes are far away from us. But we can see some of the consequences of matter falling into them by the bright flares at their poles.

Newton's formula for the energy of a moving object is $E = (1/2)\, mv^2$. At first, this looks like a huge disagreement with Einstein's formula even at the small speeds where Newton's formula is supposed to be right. But appearances are deceiving: Newton deals only in differences of energy, so his formula is really for the difference between the energy of a moving object and the energy that the same object had when it was at rest. Put another way, Newton says that the energy of a moving object is $E_0 + (1/2)\, mv^2$, where $E_0$ is some unspecified constant. Is this formula consistent with Einstein's formula? To see, we write the longhand version of $E = \gamma mc^2$.

$$E = mc^2/\sqrt{1-v^2/c^2}$$

The first two terms of the Taylor series for $1/\sqrt{1-v^2/c^2}$ are

$$1 + (1/2)\left(v^2/c^2\right)$$

So Einstein's formula says that for small velocity, the energy is

$$E = mc^2\left(1 + (1/2)\ v^2/c^2\right) = mc^2 + (1/2)\ mv^2$$

So Einstein's formula actually agrees with Newtonian mechanics despite the huge $mc^2$.

The expression $E = mc^2/\sqrt{1-v^2/c^2}$ provides a physical reason for why objects always travel slower than light. Suppose you try to make an object go faster than light by applying a force to it. The longer you apply your force, the more energy you are putting into the object's motion. But it would take an infinite amount of energy to get the speed to c, so however much energy you put in, the object is always moving at a speed less than c. This is just what occurs at particle accelerators like the Large Hadron Collider. The particles are pushed around a large ring of magnets, with each particle given a truly staggering amount of energy, and yet the particles are always just a little bit slower than light.

Nonetheless, despite this energy argument, you might think to make something go faster than light by the following method: suppose I myself am moving at (3/4)c, and I throw forward an object at a speed of (3/4)c relative to me. Since ¾+ ¾ = 3/2, surely the thrown object will be moving at a speed of (3/2)c, which is faster than the speed of light.

It turns out that this won't work either, for reasons that we will now explore. When someone moving at speed v1 throws an object at speed v2, what add are not speeds but boost parameters. That is, the object is moving at a speed $v = c \tanh(\psi1+\psi2)$, where $v1 = c \tanh(\psi1)$ and $v2 = c \tanh(\psi2)$. From the properties of the exponential function, one can show that for any $\psi1$ and $\psi2$

$$\tanh(\psi1+\psi2) = (\tanh(\psi1)+\tanh(\psi2))/(1+\tanh(\psi1)\tanh(\psi2))$$

so the speed of the thrown object is

$$v = (v1 + v2)\ /\left(1 + v1\ v2\ /c^2\right)$$

In particular, if you are moving at ¾c and throw an object at ¾c, it won't move at (3/2)c, but instead at (3/2)c/(1+9/16), which is 0.96c.

**This fact is the bane of so many SF writers. c is fast compared to everything we do in everyday life, but it's very slow if you want to travel the stars. It takes four years to get to the nearest star at lightspeed and we can't reach lightspeed. So, some SF makes use of this fact to introduce generation ships that take centuries to travel between the stars. Others just dump the physics and hand wave our way out there.**

*Also, the math is about to get much more complicated.*

## General Relativity

The relativity that we have been talking about so far is called special relativity. Now we move on to general relativity, which is Einstein's theory of gravity, which you may notice was not discussed above. For Newton, gravity and mechanics explained each other and created the elegance of his laws of motion and gravity, all of which are nice calculations with mostly algebra and bits of calculus when needed.

For Einstein, gravity opens up whole new wells of math.

One of the consequences of relativity is that not only can't objects travel faster than light, but no influence, nothing that could be used to send any kind of message, can travel faster than light.

Judged by this criterion, we can immediately see that Newton's theory of gravity is going to be trouble. Start with the formula

$$F = G\,M\,m/r^2$$

for the force that a mass M exerts on a mass m separated from it by a distance r. If we move M a little, that immediately changes the distance r, which causes a change in the force felt by m. In effect, we can send a message from M to m over a distance r in no time at all simply by adding to or subtracting from M (add rocks or remove rocks).

In fact, this is the very same "action at a distance" property that bothered Newton's contemporaries. Only now Newton's "hypotheses non fingo" dodge wouldn't work anymore. A speed limit of $3 \times 10^8$ meters per second isn't just a good idea, it's the law. Newton's theory of gravity was breaking the law and needed to be replaced.

You might worry that Coulomb's law for the force between charges (which basically has the same form as Newton's law of gravity) also has this action at a distance problem and would need to be replaced. But in fact, Coulomb's law had already been replaced by Maxwell's equations and the Lorentz force law.

Electromagnetic waves travel at the speed of light (they are light), so they have no way of sending a message faster than light. Maxwell's equations play well with special relativity: all that needed to be done with them was to find the transformation law for how electric and magnetic fields changed when viewed by a moving observer.

Given the success of Maxwell's equations and the similarity of Coulomb's law to Newton's law of gravity, you might think that Einstein would have tried to find a sort of Maxwell's equations for gravity. But oddly enough, he took a completely different route.

Einstein began by noting a striking property of Newtonian gravity: combining Newton's law of gravity with F = ma, one finds that the acceleration of mass m is given by

$$a = G\,M/r^2$$

In this expression, m is nowhere to be found. "The feather and the ten ton truck fall at the same rate" as David's high school physics teacher liked to say. Einstein called this feature the principle of equivalence. That is, there is an equivalence between gravity and acceleration.

Einstein set out to make a theory of gravity that incorporated the principle of equivalence: that is, he would dispense with the notion of gravitational force and seek a theory that directly predicted gravitational acceleration.

This is an even more radical conceptual break from classical (i.e., Newtonian) mechanics than any of the earlier Einstein caused changes. Newton's three laws of motion are founded on the idea that changes in motion (acceleration) are caused by forces. So, every problem in mechanics is framed around moving objects and forces acting on them. The principle of equivalence not only erases the forces, it causes physicists to think about motion without the idea of force. It changes the model at a deep conceptual and practical level.

Einstein began with some facts about optics: c is the speed of light in vacuum, but light is actually slowed down when it travels through a transparent substance: it moves slower in air than vacuum, slower in water than air, and slower in glass than water. Each transparent substance is characterized by what is called an index of refraction, which is the ratio of c to the speed of light in the substance.

Light travels in straight lines in a given transparent substance, but it is bent as it passes from one kind of transparent substance to another, with the angle of the bend having to do with the mismatch between the two indices of refraction. If the index of refraction were to change continuously, light would travel a curved path through this ever-changing transparent substance.

Suppose, Einstein posited, one were to replace his metric with one where c changed depending on the strength of the gravitational field. Then, light and material objects would travel on curved paths through spacetime in a way consistent with the principle of equivalence. Light rays would still be given by $ds^2=0$, and objects in free fall (i.e. acted on only by the force of gravity) would still follow curves of maximum proper time, but these paths would no longer be straight lines because of the dependence of c on space and time.

Unfortunately, this theory of Einstein did not work. It could replicate all the results of Newtonian gravity, but there was one result that it couldn't account for: the behavior of the planet Mercury. In Newtonian gravity, each planet travels in an ellipse. But astronomers had found that the orbit of Mercury was actually a precessing ellipse: the direction of the long axis of the ellipse changed a tiny amount in each orbit. These changes added up over time, but even the cumulative effect was tiny only 43 seconds of arc per century, where a second of arc is 1/3600 of a degree. Einstein's theory also gave a precession of the direction of Mercury's orbit, but it was the wrong number. It was time to go back to the drawing board.

Eventually, Einstein realized that he was going to need to have all the components of his metric (not just the time component) depend on space and time. In other words (though he didn't know at the time about their work) Einstein was going to need to change his flat spacetime metric of special relativity in just the same way that Gauss and Riemann changed the flat geometry of Euclid to produce the metric of curved space.

The mathematics of his metrics quickly became very complicated, and Einstein realized that he was going to need help. In our knight and smith terminology, Einstein was the greatest knight of the age, but unlike Newton, he was not also the age's greatest smith. Einstein consulted his friend, the mathematician, Marcel Grossmann.

Grossmann informed Einstein that he was in luck: all the mathematics he needed had been worked out by Riemann. Remarkably, the funny minus sign in Einstein's metrics made very little difference. The mathematics of curved spacetime was almost identical to the mathematics of curved space that Riemann had worked out in the previous century. Ironically, the mathematics of non-Euclidean spaces that had been considered completely unphysical was exactly what was needed to describe the physics of gravity. The light rays and the trajectories of objects in free fall were simply the geodesics of curved spacetime: the analog of the curves of shortest distance of curved space.

Grossmann the smith handed Sir Einstein the flexible curved sword of spacetime with all its manifold uses.

Using this, Einstein could make the math work. But he didn't have the physics yet. He could see that gravity curved spacetime, but gravity from a classical perspective was caused by matter, so how and why did the presence of matter affect the shape of spacetime?

To make an analogy with electromagnetism, what Einstein had found was the analog of the Lorentz force law. But he still needed to find the analog for gravity of Maxwell's equations. That is, he needed to find the gravitational field equations.

Einstein's path to the field equations was long and complicated. So instead we're going to tell you a simpler part of the story: the path to the field equations taken by David Hilbert, a 19th- and 20th-century German mathematician. General relativity was only a sideline for Hilbert, who made important contributions to many different areas of mathematics, and had particular interests in both abstract algebra and the foundations of mathematics.

It was only a matter of time before the world's greatest knight saw fit to call on the world's greatest smith. And so Einstein made the short trip from Berlin to Gottingen to talk to Hilbert, the world's greatest mathematician, about the problem of the field equations. They met in person and later corresponded by letter.

Eventually, in one of those letters, Hilbert made a most un-smith-like statement. Just as Clemenceau, the French prime minister of the time, famously said that war was too important to be left to the generals, so Hilbert seemed to think that physics was too important to be left to the physicists. Eventually, Hilbert decided to solve the problem of the field equations himself, and he informed Einstein of this decision in a letter.

All of a sudden Einstein found himself in a competition to complete his own theory! In the waning months of 1915, Einstein and Hilbert each worked furiously on the problem, and finally in a sort of Newton/Leibniz photo finish, they each independently (and approximately simultaneously) arrived at the field equations of general relativity.

Just to be clear on the nature of the competition, at the end of 1915, Einstein had done far more than just find the field equations. He had also used them to calculate the orbit of Mercury and show that general relativity gave the correct 43 seconds of arc per century for the change in orientation of the ellipse. He also used general relativity to predict the amount that light would be bent by the gravity of the Sun: a prediction that was confirmed in measurements made during the eclipse of 1919. (Einstein was fortunate in this timing: his earlier incorrect theory had predicted only one-half of the correct value of light bending. However, an earlier attempt to measure the bending during an eclipse had been derailed by the onset of World War I.) This, along with Einstein's earlier work, is why it is Einstein's theory of general relativity, not Hilbert's theory of general relativity.

Nonetheless, it is remarkable that Hilbert, who had only studied the problem of the field equations for a comparatively short amount of time, managed to find the answer at the same time as Einstein. How did he do it? By using a simpler method called the method of least action.

We will now take a slight detour to explain the method of least action, and then will explain how Hilbert used it.

We have actually encountered the principle of least action before, though not using those words. The statement "the shortest distance between two points is a straight line" is an example of this principle. Here one starts with the two points, considers all possible curves that connect the points, and calculates the length of each of those curves.

The calculus of variations considers the shape of the curve as determined by a set of parameters, and since we are looking for the curve of minimum length, we require that the derivative of the length with respect to each parameter is zero. This requirement then gives us an equation for the curve, which can then be shown to be a straight line. Exactly the same calculus of variation method can be used in a curved space, giving the geodesic equation.

Broadly speaking, the principle of least action says that laziness is inherent in motion. Or, objects never take the scenic route if they can avoid it.

Remarkably, $\mathbf{F} = m\mathbf{a}$ can also be reformulated in terms of the principle of least action. We begin with the kinetic energy $K = (1/2)mv^2$. There is also a potential energy V, which is a function of position. In physical terms, V is the amount of energy stored by putting the object in its current position. But mathematically, V is a function whose gradient is minus the force: $\mathbf{F} = -\nabla V$. For example, for a spring, where $F = -kx$, the potential is $V = (1/2)kx^2$. And for gravity, where the force is an attraction whose magnitude is $F = GMm/r^2$, the potential is $V = -GMm/r$. The total energy is $E = K + V$.

But the principle of least action makes use of a different combination of the two types of energy called a Lagrangian: $L = K - V$, named for the 18th- and 19th-century French mathematician Joseph-Louis Lagrange. Rather than looking at the total energy, the Lagrangian measures the difference between the potential and kinetic energies of an object.

The action I is the definite integral of the Lagrangian

$$I = \int_{t1}^{t2} dtL$$

That is, we consider all possible trajectories that start at point x1 at time t1 and end at point x2 at time t2. We consider the action of each possible curve and then use the calculus of variations to find the equation of motion corresponding to the condition that action be minimized. For $L = K - V$, the equation of motion that we get is $\mathbf{F} = m\mathbf{a}$.

At first sight this doesn't seem very helpful. Why not just start with $\mathbf{F} = m\mathbf{a}$ to begin with and never think about Lagrangians?

In part because this is formulation in terms of energy without recourse to forces.

But the technique becomes very useful with field theories, because sometimes the Lagrangian is much simpler than the field equations. For field theories, there is a Lagrangian density £ and the action is the integral of £ over a given region of spacetime. The field equations are then those that come from that action by using the calculus of variations. For example, for electromagnetism

$$£ = E^2 - B^2$$

This is much simpler than Maxwell's equations. But Maxwell's equations are a consequence of using the calculus of variations on this expression. So sometimes the simplest way to specify a theory is by giving an expression for £.

So in Hilbert's view, the question of "what are the field equations?" basically reduces to "what is the Lagrangian density?" Recall that for two-dimensional surfaces, Gauss showed that there is a quantity, the Gaussian curvature, that is an intrinsic property of the surface. Riemann showed that for any dimension of curved space, there is a quantity, the scalar curvature R, that is the generalization to n dimensions of the Gauss curvature of two dimensions. Einstein's minus sign in the metric doesn't change this, so there is a scalar curvature R for four-dimensional spacetime.

Hilbert reasoned that the theory of gravity that Einstein was developing was a geometric theory, so its Lagrangian density should be a geometric quantity. Riemann's scalar curvature seemed to have the appropriate properties to be the Lagrangian density, and there didn't seem to be any other appropriate geometric quantities. So Hilbert's formulation of general relativity is the simple one line

$$£ = R$$

And that's it! That gives the correct theory.

This is an example of concealing a nightmare of calculations within an elegant formulation. As a smith, Hilbert was well pleased with the result.

Of course, there's still the matter of working out the field equations, which Hilbert did using the calculus of variations and Einstein did by other methods. We aren't going to write these equations down, but they basically say that the curvature of spacetime at each event is equal to the overall density (including both matter density and energy density/ $c^2$) at that point. Written out explicitly, the field equations provide ten coupled, nonlinear partial differential equations for the ten components of the spacetime metric.

**Yup, a nightmare of calculations. Ah well. Leave the nightmares to the knights I say.**

What can one do with such complicated equations? Well, there's always the possibility that they can be solved in closed form (i.e., that there is a formula that one can write down for the solution) for particular cases of physical interest. In 1916, mere months after Einstein had written down his theory, Schwarzschild found a closed form solution for the gravitational field outside a spherical object.

Though Schwarzschild didn't know it, his solution also describes a nonspinning black hole. The closed form solution for the spinning black hole had to wait until 1963, when it was found by Roy Kerr. A closed form solution for the expanding universe was found in 1921 by Alexander Friedmann, several years before Hubble showed that the universe is expanding.

What if you can't find a closed form solution? Well, there's the Taylor series method. The first term in the Taylor series is Newtonian gravity, and the next is a correction of Newtonian gravity due to general relativity. This is how Einstein calculated the orbit of Mercury and the bending of light. Most of the tests of general relativity have been experimental measurements compared to predictions made in this way.

Mathematical methods of differential geometry can be applied to find general properties of spacetime. Mathematical methods of partial differential equations can be used to deduce general properties of solutions of the Einstein field equations.

And when all else fails, one can tackle the Einstein field equations using the same tool we use for all other intractable differential equations: a computer.

So, what emerged from all this work? A hard to calculate model of the universe that involves an equivalence between curved spacetime and a distribution of matter that is useful for looking at the universe as a whole and the motions of stars and the formation of galaxies.

And this model reduces to Newtonian mechanics when we look at the human scale of distance and velocity. It employs a sophistication of the tools of calculus and digs us into multiple differential equations that tax computers no matter how powerful and fast and memory heavy they are.

It has some technological applications (GPS systems use relativity in their calculations), but it's largely outside human scale experience and it's weird from a human perspective.

But therein lies a great deal of relativity's utility. Early mathematicians assumed that the way they observed the universe revealed the obvious truths (axioms) of that universe. Advances in mathematics opened this up to wider possibilities. But relativity showed us that our scale of thinking and measuring was only a small slice of how the universe itself was working.

Einstein revealed that our perceptions and mechanics were only narrow cases of a universe that did not work in a way convenient to our thinking, that when we looked out to the stars, we were looking at a deeper math than our sticks and stones.

Looked at on the large scale, the universe was not humanlike or human convenient at all.

What about the small scale? Surely that would still be nice to us …

## Quantum Mechanics

By 1900, cracks had begun to appear in the tidy picture of particles described by Newtonian mechanics and waves described by the fluid equations or Maxwell's equations. As time went on, these cracks became ever wider.

The first difficulty had to do with light emitted by hot objects. A hot object gives off light. The hotter the object, the more light it give off. But also the color of the light changes: hot objects give off red light, even hotter and the light becomes orange, extremely hot objects are white hot. To try to account for all this, 19th-century physicists used Fourier's trick. For an object with temperature T, they divided the light into notes with time dependence $\exp[-i\omega t]$ and calculated a formula for how much light was emitted in each note. When compared with experiment, their formula was right for small $\omega$, but gave increasingly bad answers the larger $\omega$ got. Even worse, their formula gave a nonsensical answer to the question "how much light is emitted in total?" because the sum over all $\omega$ of the answer in the formula did not converge.

This is not completely surprising since $\omega$ ranged over all the real numbers. That's a very big integral.

In 1900, German physicist Max Planck came up with the correct formula by adding one arbitrary rule to the method used to derive the old formula.

Planck had no idea that his fiddling around with formulas for heat and light would lead to the destruction of the foundations of physics, and their subsequent reconstruction in an entirely different form.

Instead of allowing light to be emitted in any old amount, Planck put in a rule that light in a note of $\exp[-i\omega t]$ could only be emitted with energy in multiples of $\hbar\omega$, where $\hbar$ was some new universal constant of nature.

This is actually a radical jump downward in a number of components and jump backward in complexity of operation.

*The complexity will be back in waves.*

Instead of summing over all possible real values (i.e., integrating over a stick), Planck was summing over a countably infinite set of energies (adding up stones).

This drops from the power of the continuum down to good old aleph-0.

Planck found that he could get his new formula to agree with the experiments if he chose the constant to have the value

$$\hbar = 1.05 \times 10^{-34} \text{kg (m)}^2/\text{sec}$$

This is an extremely tiny number, so small that each of Planck's packets would have such a tiny energy that a rain of light packets would be practically indistinguishable from a continuous beam of light.

In 1905, Einstein extended Planck's idea and used it to explain the photoelectric effect. This is the effect that allows solar cells to make electricity using light. Experiments on the photoelectric effect had shown that light could eject electrons from a piece of metal. But if the frequency of the light were too small, no electrons would be ejected. If the frequency was high enough to eject electrons, then the maximum kinetic energy of the electrons was proportional to the amount that the frequency exceeded the critical amount needed to eject electrons.

Einstein explained this effect by reconceiving Planck's light packets as actual particles of light (now called photons). For each type of metal, it takes a certain amount of energy $W_0$ to pull an electron out of the metal. Each photon would have an energy $\hbar\omega$; so if $\omega$ were too small, then $\hbar\omega$ would be less than $W_0$, and a photon would have too little energy to eject an electron.

If, on the other hand, $\hbar\omega$ were larger than $W_0$, then having used up an energy $W_0$, just to eject the electron, the photon would only have $\hbar\omega - W_0$ left to give the electron as kinetic energy. This explanation and formula fit the experimental data perfectly, even down to the fact that the value of $\hbar$, which Planck had found from the behavior of hot objects, was also precisely the right value to fit the experiment on electricity generated by light.

Light waves behave like particles, bounce electrons out of atoms, and transfer momentum and energy like colliding pool balls. These two results sounded like they might be a return to a simpler mechanics.

Not likely.

Later there were other experiments that showed that beams of electrons sometimes behaved as waves. If the electrons had momentum $\mathbf{p}$, then the beam behaved like a wave whose spatial dependence was the Fourier note $\exp[i\mathbf{k}\cdot\mathbf{r}]$, where $\mathbf{p} = \hbar\mathbf{k}$.

For our purposes, these two sets of experiments, waves acting like particles and particles acting like waves, will be enough. But there were other puzzling experiments involving atoms: like the fact that they only emitted light of particular wavelengths, or even the fact that they were stable at all when Maxwell's equations said that they shouldn't be. These experiments too were explained with somewhat ad hoc rules, somewhat loosely connected to the ideas of Planck and Einstein.

This hodgepodge of ad hoc theories (in hindsight referred to as "the old quantum theory") was finally put on a firm footing with the invention of quantum mechanics in 1926. Quantum mechanics was actually invented twice: once by Austrian physicist Erwin Schrodinger, and once by German physicist Werner Heisenberg, using entirely different methods. It was only later that these two theories were shown to be the same theory viewed from different points of view. We will start with Schrodinger's theory and later look at Heisenberg's theory.

For Schrodinger, this talk of Fourier notes of matter waves seemed backward. You're supposed to *start* with a wave equation and *then* apply Fourier's trick to it, and only *after* you've done that do you get to talk about the behavior of the individual Fourier notes. In a remarkable feat of reverse engineering, Schrodinger set out to generate the appropriate wave equation.

If you start with a wave equation, then Fourier's trick allows you to replace each $\partial/\partial t$ with a $-i\omega$. But now if we're going backward, we should replace each $-i\omega$ with a $\partial/\partial t$. Or equivalently, we should replace each $\omega$ with $i\partial/\partial t$. Then, since Planck and Einstein tell us that $E=\hbar\omega$, it means that we should replace each E with $i\hbar\partial/\partial t$.

Similarly, Fourier's trick says that we should replace each $\nabla$ with $i\mathbf{k}$. Going backward, we should replace each $i\mathbf{k}$ with $\nabla$. Or equivalently, replace each $\mathbf{k}$ with $-i\nabla$. But since $\mathbf{p} = \hbar\mathbf{k}$, this means that we should replace each $\mathbf{p}$ with $-i\hbar\nabla$. Since $p^2 = \mathbf{p}\cdot\mathbf{p}$, this means that we should replace each $p^2$ with $(i\hbar\nabla)\cdot(i\hbar\nabla) = -\hbar^2 \nabla^2$. So all we have to do is find some relation between E and p and make those replacements, and we will have the quantum wave equation.

**It's strange to take simple algebra and replace it with rough calculus and call that a solution.**

What is the formula relating E and p? We know that p=mv and that the kinetic energy is $K = (1/2)mv^2$. But that means that $K = (1/2m)m^2v^2 = p^2/(2m)$. Since E = K + V, where V is the potential energy, it means that

$$E = p^2/(2m) + V$$

Now we make our backward Fourier replacements in this formula and we get Schrodinger's equation for the wave function $\psi$.

$$i\hbar\,\partial\psi/\partial t = -\left(\hbar^2/2m\right)\nabla^2\psi + V\psi$$

Note the presence of i in this equation. This means that $\psi$ must be a complex function. This is different from all the previous physics equations. There the equations all had real number coefficients and the functions to be found were all real functions. We might have used complex functions as a convenient intermediate step, but at the end of the day we just took the real part and got a real solution.

But what is the physical meaning of the wave function $\psi$? In particular, since all measurements give real numbers as their result, there must be something that one does to $\psi$ to turn it into something used to predict the results of measurements. Recall that for any complex number z = a + bi, we have the complex conjugate $z^* = a - bi$ and the absolute square

$$|z|^2 = z^* z = (a+bi)(a-bi) = a^2+b^2 \text{ is real}$$

So $|\psi|^2$ is a convenient real quantity, but what is its physical interpretation?

This transformation of simple math to complex math was accompanied by a transformation of a simple physical concept into a more complex interpretation.

Up until this point, we've always interpreted motion as a continuous process wherein a particle moves from point to point over time along a definite path that we can measure with as much exactitude as our tools will muster.

Schrodinger's wave equation represents motion not as moving particles but as wavelike solutions. So where's the particle?

Eventually, Max Born found that $|\psi|^2$ is a probability density representing the likelihood that a given quantum particle described by the wave function $\psi$ lies within a given region. The probability of finding the particle in some region is the integral of $|\psi|^2$ over the region.

Before quantum mechanics the attitude toward probability was that physics predicted definite occurrences, but in many situations practical considerations prevented the implementation of those predictions. Probability was a tool to be used in those situations. In quantum mechanics, probability was a part of the fundamental description of nature.

Probability as a fundamental property of nature was (and remains) very disturbing to physicists. Einstein famously said "I, at any rate am convinced that He [God] does not throw dice." (Einstein was not religious but liked to use God as a metaphor for the laws of nature.) Schrodinger was even more vehement, saying "if we are going to stick with this damned quantum jumping, then I regret that I had anything to do with quantum theory."

The objections of these brilliant physicists were philosophical and aesthetic in origin. The tools of motion were geometric in origin and analytic in expression. Probability was an algebraic oddity created for the mathematics of gambling. It was as if somebody had come along and told knights that their ways of fighting were outmoded and that swords were nowhere near as effective as these chemical propellant weapons. Oh, and your metal armor won't work anymore either. But these ceramic and fabric combinations will keep you safer.

Alchemy, pottery, weaving in place of metalwork? What's next?

Dice rolling in place of curve drawing? What's next?

But the math worked, the experiments worked, so they got down to it.

Since $|\psi|^2$ is a probability density, and since there is a probability of 1 that the particle is somewhere, this means that the wave function must satisfy the condition $\int |\psi|^2 = 1$, where the integral is to be taken over all of space. For this integral to converge, the wave function must approach zero at large distances.

If quantum mechanics is so different from the Newtonian mechanics that preceded it, then how do we account for the fact that Newtonian mechanics works? That is, we must ask the question of quantum mechanics that we ask of all new physics: how does it reproduce, as an approximation, the results of the old physics? A probability density with a steep and narrow peak centered at $x = x_0$; for some, $x_0$ is a good approximation to the definite statement "the particle is at $x_0$." (Figure 15.5)

In other words, the particle could be anywhere, but you should bet that it'll be close to that curve you were going to draw anywhere to depict its standard Newtonian motion.

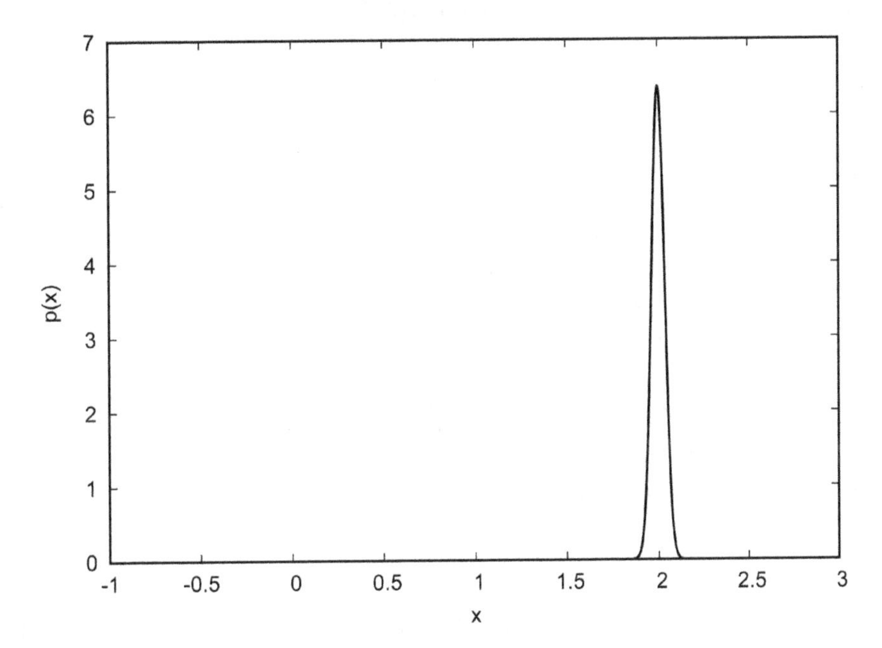


**FIGURE 15.5**
Very Steep Bell Curve

Thus, if there are solutions to Schrodinger's equation $\psi(\mathbf{r}, t)$ that at each time t have a steep and narrow peak centered on the classical trajectory $\mathbf{r}(t)$ given by Newtonian mechanics, then the wave function is saying approximately the same thing about the particle that Newton is saying. There are such trajectories but only for the objects of ordinary size mass traveling the ordinary size distances that we are used to in daily life. But for electrons with their tiny mass confined to the tiny space of an atom, there are no such approximately Newtonian solutions of Schrodinger's equation. Thus, quantum mechanics gives us the answers we are used to in daily life, while giving completely different answers on the atomic scale.

Twentieth-century physics looked up and out and the universe did not fit how humans wanted to think about it. It looked down and in and the materials of the universe did not fit how humans wanted to think about it.

But humans had the mental tools, the mathematics to model what they were not comfortable with. The smiths had forged ways to act and investigate without requiring that the humans understand first. The mathematicians had made methods that worked as well as they fit what was being modeled.

Mathematics let physics evolve well ahead of philosophy, to determine the way and how of things without yet being able to fit those ways and hows into human-comfortable thought.

The universe doesn't care. But we don't need it to care in order to develop and understand it.

And here's the thing about tools and comfort. Humans become comfortable with tools that work for them. A few generations of quantum mechanical teaching and people will be as okay with a dice rolling universe as they were with a deterministic one.

How do we solve Schrodinger's equation? The first step is to undo one of the reverse Fourier steps: we look for a solution of Schrodinger's equation of the form

$$\psi(\mathbf{r}, t) = \psi_E(\mathbf{r}) \exp[-iEt/\hbar]$$

Plugging this expression into Schrodinger's equation, we find

$$-(\hbar^2/2m)\,\nabla^2\psi_E + V\,\psi_E = E\psi_E$$

This equation is called the time-independent Schrodinger equation. There are solutions of this equation for any E, but only for particular values of E do those solutions approach zero at large distances, which is required for wave functions by their role as probability densities. This means that only certain values of energy are allowed.

The task in solving the time-independent Schrodinger equation is to find the allowed energies, and for each energy to find the corresponding wave function. We are especially interested in the lowest allowed energy, and its corresponding wave function, since systems that can give off energy to their surroundings will tend to settle down to the state of lowest energy.

Somewhat confusingly, the standard textbook notation is to leave off the subscript E and write the time-independent Schrodinger equation as

$$-(\hbar^2/2m)\,\nabla^2\psi + V\,\psi = E\psi$$

Schrodinger solved this equation for the simplest atom, the hydrogen atom, finding all the allowed energies and their corresponding wave functions. We will reproduce a piece of this result by finding the lowest energy and its wave function.

The hydrogen atom consists of a single proton in the center with a single electron in orbit around it. The potential energy V is

$$V = k_e Q_p Q_e / r$$

where $k_e$ is the constant of Coulomb's law, $Q_p$ is the charge of the proton, and $Q_e$ is the charge of the electron. Atoms are electrically neutral because the electron has an equal and opposite charge to the proton: that is, $Q_e = -Q_p$. So we can write the potential energy as $V = -k_e(Q_p)^2/r$. We now introduce the quantity $\alpha$ (called the fine structure constant) defined by

$$\alpha = k_e (Q_p)^2 / (\hbar c)$$

which allows us to rewrite the potential energy as $V = -\alpha\hbar c/r$.

There is no relativity in Schrodinger's equation, so it might seem strange to introduce the speed of light, c, into the problem. But it turns out to be convenient because with this definition the quantity $\alpha$ is a number with no units. We can see this as follows: from the formula $V = -k_e(Q_p)^2/r$, we see that $k_e(Q_p)^2$ has units of energy multiplied by length. From the formula $E = \hbar\omega$, we see that $\hbar$ has units of energy multiplied by time. The units of $\alpha$ are therefore

(Energy)(length) divided by (Energy time)(length/time)=(Energy)(length) divided by (Energy)(length) = no units: just a plain number.

It's actually a fairly small number (on a nonquantum scale), approximately equal to 1/137.

With the expression for V, we find that the time-independent Schrodinger's equation has become

$$-(\hbar^2/2m)\,\nabla^2\psi - (\alpha\,\hbar c/r)\psi = E\psi$$

Since V depends only on the distance r from the center, we can look for a $\psi$ that similarly depends only on r. Not all solutions of Schrodinger's equation for the hydrogen atom will have this form, but the solution with lowest energy (which is the only one we will find) does have this simple form.

We begin our computation with a short detour to find a formula for $\nabla^2 f$ for any function f that depends only on r. We have

$$r^2 = x^2 + y^2 + z^2$$

So, taking $\partial/\partial x$ of both sides of this equation, we find $2r\partial r/\partial x = 2x$, from which we find $\partial r/\partial x = x/r$. Now, using the chain rule, we find

$$\partial f/\partial x = (x/r)df/dr$$

Now taking another derivative and applying the product rule and the chain rule, we have

$$\partial^2 f/\partial x^2 = (1/r)df/dr + x(x/r)(d/dr)\big((1/r)df/dr\big)$$
$$= \left(x^2/r^2\right)d^2f/dr^2 + (1/r)(1 - x^2/r^2)df/dr$$

Exactly the same reasoning that we used for x applies for y and z, so we have two more formulas

$$\partial^2 f/\partial y^2 = \left(y^2/r^2\right)d^2f/dr^2 + (1/r)(1 - y^2/r^2)df/dr$$

$$\partial^2 f/\partial z^2 = \left(z^2/r^2\right)d^2f/dr^2 + (1/r)(1 - z^2/r^2)df/dr$$

Now we just add up these three formulas. On the left-hand side, the terms add up to $\nabla^2 f$. On the right-hand side, we use the fact that $x^2 + y^2 + z^2 = r^2$ and we find

$$\nabla^2 f = d^2f/dr^2 + (2/r)(df/dr)$$

Now that we know how to take $\nabla^2$ of a function of r, we apply that knowledge to Schrodinger's equation for the hydrogen atom to find

$$-\left(\hbar^2/2m\right)(d^2\psi/dr^2 + (2/r)(d\psi/dr)) - (\alpha\,\hbar c/r)\psi = E\,\psi$$

We can write this by grouping terms to keep the 1/r pieces together as

$$\left[E\psi + \left(\hbar^2/2m\right)d^2\psi/dr^2\right] + \left(\hbar/r\right)\left[\alpha\, c\psi + \left(\hbar/m\right)d\psi/dr\right] = 0$$

Buried in this we have the kind of differential equation that tends to have exponential solutions. By choosing the coefficient in the exponential just right, we can make the second term in brackets vanish, while E would take on the value that it would need to make the first term in brackets vanish. We are thus led to look for a solution of the form $\psi = \exp[-r/a]$ for some constant a. We need the minus sign so that $\psi$ will go to zero at large r. Then, $d\psi/dr = (-1/a)\exp[-r/a]$ and $d^2\psi/dr^2 = (1/a^2)\exp[-r/a]$. So Schrodinger's equation becomes

$$\left[E + \hbar^2/(2ma^2)\right]\exp[-r/a] + \left(\hbar/r\right)\left[\alpha\, c - \hbar/(ma)\right]\exp[-r/a] = 0$$

To make the quantity in the second bracket vanish, we must have

$$a = \hbar/(mc\alpha)$$

while the energy must be given by

$$E = -\tfrac{1}{2}(1/m)(\hbar/a)^2 = -\tfrac{1}{2}\, m\, c^2\alpha^2$$

We will now consider the meaning of each of these results in turn. The quantity a gives the scale for the size of atoms. Since quantum mechanics is probabilistic, atoms are fuzzy and do not have a definite size. Nonetheless, because of the properties of the exponential function, the electron is likely to be within a distance of about 1a or 2a of the nucleus and very unlikely to be much farther.

The expression for E says that the scale for the energy of electrons in atoms is around $mc^2\alpha^2$, which thus gives a rough guide to the scale of energies that can be expected in chemical reactions.

That scale can be calculated by examining chemical reactions and taking temperature readings before and after the reactions. Our quantum abstraction has measurable results in human scale events.

Oddly enough, this formula also constitutes an explanation for why the speed of light is so large. The speed of light has units, so what we mean by c is large is that it is large compared to the speeds that we are used to. But to attain speed v, a mass M must be given an amount of energy $(1/2)Mv^2$, so v is small compared to c, because the available energy is small compared to $Mc^2$. Let M be the mass of the proton. Then, since the electron is much less massive than the proton, the mass of a hydrogen atom is approximately M. So now if we compare the scale of chemical energy to $Mc^2$, we find $E/(Mc^2) = (m/M)\alpha^2$. So c is large on our scale because of the presence of two small numbers: m/M (the ratio of electron mass to proton mass, which is about 1/2000) and $\alpha$ (the fine structure constant, which is about 1/137).

The function exp[−r/a] is a solution of Schrodinger's equation, but it is not yet the complete expression for the wave function, because the absolute square of the wave function is a probability density, and therefore must have an integral over all space of 1. So there is a constant N such that the wave function is $\psi = N \exp[-r/a]$, and we must have

$$1 = N^2 \int \exp[-2r/a]$$

where the integral is to be taken over all space. This condition yields

$$N = 1/\sqrt{\pi a^3}$$

So the complete expression for the wave function is $\psi = 1/\sqrt{\pi a^3} \exp[-r/a]$

Note that this is for the simplest possible configuration of matter: a hydrogen atom. The formulae get much messier when we deal with more complicated atoms and molecules.

Probability theory gives us a toolkit that we can apply to this distribution. We should be able to derive physically meaningful quantities from the use of those tools.

First a probability density can be used to calculate expectation values. So, even though we don't know precisely where the particle is, we can figure out where its average position is. For simplicity, we will consider a particle on a line, with a wave function of a single variable $\psi(x)$. Then the average value of x, which we will denote $<x>$ is given by

$$<x> = \int_{-\infty}^{\infty} dx |\psi|^2 x$$

Since $|\psi|^2 = \psi^*\psi$, we can (for reasons that will become clear later) rewrite this expression as

$$<x> = \int_{-\infty}^{\infty} dx\, \psi^* x\psi$$

A probability density also allows one to calculate the standard deviation, which is a sort of average amount that a random variable deviates from its average. In the context of quantum mechanics, we will call the standard deviation an uncertainty and denote it by the symbol $\Delta x$. Then, the usual formula from probability theory gives

$$(\Delta x)^2 = \langle (x - <x>)^2 \rangle = \langle x^2 - 2<x>x + <x>^2 \rangle = <x^2> - <x>^2$$

These give us that peaked bell curve we were hoping for. Position in quantum mechanics is likely to be close to predicted position in classical mechanics.

But physics is all about motion. What can we do about velocity when we've thrown our dice? In Newtonian mechanics, a particle had a definite position at each time, and the velocity was the derivative of position. But in quantum mechanics, there is no definite position at any time; so it seems that we can't even get started on defining velocity.

But remember that in getting Schrodinger's equation, we made the substitution of $-i\hbar \nabla$ for **p**, or for our one-dimensional case that would be a substitution of $-i\hbar d/dx$ for p. It turns out that making that substitution in the expectation formula is what is needed. That is, the average value of p is given by

$$<p> = \int_{-\infty}^{\infty} dx\psi^*(-i\hbar)\, d\psi/dx$$

More generally, for any physical quantity A that we can measure, there is a corresponding operator, $A_{Op}$, such that the average value of A is

$$<A> = \int_{-\infty}^{\infty} dx\psi^*\, A_{Op}\psi$$

Here the term "operator" means linear map. That is, $A_{Op}$ is a rule that takes a function and produces another function. That rule satisfies the condition that given any constants a and b, and any functions f(x) and g(x), we have

$$A_{Op}\left[af(x)+bg(x)\right] = aA_{Op}\left[f(x)\right]+bA_{Op}\left[g(x)\right]$$

So, for the physical quantity x, $x_{Op}$ is the rule: multiply the function by x.

For the physical quantity p, $p_{Op}$ is the rule: take the derivative and then multiply by $-i\hbar$. Since physical quantities are real numbers, their operators must have the property that the expectation value is always a real number, no matter what the wave function. Such special operators are called Hermitian.

Operators can be multiplied: $A_{Op}\, B_{Op}$ acting on a function f(x) is the rule first act on f(x) with $B_{Op}$, then act on the resulting function with $A_{Op}$. It makes a difference in which order you multiply operators. Let's try this with $x_{Op}$ and $p_{Op}$:

$$x_{Op}\, p_{Op}\left[f(x)\right] = x_{Op}[-i\hbar\, df/dx] = x\,(-i\hbar\;\; df/dx) = -i\hbar x\; df/dx$$

$$p_{Op}\, x_{Op}\left[f(x)\right] = p_{Op}[x\; f] = -i\hbar\,(d/dx)[x\; f] = -i\hbar x\; df/dx - i\hbar\; f$$

Subtracting the second equation from the first, we find

$$\left(x_{Op}p_{Op} - p_{Op}x_{Op}\right)\left[f(x)\right] = i\hbar\, f(x)$$

Since this holds for any f(x), we have the operator equation

$$\left(x_{Op}p_{Op} - p_{Op}x_{Op}\right) = i\hbar$$

Here the number on the right-hand side is the operator rule "multiply by this number."

This is an odd formula. It connects position and momentum operators in a way where doing one then the other differs from the reverse order by an imaginary constant value. What does this mean?

From this operator equation, Heisenberg was able to show an inequality involving the uncertainty in x and the uncertainty in p

$$\Delta x \, \Delta p \geq \hbar/2$$

This is the famous Heisenberg uncertainty principle.

Despite the formal abstraction of the operator approach, Heisenberg emphasized that the uncertainty principle has direct physical meaning: whatever steps you take to make an especially accurate measurement of position will limit your ability to accurately measure momentum. And if, on the other hand, you take pains to accurately measure momentum, your measurement procedure will limit your ability to accurately measure position.

At this point, we come to a temporary screeching halt. Back at the beginning, we made sticks to measure things and thought that we could just use them casually and write down the results. The idea that using one stick would interfere with using another was twitch-inducing.

Both the Schrodinger picture and the Heisenberg picture are the same theory: quantum mechanics. In our terminology, the Schrodinger picture is quantum mechanics as seen by the functional mindset: the focus is on wave functions, with operators playing a subordinate role. The Heisenberg picture is quantum mechanics as seen by the algebraic mindset: the focus is on operators, with the wave function playing a subordinate role.

Both employ unexpected tools to produce unexpected results that actually reveal themselves in macroscopic results. By creating an understanding of atomic structure and the physics of bond formation and breaking, chemistry acquired a foundation in physics that it had been lacking since the first time people stuck stuff together in a fire.

## Quantum Field Theory

Schrodinger's equation implements the relation between momentum and energy of Newtonian mechanics. So just like everything else in Newtonian mechanics, it needed to be rewritten to be compatible with special relativity. This merging of quantum mechanics with special relativity is called quantum field theory. The quantum field theory of electrons and photons is called quantum electrodynamics, or QED for short (not to be confused with the other QED). The formulation of this theory was a monumental effort started by Paul Dirac and completed by Richard Feynman.

Dirac devised a more complicated version of Schrodinger's equation that incorporated the correct special relativistic relation between momentum and energy. As a nice bonus, the theory automatically incorporated the experimentally known fact that electrons were spinning particles that acted like tiny bar magnets.

But the theory also made a very puzzling prediction: antimatter. In addition to photons and electrons, the theory predicted a new particle (eventually to be called the positron) with the same mass as the electron but with opposite electric charge. Soon after this prediction, the positron was found in cosmic ray experiments. When a positron hits an electron, both particles disappear to be replaced with two photons: direct and complete conversion of mass into energy.

Unfortunately, Dirac's theory was not enough for compatibility between quantum theory and special relativity. What was needed was quantum fields: these are operator-valued fields, which essentially means a different operator for each point of spacetime. This is enormously complicated, but it was eventually made tractable by Feynman, who used three tricks: Taylor's trick, Fourier's trick, and a new trick of his own.

QED has a small parameter: $\alpha$ the fine structure constant, which we encountered in our treatment of the hydrogen atom. Feynman wrote QED as a Taylor series in powers of $\alpha$, and gave a method for doing calculations at each order. Since $\alpha$ is small, most calculations could be done to excellent accuracy using only a small number of terms in the Taylor series.

Using Fourier's trick, Feynman wrote things in terms of the behavior of individual Fourier notes. Each Fourier note represented a particle (electron, positron, or photon) of a particular momentum.

The first term in the Taylor series was reasonable, but boring. Each particle maintained its own momentum, not affecting, or being affected by, any of the other particles.

The next term in the Taylor series was better: electrons repelled each other by an amount that agreed fairly well with experiments. In the Fourier note picture of this process, the repulsion was accomplished by a "virtual photon" which took some of the momentum of one electron and transferred it to the other electron.

At the next term in the Taylor series, all hell broke loose. The calculation required an integral over all k values in the Fourier notes, and the integral did not converge. In what was supposed to be a small correction to the previous terms in the Taylor series, the theory said that the answer was infinity!

Feynman found a method, now called renormalization, to sequester these infinities in such a way that they didn't appear in the final calculation. With Feynman's method, the answers were finite and in excellent agreement with the experiment.

This is a disturbing way to do physics, but there is also a sense of déjà vu. Feynman's subtracting infinity from infinity to get a finite answer has a similar flavor to the way that Newton and Leibniz would divide an infinitesimal by an infinitesimal to get a finite answer. Perhaps at some point, some latter day Riemann will appear to put quantum field theory on a sound mathematical footing.

QED is a quantum theory of the electromagnetic force. But its methods were generalized to the strong and weak forces that act inside the nucleus of an atom. The list of matter particles treated by the theory was expanded from the electrons treated in QED to include neutrinos and quarks, as well as heavier versions of the electron called the muon and the tau. The list of force particles was expanded from the photons treated in QED to include the gluons of the strong force, the W and Z bosons of the weak force, and finally the Higgs boson that gives mass to the other particles.

Special relativity added to quantum mechanics gave us modern particle physics.

But, as with relativity itself, things get complicated when gravity pulls itself in.

So what about a quantum field theory of gravity? Well … there's a problem. Feynman's renormalization trick only works on a certain class of theories, which in the jargon of particle physics are called renormalizable. General relativity is not renormalizable.

So we have two options:

A: we need a better theory of gravity that is renormalizable to replace general relativity (and this better theory has to somehow predict the same things as general relativity for all the cases in which gravity has been tested by experiment).

B: we need a better way to do quantum field theory, hopefully one that doesn't need to rely on the process of renormalization.

Unfortunately, we have no experiments to test the quantum properties of gravity, and so whether we pursue option A or option B, we have no guidance from nature as to whether we are on the right track. (We could stride boldly and confidently into the unknown without experimental guidance, but only if we forget all those warnings from Greek tragedy about the danger of hubris.)

In any case, even without quantum gravity, there are plenty of interesting problems in quantum field theory to tackle, like the nature of the dark matter and dark energy that pervade the universe.

This is where we stand in the theoretical universe. We have expanded in many directions by applying tools that did not look like they belonged, but always maintaining the need for testable results.

When physicists cannot test, they sometimes become like mathematicians, picking up and elaborating tools and theories based on elegance and personal aesthetics.

If they were left to do that too long, they'd end up generating new branches of math or elaborating old ones and find out that they had hung up their knightly armor and weapons and become smiths until testable ideas emerged and they armed up and got back to the field.

# 16

## *Math Education and Math in Education*

For most people mathematical inheritance is found or lost in school. This is where those who learn math take it up eagerly, but it's also where a much larger number of people learn to give up on math.

Why does this happen? Part of it has to do with the nature of the subject; but an even greater part has to do with the way math is taught. By way of explanation, we are going to start by saying a few general things about teaching, and we will then contrast the teaching of reading with the teaching of math.

Teaching has four different purposes, which intermingle and intertwine:

1. Teaching people what they need to get along in life
2. Teaching people what they might be interested in
3. Teaching people what is needful for particular jobs
4. Teaching people what is needful to expand the field being taught

Reading and writing, to take the other two R's away from arithmetic, are taught for all four purposes:

1. In a literate society, reading and writing are necessary. Therefore, we strive for universal literacy so that each person can read bills and write letters/e-mails as needed.
2. There are a lot of things to read out there, which contain information and ideas about a vast range of subjects, some shallow, some deep, all written down.
3. A great deal of reading and writing is necessary in a large number of jobs. Every field teaches the specific ways of reading and writing what is necessary for that field.
4. People who learn how to write can expand what is available to be read.

But what about math?

1. Teaching people what they need to get along in life. The standard usually set here is that people should be able to do arithmetic and maybe work with fractions and percentages. Little more is regarded as necessary, despite the presence of statistics in the news every day, the scientific issues people might need to vote on, and the computers sitting on people's desks, laps, belts, wrists, etc.
2. Teaching people what they might be interested in. Mathematical recreations are only given to people who are interested in math already. There is no attempt to make math interesting to the nonmathematically inclined. Furthermore, the mathematically dependent fields of interest, such as science, economics, and architecture, are sanitized of math when presented as everyday interests.

3. Teaching people what is needful for particular jobs. Jobs that require more math than basic arithmetic select their new hires from among those who have learned more math, rather than teaching enough math to those who might find the jobs themselves interesting enough to be worth the effort to learn the math.
4. Teaching people what is needful to expand the field being taught. Those who persevere and learn through all the obstacles in learning math, passing through all the barriers to understanding, are deemed capable of becoming mathematicians and therefore are given the honor of adding to the field. They have been selected for the correct mindset and given the background and practice necessary to become contributors to the field.

What has happened from this distribution of goals is three levels of mathematical knowledge, each of which is taught with its own purpose, methods, and orientation:

*Math to get by*: Needed for everyone. Concentrates on basic skills centered on arithmetic.

*Math for math-using jobs*: Needed for architects, engineers, scientists, economists, etc. Selected for by those who can get through the appropriate classes. Concentrates on problem-solving with real-world emphasis.

*Math for budding mathematicians*: Needed for mathematicians. Concentrates on proofs.

The gap between math for jobs and math for mathematicians is the gap of the knight and the smith, which we've already talked about. The gap in which the inheritance is lost is the one between the math to get by and the rest of it.

Actually, gap is too small a word. This is a chasm the size of the Grand Canyon. On the one side of the vast abyss are those culled from math learning, who come to think of themselves as disliking or fearing math and mistrusting those who can learn it. On the other side are those in groups two and three who see themselves as capable and interested and tend to have a certain pity toward those on the other side but who also know that their ability makes it difficult for them to fit in with the majority.

To understand how and why this chasm opens up, it is helpful to contrast the teaching of math with the teaching of reading. There is an old quote from Plutarch to the effect that the mind is a fire to be kindled, not a bucket to be filled.

The first and most important task of modern schools is to teach reading. When teaching reading works, which it often does, a fire is kindled in the mind. At the beginning, there is a lot of hard work to be done by both the teacher and the student: letters and their sounds to be learned, words to be sounded out and eventually recognized by sight.

The rewards for this initial learning are meager. The reading primers of our youth contained such phrases as "See Dick run. See Jane run. Run Dick, run." It's a wonder that anyone of our generation was motivated to read after being initially exposed to such insipid stuff. Indeed, the most motivated are those who are frequently read to. Those students are shown that there is a world of reading beyond Dick, Jane, and running.

Eventually, the students acquire enough skill that they can read with a certain amount of ease material that interests and entertains them. And at that point the fire is burning! Each book that they read increases their skill and ease with reading, as well as increasing their vocabulary. This brings more and more books within their reach, which in turn tend to increase skill, ease, and vocabulary.

Once a fire starts burning in earnest, all that is needed to keep it going is a steady supply of fuel. Through school, home, library, and bookstore, there are plenty of sources of books on a variety of subjects and in a variety of styles, so that each student can continue to grow and progress as a reader of whatever it is that they want to read.

Let's now look at what happens to classroom instruction after reading is taught. It is fortunate that reading once started is somewhat self-perpetuating, because classrooms are challenging places and the challenges don't get easier as the students get older. A host of emotional and behavioral problems not addressed at home spill over into the classroom, not to mention any particular cognitive or attention difficulties that each student might have. Much of the effort and energy of the teacher is devoted to "crowd control."

There is an additional challenge having to do with teaching anything cumulative, that is, anything that needs to build on previous learning. Each teacher teaches only one grade and can only teach a cumulative subject if the students have been adequately prepared by the previous teacher. This challenge is especially severe for math, which by its nature is entirely cumulative.

Perhaps in response to these challenges, many schools have adopted, after teaching reading, a lesson style that is piecemeal, "one size fits all," and emphasize memorization rather than skill or understanding. In this lesson strategy, whatever the subject, one picks some vocabulary words or facts to be memorized. They could be state capitals, or terms for different cloud formations, or names of polygons.

This is definitely teaching as a bucket to be filled. Furthermore, the bucket is very leaky. After the lesson has been taught and homework has been done, there might be some kind of quiz or test. But once the test is done, there is no expectation that any of the new knowledge will be retained. Some small part of it might be, but that is not required.

*It seems that all that most people remember of the astronomy they learned in school is the names of the planets, and then they're miffed that Pluto is no longer on the list.*

Basic arithmetic skills are taught, but there is not enough time set aside to practice those skills (because we need all that time to have them memorize the name of a polygon with eight sides). Certainly not enough time for them to gain the ease and facility with those skills that are needed to move on to more advanced math.

Because in the piecemeal approach to math instruction there is no expectation that lessons will be remembered, this approach constantly circles back, with small variations, to teach the same math over and over again. A small number of the students remember what they learned the previous time that math lesson was taught and are bored by its repetition.

These students are labeled as good at math. Eventually they may be encouraged to skip one or more grades in math. If there are multiple sections of a given math course, these good at math students will be encouraged to enroll in the more "mathy" section. One spillover effect of this labeling is that all of the other students in the class are encouraged to think of themselves as bad in math.

To see why the piecemeal teaching style is such a disaster for math instruction, it is helpful to look at the concept of facility, which is the ease of use that comes with enough practice. Yes, practice can be boring. But in every field practice is necessary. Musicians do not just pick up instruments and play them. They practice. Sports team members practice all the time. Actors practice, writers practice.

Everyone needs to practice to learn things. That's just a fact of the way humans work.

There has to be enough practice time and exercise in learning math because math builds up. Everything learned is based on facility with earlier things. You can't learn to do algebra without being able to do arithmetic easily. You can't do calculus unless you can do algebra easily.

This learning-built-upon-learning aspect of facility is the most crucial hidden cause of lost inheritance. Persons facile with a certain stage of mathematical learning (let's say arithmetic) can rely upon that facility when they need to learn practices that rely upon that learning (fractions build upon arithmetic in exactly this way). Metaphorically, this person has a solidly built first floor in a building upon which the next story is built. But a person without that facility who tries to climb to the next floor more resembles someone building up a house of cards. Each level is less stable than the last, until at some point the whole structure collapses and they lose their interest in and connection to mathematics.

It might seem that it is difficulty with understanding that most causes the collapse of these cardhouses, but lack of facility is at least as much of a contributor. Understanding without facility does not impart the ability to know what the people inside the field are actually doing to produce their work, only what they are producing.

This is considerably better than not understanding at all, and indeed reclamation of that understanding is the primary purpose of this book, but if one is teaching someone to actually practice mathematics, then facility must accompany understanding in order to build practice upon practice. Taken together, understanding and facility support each other in a manner similar to the way the theoretical and detected universes support each other.

The piecemeal, memorization-intensive approach to math doesn't end with elementary school. It goes on through algebra, only now what is being memorized are not math names or math facts, but self-contained methods disconnected from all other methods. As an example, we now discuss the FOIL method, which is a method to multiply $(a + b)$ by $(c + d)$. The students have already learned the distributive property, which tells them that $a(b + c) = ab + ac$, and since multiplication is commutative, this also means that $(b + c)a = ab + ac$. Now suppose a student who understands the distributive property is told to compute $(a+b)(c+d)$. This can be done just by using the distributive property twice. That is,

$$(a+b)(c+d) = a(c+d) + b(c+d) = ac + ad + bc + bd$$

But instead the students are taught that they should first memorize the acronym FOIL, standing for First, Outer, Inner, Last. In the product $(a+b)(c+d)$, the two "first" numbers are a and c, the two "outer" numbers are a and d, the two "inner" numbers are b and c, and the two "last" numbers are b and d. Thus, the product of the two first numbers is ac, the product of the two outer numbers is ad, the product of the two inner numbers is bc, and the product of the two last numbers is bd. Thus, when we "FOIL," we write down those products in that order:

$$ac + ad + bc + bd$$

Why is it taught this way instead of using the distributive property? Well, first, we can't rely on the students remembering the distributive property (that was on last week's test, so they shouldn't have to remember it for this week's lesson. That wouldn't be fair). Second, even if they do remember the distributive property, we can't expect them to think of it as a tool to be used beyond the confines of whatever exercises were given to them in that section of the book. And what if they need to calculate $(a+b)(c+d+e)$? Oh no. That's too complicated. We would never give them anything like that.

Disconnected algorithms like this are crutches for people who we never expect to be able to walk. In other words, by the time this sort of math instruction is used, we have given up on the possibility of the students ever learning to use or understand math and we're not even going to try to teach them.

We have emphasized that math is a toolkit for understanding the world. So surely the application of math to real-world problems should form a central part of math education. And it does … sort of, in an odd and artificial way that we will now explore.

## Welcome to Math Problem World

Yes, come to Math Problem World, where the trains always run at the same speed until they collide because they are always on the same track, where the farmers have uniform fertility in their perfectly rectangular fields and never suffer crop failures, where interest rates never change, and where price … . Well, let us illustrate, or rather twist, someone else's illustration.

Allow us to digress about the hilarious comic strip "Foxtrot," written by Bill Amend. "Foxtrot" is about a family named Fox. (They're human beings. Fox is just their name.) Two of the kids in this family are Paige, aged 14, who is among other things a fashionista, and her brother Jason, aged 10, who is a math genius. One of the running gags in the strip concerns Jason helping Paige with her math homework.

In one strip, Paige was confronted with a word problem in which a girl is trying to choose among three sweaters. The sweaters were specified as being of three different colors, and they were given list prices and sale percentages (such as 20%) off, and the question was "which sweater should she buy?"

Jason asked what the difficulty was, since it was a simple matter of calculating the actual prices using the percentage reductions in the list prices and then comparing the resulting prices to see which sweater was cheapest. But Paige complained that she did not know what color the girl's hair or eyes were. We could add to this the lack of knowledge of style, personal preferences, heaviness or lightness, season, and just how the sweaters looked.

The joke implies that Paige is being distracted from solving the problem by her fashion attitudes, but the truth is that Paige is right. The problem asked which sweater she should buy, not which sweater is cheapest.

This, by the way, is not isolated. While Jason is always right about the math, Paige is often right about the real-world considerations to the problems. But Jason is so used to the Math Problem World that he doesn't even notice the missing real-world elements.

*That's a lot of analysis for one comic strip.*

**You take your insights where you find them.**

The central problem with Math Problem World is that the problems are constructed artificially to help reinforce and test whatever lesson has just been given. The problem above would show up in a lesson on percentages. Such a lesson would cover the idea that 1% = 0.01 and talk about sale prices and tax rates and interest rates and other things that seem to relate to the real world. But at no point in such a lesson would anyone explain what percentages are for, or why they exist. What real-world matter caused % to be created as mathematical concept?

The answer is a little embarrassing, given how many kids have stumbled over the idea. % is just convenience notation. It's a longer stick, like meters and centimeters. It turns 1/100ths into 1s. 0.35 = 35%. It's just like calculating costs in cents instead of dollars. It's just a way of writing things, nothing more. And yet it is belabored as if it were a deep concept, while other ideas that are deep (like the compromises necessary in creating math problems from the real world and the implications to accuracy of result) are ignored completely.

One difficulty with percentages is that the terminology for their use is somewhat misleading. "20% off" sounds like subtraction, but it really means "multiply by 0.8" just as "plus an additional 20%" sounds like addition but really means "multiply by 1.2." Also, "plus an additional 20%" followed by "20% off" sounds like it should get you back where you started, but since it is equivalent to multiplying by $(1.2)(0.8) = 0.96$, it's actually equivalent to "4% off."

If instead of saying "we're going to cover percentages," these lessons were introduced by saying "we're going to help you navigate the misleading math terminology used by stores and banks and the government," the real-world application of the lesson would be much clearer.

The serious difficulty of Math Problem World is that its inhabitants and the things that happen to them are created artificially to test the ability to do one or two calculations. It is a substitute for abstraction, and therefore teaches nothing to the people who cannot automatically connect math with reality.

## How to Teach It

After all this carping, do we have any practical suggestions about how math should be taught? We do.

First and foremost, this is not rocket science: if we want kids to learn to do math, then we need to teach it to them the way we teach those things that we, as a society, care about people doing (e.g., reading, playing sports, playing a musical instrument). After learning basic skills, students must be given enough time to practice those skills and gain facility with them. Then, once those basic skills are second nature, the students can be taught the more advanced skills that need the basic skills as a foundation. Then the students must in turn be given enough time to practice those more advanced skills and gain facility with them. And so on.

*If this were rocket science (which should really be called rocket engineering), we would have a bunch of diagrams of rocket engines, numbers for the proper fuel to oxygen ratio in fuel mixtures, and the value of the speed the rocket needs to escape from Earth's gravity (which, by the way, is about 11 km/sec).*

Second, we need to put more care and effort into how we design math word problems. We're fully aware that keeping the complexity of the word problem down requires a certain amount of artificiality, and therefore that math word problems will never be able to fully escape from Math Problem World. But much could be done to design problems that do a better job of showcasing why math is a set of tools whose purpose is to solve real-world problems.

It is also important to recognize that the math word problem skill is itself an important skill to learn and cultivate. This skill involves looking at the verbal description of the problem, figuring out what mathematical tool is needed for that situation, solving the appropriate equation, and stating the result as the answer to the original question.

*First semester college physics can be viewed as nothing more than an application of the math word problem skill to $F = ma$ and formulas that can be derived from $F = ma$.*

One difficulty with the piecemeal approach to math instruction is that students don't really learn the "pick the right tool" skill if we're always implicitly saying, "Pro tip: since this is the section of the book on number 2 Phillips head screw drivers, that means that the right tool for this problem is a number 2 Phillips head screwdriver."

The solution to this difficulty, which is also an opportunity to underscore the cumulative nature of math, is the following: make word problems that use more than one tool. This way, in addition to the tool being taught in that particular section, students will need to reach into their mental toolkit to figure out which of the previously learned math is needed for this particular problem.

Solving a word problem that needs multiple tools is excruciatingly difficult for students taught entirely by piecemeal methods (*which is one reason students have difficulty with college physics*). But that's not an argument against word problems. It is in fact another urgent reason why the piecemeal approach to instruction has to be discarded as soon as possible.

Third, we need to present the same parts of math in multiple ways so as to appeal to the understanding of different mindsets. The wonderful thing about analytic geometry is that it turns equations into pictures and pictures into equations. This gives people with a geometric mindset a second way to understand algebra, and people with an algebraic mindset a second way to understand geometry.

When two linear equations with two unknowns is taught, the algebraic method for solving the equations should be presented, but it should be accompanied with the statement (along with the corresponding picture) that each equation is a line, and the solution is the point where the two lines cross.

When circles are taught, the pictures and formal definitions should be accompanied by the equation $x^2 + y^2 = R^2$. When the quadratic equation is taught, the derivation of the quadratic formula should be accompanied by pictures of parabolas that do or don't cross the x-axis.

## Math in Education

The problems of teaching math are accompanied by the questionable nature of many of the numbers generated by and used in modern education.

While many people shy away from math, they are sometimes still enamored of numbers. They have the sense that a characteristic of something that is put against a stick and measured is somehow more real than one that is unmeasured.

But while it's possible to generate numbers by all sorts of different methods, not all of those numbers contain meaningful information. You can count the number of pieces of fruit in your refrigerator. But if the refrigerator contains blueberries, cherries, apples, and watermelons, how meaningful a piece of information is the overall number of pieces of fruit?

The numbers generated in education are test scores, grades, and the grade point average (GPA) formed by taking the numerical average of all the grades of a given time period (e.g., one semester, or all of high school, or all of college).

Why are these numbers generated? And how meaningful are they? In addition to teaching, the job of the teacher requires evaluation: at the end of each course, the teacher must produce for each student a grade – the number (or corresponding letter) that is the teacher's overall evaluation of how the student did in the course.

Tests can be used for both parts of the teacher's job. An individual test can be used as a diagnostic aid to teaching: by looking question by question at the completed test, the teacher can get an idea of what parts of the student's understanding are strong, and what parts are shaky and thus are in need of shoring up through more teaching.

All that detailed information about strengths and weaknesses is lost when looking solely at the overall test score. Even more information is lost when the scores of all the tests and homework assignments are used to produce the overall grade. And then yet more information is lost when the grades are used to produce the GPA.

At some times and in some cultures, tests ended up being very serious indeed. In Imperial China, tests were given for accomplishment in various areas of knowledge and skill. Those who did best became the bureaucrats who ran the country. Knowledge of the Book of Odes could make or break a person's ability to determine whether or not the dikes on the Yellow River needed to be shored up.

Comparable make or break tests in our culture include the MCAT for admission to medical school, and the Bar Exam for certification to practice law.

In the early years of a student's education, tests and grades are not taken very seriously. But this changes over time. In particular, at the end of high school both GPA and scores on standardized tests are used as important criteria for admission to college. As we noted before, both measures are problematic because of the enormous loss of information that comes in producing a single number.

But high school GPA has an additional problem of the "comparing apples and oranges" sort. Even for what is nominally the same course, two different high schools may not teach the course at the same level of difficulty, nor grade the course at the same level of difficulty. Thus, even for a single course, the grade at high school 1 may not mean the same thing as the grade at high school 2. And by extension, the GPA at high school 1 may not mean the same thing as the GPA at high school 2.

This "apples and oranges" problem is alleviated by standardized tests, since at least then all the students are taking the same test. But if too much reliance is placed on standardized tests, then enormous social havoc can be wreaked by whatever cultural bias the test (or different amounts of preparation for the test) contains.

Finally, we want to point out ways in which both tests and GPA can be counterproductive: that is, ways in which they can have a tendency toward measuring (and thus rewarding) the opposite of what they are supposed to measure.

David's Ph.D. advisor liked to say about physics research projects that "if you make all the bases you steal that means that you're not stealing enough bases." What he meant by this baseball metaphor is that a harder research project means both a greater chance of failure and a more important result if it succeeds. In the hope of doing important work, one should not play it safe (i.e., one should not choose only projects that are assured of success).

This advice about research also applies to GPA. One route to a high grade is to take an easy course. Thus, when we see a high GPA, should we say, "this is a good student" or "this is a student who played it safe"? Furthermore, if we judge students by GPA alone, then we are encouraging them to play it safe and thus keeping them from realizing their potential as students.

There is a different kind of counterproductive tendency of tests: having to do with the fact that the tests are timed. We are all prone to make errors in mathematical calculations, with the chance of error growing for calculations that are longer or more complicated. We therefore emphasize, whenever we teach math, that calculations should be done in a slow, careful and methodical way; and that even after the answer is obtained, that answer should be checked in a careful and methodical way. This is a fundamental teaching of mistrust needed to learn math.

But that advice is incompatible with having enough time to finish all the questions on a test (especially the sort of standardized tests that are so much relied on for admission to college). To cope with this situation, students must develop a finely calibrated "math test taking skill" that involves going fast enough to finish the test, but within that constraint doing things in a sufficiently slow, careful, and methodical way to avoid as many errors as possible. This math test taking skill must be kept scrupulously separate from all math tasks outside of test taking, since otherwise it would wreak havoc by causing a whole host of unnecessary errors.

We realize that despite all these qualms, tests (both classroom and standardized) will need to be given. Grades will need to be assigned and used to calculate GPA. We're not arguing against any of that. However, it's important to be aware of just how problematic these numbers are and how little they should be taken at face value.

It's helpful to make the distinction between the numbers that arise in the physical sciences and the numbers that result from these arbitrary educational testing procedures.

In the physical sciences, a number that is measured tells something directly about nature. If a theory in science is well tested by such measurements (and it is being used in an area where the measurements say that it can be trusted), then the calculations of that theory also have direct physical meaning.

In contrast, the numerical result of some classroom test (or the average of several such tests) has no direct meaning and may not have any useful meaning at all. We should never take such numbers at face value and always ask, "what, if anything, does this number mean?"

*If we were going to adopt Einstein's use of God as a metaphor for nature, we could use the old joke grocery store sign: "in God we Trust: all others pay cash."*

## Reclaiming the Lost

The process outlined above for reforming mathematical teaching may work for kids learning math for the first time, but what about the orphans of mathematics, those cast adrift in the sink-or-swim teaching, those who grew up thinking they could not do math and who therefore gave up on the idea. What can they do to reclaim the lost mathematical knowledge? (And do they have sufficient reason to do so?)

As we said at the beginning, math is part of our shared heritage. One should not let accidents or errors of education keep one from one's inheritance. Math, as we've shown in this book, is used all around us, in matters that affect our lives and in the tools we use to do our jobs. Yes, one can say that one was robbed, but does that mean one should not work to get back what was taken?

We're not saying to the disinherited that they should pick up math texts and go through them. We are saying that a lot of mathematical understanding can be learned without needing to do the practice necessary to gain facility with it. This is not as easy with math as it is with science. There are fewer math popularizations than books on science. But the books on science point toward the math (while mostly omitting it). You find the math in these books hidden in the discussions of the uses of the math.

There are books written by expert gamblers that give a sense of probability. There are books on lying with statistics that tell about that. There are books on architecture and drafting which give some of the geometry. There are books on astronomy that give more of it. There are books about computer imaging and animation that show something of fractals. There are books about weather and chaos that are not all wet.

A person seeking to reclaim what is lost can find the tools in the use of the tools, and from there can stretch out if desired to get a better sense of the tools themselves.

Then there is that ancient human activity, talking to people, the tool users, and the tool makers. Many, if not most, businesses have such people: accountants, programmers, engineers, people whose learning involves a fair amount of math. Many such people are actively ignored or separated into their own departments, shunted off in some way, creating a social manifestation of the gap.

In many respects they're like the crew at a theater, the stagehands who make sure that things work so the actors can spend their time saying their lines in front of an audience. Sometimes the stagehands seem to butt in for no clear reason, pointing out a loose sandbag or a weak trap door or a fire hazard, but otherwise they just seem to be behind the scenes.

It's worth it for everyone else to find out what the stagehands are doing there and how they're doing what they're doing. Who knows? They might even talk back and give the asker some idea of how things work behind the scenes where the tools are used and made.

## A Stone's Throw

In the modern world, nearly everything we use has been shaped by mathematics. Every tool is made with measurement and calculation. Everything we eat has been weighed and measured. Our clothes are woven using machines that are the precursors to computers or are themselves computerized. Our phones are computers, our toys are computers, our computers are computers. Our buildings are designed and built with geometry, their materials analyzed for stress and tolerance.

And yet, like the actors on a stage, most people behave as if these things are simply props to pick up, use, and put down, appearing when they need them, to be used as they need them. Behind the scenes are those who use the tools that make the props and keep the stage running and those who make those tools and those who make the tools to make those tools. In the background can still be found the stick and the stone out of which we have made our world.

Reach lazily down, abstract the tools, and through them look mistrustfully around to reclaim the lost world.

# Index

## D

## E



## F



## G

## M



## N



## O

S



T

CPSIA information can be obtained
at www.ICGtesting.com
Printed in the USA
LVHW050937190423
744686LV00005B/292